Nonlinear, Nonlocal and Fractional Turbulence

Peter William Egolf • Kolumban Hutter

# Nonlinear, Nonlocal and Fractional Turbulence

## Alternative Recipes for the Modeling of Turbulence

Springer

Peter William Egolf
Institute of Theoretical Turbulence Research
Niederlenz, Aargau, Switzerland

Kolumban Hutter
VAW
ETHZ
Zürich, Switzerland

ISBN 978-3-030-26032-3 ISBN 978-3-030-26033-0 (eBook)
https://doi.org/10.1007/978-3-030-26033-0



The book cover image shows the ion and electron streamlines, colored by the magnitude of the flow vorticity, of the solar wind. This visualization reveals the hierarchical development of coherent structures in formations of current sheets and magnetic islands in fully developed turbulence. From Phys. Plasma, 20, 012303 (2013) with permission. © 2019 American Institute of Physics

This Springer imprint is published by the registered company Springer Nature Switzerland AG.
The registered company address is: Gewerbestrasse 11, 6330 Cham, Switzerland

# Preface

Chaos and turbulence are closely related terms, describing "irregular motion." This kind of irregularity, with all its facets, is a main topic discussed throughout this book. Chaos is omnipresent in the entire universe, in and on the earth, especially also in the atmosphere and oceans. The two terms "chaos" and "turbulence" are tightly related to order and disorder, the main ingredients in life, art, music, etc. Music of a single frequency and that containing the entire frequency spectrum (white noise) would be annoying and boring. The highly varying mixture of order and disorder, with different tempi and intensities, gives our lives and us human beings the wonderful richness observed in so many features in the dynamics of nature. Isn't a visualized turbulent flow above a candle one of the most beautiful examples of volatizing coherent structures? Furthermore, especially at high excitation (energy input), it gives the unnatural impression of an object (in this example it is a plume), which is a simplifying interpretation of a very complex reality experienced by us human beings.

In our book, the terms "chaos" and "turbulence" are mainly used in the original Greek manner as phenomena with states located somewhere between order and disorder. Greek χάος denotes gaping void, infinite darkness or the primordial state of any existence. In the Greek terminology, it meant the very early matter, which—before the expansion of the universe—already contained all the ingredients and information necessary for its entire wonderful development. This implied not only the originating of the cosmos (from its big bang), but even the creation of the Gods. Furthermore, it is interesting that in the Greek mythology later chaos was also personified as the God of air. Chaos originated in Ancient Greek, but also in related religions of the Ancient Near East, this term played a crucial role.

Whereas in the Greek myths chaos was a very positively occupied expression, in the middle age, for example, the motif "Chaoskampf" (German), which means the battle against chaos, symbolized by God defeating a dragon (chaos monster), got a slightly negative occupied meaning (see Picture 1). Then in older religious Christian books, a separation of heaven and earth was believed. In this time, the separation was seen as a cosmogonic act of a deity creator. It was a creation "ex nihilio" (out of nothing) by a single almighty God.

**Picture 1** A cultural hero deity with a chaos monster, shown as dragon, symbolizes Chaoskampf, the German denotation of struggle against chaos. He is also known as the God of darkness. In the picture, a Christian chaoskampf statue is shown where Archangel Michael is slaying Satan represented as a wild dragon. From reference: https://en.wikipedia.org/wiki/Chaos_(cosmogony)#Chaoskampf

In Elizabethan England, the slightly altered meaning of chaos was even changed to describe complete disorder and total confusion, which naturally was a new terming of highest exaggeration. Since then men and women on the road use this term in the sense of "order and chaos," where now chaos describes practically the highest state of disorder and even disasters.

It would be advantageous to go back to the ancient Greek usage of the word "chaos" in daily life! Then instead of order and chaos, *order and disorder* would be the correct opposite terms. The vast range between these two would be reserved for the terms chaos (and turbulence). Scientists also use the term "deterministic chaos" in this more positive sense. Physical systems showing deterministic fluctuations are called chaotic if their phase space is low-dimensional and turbulent if a higher degree of freedom occurs. Moreover, physical theories in this book reveal that a more chaotic and turbulent system, after several bifurcations and symmetry breaking processes, shows a lower entropy and higher order!

An early attempt of picturing turbulent processes in water is attributed to Leonardo da Vinci (1452–1515), who was an excellent observer of nature, human beings, animals, etc., and a highly gifted designer of new technical discoveries. He, at his time, did not distinguish between art and science. He realized that flowing water at high excitation contains two kinds of motion, a calmer stream and a more agitated fluctuating and whirling part. In a study of turbulence, he asked the following important questions: (1) Where is the turbulence in the water generated? (2) Where does turbulence in the water persist for a long time? (3) Where does the turbulence in the water come to rest? Sophisticated scientific answers to his questions are given in this book. His pictures of turbulent air and mostly water flows are very realistic and show eddying motion of eddies of different sizes (see Picture 2), an important idea taken up by different theories also presented in this book.

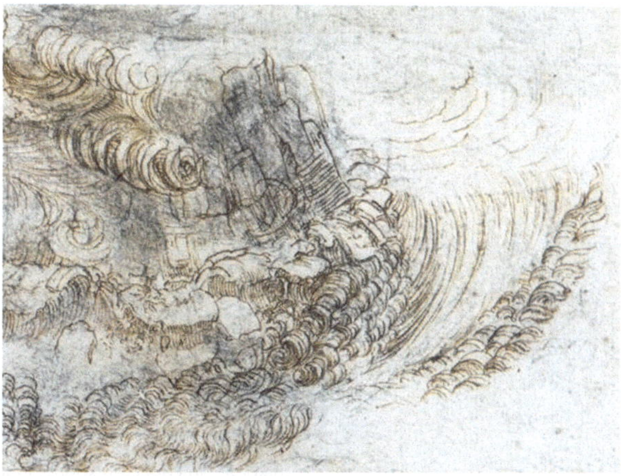

**Picture 2** Drawing of a turbulent flow as it occurs in stormy weather by Leonardo da Vinci. From reference: https://www.researchgate.net/figure/Turbulent-flow-by-Leonardo-da-Vinci-Storm_fig 1_282292405

**Picture 3** In June 1889, the famous impressionist artist Vincent van Gogh painted the picture "starry night" by oil on canvas. It shows among glowing stars' large-scale turbulent air movements in the atmosphere. They are characterized by large eddies of high vorticity. From reference: https://en.wikipedia.org/wiki/The_Starry_Night

In the literature on turbulence, Leonardo da Vinci's drawings are omnipresent, but also the picture "starry night" of Vincent van Gogh (1853–1890) is presented from time to time (see Picture 3).

Furthermore, in Asia famous artists were observing and painting turbulent flows. One of the most beautiful woodblock print of turbulent movements in water is "The Great Wave of Kangawa" that was produced by the famous Japanese artist Katsushika Hokusai (1760–1849) and shows Mount Fuji. In Hiroshige Utagawa's (1797–1858) wooden print "Vortices in the Konaruto River" (see Picture 4) two important elements of fluid dynamics research are addressed, namely self-similar fractal-like surface waves with their typical fingering at the crest. Furthermore, in a trilogy of prints by the same artist, produced in 1857, an ensemble of eddies of different sizes is shown (Picture 5). In this beautiful picture, the eddy boundaries were printed fuzzily. The fine structure is fractal, which, because of the distant view, is less observable than in the eddies of Picture 4.

**Picture 4** Whirlpools (vortices) in the woodblock print by Hiroshige Utagawa: "Vortices in the Konaruto stream." From: https://www.math.waikato.ac.nz/~seano/research/turbulence-pictures.html

**Picture 5** View of the whirlpools at Awa triptych, part of the series "Snow, Moon and Flowers." From: https://www.bing.com/images/search?q=Hiroshige+Utagawa&id=A9ED2E02901A5B1C6C319FqB3FA66106327CA6DBA&FORM=IQFRBA

Artists were attracted by turbulent phenomena also in the following years. In contemporary art, for example, out of mainstream art activities, Haltiner-Lüscher and Neukomm created drawings with lattice-like configurations surrounded by coherent turbulent flow structures (see Picture 6).

Turbulence is evident in many manifestations, e.g., it finds applications in *astrophysics* in the mathematical description of planetary rings and the formation of stars from stellar nebulae and spiral galaxies with turbulent star distributions, in *meteorology* in the jet stream, tornados, as wind gusts, cloud motions, etc. (see, e.g.,

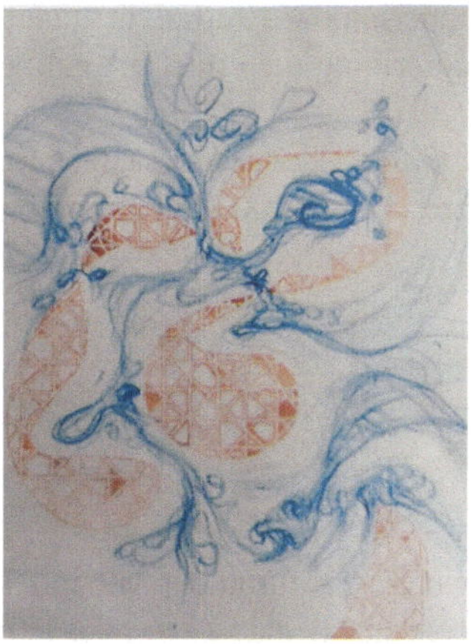

**Picture 6** A wax drawing with Tjanting on canvas by Esther Haltiner-Lüscher and Jacquy Philipp Neukomm. In this joint presentation, the artists seeked for an *equilibrium* between their single personal contributions of the *nonequilibrium* physical phenomenon *turbulence*. Printed with permission by Haltiner and Neukomm

Picture 3), in particular in early storm formations (Picture 2), in *oceanography* in the oceans, especially also the Gulf stream, tsunamis, lakes, and rivers, in *hydraulics* in the flow of river waters, e.g., the motion of the river water past bridge piers, in *fluid dynamics* as turbulent pipe flows, in boundary layers adjacent to forcing outer laminar or turbulent parallel flows, etc. Technically, turbulent phenomena arise in many fluid-related processes in almost every field of applications in engineering disciplines (mechanical, civil, geotechnical, environmental, etc.) within motions of liquids, gases, or composite materials of fluid-like behavior (particle suspensions and avalanching flows). Finally, yet importantly, there is turbulence also at smaller scales in daily life, e.g., the flow across wings of birds, the leaves falling from a tree visualizing the chaotic or turbulent nature of this peculiar physical flow phenomenon (singular!) in the nearest earth boundary layer or just the swirling flow in a spoon-agitated cup of our morning tea.

The scientific literature on turbulence is abundant even though it is only part of that of fluid dynamics, but it is a significant example of the physical appearance of deterministic fluctuations, often carelessly named randomness, which after Osborne Reynolds (1842–1912) may be considered as fluctuations, which in most cases is superposed on an otherwise rather regular averaged overall fluid process.

Mathematical attempts of the description of turbulence began in the nineteenth century with Claude Louis Marie Henry Navier's (1785–1836) and George Gabriel Stokes's (1819–1903) development of the Navier–Stokes equations (NSE) for density preserving viscous fluids and the observations and description of two characteristic flow regimes—seemingly orderly laminar flows and randomly oscillating turbulent motions. It needed Reynolds' ingenuity to recognize the laminar-turbulent transition as an instability process and to interpret the turbulent motion as a perturbation above the "laminar process" of chaotic quality, described by fluctuating quantities. This led, by statistical averaging techniques, to the Reynolds equations being identical to averaged evolution equations, describing the smooth orderly "ground motion," composed of what was the regular motion before, with superimposed turbulent quantities as correlation averages of fluctuating products describing the effect of "sub-grid fluctuations."

This is all well known, as is the fact that the most popular constitutive equations for the viscous stress tensor $\mathbf{t}^{(\text{visc})}$and its turbulent analogue $\mathbf{t}^{(\text{turb})}$ in a density preserving fluid are parameterized by:

$$\mathbf{t}^{(\text{visc})} = \rho\nu^{(\text{visc})}\left[\text{grad}\left(\mathbf{u}+\mathbf{u}^T\right) - \frac{2}{3}\text{tr}\mathbf{D}\right] = 2\rho\nu^{(\text{visc})}\left(\mathbf{D} - \frac{1}{3}\text{tr}\mathbf{D}\right) = 2\rho\nu^{(\text{visc})}\mathbf{D},$$

$$\mathbf{t}^{(\text{turb})} := -\rho\langle\mathbf{u}' \otimes \mathbf{u}'\rangle = 2\rho\left(\nu^{(\text{turb})}\langle\mathbf{D}\rangle - \frac{1}{3}k\,\mathbf{I}\right), \tag{P1a – e}$$

where the turbulent version is called Boussinesq closure and in which $\rho$ is the constant density, $\langle\mathbf{u}' \otimes \mathbf{u}'\rangle$ the second-order correlation term, the inner product, $k = 1/2\ \langle\mathbf{u}' \cdot \mathbf{u}'\rangle$, the turbulent kinetic energy, and $\nu^{(\text{visc})}$, $\nu^{(\text{turb})}$ the "laminar kinematic viscosity" and the "turbulent kinematic viscosity," also called "eddy viscosity." For a brief derivation of these formulas, see, e.g., Hutter & Jöhnk (2004, Chap. 10, p. 459).

In these early times of turbulence research, it was Ludwig Prandtl who had the ingenious idea to introduce a flux law, in analogy to Fick's law and Newton's shear layer gradient law (see Eq. (P2a)), leading to the second of the following two equations:

$$\frac{\tau_{21}^{(\text{visc})}}{\rho} = \nu^{(\text{visc})}\frac{\mathrm{d}u}{\mathrm{d}y} \quad\rightarrow\quad \frac{\tau_{21}^{(\text{turb})}}{\rho} = \nu^{(\text{turb})}\frac{\mathrm{d}\overline{u}}{\mathrm{d}y}, \tag{P2a, b}$$

where $y$ is the transverse coordinate, $u$ the velocity in the main flow direction and $\overline{u}$ its average. The viscosity $\nu^{(\text{visc})}$ is proportional to a characteristic length $l$ and a characteristic velocity scale $u$, a result that also follows from molecular theory. Therefore, $\nu^{(\text{visc})} \propto lu$ is a constant. The gradients in Eq. (P2a,b) are operators at a single space location and, therefore, they are local. On the other hand, $l$, which is identical to the mean free length of the molecules in the fluid, has a finite extension and would actually be nonlocal. However, because it is a constant, an equation as (P2a) is mathematically still called a local form. The application of Prandtl's analogy to the averaged velocity field of a turbulent flow extends this length scale to the size of a large-scale eddy or even the entire overall fluid domain. Therefore, its virtual nonlocality becomes even more evident. Important in this book is that the above closure conditions are *linear* (they contain $\langle\mathbf{D}\rangle$ only in a linear fashion) and they are *local*. This means that the viscous and turbulent stresses at position $\vec{x}$ and time $t$ only depend on values of the unknown, evaluated at the same spatial point $\vec{x}$ and time $t$.

It was Ludwig Prandtl's good idea to make the turbulent viscosity dependent on space or/and averaged velocity (in his famous mixing-length turbulence model on the derivative of the averaged velocity), a generalization that makes the turbulent shear stress, also called Reynolds shear stress, *nonlinear*. With this improvement in the prediction of turbulent shear flows fairly better results could be obtained.

Interestingly, a systematic treatment of *nonlocal* behavior was attacked much later (even though it remained untouched in the twentieth century for several decades). In April 1985, one of us (Peter William Egolf – PWE), deriving the *Difference-Quotient Turbulence Model* (DQTM), got automatically involved into the problem of nonlocality and started to tackle it. Significant results were obtained (partly with associate Daniel Weiss) for a number of simple fluid dynamical configurations. It was demonstrated that the theoretical results matched experimental results more convincingly than the results that were developed by the early mathematical, physical, and engineering experts, applying often stronger phenomenological local implications.

In these papers, the turbulent shear stress $\tau_{21}$ does not only depend on $\nu^{(\text{turb})}$ and $\langle \text{D} \rangle_{21}$ at a given point $x^{(\text{local})}$ and an initial time $t^{(\text{present})}$, but also on a possibly far-distant (neighboring) position $x^{(\text{neigh})}$ and in dynamic considerations at an earlier time $t^{(\text{earlier})}$, which is prior to the time $t^{(\text{present})}$ and, thereby, describes a *hereditary* (or *history*) *effect*, which is also called *memory effect*. However, P.W.E's main finding was that in a fluid domain with shear flow only three points are of significance, namely two boundary values and the actual position varying in between them. This variable location may change from one boundary value to the other, e.g., in a plane turbulent Poiseuille flow from the lower plate to the upper one. This was equivalent in a replacement of the local "gradient" in Boussinesq's approximation by a "difference quotient" as it occurs in the nonlocal DQTM

$$\frac{\text{d}\overline{u}}{\text{d}y} \quad \rightarrow \quad \frac{\overline{u}^{(\text{local})} - \overline{u}^{(\text{neigh})}}{x^{(\text{local})} - x^{(\text{neigh})}}. \tag{P3}$$

It is fascinating that there is a certain potentiality in this approach, because in the mathematical formulation the positions in between these three distinct points are of no importance. Physically they are very relevant; however, in the mathematical description, because of the assumed self-similarity of the turbulent structures, they do not appear. P.W.E. was also able, in analogy to the molecular theory of gases, to explain momentum transfer: not by assuming a single species of molecules of the same size, but by introducing an infinite set of different sized whirling eddies. The idea of momentum transfer by eddies of a single size also goes back to Prandtl and was correspondingly extended and generalized.

Based on these new ideas, the first author, partly together with Daniel Weiss, solved several elementary turbulent shear flows with success. Initiated by climatologist Thomas Stocker, Peter William Egolf and Kolumban Hutter met and discussed synergies of their works. In 2012, they decided to collaborate on the nonlocality and fractionality of turbulence and to further develop the corresponding ideas that came out of the DQTM. Furthermore, Kolumban Hutter, who with coauthor Yongqi Wang was writing an advanced fluid dynamics book series, in his second volume dedicated a full chapter to the above-described elementary turbulent flow solutions and made some additions and improvements (Hutter & Wang 2016).

The two authors then worked on nonextensive nonequilibrium thermodynamics of turbulence and were able to generalize the important energy-enstrophy spectrum of Robert Kraichnan (the last assistant of Albert Einstein). He had applied equilibrium Gibbs–Boltzmann thermodynamics to isotropic turbulence and obtained the correct power law for small-scale eddies that are in a quasi-equilibrium. The large eddies, on the other hand, show a different power law exponent that Kraichnan (at that time Tsallis' nonextensive thermodynamics did not exist) could not predict. Egolf and Hutter's new calculus explains the entire spectrum and, for large wave numbers, contains Kraichnan's correct part as a special case. They also developed a generalized temperature that for the quasi-equilibrium part of the spectrum, which relates to small eddies, reveals an identical temperature for all the different eddy classes with eddies of different sizes. However, for the small wave number nonequilibrium region of the spectrum these temperatures decrease for eddies of decreasing eddy diameter.

A further contribution of us to the development of turbulence is an analogy of turbulence with other physical systems showing critical or cooperative phenomena. By comparison of the different physical systems, a Curie and a Curie–Weiss law of turbulence could be developed. The first leads to the right response function of turbulence, which is the turbulence intensity, and in analogy to the "magnetization curve" a "vortization curve" is derived, which is a measure of the order in the turbulent system.

On this basis, quite naturally, this book provides an approach to turbulence that differs from usual standard textbooks published in the last few decades. So, we do not elaborate on, say, *local* higher order Reynolds stress parameterizations and how such procedures may better resolve the statistical nature of a particular turbulent process. We stay rather modest on the early simple *nonlocal* turbulent stress and deformation configurations, and we hope to convince the reader that significant fundamental results can be obtained in this way. One reason for this deviating presentation is that it addresses three main objectives:

(a) One goal of this presentation is to fully avoid empiricism and phenomenology and to present as much basic physics as possible. This was a rather difficult task, because most of the material of turbulence is of empiric or at least of semi-empiric nature. On the other hand, we did not hesitate to also base our considerations on rather simple physical models of turbulence invented by other authors. Models in our view are usually rather crude approximations to reality, and if a model mimics to a certain extent the behavior of a real-life system, it may lead to positive physical insight(s) and may permit some successful use in lecturing, research, and applications.
(b) We tend to strictly avoid linear and local concepts in favor of nonlinear, nonlocal, and fractional ones. In this treatise, it will be demonstrated that the terms "nonlocal" and "fractional" behavior will have many of their features in common. There is a large consensus that turbulence is "nonlinear" and

"nonlocal." However, especially in the field of its numerical simulations, there is an enormous resistance to abandon the conventional linear and local methods. This may be so, because at present especially nonlocal theories seem to be esoteric in their procedure of exploration and somewhat distant from showing a form that can be directly transformed to useful numerical algorithms set up to calculate complex flows in complex geometries.

(c) The third objective is to demonstrate that the desired belief that turbulence can hardly be treated analytically and that one must imperatively switch to sophisticated numerical methods is at least not correct in the area of solving fundamental turbulent flows of incompressible Newtonian fluids. This part of the book may bring forth motivation that future standard textbooks can also contain "clean" physical examples of beauty as do reviews and books of mechanics, electromagnetism, thermodynamics, and such in other physical fields.

In more specific details, the various chapters cover the following topics:

*Chapter 1:* The first chapter contains a tour d'horizon through today's turbulence field and its modeling. It lists all the scientific domains from mathematics to physics and geophysical sciences, etc., where turbulence plays a crucial role and is further investigated. A large number of references going back to the nineteenth century give the reader a broad view also of older scientists and their valuable contributions to turbulence.

*Chapter 2:* Here a brief, however serious and complete, mathematical derivation of the *R*eynolds *A*veraged *N*avier–*S*tokes (RANS) equations for incompressible Newtonian fluids is outlined. As a result, the Reynolds shear stress tensor is derived and its properties, e.g., symmetries, are discussed.

*Chapter 3:* The origin and the idea of the closure problem are explained in detail; we support the idea that (because of self-similarity) only closure on the lowest level is necessary and, therefore, also sufficient. Our hypothesis is that higher-order turbulence models are unnecessarily complex and can easily be discarded. This somewhat extreme idea is supported by numerous modern developments throughout this book.

*Chapter 4:* In this part of the book, the work on turbulent momentum and vorticity transport from the 1870s, when Boussinesq published his "constitutive equation", is reviewed and presented in detail.

*Chapter 5:* It is shown that Ludwig Prandtl, in his modeling of turbulent momentum transport, was on a successful path of discovery that he could not go to the end with a full harvesting of results. Because of this, even if today his mixing-length theory still finds many applications in numerous areas of turbulence research, e.g., in fluid machinery, air-conditioning, refrigeration, atmospheric physics, oceanography, and astrophysics, the power of simple zero-equation turbulence models is very much underestimated. Many zero-equation turbulence models of Prandtl and his contemporary scientists and engineers are reviewed, and their advantages and disadvantages are discussed. Furthermore, critics on present turbulence modeling by Corsin, Schmitt, etc., are presented and deficiencies and fallacies of existing turbulence models are listed and thoroughly explained. It is also demonstrated that a generalization of the "logarithmic law of the wall" fails with a tremendous error when compared with experiments.

*Chapter 6:* In this chapter, nonlocality in phase space is explained. Furthermore, it contains a review of existing nonlinear and nonlocal turbulence models given, e.g., by Kraichnan, Hinze, and Hamba. In the chapter at first atomic and continuum theories are explained and it is then demonstrated how in analogy to the molecular dynamic theory of ideal gases history dependent and nonlocal Reynolds stressses may be developed. Furthermore, in this chapter numerical proofs are given that demonstrate the failure of local and linear turbulence models.

*Chapter 7:* The main model of this book, the *D*ifference-*Q*uotient *T*urbulence *M*odel (DQTM), is derived by ideas of momentum transfer by whirling eddies. These considerations are based on the Lévy flight statistics and the fractal-$\beta$ model. After its development in the middle of the 1980s and its first publication in 1991, the DQTM turned out to be a nonlocal and fractional extension of Prandtl's former mixing-length model and shear-layer model. Finally, a comparison of fundamental aspects of shearing in laminar and turbulent flows is outlined.

*Chapter 8:* The technique of self-similarity is explained and a self-similar version of the *R*eynolds *A*veraged *N*avier-*S*tokes (RANS) equations is developed. This equation is the basis for developing solutions of elementary turbulent flow problems. This is important, because it transforms partial differential equations into a single ordinary differential equation. This simplification is essential to derive analytical solutions of the occurring linear and nonlinear differential equations.

*Chapter 9:* The application of the DQTM leads to analytical solutions of elementary turbulent shear flows of great simplicity and beauty. For example, the infinite Reynolds number plane turbulent Poiseuille flow reveals a circular mean velocity profile, in good agreement with observations. Moreover, "wall"-turbulent flow problems, already solved with the DQTM, reveal a critical phenomenon with the inverse Reynolds number acting as stress parameter and turbulence intensity being the order parameter. A new law of the wall, which is a deficit power law, will serve as an alternative of the periodically criticized logarithmic law of the wall (see Chap. 5).

*Chapter 10:* With the nonextensive thermodynamics of Tsallis, the authors have derived a fractional generalization of Kraichnan's energy and enstrophy spectrum of two-dimensional turbulence. For small wave numbers, it reveals the nonequilibrium energy spectrum of Kolmogorov–Obukhov, with an intermittency correction, and for large wave numbers the thermal equilibrium power law enstrophy spectrum with its "minus three exponent." A generalized Kelvin temperature of turbulence, $T_q$, is discovered, satisfying the $q$-thermodynamic relation $dS_q/dU_q = 1/T_q$, which in the nonequilibrium energy transfer range toward larger wave numbers monotonically decreases and rests constant in the thermal quasi-equilibrium enstrophy range.

*Chapter 11:* The authors of this book developed, in analogy to magnetism, a mean field theory of turbulence. By this, in analogy to their counterparts in the theory of magnetism, the "Curie law of turbulence" and the "Curie–Weiss law of turbulence" were discovered. The first leads directly to the right response function of a turbulent system, which—in analogy to the compressibility of a gas and the susceptibility of a ferromagnetic specimen—is a differential form of the turbulence intensity. Finally, all these findings propose incorporation of the presented theory of turbulence into the class of physical systems showing a cooperative phenomenon possibly with a dynamical phase transition exhibiting critical slowing down.

Possibly, this book with many theories, models, and numerous applications might bring support in stepping into the likely right direction toward a nonlocal and fractional structure of the physics of turbulence. This will lead to a better mathematical-physical understanding of the phenomenon "turbulence" and, finally, to improved numerical predictions of its temporal-spatial behavior. However, it is a long way to that end. Much must still be discovered by following generations of passionate younger turbulence researchers. We hope that this book may yield a serious launching pad for their future work.

Peter William Egolf is grateful to Daniel Weiss for the constructive collaboration (1995–2000) on some of the elementary turbulent shear flow problems and to Louis-Nicolas Douce and Fatou Samba for their valuable Bachelor diploma works. Furthermore, he thanks Francois Schmitt, Joachim Peintke, and Friedrich Busse for helpful discussions on different modern aspects of turbulence.

Kolumban Hutter is thankful to the Director of the Laboratory of Hydraulics, Hydrology and Glaciology at ETH-Zurich, R. Boes, and to ETH-Zürich in general, for the permission to work within the institute as a free researcher from outside.

Furthermore, we are thankful to our wives Hildegard and Barbara for their care and positive influence on our work–life balances, which is crucial to keep creativity awake and which was important for the production of this book.

Finally, the authors are grateful to Lisa Scalone (Springer editor, Heidelberg) for the friendly and competent accompaniment and her helpful advices throughout the production process of this physics book. Good support of production was also given by Shivakumar Krithika.

Niederlenz, Switzerland — Peter William Egolf
Zürich, Switzerland — Kolumban Hutter
May 2019

## References

Hutter, K., Jöhnk, K.: Continuum Methods of Physical Modeling. Springer, Berlin (2004). ISBN 13-978-364-205-831-8

Hutter, K., Wang, Y.: Fluid and thermodynamics, vol. 2. Advanced Fluid Mechanics and Thermodynamic Fundamentals, Springer International Publishing, Switzerland (2016). ISBN 978-3-319-33635-0

# Contents

# List of Symbols

## Roman Symbols

| | |
|---|---|
| $a$ | General physical quantity |
| $a$ | Shearing coefficient |
| $a$ | Space location |
| $a$ | Inverse ratio of jump probabilities of neighboring classes |
| $a$ | Location of the upper plate of plane turbulent Couette flows |
| $a$ | Integer number in a quadratic irrational number |
| $a$ | Parameter of a Reiner–Riwlin-type parameterization of the Reynolds stress |
| $a_1, a_2, \ldots$ | Integer numbers in general formula of a continued fraction |
| $a_k$ | Constants in a polynomial tensor development |
| $\overline{a}_0$ | Downstream acceleration of a fluid lump in a turbulent axisymmetric jet in and at the end of the theoretical core |
| $\overline{a}(\overline{t})$ | Mean acceleration of a fluid lump on the axis of a two-dimensional turbulent jet |
| $A$ | Surface |
| $A$ | Constant in general solution of plane turbulent Couette and Poiseuille flows |
| $A$ | Multiplicative parameter in the (universal) logarithmic law |
| $A$ | Constant in the derivation of the deficit power law of the wall |
| $A$ | Identifier for a thermodynamic subsystem |
| A, A* | Spatial locations related to eddy momentum transport |
| Ai | Abbreviation for Airy function |
| $A_j$, $A_n$ | Constants in a polynomial turbulent viscosity parameterization |
| $A_{imn}$ | Spectral coupling constants in a Galerkin–Ritz projection method |
| $\overline{\mathbf{A}}_2$ | Averaged second Rivlin-Ericksen tensor |
| $\overline{\mathbf{A}}_{2\mathrm{D}}$ | Deviator of tensor $\overline{\mathbf{A}}_2/2$ |
| $b$ | General physical quantity |

| | |
|---|---|
| $b$ | Space location |
| $b$ | Parameter of a Reiner–Riwlin-type parameterization of the Reynolds stress |
| $b$ | Half-width of turbulent shear flows, e.g., axi-symmetric jets |
| $b$ | Non-dimensional basic step size in a random walk (Brownian or Lévy walk) |
| $b$ | Integer number in a quadratic irrational number |
| $B$ | Additive parameter in the (universal) logarithmic law |
| $B$ | Constant in the derivation of the deficit power law of the wall |
| $B$ | Constant in general solution of plane turbulent Couette and Poiseuille flows |
| $B$ | Identifier for a thermodynamic subsystem |
| $\mathrm{B}(\vec{u}, \vec{u})$ | Fluid acceleration operator |
| $B_{ij}$ | Eulerian time covariance |
| $c$ | General constant |
| $c$ | Integer number in a quadratic irrational number |
| $c$ | Parameter of a Reiner–Riwlin-type parameterization of the Reynolds stress |
| $c$ | Constant arising in the DQTM |
| $c$ | Mass concentration, e.g., of an air pollutant |
| $c$ | Derivative of the width of a turbulent flow as a function of the downstream coordinate $x_1$ |
| $c_0$ | Convection term of the kinetic energy |
| $c_{\mathrm{m}}$ | General constant |
| $\breve{c}$ | General constant |
| $\hat{c}$ | General constant |
| $C$ | Constant in the derivation of the deficit power law of the wall |
| $C$ | Constant in the general solution of plane turbulent Couette and Poiseuille flows |
| $\mathrm{C}_{\alpha}$ | Nonlocal operator |
| $C_{H_0}$ | Specific heat at constant external magnetic field |
| $C_V$ | Specific heat at constant volume |
| $C_p$ | Specific heat at constant pressure |
| $C_q$ | Tsallis or $q$-generalized specific heat |
| $C_{\mu}$ | Universal constant in the diffusivity parameterization of the $k$-$\varepsilon$ turbulence model |
| $C_{\lvert u \rvert}$ | Specific heat of a turbulent system at constant absolute velocity $\lvert u \rvert$ |
| $d$ | General constant in the turbulent kinetic energy in critical theory of turbulence identified to be $\rho/2$ |
| $d$ | Euclidian space dimension |
| $d$ | Diameter of a cylindrical obstacle in wake flows |
| $d_0$ | Diameter of the orifice of a cylindrical pipe |
| $da$ | Differential surface element |

| | |
|---|---|
| $dv$ | Differential volume element |
| $d_f$ | Diffusion of the turbulent kinetic energy |
| $d_i$ | Dissipation of the turbulent kinetic energy |
| $D$ | von Kármán turbulent diffusion coefficient |
| $D$ | Diffusivity (constant) |
| $D$ | Hausdorff–Besicovitch dimension of a fractal set |
| $D_1$ | Hausdorff–Besicovitch dimension of a fractal set in a one-dimensional space |
| $\mathrm{D}_\alpha$ | Nonlocal operator |
| $D_\mathrm{m}$ | Molecular diffusion term |
| $D_{ij}^{(s)}$ | Transfer function of turbulent diffusion |
| ${}_L\mathrm{D}_+^\alpha$ | Left-handed Liouville fractional derivative of order α |
| ${}_L\mathrm{D}_\pm^\alpha$ | Left- and right-handed Liouville fractional derivative of order α |
| $\mathbf{D}$ | Stretching tensor, strain rate tensor of the symmetrized velocity gradient |
| $\overline{\mathbf{D}}$ | Average value of the stretching tensor or strain rate tensor of the symmetrized velocity gradient |
| $\mathbf{D}'$ | Fluctuation of the strain tensor |
| $e$ | Entrainment rate of axisymmetric turbulent jet |
| $e$ | Specific energy (per unit of mass) |
| $\hat{e}(k)$ | Energy density in Fourier space at wave number $k$ |
| $\hat{\breve{e}}(k)$ | Energy density in Fourier space of $\breve{e}$ at wave number $k$ |
| $e_q^x$ | Tsallis generalized $q$-exponential function |
| $E$ | Turbulent kinetic energy |
| $E$ | Enstrophy of a fluid |
| $E_\varepsilon$ | Turbulent kinetic energy of a two-dimensional flow |
| $E_H$ | Turbulent kinetic energy of a three-dimensional flow |
| $E_n$ | Turbulent kinetic energy of the eddies of class $n$ |
| $E_q$ | Tsallis $q$-averaged energy $E$ |
| $\breve{E}_\chi$ | Generalized energy |
| $\check{E}_\varepsilon$ | Generalized energy of a two-dimensional flow |
| $\hat{\check{E}}_\varepsilon(k)$ | Fourier transform of the generalized energy $\check{E}_\varepsilon$ |
| $\check{E}_H$ | Generalized energy of a three-dimensional flow |
| $\widetilde{E}$ | Kraichnan's discretized turbulent kinetic energy |
| $\langle\breve{E}\rangle$ | Averaged generalized energy $\breve{E}$ |
| $f$ | Function, e.g., elevation or amplitude of a wave |
| $f$ | Size of step in a random walk |
| $f(b)$ | Power law description of Lévy walks |
| $\vec{f}$ | Body force density |
| $f_\mathrm{T}$ | Turbulent fluidity |
| $f_n$ | Frequency of creation of eddies of eddy class $n$ |

| | |
|---|---|
| $f_k$ | Dimensionless average velocity of quasi-two-dimensional flows |
| $f_{kl}$ | Dimensionless Reynolds shear stress of quasi-two-dimensional flows |
| $f_1(\eta)$ | Self-similar function of mean downstream velocity |
| $f_2(\eta)$ | Self-similar function of mean spanwise velocity |
| $f_{11}(\eta)$ | Normal turbulent intensity of the downstream velocity fluctuation |
| $f_{21}(\eta)$ | Self-similar function of Reynolds shear stress |
| $f_{22}(\eta)$ | Normal turbulent intensity of the spanwise velocity fluctuation |
| $\hat{f}$ | Fourier transform of function $f$ |
| $F$ | Fourier transform of function $f$ |
| $F$ | Force |
| $F$ | Functional |
| $\overline{F}$ | Average of a flux quantity $F$ |
| $F_q$ | Tsallis $q$-averaged function $f$ |
| $\mathbf{F}$ | Large-scale solenoidal forcing field |
| $g$ | Specific physical quantity |
| $g$ | Functional |
| $g$ | The golden mean |
| $\vec{g}$ | Vectorial physical quantity |
| $g(x, x')$ | Nonlocality function, response function |
| $g_H(x, x')$ | Hamba's response function |
| $g_K(x, x')$ | Kraichnan's response function |
| $g_1(\eta)$ | Normalized longitudinal mean velocity in plane Couette or Poiseuille flows |
| $g_{21}(\eta)$ | Normalized tangential turbulent intensity of turbulent axisymmetric jets and plane Couette or Poiseuille flows |
| $G$ | Shear modulus |
| $G$ | Characteristic function in wave number spectrum |
| $G(\vec{r}, \vec{r}')$ | Density–density correlation function |
| $G(r)$ | Spatial correlation function |
| $G_{21}$ | Shear modulus of plane shear flows |
| $G_q$ | Tsallis $q$-averaged Gibbs free energy $G$ |
| $h$ | Auxiliary function in the theory of plane turbulent Poiseuille flows |
| $h$ | Hurst exponent of fractal and superdiffusive systems |
| $h$ | Exponent of the structure function in the multi-fractal model |
| $h_1 \ldots h_4$ | Bessel and Weber functions of first and second kind |
| $h_1$ | First exponent of the structure function in the bi-fractal model |
| $h_2$ | Second exponent of the structure function in the bi-fractal model |
| $H$ | Auxiliary function in the theory of plane turbulent Poiseuille flows |
| $H$ | Internal magnetic field |
| $H_0$ | External magnetic field |
| $H$ | Helicity |
| $H_q$ | Tsallis $q$-averaged Helmholtz free energy $H$ |

| | |
|---|---|
| $\mathbf{H}$ | Hessian tensor of the pressure field |
| $\widetilde{\mathbf{H}}$ | Traceless symmetric Hessian tensor |
| $i$ | Imaginary number |
| $\mathbf{I}$ | Unit tensor, unit matrix in three dimensions |
| $j_2$ | Transverse specific eddy momentum flux |
| $\vec{j}$ | Mass flux vector |
| $\vec{j}_{\mathrm{t}}$ | Turbulent mass flux vector |
| $J_\kappa$ | Bessel function of the first kind and of order $\kappa$ |
| $k$ | Wave number |
| $k$ | Generalized Boltzmann constant in the thermodynamics of nonlinear or turbulent systems |
| $k$ | von Kármán constant |
| $k$ | Turbulent kinetic energy |
| $k$ | Running or counting index of eddy classes |
| $k$ | Exponent in deficit power law of the wall |
| $k_{\mathrm{B}}$ | Boltzmann constant |
| $k_c$ | Critical value of the exponent of the deficit power law of the wall |
| $k_p$ | Pole value of the exponent of the deficit power law of the wall |
| $k$-$\varepsilon$ | Two-equation turbulence model for the transport of kinetic energy $k$ and dissipation rate $\varepsilon$ |
| $k$-$l$ | Two-equation turbulence model for the transport of turbulent kinetic energy $k$ and Kolmogorov length $l$ |
| $k$-$\omega$ | Two-equation turbulence model for the transport of turbulent kinetic energy $k$ and vorticity $\omega$ |
| $k'$ | Lowest occurring wave number modulus |
| $k''$ | Highest occurring wave number modulus |
| $K$ | Diffusivity constant |
| $K_1$ | Abbreviation for a constant in the deficit power law of the wall |
| $K_2$ | Abbreviation for a constant in the deficit power law of the wall |
| $K_3$ | Abbreviation for a constant in the deficit power law of the wall |
| $K_i$ | Coefficient in a higher order flux parameterization |
| $K_z$ | Vertical eddy diffusivity of contraction profile $c$ |
| $K_{ij}$ | Tensorial diffusivity |
| $\mathbf{K}$ | Bachelor's diffusivity tensor ("coefficient") |
| $l$ | Characteristic length |
| $l$ | Turbulent eddy length or diameter |
| $l_0$ | Characteristic length of a flow domain |
| $l_n$ | Characteristic size (height) of a workman or eddy of class $n$ |
| $l_K$ | Kolmogorov's dissipation scale |
| $\ln_q$ | Tsallis generalized $q$-logarithmic function |
| $\ell_{\mathrm{m}}$ | Prandtl's (first) turbulent mixing length |
| $\ell'_{\mathrm{m}}$ | Prandtl's (second) turbulent mixing length |
| $\ell_\omega$ | Taylor's vorticity mixing length |
| $L$ | Characteristic length scale of a flow domain |

| | |
|---|---|
| $L_0$ | Characteristic length scale of a flow domain |
| $\mathbf{L}$ | Turbulent velocity gradient tensor |
| $\overline{\mathbf{L}}$ | Average turbulent velocity gradient tensor |
| $m$ | Exponent of deficit power law of the wall |
| $m$ | Running or counting index of eddy classes |
| $m$ | Specific magnetization (1-d, scalar) |
| $m$ | Mass flux of axisymmetric turbulent jets |
| $m_c$ | Critical value of the exponent of the deficit power law of the wall |
| $m_p$ | Pole value of the exponent of the deficit power law of the wall, called marvelous proportion (new naming by authors) |
| $m_{\mathrm{eddy,min}}$ | Mass of the large-scale eddy on the lower side of $x_2^*$ at its maximum size; see Fig. 7.5 |
| $m_{\mathrm{eddy,max}}$ | Mass of the large-scale eddy on the upper side of $x_2^*$ at its maximum size; see Fig. 7.5 |
| $m_{\mathrm{mol}}$ | Molecular mass |
| $m_0$ | Mass flux of axi-symmetric turbulent jet in the orifice |
| $M$ | Total number of eddies of all classes $n$ created during the time period $T$ |
| $M$ | Memory kernel in turbulence model |
| $M$ | Magnetization |
| $\overline{M}^{(\nu)}_{\nu_1 \ldots \nu_d}$ | Average of the velocity moment of order $\nu$ |
| $\vec{\mathcal{M}}$ | Magnetic moment |
| $n$ | Number of the workmen or turbulent eddy classes, respectively |
| $n$ | Number of particles in a volume element, particle number density |
| $n$ | Exponent of $\overline{\mathbf{D}}$ in a parameterization of $\nu_{\mathrm{T}}$ |
| $n$ | Integer number in a quadratic irrational number |
| $n_j$ | Exponents of $\overline{\mathbf{D}}$ in a parameterization of $\nu_{\mathrm{T}}$ ($j = 1,2,\ldots,N$) |
| $n_{min}$ | Eddy density of the eddies on the lower side of $x_2^*$; see Fig. 7.5 |
| $n_{max}$ | Eddy density of the eddies on the upper side of $x_2^*$; see Fig. 7.5 |
| $\vec{n}$ | Normal vector |
| $\tilde{n}$ | Number of molecules transporting momentum over the plane $\partial\Gamma$ with a unit surface $\Delta A_2$ per unit of time $\Delta t$ |
| $N$ | Number of Lévy flight steps or turbulent eddy sizes (classes), respectively, minus one |
| $N$ | Nonlocal kernel in a turbulence model |
| $N$ | Number of partial systems in a thermodynamic system |
| $N(\lambda, \xi_2)$ | Nonlocal kernel in a DQTM parameterization |
| $o$ | Order parameter (e.g., density, magnetization) |
| $o_c$ | Critical value of the order parameter $o$ |
| $o_n$ | Number of eddies of class $n$ (occupation number) |
| $o_p$ | Pole value of the order parameter $o$ |
| $O$ | Order parameter (normalized) |
| $O$ | Total number of eddies of all existing classes |
| $\mathrm{O}(\ldots)$ | Attenuation behavior (order of magnitude) |

| | |
|---|---|
| $\hat{\mathrm{O}}$ | Operator in fractional calculus |
| $p$ | Pressure |
| $p$ | Abbreviation for a derivative in the derivation of the logarithmic law |
| $p$ | Pole distance of turbulent flows behind obstacles and openings |
| $p$ | Probability distribution (e.g., of a Lévy flight process) |
| $p$ | Pole distance |
| $p_\mathrm{d}$ | Pressure diffusion |
| $p_i$ | Probability of state $i$ |
| $p_i^q$ | Escort probability of state $i$ |
| $p_i^\mathrm{A}, p_i^\mathrm{B}$ | Probability of state $i$ in partial system A or B, respectively |
| $p_{ij}^{\mathrm{A+B}}$ | Probability of state $i$ in partial system A and state $j$ in partial system B |
| $p_n$ | Probability to have a creation of an eddy of class $n$ |
| $p_n^{(\mathrm{active})}$ | Active space ratio for turbulent patches of eddies of class $n$ |
| $p_r$ | Production of turbulent kinetic energy |
| $\overline{p}_\mathrm{min}$ | Momentum of the lower eddy (including the momentum of the smaller scale eddies); see Fig. 7.5 |
| $\overline{p}_\mathrm{max}$ | Momentum of the upper eddy (including the momentum of the smaller scale eddies); see Fig. 7.5 |
| $\hat{p}$ | Average specific momentum of a single eddy |
| $\hat{p}_n$ | Specific momentum of the eddies of class $n$ |
| $p_\mathrm{mol}$ | Momentum density of a molecule |
| $p(s)$ | Jump probability distribution |
| $p', p''$ | Impact factors or probabilities |
| $\vec{p}$ | Momentum (in phase space) |
| $\vec{p}_\mathrm{A}$ | Momentum of particle A on trajectory $x_\mathrm{A}$ |
| $\vec{p}_\mathrm{B}$ | Momentum of particle B on trajectory $x_\mathrm{B}$ |
| $p_0$ | Exponent of self-similar width of a turbulent shear flow |
| $p_0$ | Initial pressure |
| $p_1$ | Exponent of self-similar mean downstream velocity of turbulent shear flows |
| $p_2$ | Exponent of self-similar mean spanwise velocity of turbulent shear flows |
| $p_{21}$ | Exponent of self-similar Reynolds stress of turbulent shear flows |
| $P$ | Pressure |
| $P$ | Transition probability rate of a fluid lump |
| $P_i$ | Escort probability |
| $P_n$ | Projection operator in Galerkin–Ritz spectral method |
| $P_{r\mathrm{T}}$ | Turbulent Prandtl number |
| $P_{r\mathrm{T}}^{(\mathrm{B})}$ | Batchelor's generalized turbulent Prandtl number |
| $\hat{P}$ | Total specific momentum of all eddies |
| $\hat{P}_{m,N}$ | Total linear momentum of all eddies of the classes $m$ to $N$ |
| $q$ | Index to identify Tsallis's thermodynamic quantities |
| $q_n$ | Occupation probability of eddies of class $n$ |

| | |
|---|---|
| $\vec{q}$ | Location (in phase space) |
| $\vec{q}_{\mathrm{t}}$ | Turbulent heat flux vector |
| $\vec{q}$ | Heat flux vector |
| $r$ | Parameter for measuring the validity of the Boussinesq-type turbulence closure |
| $r$ | Similarity ratio in a fractal set |
| $r$ | Decadal ratio of mean velocity profiles |
| $r$ | Radius of a turbulent eddy |
| $r$ | Variable position in a solid, liquid, or gaseous body |
| $r_k$ | Position of particle $k$ in a solid, liquid, or gaseous body |
| $R$ | Radius of a pipe |
| $R_{u_1 u_2}$ | Second-order Lagrangian correlation function |
| Re | Overall flow Reynolds number |
| $\mathrm{Re}_c$ | Critical Reynolds number |
| $\mathrm{Re}_\lambda$ | Taylor Reynolds number |
| $\mathrm{Re}_\eta$ | Kolmogorov's dissipation Reynolds number |
| $\mathrm{Re}_{0,\mathrm{channel}}$ | Reynolds number related to channel flows |
| $\mathrm{Re}_{0,\mathrm{wall}}$ | Reynolds number related to flows along a wall |
| $\mathrm{Re}^*$ | Reynolds number of plane turbulent Couette and Poiseuille flows related to the overall width |
| $\mathrm{Re}_c^*$ | Critical Reynolds number of plane turbulent Couette and Poiseuille flows related to the overall width |
| $\mathrm{Re}^+$ | Reynolds number of a flow along a boundary layer related to the boundary layer thickness |
| $\mathbf{R}$ | Reynolds stress tensor |
| $\overline{\mathbf{R}}$ | Averaged Reynolds stress tensor |
| $\overline{\mathbf{R}}_{\mathrm{D}}$ | Averaged Reynolds stress deviator |
| $\overline{\mathbf{R}}_{\mathrm{DL}}$ | Linearized averaged Reynolds stress deviator |
| $\overline{\mathbf{R}}_{\mathrm{DN}}$ | Nonlinear (e.g., polynomial) averaged Reynolds stress deviator |
| $s$ | Integration variable |
| $s$ | Step size in a Lévy flight distribution |
| $s$ | Specific entropy (per unit of mass) |
| $s$ | External stress parameter |
| $s_c$ | Critical value of the external stress parameter |
| $s_\uparrow$ | Spin up (+1/2 or +1) |
| $s_\downarrow$ | Spin down (−1/2 or −1) |
| $\hat{\mathrm{s}}_\pm$ | Positive and negative shift operator |
| $S$ | Dimensionless stress parameter |
| $S$ | Surface area |
| $S_p$ | Velocity structure function of order $p$ |
| $S_q$ | Tsallis generalized entropy |
| $S_\varepsilon$ | Turbulent energy spectrum |
| $S_{\mathrm{BG}}$ | Entropy of Boltzmann–Gibbs type |
| $S_q$ | Tsallis or nonextensive entropy |

| | |
|---|---|
| $\mathbf{S}$ | Vorticity tensor |
| $\overline{\mathbf{S}}$ | Averaged vorticity tensor |
| $t$ | Time |
| $t$ | Turning time of an eddy (for a single rotation) |
| $t^-$ | Time smaller than $t_0$ (backward in time) |
| $t^+$ | Time larger than $t_0$ (forward in time) |
| $\bar{t}$ | Time of flight of a fluid lump in a turbulent axisymmetric jet starting in the orifice and moving on the jet axis |
| $t_0$ | Present or initial time |
| $\bar{t}_0$ | Time of flight of a fluid lump in a turbulent axisymmetric jet at the end of the theoretical core |
| $t_n$ | Characteristic time in which the energy of eddies of class $n$ is transferred to that of class $n + 1$, also: turnover time of eddy of class $n$ |
| $\langle \mathbf{t}_\mathrm{D} \rangle$ | Averaged dissipation tensor |
| $\langle \mathbf{t}_\mathrm{R} \rangle$ | Averaged Reynolds stress tensor |
| $T$ | Overall characteristic time scale |
| $T$ | Total time of eddy creation process |
| $T$ | Kinetic energy of a harmonic oscillator |
| $T$ | Absolute temperature, also called Kelvin temperature |
| $T_1$ | Kinetic energy of a first harmonic oscillator coupled to a second one |
| $T_2$ | Kinetic energy of a second harmonic oscillator coupled to a first one |
| $T_\mathrm{c}$ | Curie temperature or critical temperature |
| $T_n^{(\mathrm{eddy})}$ | Time when an eddy of class $n$ is present |
| $T_q$ | Tsallis generalized temperature |
| $T_\mathrm{W}$ | Freezing/melting temperature of water |
| $\overline{T}$ | Mean absolute temperature |
| $\langle T_k \rangle$ | Average flux of kinetic energy |
| $\overline{\mathbf{T}}_k$ | Traceless tensor |
| $\langle \vec{T}_p \rangle$ | Pressure contribution to the averaged turbulent enthalpy flux |
| $u$ | General velocity |
| $u$ | Velocity in downstream direction |
| $u$ | Tangent in phase space of the trajectory |
| $\bar{u}$ | Downstream velocity of a fluid lump moving on the axis of a turbulent axisymmetric jet |
| $u_\mathrm{mol}$ | Flight velocity of molecules |
| $u_\mathrm{mol}^{(\mathrm{rms})}$ | Root-mean-square velocity of molecules |
| $u_n$ | Characteristic velocity of eddies of class $n$ |
| $u_\eta$ | Kolmogorov's micro-velocity scale or dissipation velocity scale |
| $\bar{u}_0$ | Downstream velocity of a fluid lump moving on the axis of a turbulent axisymmetric jet |
| $\vec{u}_0$ | Initial velocity vector |
| $\hat{u}_n$ | Fourier components of the velocity vector |
| $\vec{u}$ | Three-dimensional velocity vector |

$\widehat{\vec{u}}^{(\kappa')}_{\vec{n}}$ Fourier component of velocity $\vec{u}$ with dimensionless integer wave number $\vec{n} = (n_1, n_2, n_3)$ and lowest wave number $\vec{k}' = \kappa'(n_1, n_2, n_3)$
$\widehat{\vec{u}}^{(\kappa',\kappa'')}_{\vec{n}}$ Fourier component of velocity $\vec{u}$ with dimensionless integer wave number $\vec{n} = (n_1, n_2, n_3)$, lowest wave number $\vec{k}' = \kappa'(n_1, n_2, n_3)$ and highest wave number $\vec{k}'' = \kappa''(n_1, n_2, n_3)$
$\widehat{\vec{u}}_{\vec{k}}$ Fourier transformed velocity of wave number $\vec{k}$
$\langle \vec{u} \rangle$ Averaged value of $\vec{u}$
$\sqrt{\overline{(u'_1)^2}}$ Root mean square of the downstream fluctuation velocity $u'_1$
$\sqrt{\overline{(u'_2)^2}}$ Root mean square of the spanwise fluctuation velocity $u'_2$
$\sqrt{\overline{(u'_3)^2}}$ Root mean square of transversal or azimuthal fluctuation velocity $u'_3$
$u_1$ Cartesian or cylindrical streamwise component of the vector $\vec{u}$
$u_2$ Cartesian spanwise or cylindrical radial component of the vector $\vec{u}$
$u_3$ Cartesian orthogonal or cylindrical azimuthal component of the vector $\vec{u}$
$u_{1a}$, $u_{1b}$ Streamwise velocities at locations $a$ and $b$
$u_{2a}$, $u_{2b}$ Spanwise velocities at locations $a$ and $b$
$\overline{u}_1$ Mean velocity in streamwise direction
$\overline{u}_2$ Mean velocity in spanwise direction
$\overline{u}_{1_{\min}}$ Minimum mean streamwise velocity in a shear flow
$\overline{u}_{1_{\max}}$ Maximum mean streamwise velocity in a shear flow
$\widehat{u}_1$ Dimensionless (normalized) streamwise velocity
$\overline{u}_1^*$ Mean perturbation part of the streamwise mean velocity of a shear flow
$\vec{\widetilde{u}}$ Projected vector in Galerkin–Ritz truncation and projection method
$u^*$ Boundary shear velocity
$u^+$ Dimensionless longitudinal velocity in "wall"-turbulent flows
$\overline{u}^*{}_{\alpha\beta}$ Root-mean-square velocities, $\alpha \in \{1, 2\}$, $\beta \in \{\min, \max\}$
$\overline{u'_i u'_j}$ Second-order velocity fluctuation correlation
$u_1^*$ Average turbulent streamwise velocity
$u_2^*$ Average turbulent spanwise velocity
$U$ Central velocity in a turbulent pipe flow
$U$ Outlet velocity in a round orifice
$U$ Free stream velocity in "wall"-turbulent flows
$U$ Potential energy of a harmonic oscillator
$U_G$ Basic and constant downstream velocity
$U_q$ Tsallis generalized internal energy
$U_0$ Constant reference velocity
$v$ Specific volume (per unit of mass)
$V$ Volume
$w$ Width of a channel

| | |
|---|---|
| $\vec{w}_n$ | $n$-th Eigenvector of the Stokes operator |
| $\vec{w}_{i,n}$ | $i$-th component of the $n$-th Eigenvector of the Stokes operator |
| $W$ | Work |
| $W_n$ | Work performed by a workman of class $n$ |
| $\overline{w'c'}$ | Diffusive flux of a boundary layer concentration profile $c$ due to the normal velocity fluctuation $w'$ |
| $x$ | Coordinate in the streamwise direction |
| $\overline{x}$ | Downstream position of a fluid lump moving on the axis of a turbulent axisymmetric jet |
| $\vec{x}$ | Space or velocity vector in Cartesian or cylindrical coordinates |
| $\widetilde{x}$ | Cartesian replica of the space field in a Galerkin–Ritz projection method |
| $x_{\mathrm{A}}$ | Trajectory of a fluid lump A |
| $x_{\mathrm{A}}$ | Variable of a subsystem A as part of a larger system |
| $x_{\mathrm{B}}$ | Trajectory of a fluid lump B |
| $x_{\mathrm{B}}$ | Variable of a subsystem B as part of a larger system |
| $x_0$ | Initial coordinate value at time $t = 0$ |
| $x_0$ | Core length of a turbulent cylindrical jet |
| $\overline{x}_0$ | Streamwise position of a fluid lump moving on the axis of a turbulent axisymmetric jet at the end of the theoretical core |
| $x_1$ | Coordinate in streamwise direction |
| $x_2$ | Coordinate in spanwise direction |
| $x_3$ | Coordinate in rectangular or azimuthal direction |
| $\overline{x}_1$ | Streamwise position of a fluid lump moving on the axis of a turbulent axisymmetric jet |
| $\widehat{x}_i$ | Stretched coordinate, stretched with a constant and the angular velocity of the largest eddies |
| $x_2^*$ | Actual spanwise spatial location |
| $x_{2_{\min}}$ | The $x_2$-position in a turbulent shear flow, where $u_1$ reaches its minimum |
| $x_{2_{\max}}$ | The $x_2$-position in a turbulent shear flow, where $u_1$ reaches its maximum |
| $x'$ | Free parameter for a space location |
| $X_q$ | Tsallis $q$-averaged position $x$ |
| $y$ | Radial coordinate |
| $Y_k$ | Bessel function of the second kind and of order $k$ |
| $y^+$ | Dimensionless boundary layer variable in the $x_2$-direction of "wall"-turbulent flows |
| $z$ | Vertical or azimuthal component of a Euclidian coordinate system |
| $Z_{\mathrm{BG}}$ | Boltzmann–Gibbs partition function |
| $Z_q$ | Tsallis $q$-generalized partition function |
| $Z_{\mathrm{ES}\varepsilon}$ | Partition function derived with the Escort probability distribution |

## Greek Symbols

$\alpha$ General constant
$\alpha$ Constant in a polynomial tensor development
$\alpha$ Exponent in the power law parameterization of turbulent boundary layer flows
$\alpha$ Order of Liouville fractional derivative
$\alpha$ Exponent of a power law describing the workman's sizes
$\alpha$ Spreading angle of round turbulent jet
$\alpha$ Constant in Kraichnan's energy-enstrophy spectrum
$\alpha$ Exponent characterizing diffusion (subdiffusion, normal diffusion, superdiffusion)
$\alpha$ Stress parameter in the DQTM formulation of turbulent shear flows
$\alpha$ Critical exponent for a quantity that approaches criticality from above
$\alpha$ Exponent of time behavior of mean quadratic separation distance in Brownian motion
$\alpha_c$ Critical stress parameter in the DQTM formulation of turbulent shear flows
$\alpha'$ Critical exponent for a quantity that approaches criticality from below
$\beta$ Constant in a polynomial tensor development
$\beta$ Reduction factor of space filling (this symbol gives the random beta-model its name)
$\beta$ Scale parameter in the self-similar parameterization of turbulent shear flows
$\beta$ Factor for normal turbulence intensity in streamwise direction of turbulent axisymmetric jets
$\beta$ Non-normalized order parameter of plane turbulent Couette and Poiseuille flows
$\beta$ Coefficient of $\overline{T}_2$ in the nonlinear deviatoric Reynolds shear stress tensor
$\beta$ Boltzmann factor
$\beta$ Constant in Kraichnan's energy-enstrophy spectrum
$\beta$ Critical exponent for a quantity that is approached from above
$\beta'$ Critical exponent for a quantity that is approached from below
$\beta_q$ Coldness in Tsallis non-extensive thermodynamics
$\gamma$ Counter gradient term in a mass concentration flux profile
$\gamma$ Critical exponent for a quantity that approaches criticality from above
$\gamma$ "Identical" parts in a fractal set
$\gamma$ Reduction factor for normal turbulence intensities in transverse and azimuthal directions of turbulent axisymmetric jets
$\gamma$ Coefficient of $\overline{\mathrm{T}}_3$ in the nonlinear deviatoric Reynolds stress tensor
$\gamma$ Scale factor in orthotropic turbulence intensity parameterization
$\gamma_{\mathrm{L}}$ Exponent in the asymptotic Lévy distribution
$\gamma_{\mathrm{T}}$ Diffusivity in the flux parameterization of the turbulent kinetic energy
$\gamma_{21}$ Shear angle

| | |
|---|---|
| $\gamma'$ | Critical exponent for a quantity that approaches criticality from below |
| $\Gamma$ | Gamma function |
| $\Gamma$ | Space set or subset |
| $\overline{\Gamma}$ | Mean conserved physical quantity $\Gamma$ |
| $\mathbf{\Gamma}_{\mathrm{T}}$ | Turbulent diffusivity tensor of the anisotropic flux parameterization of the turbulent kinetic energy |
| $\delta$ | Dirac's delta distribution |
| $\delta$ | Abbreviation for order parameter combination in plane turbulent Couette flows |
| $\delta$ | Critical exponent |
| $\delta$ | Boundary layer "thickness," i.e., position where the velocity in the boundary layer is at its maximum |
| $\delta a$ | Small but not infinitesimal change of $a$ |
| $\delta_{21}$ | Shear deformation angle |
| $\delta^{+}$ | Normalized boundary layer thickness of "wall"-turbulent flows |
| $\delta\vec{u}(\vec{r})$ | Nonlocal velocity increment |
| $\Delta a$ | Small but not infinitesimal change of $a$ |
| $\Delta N$ | Number of particles in a small spatial element |
| $\Delta V$ | Small volume element |
| $\Delta t$ | Characteristic time in the eddy creation process (adjustable parameter) |
| $\overline{\Delta M}$ | Total average linear excess momentum |
| $\varepsilon$ | Abbreviation for order parameter combination in plane turbulent Couette flows |
| $\varepsilon$ | Enstrophy |
| $\varepsilon$ | Energy transfer rate of turbulent eddies |
| $\varepsilon$ | Specific turbulent dissipation rate |
| $\varepsilon_{\mathrm{m}}$ | Eddy viscosity of turbulent momentum transfer |
| $\varepsilon_{\mathrm{m,ml}}$ | Eddy diffusivity of Prandtl's mixing length turbulence model |
| $\varepsilon_{\mathrm{m,mg}}$ | Eddy diffusivity of Prandtl's mean gradient turbulence model |
| $\varepsilon_{\mathrm{m,sl}}$ | Eddy diffusivity of Prandtl's shear layer turbulence model |
| $(\varepsilon_{\mathrm{m}})_{ij}$ | Eddy diffusivity tensor |
| $\varsigma$ | Generalized first coordinate of circle profile of plane turbulent Poiseuille flows |
| $\zeta_2$ | Correlation of the two-particle number density |
| $\zeta_p$ | Exponent of the velocity structure function of order $p$ |
| $\eta$ | Kolmogorov's micro-length scale |
| $\eta$ | Generalized second coordinate of circle profile of plane turbulent Poiseuille flows |
| $\eta$ | Critical exponent of density–density correlation $G$ |
| $\eta$ | Dimensionless spanwise coordinate |
| $\theta$ | Scalar physical quantity |
| $\theta$ | Heaviside distribution |
| $\kappa$ | Constant arising in Prandtl's parameterization of $\varepsilon_{\mathrm{m}}$ |
| $\kappa$ | Factor in multi-eddy description of turbulence |

| | |
|---|---|
| $\kappa$ | Normalized dimensionless turbulent kinetic energy |
| $\kappa_{\mathrm{m}}$ | Molecular diffusion coefficient |
| $\kappa_{m,N}$ | Ratio of eddy momentum of eddies of classes $m$ to $N$ to that of classes 0 to infinity |
| $\kappa_S$ | Compressibility at constant entropy |
| $\kappa_T$ | Compressibility at constant temperature |
| $\kappa_{T,0}$ | Compressibility of an ideal gas at constant temperature |
| $\kappa'$ | Lowest occurring wave number of the turbulent spectrum of eddies |
| $\kappa''$ | Highest occurring wave number of the turbulent spectrum of eddies |
| $\lambda$ | Mean free path length between two molecules |
| $\lambda$ | Internal characteristic length in "wall turbulence" |
| $\lambda$ | Coefficient in the DQTM that describes reduced eddy momentum transfer by dissipation of the small-scale eddies |
| $\lambda$ | Scaling factor in random motion |
| $\lambda$ | Parameter in weighting function of fractional calculus |
| $\lambda$ | Stretching factor of normalized transversal coordinate in plane turbulent Poiseuille flows |
| $\lambda_{k,N}$ | Coefficient and ratio in the DQTM parameterization |
| $\lambda_1$ | First Lagrange parameter |
| $\lambda_2$ | Second Lagrange parameter |
| $\Lambda_{\mathrm{T}}$ | Coefficient in the DQTM parameterization (turbulent mixing length) |
| $\Lambda_{\mathrm{T},2}$ | Coefficient in the DQTM parameterization or turbulent mixing length in the flow transverse direction |
| $\mu$ | Dynamic viscosity of the fluid or laminar dynamic viscosity |
| $\mu$ | Intermittency exponent |
| $\mu$ | Ratio of the logarithms of constants $a$ and $b$ characterizing Lévy flights |
| $\mu(h)$ | Distribution function of exponents $h$ in the multi-fractal model |
| $\mu_{\mathrm{T}}$ | Turbulent dynamic viscosity |
| $\nu$ | Kinematic viscosity |
| $\nu$ | Order of a moment |
| $\nu$ | Critical exponent for a quantity that approaches criticality from above |
| $\nu_{ij}$ | Kernel related to the eddy diffusivity |
| $\nu_{\mathrm{T}}$ | Turbulent kinematic viscosity |
| $(\nu_{\mathrm{T}})_{ij}$ | Eddy diffusivity tensor |
| $\nu'$ | Critical exponent for a quantity that approaches criticality from below |
| $\nu'$ | Fluctuation quantity averaged over some neighboring sites of a grid |
| $\xi$ | Dimensionless downstream coordinate |
| $\xi$ | Integration variable in Liouville fractional derivative |
| $\xi$ | Integration variable in the theory of turbulent axisymmetric jets |
| $\xi$ | Correlation length of the density fluctuations |
| $\xi(x,t)$ | General physical quantity |
| $\xi(x,t,\rho)$ | Probability density of quantity $\xi(x,t)$ |
| $\langle\xi\rangle$ | Statistical average of quantity $\xi$ |
| $\xi'$ | Fluctuation of quantity $\xi$ |

| | |
|---|---|
| $\xi_1$ | Self-similar streamwise dimensionless coordinate in turbulent shear flows |
| $\pi^{(g)}$ | Production rate density of quantity $g$ |
| $\vec{\pi}^{*(g)}$ | Vectorial production rate density of quantity $g$ |
| $\pi(\eta)$ | Normalized production of turbulent kinetic energy |
| $\Pi$ | Dimensionless average pressure |
| $II_{\mathbf{A}}$ | Second invariant of tensor **A** |
| $I_{\overline{\mathbf{D}}}$ | First invariant of tensor $\overline{\mathbf{D}}$ |
| $II_{\overline{\mathbf{D}}}$ | Second invariant of tensor $\overline{\mathbf{D}}$ |
| $III_{\overline{\mathbf{D}}}$ | Third invariant of tensor $\overline{\mathbf{D}}$ |
| $\rho$ | Probability measure |
| $\rho$ | Density (mass density) |
| $\rho_k$ | Ratio of the alignment of the total flux $\vec{T}$ with the gradient of $k$ |
| $\rho_{\mathrm{RD}}$ | Measure of the alignment of $\overline{\mathbf{R}}_{\mathbf{D}}$ and $\overline{\mathbf{D}}$ |
| $\sigma$ | Constant arising in the DQTM |
| $\sigma$ | Constant arising in the molecular theory of gases |
| $\sigma$ | Standard deviation of a Gaussian probability distribution |
| $\sigma_q$ | Standard deviation of $q$-Gaussian distribution |
| $\sigma^{(g)}$ | Supply rate density of quantity $g$ |
| $\vec{\sigma}^{(g)}$ | Vectorial supply rate density of quantity $g$ |
| $\tau$ | Internal time scale for intermolecular flights (mean time between a collision of two molecules) |
| $\tau$ | Typical internal time scale |
| $\tau$ | Time parameter in temporal integration |
| $\tau$ | Second-order correlation |
| $\tau$ | Variable of weighting function in fractional calculus |
| $\tau$ | Shift parameter in fractional calculus |
| $\tau_n$ | Time between the creation of two succeeding eddies of class $n$ |
| $\tau_{u_0}$ | Vorticibility (new naming), which is the response function of a turbulent system |
| $\tau_{tot}$ | Viscous shear stress plus Reynolds shear stress in plane Poiseuille flows |
| $\tau_0$ | Viscous shear stress of flow at the wall |
| $\tau_{21}$ | Reynolds shear stress component in plane turbulent shear flows |
| $\tau_{\mathrm{T},21}$ | Turbulent shear stress component in plane turbulent Couette flows |
| $\phi$ | Diverging function describing critical phenomenon |
| $\Phi$ | General function or operator of moments |
| $\Phi$ | Lagrange function |
| $\Phi_{\mathrm{D}}$ | Dissipation operator |
| $\Phi_{\mathrm{L}}$ | Local and linear operator |
| $\Phi_{\mathrm{NL}}$ | Nonlinear and nonlocal operator |
| $\Phi_{\mathrm{P}}$ | Pressure operator |
| $\vec{\Phi}^{(g)}$ | Vectorial flux quantity of $g$ |
| $\chi$ | Symbol for an unspecified function |

| | |
|---|---|
| $\chi$ | Order parameter |
| $\chi$ | Normalized turbulent convection |
| $\chi$ | Order parameter of cooperative phenomenon in turbulent plane Couette flows |
| $\chi$ | Dimensionless convection term of the turbulent kinetic energy |
| $\chi_2$ | Overall constant length scale or variable (difference in space) in the DQTM |
| $\chi_T$ | Differential susceptibility at constant temperature of a magnetic system |
| $\psi$ | General physical quantity |
| $\psi$ | General symbol for a response function |
| $\psi$ | Stretched normalized spanwise variable of plane turbulent Poiseuille flows |
| $\psi$ | A differentiable function |
| $\psi$ | Slope of $O$ relative to $s_2$ at fixed $s_1$ |
| $\psi$ | Pressure function |
| $\omega$ | Characteristic frequency |
| $\omega$ | Abbreviation symbol in the theory of eddy occupation numbers |
| $\omega$ | Weighting function in fractional calculus |
| $\omega$ | Angular velocity of a turbulent eddy |
| $\omega_0$ | Angular velocity of a largest eddy |
| $\omega(x')$ | Probability density function; also weighting function |
| $\vec{\omega}$ | Vorticity vector |
| $\vec{\omega}'$ | Fluctuation quantity of the vorticity vector $\vec{\omega}$ |
| $\omega_3$ | Third component of the vorticity vector $\vec{\omega}$ |
| $\overline{\omega}_3$ | Average of the third component of the vorticity vector $\vec{\omega}$ |
| $\omega_3'$ | Fluctuation of the third component of the vorticity vector $\vec{\omega}$ |
| $\mathit{\Omega}$ | Domain $\in \mathrm{IR}^n$, generally $n \leq 3$ |
| $\mathit{\Omega}$ | Norm of a weighting function |
| $\Omega_\varepsilon$ | Renamed enstrophy |
| $\Omega_{\mathrm{H}}$ | Renamed helicity |
| $\widetilde{\Omega}_\varepsilon$ | Kraichnan's discretized $\Omega_\varepsilon$ |
| $\widetilde{\Omega}_{\mathrm{H}}$ | Kraichnan's discretized $\Omega_{\mathrm{H}}$ |

## Special Symbols

| | |
|---|---|
| $\|\|\mathbf{A}\|\|$ | Norm of the 3 × 3 tensor or matrix **A** |
| $\vec{a} \otimes \vec{b}$ | Exterior product of vectors $\vec{a}$ and $\vec{b}$, $\left(\vec{a} \otimes \vec{b}\right)_{ij} = a_i b_j$ |
| curl | Rotation operator |
| $\delta_{ij}$ | Kronecker delta |
| $\partial$ | Partial differential operator |
| $\partial\Gamma$ | Subregion (boundary) of $\Gamma$ |
| $\partial\Omega$ | Closed surface of the open set $\Omega$ |
| D/D$t$ | Total time derivative or substantial derivative |
| Det(**D**) | Determinant of **D** |
| $\Delta$ | Increment, difference |
| $\Delta$ | Laplace operator |
| $\vec{\nabla}$ | Gradient, Nabla operator |
| $\mathfrak{J}$ | Symbol for first and second kind of Bessel functions |
| tr(**A**) | Trace operator of the second rank tensor **A** |
| $\ldots^{\bullet}$ | Derivative related to time |
| $\ldots'$ | Fluctuation quantity |
| $\langle\ldots\rangle$ | Statistical average |
| $\|\ldots\|$ | Absolute value |
| $\|\ldots\|$ | Frobenius norm of a matrix |

# Chapter 1
# Introduction

## 1.1 Aims and Scopes of This Book

In this book, our intentions are to review early methods of turbulent closure proposals with a view to generalize these, not so much by a hierarchy of first, second, or even higher order closure schemes, but by scrutinizing the basic ideas that were likely the underlying concepts of the pioneers—de Saint-Venant, Boussinesq, Prandtl, von Kármán, Taylor, and others. Their models did not seem to have been fully rationalized at those times of nascent development. What we mean is that the pioneers were highly influenced by Boussinesq's turbulent viscosity concept, in which the turbulent stress tensor at a point in space was set proportional to the local symmetrized mean velocity gradient at the same point, and the turbulent viscosity was a multiplicative coefficient, which by Prandtl could in some way depend on the state of the mean flow at that point. Only Prandtl (1942) showed signs of abandoning this locality concept by the introduction of an extended mixing length that was treated in agreement with Boussinesq as a constant. Furthermore, Prandtl (1925) had already applied an approximation of this model that depends on the modulus of the mean rate of strain—the second mean rate of strain invariant as we would say today—and thereby is a pure local model! The early struggle with improvements, essentially parameterizations of the mixing length, e.g., the introduction of second spatial derivatives of the mean velocity, were of mixed success and maintained the "credo" of locality. This proposal contained a second length parameter, culminating in the statement that it would resolve at least the momentum transfers of two eddy classes with eddies of different sizes out of all classes describing a whole cascade. Success was rather limited, because the model is still local and the two parameters are not much more than fudging factors to match experimental facts. Therefore, only Prandtl proposed a parametrization, in which the mixing length was larger than a differential length, thus, introducing a weak concept of nonlocality: weak, because his Reynolds stress was still of Boussinesq type; it was still collinear with the average rate of strain tensor at the same point $(\vec{x}, t) \in \mathbb{R}^3 \cup \mathbb{R}$.

---

The original version of this chapter was revised. A correction to this chapter can be found at https://doi.org/10.1007/978-3-030-26033-0_14



P. W. Egolf, K. Hutter, *Nonlinear, Nonlocal and Fractional Turbulence*,
https://doi.org/10.1007/978-3-030-26033-0_1

At the level of zero-order turbulence closure, virtually no breathtaking novel fundamental ideas were introduced until the early 90s of the last century. An earlier approach similar to Prandtl's was making the Reynolds stress tensor depending on other variables (e.g., mean vorticity) (see Taylor 1932), not leaving the concept of locality, but at least guaranteeing frame indifference (for this issue see Speziale 1979). More successful is a consequent extrapolation of Prandtl's mixing length concept (Prandtl 1925) to the Reynolds stress from a dependence on the mean rate of strain (mean velocity gradient) at the same point and to replace it by a difference quotient in the nonlocal Difference Quotient Turbulence Model (DQTM), presented in the second part of this book.

For more than two decades (~1970–1995), the efforts in amending the zeroth closure models concentrated in first- and higher order gradient-type proposals and in the consistency of these proposals with the second law of thermodynamics. Such efforts have been done by Ahmadi (1985, 1987, 1988, 1989, 1991a, b), Abu-Zaid and Ahmadi (1991), Ahmadi and Chowdhuri (1991), Hutter and Jöhnk (2004), Sadiki and Hutter (1996), Sadiki (1998), Sadiki et al. (1999), Speziale (1987, 1989), Speziale et al. (1990, 1991), Speziale and Gatski (1997), and others. All these models, in spite of their complexity, are of local and extended gradient type.

Here in this book, it is our intention to demonstrate that Egolf's DQTM zero-order turbulent closure proposal of the early 1990s is, for the classical quasi-two-dimensional turbulent dispersion configurations—as, e.g., quasi-two-dimensional wakes, two types of axisymmetric jets, plane Couette and Poiseuille flows, plane "wall" turbulent flows—for which it was applied and thoroughly tested by comparisons with experiments, in better agreement with measured data than any other earlier (and any to the authors known later) proposals. This will be demonstrated in the second part of this memoir in Chaps. 7–9.

The turbulence closure scheme of Egolf, reviewed and its explanations slightly extended in this book, is clearly of spatial nonlocal nature. Its superior performance over the early local closure models, including those by Prandtl, is impetus for basing first- and second-order closure schemes also on a nonlocal (continuum) formulation, in which gradient-type postulations for the fluxes of turbulent kinetic energy and its dissipation rate, as well as Fourier-type heat flux and Fickian-type mass fluxes are possibly altered to be more general. To this end, nonlocal theories of continuum mechanics and thermodynamics need to be developed and applied (see, e.g., Eringen 2002).

In this chapter, we shall present a somewhat broader view of the significance of turbulence in natural philosophy, science, physics, and engineering. This is done by embedding newer developments in a broader environment of the present knowledge. Chapter 3 then quickly deals with the turbulent closure scheme akin to a statistical formulation of the theory of a kinetic gas. It is shown that the system of evolutionary equations for the moments (correlation terms) of the momentum equations exhibit a forward coupling. However, it is also shown that this structure of the evolutionary equations for the velocity moments can theoretically (but not computationally) be functionally related to the lowest order moment. This means that a lowest order closure is, in principle, sufficient. Chapter 4 provides a modern presentation of

Boussinesq's closure law. With Chap. 5, a detailed review of the early turbulent Reynolds stress parametrizations are given. The presentation of the relevant literature is extensive but not exhaustive. This section is terminated by a critical overview and description of the deficiencies of these early models and their reasons and a serious questioning of the logarithmic mean velocity profile along a plane boundary, as it was derived with Prandtl's mixing-length model, in the last subchapter comparing its performance with that of power laws.

## 1.2 A Brief Tour d'Horizon Through Today's Turbulence Field and Modeling

After a statement of Richard Feynman (see Vergano 2006) turbulence, during his working lifetime prior to 1988, was the most important unsolved problem of classical physics. It is broadly accepted that this statement is still valid today. Turbulence, by its field quantities, e.g., the velocity field, exhibits nonlinearity and irreversibility leading to great "irregularity" in space and time. Chaos and turbulence, in all its facets, fascinates and motivates not only, e.g., artists and musicians, but equally scientists, applying a broad spectrum of different methods and concepts, to search for a deeper comprehension of this complex phenomenon that withstood for centuries a comprehensive and complete mathematical-physical description.

Pure and applied mathematicians, for instance, try to analyze the existence, regularity, and smoothness of solutions of the three-dimensional incompressible Navier–Stokes Equations (NSE) (see, e.g., Constantin and Foias 1988; Sohr 2001; Khmelnik 2011; Galdi 2011; Boyer and Fabrie 2014; Seregin 2014)[1]. For their advances they employ (e.g., for an existence proof of an initial value problem of the NSE) sophisticated mathematical tools such as operator theory, differential geometry, and functional analytic methods in complex inner product spaces (Hilbert, Banach, Sobolev). The single constituent NSE, which are a generalization of the inviscid Euler equations for viscous fluids, yield, in non-isothermal processes, a system of five coupled (nonlinear) partial differential equations (see Chorin and Marsden 1992). This system is believed to form the basic equations for the description of viscous flow problems, also at higher excitations involving turbulent states (see, e.g., Foias et al. 2001). In the natural and technical sciences, fluid turbulence is the most frequently observed flow type; it is more frequent than the damped and regular laminar and transitional flow configurations. The above formulated set of equations is also the third problem out of the seven Millenium prize problems stated by the Clay Institute of Cambridge at the University of Massachusetts (see Clay 2000). It is not surprising that the natural-philosophical description of turbulence still remains among the universally unsolved problems. The Russian mathematician,

[1] In what follows the physical balance laws of mass, linear momentum (and energy and species mass if necessary) for linearly viscous fluids will be referred to as Navier-Stokes Equations (NSE).

Kolmogorov (1941a, b), (see Hunt et al. 1991; Frisch 1995), worked intensely on a stochastic description of turbulent processes. He improved Richardson's eddy cascade model, proposed the Kolmogorov microscales as well as the dissipation length scale (see, e.g., Tennekes and Lumley 1972). Furthermore, he contributed to the theorem of Kolmogorov, Arnold, and Moser (KAM) (see Kolmogorov 1954; Arnold 1963; Moser 1967), today extended to a theory that describes the persistence of low-dimensional Hamiltonian systems with quasi-periodic motions. Kolmogorov introduced the inertial range, studied and contributed to local isotropy, worked on fluid dynamic attractors, intermittency, and global scaling, providing much impetus to the work of another group of researchers in the field of turbulence (see the ensuing paragraph).

Some problematic views are found in the early literature on the randomness of turbulence. It was the concept of deterministic chaos (see, e.g., Poincaré 1899; Lorenz 1963; Smale 1967; May 1976; Feigenbaum 1978) that brought a deeper understanding of basic mechanisms into deterministic chaotic and turbulent systems, and the difference between irregular stochastic (random) and chaotic or turbulent motion (Lichtenberg and Liebermann 1983). Today, it is well known that chaos and turbulence manifest themselves in signals that roughly look like those of random processes; however, they are fully deterministic and enjoy special features, e.g., an extreme sensitivity to initial conditions and a less regular occupation of the phase space, e.g., with quasi-steady states on strange attractors of fractal or multifractal nature (see Mandelbrot 1977), than random systems reveal. Hopf (1948) contributed to the understanding of the onset of chaos and turbulence of an entire class of systems, which obey the Hopf bifurcation scenario.

Similarly, applied mathematicians also performed important work on the theory of homogeneous turbulence, e.g., Obukhov (1949) and Batchelor (1953).

Turbulence specialists of a different class are to be found among theoretical physicists who are concerned with a great variety of technical aspects of turbulence, e.g., field-theoretical descriptions (see also references in Egolf and Weiss 1998), effects of nonlinearity (Burgers 1948), statistical descriptions of turbulence (see Monin and Yaglom 1971), the study of structure functions, e.g., by Anselmet et al. (1984), multifractal behavior (Mandelbrot 1977; Benzi et al. 1984), the elimination of infrared divergences and renormalization (L'vov and Procaccia 1995), anomalous scaling (Yakhot 1994), universality (Grossmann and Lohse 1994), turbulent cascades and intermittency (Frisch 1995 and references therein), scaling, self-similarity and strong interactions (Barenblatt 1996), the Lévy flight concept (Lévy 1948; Shlesinger et al. 1986, 1993), and its application to turbulence (Egolf 2009). Substantial work was also performed by Kraichnan (1961), who, by developing the direct interaction approximation, applied second quantization ideas to the description of turbulence.

Other giants of theoretical physics from the end of the second half of the nineteenth, the twentieth, and beginning of the twenty-first century also dedicated substantial efforts to turbulence. Among these, only those with exceptional merits are here mentioned (for more information, please consult the given references below): Lord Kelvin (1871, 1887), who, by the way, introduced the phrase "turbulence"; Lord Rayleigh (1892), Sommerfeld (1908), Wiener (1938), Landau (1944), von Weizsäcker (1948), Heisenberg (1948), and Onsager (1949) (see also Eyink and Sreenivasan (2006) and references therein); Feynman (1955) and Chandrasekhar (1981), who, besides his devotion to astrophysics also worked on stochastic problems and the stability of hydrodynamic systems.

Similar as Hopf, also Ruelle (a physicist) and Takens (a mathematician) contributed to the onset of chaos and turbulence by discovering another route to chaos. Before these developments, the Landau picture (Landau 1944) was the established understanding. It states that toward higher excitations, the number of occurring frequencies that are created in the power spectrum of the system continuously increases up to infinity. However, this picture was revised by an alternative route, where only three different frequencies are created, before at a critical point the three-dimensional torus in the system's phase space decays into a strange attractor. Honoring its inventors, this scenario is called *Ruelle-Takens route to chaos* or turbulence (for more information see Ruelle and Takens 1971a, b).

In recent years, experimentalists in physics and engineering contributed to the advancement of turbulence understanding with the performance of quality measurements conducted with optical systems, e.g., Laser-Doppler Anemometry (LDA) (Durst et al. 1976), digital image techniques, such as Particle Image Velocimetry (PIV) (see Adrian and Westerweel 2010) and Particle Tracking Velocimetry (PTV) (see, e.g., Dracos 1996). Moreover, important experiments were reported within the last 20 years on highest Reynolds-number flows achieved in a laboratory at Princeton University in the Superpipe (see Zagarola et al. 1997 and later in this book).

Since in 1947 at the University of California the first institute for Numerical Analysis was founded (Egolf 2017, Watson 2015) and electronic computing was developed, a further group of experts became equally important in turbulence research who were the numerical mathematicians and engineers with dedication to high-performance supercomputing (Hoffman 1990; van der Steen 1997; Lanzagorta et al. 2010, etc.). Sophisticated numerical calculations of turbulent flows are mainly grouped into three categories with increasing demand of computer calculation time, called Central Processing Unit (CPU) time:

1. Turbulence modeling by *R*eynolds *A*veraged *N*avier–*S*tokes equations (RANS),
2. *L*arge *E*ddy *S*imulations (LES),
3. *D*irect *N*umerical *S*imulations (DNS).

Turbulence models (see Chap. 3) reduce the computing time as compared to LES and DNS; however, for complex flows the numerical quality of the results is not comparable with the more sophisticated methods.

A treatment of the equations (NSE) for larger scales which additionally model the influence of a background, defined by the actions of eddies of smaller scales, is a method called LES (Fröhlich 2006; Grinstein and Margolin 2010).

The highest accuracy is achieved by fully solving the NSE by DNS (Metcalfe and Riley 1981). Soon it became evident that this area (just as molecular dynamics) is excessively computation time consuming, so that till date only moderate Reynolds number flows are treatable by DNS; consequently, turbulence modeling remains a very important tool of science and engineering. DNS results have today the quality to be able to replace (in some cases) experiments, which are also very challenging in the performance, especially just above the region of the transition from the laminar to the turbulent state. New problems occur by the harvested overwhelming amount of DNS data and its appropriate evaluation and presentation, which are well addressed by George (2013).

Turbulence researchers were/are also important in environmental and geophysical fluid dynamics, e.g., among geophysicists, atmospheric physicists, dynamical meteorologists, hydrologists, hydraulicians, oceanographers, physical limnologists, glaciologists, and climatologists, who have to deal with very complex flows (see, e.g., Holton 1979; Phillips 1980; Pichler 1984; Stocker and Hutter 1987; Hutter and Jöhnk 2004; Hutter and Wang 2016a, b, 2018). This is so, because in the atmosphere, oceans, and in lakes and rivers turbulent movements are ubiquitous and often paired with stratifications (Voorrips et al. 1995). On a large scale, the Coriolis force field, initiated by the rotating Earth, leads not only to geostrophic flows and winds, which are very high Reynolds number (turbulent) flows, but also to complexer Navier–Stokes–Fourier–Fick equations (see Phillips 1963; Hutter and Jöhnk 2004). Buoyancy effects are significantly influencing atmospheric and oceanic flows, e.g., by local and global weather conditions (Turner 1973; Hutter and Jöhnk 2004). Furthermore, lava flows, avalanches (see Pudasaini and Hutter 2007; Hutter and Baillifard 2011), etc. are modeled as granular non-Newtonian materials with analogies to pulsating turbulent flows. On a smaller scale, the dynamics of lakes was also investigated showing linear and nonlinear waves related to regular and turbulent motion (Hutter et al. 2011a, b, 2014; Hutter 2012). Thus, important contributions to the understanding of turbulence came from the environmental and geophysical sciences. An early significant discovery in this field and a contribution to the fundamental theory of turbulence was the important observation, made by Richardson (1926), on the energy cascade of turbulent eddies and the famous derived $l^{4/3}$-law (Richardson 1922, see also Davidson et al. 2011). It is known that Poincaré, already in the late nineteenth century, had deep insights into mechanisms of nonlinear systems (see Poincaré 1899). However, because of his extraordinary advance, it was difficult for him to communicate this to his contemporary scientific colleagues. Besides this, more than half a century later, it was Lorenz (1963), who discovered deterministic chaos in numerically produced time series of a three-dimensional weather forecasting model that he solved with an earlier computer version. This discovery led to a scientific revolution and conceptual revisions of philosophical and scientific concepts also triggering turbulence research.

Other important groups contributing to turbulence are mechanical, chemical, and environmental engineers; they are, e.g., deeply concerned with mixing processes. Turbulent momentum and particle transport plays a crucial role in improving the efficiency of mixing processes. This is very important in judging about production processes of food, such as chemicals and plastics (see, e.g., Dimotakis 2005). Moreover, in the first half of the twentieth century, the development of ships and aircrafts led to intense design activities, which accelerated the development of fluid flows under turbulent performance in the mechanical and aerospace engineering industry. Experts in this area were Prandtl (1925), von Kármán (see von Kármán and Howarth 1938), etc. mainly in Germany and Taylor (1938) in England, an applied mathematician and physicist who fundamentally contributed to the basic understanding of turbulence phenomena, as well as to the design of technical devices associated with turbulent flows (Taylor 1937). Works of additional scientists and engineers, who studied fluid flows theoretically, numerically or at laboratory scale and contributed significantly to the field of turbulence, will be cited below.

This book will focus on the period of time when first turbulence models were developed, and which are based on previous works by Boussinesq (1877, 1897) and Reynolds (1883, 1894). Exclusively, zero-order turbulence models will be analyzed, simply for reasons of space limitation and our earlier personal experiences. More subtle, higher-order turbulence models will only be discussed to shed light on the hierarchical structure of today's higher-order turbulence models, which in their closure schemes are usually also based on the criticized gradient-based "constitutive equation" of Boussinesq.

This rather extensive introduction of the topic "turbulence" by scientific areas is meaningful because it is astonishing how the simple model, the Difference-Quotient Turbulence Model (DQTM), to be introduced in Chap. 7 [which is a natural continuation of Prandtl's and Taylor's zero-equation turbulence models], relates so much to different features of turbulence. This is so, e.g., because of an alteration from one, respectively two, to an infinite number of (eddy) scales and makes the DQTM a link and merger with many important developments of turbulence presented in the past by mathematicians and physicists: we mention, e.g., (multi-) fractal behavior, anomalous scaling, cascading, and self-similarity. Moreover, it reveals the right solutions to elementary turbulent shear flows, which were the subject of applied mathematicians, engineers, and physicists early on. And at last, DQTM captures all the treated problems of "wall-turbulent" shear flows as it reveals a critical phenomenon with a phase transition as occurring in other physical domains (Sects. 11.4–11.6). In the future links to geophysics, numerical and computational fluid dynamics will likely also be established, and with high probability, based on ideas presented in the second part of this book, better turbulence modeling might be achieved.

## References


Abu-Zaid, S., Ahmadi, G.A.: Thermodynamically consistent stress transport model for rotating turbulent flows. Geophys. Astrophys. Fluid Dyn. **61**, 109 (1991)

Adrian, R.J., Westerweel, J.: Particle Image Velocimetry. Cambridge University Press, Cambridge, UK (2010). ISBN 978-805-214-400-80

Ahmadi, G.: A generalized continuum theory for multiphase suspension flows. Int. J. Eng. Sci. **23** (8), 1 (1985)

Ahmadi, G.: On material frame-indifference of turbulence closure models. Geophys. Astrophys. Fluid Dyn. **38**, 131 (1987)

Ahmadi, G.: Thermodynamically consistent *k-Z* models for compressible turbulent flows. Appl. Math. Model. **12**, 391 (1988)

Ahmadi, G.: A two-equation turbulence model for compressible flows based on the second law of thermodynamics. J. Non-Equilib. Thermodyn. **14**, 45 (1989)

Ahmadi, G.: A thermodynamically consistent rate-dependent model for turbulence, part I–formulation. Int. J. Non-Linear Mech. **26**, 595 (1991a)

Ahmadi, G.: Thermodynamics of turbulence. Iran. J. Sci. Tech. **15**, 67 (1991b)

Ahmadi, G., Chowdhury, S.J.: A rate-dependent algebraic stress model for turbulence. Appl. Math. Model. **15**, 516 (1991)

Anselmet, F., Gagne, Y., Hopfinger, E.J., Antonia, R.A.: High-order velocity structure functions in turbulent shear flows. J. Fluid Mech. **140**, 63 (1984)

Arnold, V.I.: A proof of the A.N. Kolmogorov's theorem on the conservation of conditional-periodic motions in a small change of the Hamiltonian function. Uspehy Math. Nauk. **18**(5), 13 (1963)

Barenblatt, G.I.: Scaling, Self-Similarity, and Intermediate Asymptotics. Cambridge University Press, Cambridge, UK (1996). ISBN 978-0-521-43522-2

Batchelor, G.K.: The Theory of Homogeneous Turbulence. Cambridge University Press, Cambridge, UK (1953). ISBN 0-521-04117

Benzi, R., Paladin, G., Parisi, G., Vulpiani, A.: On the multifractal nature of fully developed turbulence and chaotic systems. J. Phys. A. **17**, 3521 (1984)

Boussinesq, J.: Essai sur la théorie des eaux courants. Mémoires présentés par divers savants à l'Académie des Sciences. **23**, 1 (1877). (in French)

Boussinesq, J.: Théorie de l'écoulement tourbillonnant et tumulteux des liquids. Gaulthier-Villars et fils, Paris (1897)

Boyer, F., Fabrie, P.: Mathematical Tools for the Study of the Incompressible Navier-Stokes Equations and Related Models. Springer, New York (2014). ISBN 13-978-146-145-974-3

Burgers, J.M.: A mathematical model illustrating the theory of turbulence. Adv. Appl. Mech. **1**, 171 (1948)

Chandrasekhar, S.: Hydrodynamic and Hydromagnetic Stability. Dover Press, Mineola, NY (1981). ISBN 0-486-64071-X. 978-0-521-19868

Chorin, A.J., Marsden, J.E.: A Mathematical Introduction to Fluid Mechanics. Springer, New York (1992). ISBN 0-387-97918-2

Clay, 2000: http://www.claymath.org/millennium-problems/millennium-prize-problems (2000)

Constantin, P., Foias, C.: Navier-Stokes Equations, Chicago Lectures in Mathematics, Chicago, IL (1988). ISBN 13-978-022-611-549-8

Davidson, P.A., Kaneda, Y., Moffat, K., Sreenivasan, K.R.: A Voyage Through Turbulence. Cambridge University Press, Cambridge, UK (2011). ISBN 978-0-521-19868-4

Dimotakis, P.E.: Turbulent mixing. Ann. Rev. Fluid Mech. **37**, 329 (2005)

Dracos, T.: Particle Tracking Velocimetry (PTV), Three-Dimensional Velocity Measuring and Image Analysis Techniques, ERCOFTAC Series, vol. 4. Springer, Dordrecht (1996). ISBN 978-90-481-4757-1

Durst, F., Mehling, A., Whitelaw, J.H.: Principles and Practice of Laser Doppler Anemometry. Academic Press, London (1976). ISBN 0-12-225250-0

Egolf, P.W.: Lévy flights and beta model: a new solution of "wall" turbulence with a critical phenomenon. Int. J. Refrig. **32**, 1815 (2009)

Egolf, P.W.: Numerical simulations for engineers. In: Lecture Notes of the University of Applied Sciences of Western Switzerland, Yverdon-les-Bains (2017)

Egolf, P.W., Weiss, D.A.: Difference-quotient turbulence model: the axisymmetric isothermal jet. Phys. Rev. E. **58**(1), 459 (1998)

Eringen, A.C.: Nonlocal Continuum Field Theories. Springer, New York (2002). ISBN 0-387-95275-6

Eyink, G.L., Sreenivasan, K.R.: Onsager and the theory of hydrodynamic turbulence. Rev. Mod. Phys. **78**, 87 (2006)
Feigenbaum, M.J.: Quantitative universality for a class of nonlinear transformations. J. Stat. Phys. **19**(1), 25 (1978)
Feynman, R.P.: Application of quantum mechanics to liquid helium. Prog. Low Temp. Phys. **1**, 17 (1955)
Foias, C., Manley, O., Rosa, R., Temam, R.: Navier-Stokes equations and turbulence. In: Encyclopedia of Mathematics and its Applications. Cambridge University Press, Cambridge, UK (2001). ISBN 0-52136032-3
Frisch, U.: Turbulence – The Legacy of A.N. Kolmogorov, 1st edn. Cambridge University Press, Cambridge, UK (1995). ISBN 0-85-403-441-2
Fröhlich, J.: Large Eddy Simulation Turbulenter Strömungen. Teubner Verlag, Stuttgart (2006). ISBN 978-3-319-24631-4
Galdi, G.P.: An Introduction to the Mathematical Theory of the Navier-Stokes Equations: Steady-State Problems, Springer Monographs in Mathematics. Springer, New York (2011). ISBN 13-978-038-709-619-3
George, W.K.: Lectures in Turbulence for the 21st Century. www.turbulence-online.com (2013)
Grinstein, F.F., Margolin, L.G.: Implicit Large Eddy Simulation: Computing Turbulent Fluid Dynamics. Springer, Berlin (2010). ISBN 9780521869829
Grossmann, S., Lohse, D.: Universality in fully developed turbulence. Phys. Rev. E. **50**, 2784 (1994)
Heisenberg, W.: On the theory of statistical and isotropic turbulence. Proc. Roy. Soc. Ser. A. **195**, 402 (1948)
Hoffman, A.R.: Supercomputer: Directions in Technology and Applications. National Academics, Washington, DC (1990). ISBN 0-309-04088-4
Holton, J.R.: An Introduction to Dynamic Meteorology. Academic Press, New York (1979). ISBN 0-12-354360-6
Hopf, E.: A mathematical example displaying features of turbulence. Commun. Pure Appl. Math. **1**, 303 (1948)
Hunt, J.C.R., Phillips, O.M., Williams, D.: Turbulence and Stochastic Processes: Kolmogorov's Ideas 50 Years on, The Royal Society, London. University Press, Cambridge, UK (1991). ISBN 0-854-03-441b-2
Hutter, K.: Nonlinear Internal Waves in Lakes. Springer, Berlin (2012). ISBN 978-3-642-23438-5
Hutter, K., Baillifard, O.A.: Continuum formulation of lava flows from fluid ejection to solid deposition. In: Hutter, K., Wu, T.T., Shu, Y.-C. (eds.) From Waves in Complex Systems to Dynamics of Generalized Continua. World Scientific, Singapore (2011)
Hutter, K., Jöhnk, K.: Continuum Methods of Physical Modeling. Springer, Berlin (2004). ISBN 13-978-364-205-831-8
Hutter, K., Wang, Y.: Fluid and thermodynamics. Basic Fluid Mechanics, vol. 1. Springer, Berlin (2016a). ISBN 978-3-319-33632-9
Hutter, K., Wang, Y.: Fluid and thermodynamics. Advanced Fluid Mechanics and Thermodynamic Fundamentals, vol. 2. Springer, Berlin (2016b). ISBN 978-3-319-33635-0
Hutter, K., Wang, Y.: Fluid and thermodynamics. Structured and Multiphase Fluids, vol. 3. Springer, Berlin (2018). ISBN 978-3-319-77745-0
Hutter, K., Wang, Y., Chubarenko, I.: Physics of lakes. Foundation of the Mathematical and Physical Background, vol. I. Springer, Heidelberg (2011a). ISBN 978-3-642-15-177-4
Hutter, K., Wang, Y., Chubarenko, I.: Physics of lakes. Lakes as Oscillators, vol. II. Springer, Berlin (2011b). ISBN 978-3-642-19111-4
Hutter, K., Wang, Y., Chubarenko, I.: Physics of lakes. Methods and Understanding Lakes as Components of the Geophysical Environment, vol. III. Springer, Berlin (2014). ISBN 978-3-319-00472-3
Khmelnik, S.: Navier-Stokes Equations: On the Existence and the Search Method for Global Solutions, Mathematics in Computers. MiC, Bene-Ayish (2011). ISBN 978-0-557-54079-2

Kolmogorov, A.N.: The local structure of turbulence in incompressible viscous fluid with very large Reynolds numbers. Dokl. Akad. Nauk. SSSR Seria fizichka. **30**, 301 (1941a). (in Russian). English translation in Proc. R. Soc. Lond. A **434,** 9
Kolmogorov, A.N.: Dissipation of energy in the locally isotropic turbulence. Dokl. Akad. Nauk SSSR. **32**, 16 (1941b). (in Russian). English translation in Proc. R. Soc. Lond. A **434,** 15
Kolmogorov, A.N.: On the conservation of conditionally periodic motions for a small change in Hamilton's function. Dokl. Akad. Nauk. SSSR. **98**, 527 (1954). (in Russian). English translation by Helen Dahlby, Los Alamos Scientific Laboratory Translation LA-TR-71
Kraichnan, R.H.: The Closure Problem of Turbulent Theory. Research Report of Office of Naval Research (1961). No. HSN-3, USA, pp. 1–48
Landau, L.D.: On the problem of turbulence. Akad. Nauk. **44**, 339 (1944)
Lanzagorta, M., Bique, S., Rosenberg, R.: Introduction to Reconfigurable Supercomputing, Synthesis Lectures on Computer Architecture. Morgan and Paypool, San Rafael, CA (2010). ISBN 978-160-845-336-8
Lévy, P.: Processus stochastique et mouvement brownien. Gauthier-Villars, Paris (1948)
Lichtenberg, A.J., Liebermann, M.A.: Regular and Stochastic Motion, Applied Mathematical Sciences, vol. 38. Springer, New York (1983). ISBN 0-387-90707-6
Lord Kelvin, Thomson, W.: Hydrokinetic solutions and observations. Phil. Mag. **42**, 362 (1871)
Lord Kelvin, Thomson, W.: On the propagation of laminar motion through a turbulently moving inviscid liquid. Phil. Mag. **24**(149), 342 (1887)
Lord Rayleigh: On the instability of cylindrical fluid surfaces. Phil. Mag. Series 5. **34**(207), 177 (1892)
Lorenz, E.N.: Deterministic nonperiodic flow. J. Atm. Sci. **20**(2), 130 (1963)
L'vov, V., Proccacia, I.: Turbulence: a universal problem. Phys. World. **9**, 35 (1995)
Mandelbrot, B.B.: The Fractal Geometry of Nature. W.H. Freeman and Company, New York (1977). ISBN 3-7643-2646-8
May, R.: Simple mathematical models with very complicated dynamics. Nature. **261**, 459 (1976)
Metcalfe, R.W., Riley, J.J.: Direct Numerical Simulations of Turbulent Flows, Lecture Notes in Physics, vol. 141, p. 279. Springer, Berlin (1981). ISBN 978-3-540-10694-4
Monin, A.S., Yaglom, A.M.: Statistical Fluid Mechanics: Mechanics of Turbulence, vol. I and II. MIT Press, Massachusetts (1971). ISBN 0-262-1-3062-9
Moser, J.: Convergent series expansions for quasi-periodic motions. Math. Ann. **169**, 163 (1967)
Obukov, A.M.: Temperature field structure in a turbulent flow. Izv. AN SSSR Ser. Geogr. Geofiz. **13**, 58 (1949)
Onsager, L.: Statistical hydrodynamics. Neuvo Cimento. **6**(2), 279 (1949)
Phillips, N.A.: Geostrophic motion. Rev. Geophys. **1**, 123 (1963)
Phillips, O.M.: The Dynamics of the Upper Ocean, Cambridge Monographs on Mechanics and Applied Mathematics, 2nd edn. Cambridge University Press, Cambridge, UK (1980)
Pichler, H.: Dynamik der Atmosphäre. Bibliographisches Institut, Zürich (1984). Wissenschaftsverlag. ISBN 3-411-01685-X (in German)
Poincaré, H.: Les méthodes des nouvelles de la mécanique céleste. Gaultier-Villars, Paris (1899). (in French)
Prandtl, L.: Bericht über Untersuchungen zur ausgebildeten Turbulenz. ZAMM. **5**(2), 136 (1925). (in German)
Prandtl, L.: Bemerkungen zur Theorie der freien Turbulenz. ZAMM. **22**(5), 241 (1942). (in German)
Pudasaini, P., Hutter, K.: Avalanche Dynamics: Dynamics of Rapid Flow of Dense and Granular Avelanches. Springer, Heidelberg (2007). ISBN 978-3-540-32686-1
Reynolds, O.: An experimental investigation of the circumstances which determine whether the motion of water shall be direct or sinuous, and of the law of resistance of parallel channels. Phil. Trans. R. Soc. Lond. A. **174**, 935 (1883)
Reynolds, O.: On the dynamical theory of incompressible viscous fluids and the determination of the criterion. Phil. Trans. R. Soc. Lond. A. **186**, 123 (1894)

Richardson, L.F.: Weather Prediction by Numerical Methods. Cambridge University Press, Cambridge, UK (1922)

Richardson, L.F.: Atmospheric diffusion shown on a distance-neighbour graph. Proc. Roy. Soc. Lond. A. **110**, 709 (1926)

Ruelle, D., Takens, F.: On the nature of turbulence. Commun. Math. Phys. **20**, 167 (1971a)

Ruelle, D., Takens, F.: On the nature of turbulence. Commun. Math. Phys. **23**(4), 343 (1971b)

Sadiki, A.: Thermodynamik und Turbulenzmodellierung. Habilitationsschrift, Fachbereich Mechanik, Technische Universität Darmstadt (in German) (1998)

Sadiki, A., Hutter, K.: On the frame dependence and form invariance of the transport equations for the Reynolds stress tensor and the turbulent heat flux vector: its consequences on closure models in turbulence modelling. Continuum Mech. Thermodyn. **8**, 341–349 (1996)

Sadiki, A., Hutter, K., Janicka, J.: In: Rodi, W., Laurence, D. (eds.) Engineering Turbulence Modelling and Experiments, 4th edn, pp. 99–102. Elsevier (1999). ISBN 978-0-08-043328-8

Seregin, G.: Lecture Notes on the Regularity Theory for the Navier-Stokes Equations. World Scientific, New Jersey (2014). ISBN 13-978-981-462-340-7

Shlesinger, M.F., Klafter, J., West, B.J.: Lévy walks with applications to turbulence and chaos. Physica. **140A**, 212 (1986)

Shlesinger, M.F., Zaslavsky, G.M., Klafter, J.: Strange kinetics. Nature. **363**, 31 (1993)

Smale, S.: Differential dynamical systems. Bull. Amer. Math. Soc. **73**(6), 747 (1967)

Sohr, H.: The Navier-Stokes Equations: An Elementary Functional Analytic Approach. Birkhäuser, Basel (2001). ISBN 978-3-0348-0550-6

Sommerfeld, A.: Ein Beitrag zur hydrodyamischen Erklärung der turbulenten Flüssigkeitsbewegung. In: Proceedings of the 4th International Congress of Mathematicians, vol. 3, p. 116, Rome (1908)

Speziale, C.G.: Invariance of turbulent closure models. Phys. Fluids. **22**(6), 1033 (1979)

Speziale, C.G.: On nonlinear *k*-*l* and *k*-*ε* models of turbulence. J. Fluid Mech. **178**, 459 (1987)

Speziale, C.G.: Turbulence modeling in noninertial frames of reference. Theor. Comput. Fluid Dyn. **1**, 3 (1989)

Speziale, C.G., Gatski, T.B.: Analysis and modelling of anisotropies in the dissipation rate of turbulence. J. Fluid Mech. **344**, 155 (1997)

Speziale, C.G., Gatski, T.B., Mhuiris, M.G.: A critical comparison of turbulence models for homogeneous shear flows in a rotating frame. Phys. Fluids A. **2**(9), 1678 (1990)

Speziale, C.G., Sarkas, S., Gatski, T.B.: Modelling the pressure–strain correlation of turbulence: an invariant dynamical systems approach. J. Fluid Mech. **227**, 245 (1991)

Stocker, T., Hutter, K.: Topographic Waves in Channels and Lakes on the *f*-Plane, Lecture Notes on Coastal and Estuarine Studies. Springer, Berlin (1987). ISBN 3-540-17623-3

Taylor, G.I.: The transport of vorticity and heat through fluids in turbulent motion. Proc. R. Soc. Lond. **135A**, 685 (1932)

Taylor, G.I.: Flow in pipes and between parallel planes. Proc. R. Soc. Lond. **A159**, 496 (1937)

Taylor, G.I.: The spectrum of turbulence. Proc. R. Soc. Lond. **A164**, 476 (1938)

Tennekes, H., Lumley, J.L.: A First Course in Turbulence. MIT Press, Cambridge, MA (1972). ISBN 0-262-200-19-8

Turner, J.S.: Buoyancy Effects in Fluids. Cambridge University Press, London (1973). ISBN 9780511608827

van der Steen, A.J.: Overview of Recent Supercomputers. Publication of the NCF, Stichting Nationale Faciliteiten (1997)

Vergano, D.: Turbulence theory gets a bit choppy. USA Today, September 10 (2006)

von Kármán, T., Howarth, L.: On the statistical theory of isotropic turbulence. Proc. R. Soc. Lond. A. **164**, 192 (1938)

von Weizäcker, C.F.: Das Spektrum der Turbulenz bei grossen Reynold'schen Zahlen. Z. Physik. **124**, 614 (1948). (in German)

Voorrips, A.C., Makin, V.K., Komen, G.J.: The influence of atmospheric stratification on the growth of water waves. Bound.-Layer Meteorol. **72**(3), 287 (1995)

Watson, G.A.: http://www.maths.dundee/~gawatson/nahistory.pdf (2015)
Wiener, N.: The homogeneous chaos. Am. J. Math. **60**(4), 897 (1938)
Yakhot, V.: Large-scale coherence and "anomalous scaling" of high-order moments of velocity differences in strong turbulence. Phys. Rev. E. **49**, 2887 (1994)
Zagarola, M.V., Perry, A.E., Smits, A.J.: Log laws or power laws: the scaling in the overlap region. Phys. Fluids. **9**(7), 2094 (1997)

# Chapter 2
# Reynold's Averaging of the Navier–Stokes Equations (RANS)

Turbulent flows fluctuate in space and time with a variety of wavelengths and periods. If somewhat calmer motion occurs between irregularly appearing signals or in subregions as laminar regions neighboring whirling patches, then the processes are called intermittent. In such situations one speaks of intermittency (see, e.g., Sreenivasan and Antonia 1997). Even though this irregular behavior is fully deterministic, to evaluate turbulent signals, one can also use statistical methods applied to fully random processes. Therefore, Reynolds' idea to split a physical quantity $\xi$ into an average quantity $\langle \xi \rangle$ and a fluctuation quantity $\xi'$, so that

$$\xi = \langle \xi \rangle + \xi', \tag{1}$$

yields a successful and frequently applied basic method of separating smooth and pulsating components of a variable. Three types of filters to determine the average value have been preferentially established, namely the statistical, temporal, and spatial filters. For a detailed introduction and treatment of this subtle theme and the three filters, see, e.g., Hutter and Wang (2016b). Following Reynolds, let us apply the statistical filter. It is based on the assumption that, on a local scale, the fluctuations have the properties of a quasi-stationary random process. If $\xi(\vec{x}, t)$ is the variable in question, it assumes a certain value with a given probability $\rho$. Then $\xi(\vec{x}, t, \rho)$ denotes the probability density, i.e., just that value of the function $\xi(\vec{x}, t)$, which belongs to the assigned value of $\rho$. The statistically most probable value of $\xi(\vec{x}, t)$ is obtained as the integration over all probable specific values,

---

The original version of this chapter was revised. A correction to this chapter can be found at https://doi.org/10.1007/978-3-030-26033-0_14



P. W. Egolf, K. Hutter, *Nonlinear, Nonlocal and Fractional Turbulence*,
https://doi.org/10.1007/978-3-030-26033-0_2

$$\langle\xi\rangle := \frac{\int_0^\infty \xi(\vec{x}, t, \rho)\mathrm{d}\rho}{\int_0^\infty \mathrm{d}\rho} = \int_0^\infty \xi(\vec{x}, t, \rho)\mathrm{d}\rho, \tag{2a, b}$$

where $\rho$ denotes a probability density distribution of norm one that leads to Eq. (2b). The statistical filter has been the only one used in early turbulence research; it has the following properties, which the reader may easily prove by applying Eqs. (2a,b):

1. *Linearity*

Let $\xi$ and $\psi$ be two quantities of a turbulent field, and $a$ a real number. Then,

$$\langle\xi + a\psi\rangle = \langle\xi\rangle + a\langle\psi\rangle. \tag{3}$$

2. *Commutability with differentiation*

$$\langle\partial\xi\rangle = \partial\langle\xi\rangle. \tag{4}$$

3. *Invariance under multifold averaging*

$$\langle\langle\xi\rangle\rangle = \langle\xi\rangle. \tag{5}$$

If all conditions (2b) and (3) to (5) are fulfilled (ergodic hypothesis), the following computation rules are valid and may be easily verified:

4. *Further computational rules*

$$\langle\xi'\rangle = 0, \tag{6}$$

$$\langle\langle\xi\rangle\psi\rangle = \langle\xi\rangle\langle\psi\rangle, \tag{7}$$

$$\langle\xi\psi\rangle = \langle\xi\rangle\langle\psi\rangle + \langle\xi'\psi'\rangle. \tag{8}$$

These rules are dealt with in every book on turbulence (e.g., Hutter and Jöhnk 2004). Furthermore, it must be noted that the derivatives in Eq. (4) are taken with regard to time and/or space coordinates.

In all the fields, which are described in the introduction, and for which turbulence investigations are performed, the objective is to describe the distributions of field variables and their evolutions. Examples can be found in fluid elements of machines under accelerating and decelerating (quasi-steady) modes of operation, in hydraulic, meteorological and climatological contexts, etc. Turbulent flows are never stationary because they are formed by fluid flow instabilities. Moreover, in these flows, turbulent fluctuations occur always as three-dimensional perturbations, even if the problem setting has lower dimensionality. On the other hand, the averages may be constant. Such special case flows are called quasi-stationary flows. If a flow has an average velocity in one direction that is equal to zero, it is called two-dimensional. In this book, these flows are named quasi-two-dimensional flows. Such a flow is, e.g., a swirl-free axi-symmetric jet, whose mean velocity in the azimuthal direction vanishes.

Ignoring electromagnetic effects, the most general case requires evolution equations for the field quantities: mass, linear and angular momentum, internal energy, and for scalar constituents (tracers) to the fluid. In this book, we will restrict attention to the mass and linear momentum balance equations. For a density preserving material, the first is called continuity equation and the second sometimes bears the name Navier–Stokes Equation (NSE). It is, however, more logical to call the union of the two NSEs. It is clear that strong analogies in turbulent transport also exist for the neglected conservation quantities (e.g., heat and constituents).

The NSEs belong to the class of integral balance laws of the global form

$$\frac{\mathrm{d}}{\mathrm{d}t}\int_{\Omega} g\,\mathrm{d}v = -\int_{\partial\Omega} \vec{\Phi}^{(g)}\vec{n}\,\mathrm{d}a + \int_{\Omega}\left(\pi^{(g)} + \sigma^{(g)}\right)\mathrm{d}v. \tag{9}$$

Here, $g$ is a specific physical quantity, defined over the material volume $\Omega$, $\vec{\Phi}^{(g)}$ is its flux density evaluated on the surface $\partial\Omega$ of $\Omega$ and $\vec{n}$ is the unit normal vector pointing to the exterior (complement) of the body $\Omega$, $\pi^{(g)}$ is the density of the production of $g$ within $\Omega$, and $\sigma^{(g)}$ the supply to $\Omega$ from the complement of $\Omega$. For the case that $g$ is identical to the mass or momentum, one has Table 2.1 in which $\rho$ = mass density; $\rho\,\vec{u}$ = momentum density, furthermore, $\boldsymbol{\tau}_D$ = Cauchy stress tensor; $p$ = pressure; and $\mathbf{I}$ the unit tensor.

For differentiable fields, the global balance laws can be put into their local forms. To obtain this, the Reynolds transport theorem is applied to the left-hand side of Eq. (9) and the divergence theorem is imposed to the first term on the right-hand side of this equation. For a scalar quantity $g$ this yields

**Table 2.1** Density of a physical quantity $g$, respectively $\vec{g}$, and its flux, production, and supply terms as described in the text

| | $g$ | $\vec{\Phi}^{(g)}/\mathbf{\Phi}^{(g)}$ | $\pi^{(g)}$ | $\sigma^{(g)}$ |
|---|---|---|---|---|
| Mass | $\rho$ | 0 | 0 | 0 |
| Momentum | $\rho\,\vec{u}$ | $-\boldsymbol{\tau}_D + p\mathbf{I}$ | 0 | $\vec{f}$ |

The quantity $\vec{f}$ denotes the specific external body force, which in all cases treated in this book is assumed to be zero

$$\frac{\partial g}{\partial t} + \operatorname{div}\left( g\vec{u} + \vec{\Phi}^{(g)} \right) - \pi^{(g)} - \sigma^{(g)} = 0 \tag{10}$$

and for a vector quantity $\vec{g}$

$$\frac{\partial \vec{g}}{\partial t} + \operatorname{div}\left( \vec{g} \otimes \vec{u} + \mathbf{\Phi}^{(g)} \right) - \vec{\pi}^{(g)} - \vec{\sigma}^{(g)} = 0, \tag{11}$$

where the symbol $\otimes$ denotes the dyadic product.

If one applies the Reynolds averaging technique by substituting for all physical quantities the expressions given by Eq. (1) into the basic equations and then continues by applying the averaging operator technique and its rules (3) to (8) to the entire set of equations, the averaged general balance equations for a scalar field $g$ become

$$\begin{aligned} &\frac{\partial \langle g \rangle}{\partial t} + \operatorname{div}\left( \langle g \rangle \left\langle \vec{u} \right\rangle + \left\langle \vec{\Phi}^{(g)} \right\rangle \right) \\ &-\left\langle \pi^{(g)} \right\rangle - \left\langle \sigma^{(g)} \right\rangle + \operatorname{div}\left( \left\langle g' \vec{u}' \right\rangle \right) = 0, \end{aligned} \tag{12}$$

and that for a vector field $\vec{g}$ takes the form

$$\begin{aligned} &\frac{\partial \left\langle \vec{g} \right\rangle}{\partial t} + \operatorname{div}\left( \left\langle \vec{g} \right\rangle \otimes \left\langle \vec{u} \right\rangle + \left\langle \mathbf{\Phi}^{(g)} \right\rangle \right) - \\ &\left\langle \vec{\pi}*^{(g)} \right\rangle - \left\langle \vec{\sigma}^{(g)} \right\rangle + \operatorname{div}\left( \left\langle \vec{g}' \otimes \vec{u}' \right\rangle \right) = 0. \end{aligned} \tag{13}$$

Notice that in (12) and (13), all quantities in angular brackets, $\langle \ldots \rangle$, correspond to those without brackets in (10) and (11) with the following exceptions:

1. The two terms describing the divergence of $\left\langle g' \vec{u}' \right\rangle$ and of $\left\langle \vec{g}' \otimes \vec{u}' \right\rangle$ represent averages of the product of $g'$ with $\vec{u}'$ or $\vec{g}'$ with $\vec{u}'$.
2. The production term $\left\langle \vec{\pi}*^{(g)} \right\rangle$ arises only if $\pi^{(g)}$ can be represented as the product of at least two quantities, which both may be subjected to turbulent fluctuations.

If the turbulence filtering is performed as a time average, the time derivative is associated with a longer time period than the averaging process. This is a consequence of Eq. (4) [compare, e.g., with adiabatic invariance, presented by Landau and Lifshitz (1979)].

Conservation of mass is obtained from Eq. (12) by substituting the assigned variables in row (1) of Table 2.1,

$$\frac{\partial \langle \rho \rangle}{\partial t} + \operatorname{div}\left( \langle \rho \rangle \left\langle \vec{u} \right\rangle \right) + \operatorname{div}\left( \left\langle \rho' \vec{u}' \right\rangle \right) = 0, \tag{14}$$

and involves a flux term $\left\langle \rho' \vec{u}' \right\rangle$ which is inconvenient, because it destroys the source-free structure of Eq. (14). It can be removed by replacing Reynolds averaging of the velocity by Favre averaging (see Hutter and Jöhnk 2004). Fortunately, this complexity is irrelevant for a density preserving fluid, $\langle \rho \rangle^{\bullet} = 0$ and $\rho' = 0$, for which (14) reduces to

$$\operatorname{div}\left\langle \vec{u} \right\rangle = 0, \tag{15}$$

which is commonly called continuity equation.

In non-isothermal flows, the density is also often kinematically considered to be a constant, but in the momentum supply term a varying density, describing buoyancy effects, is accepted. Such limiting cases are known as Boussinesq fluids or free convection flows. However, they are not the subject of this work (see Hutter and Wang 2016a, b; Egolf and Hutter 2015), in which $\rho$ is a constant.

The conservation of linear momentum is presented in two different forms, which for differentiable fields, are mathematically equivalent; whereas the second form contains the substantial derivative D/D$t$

$$\left. \begin{array}{l} \dfrac{\partial \left\langle \vec{u} \right\rangle}{\partial t} + \operatorname{div}\left( \left\langle \vec{u} \right\rangle \otimes \left\langle \vec{u} \right\rangle \right) \\ \dfrac{\mathrm{D}\left\langle \vec{u} \right\rangle}{\mathrm{D}t} = \dfrac{\partial \left\langle \vec{u} \right\rangle}{\partial t} + \left( \left\langle \vec{u} \right\rangle \cdot \operatorname{grad} \right) \left\langle \vec{u} \right\rangle \end{array} \right\} = \tag{16a, b}$$

$$-\frac{1}{\rho} \operatorname{grad}\langle p \rangle + \frac{1}{\rho} \operatorname{div}\langle \mathbf{t}_D \rangle + \frac{1}{\rho} \operatorname{div}\langle \mathbf{t}_R \rangle.$$

Reynold's averaging produces an additional stress tensor of the second rank, the so-called Reynolds stress tensor $\langle \mathbf{t}_R \rangle$ (Reynolds 1894), related to $\langle \mathbf{T}_R \rangle$ by

$$\langle \mathbf{t}_R \rangle = -\rho \langle \mathbf{T}_R \rangle. \tag{17}$$

In Cartesian notation $\langle \mathbf{T}_R \rangle$ takes the form

$$\langle \mathbf{T}_R \rangle = \left\langle \vec{u}' \otimes \vec{u}' \right\rangle = \begin{pmatrix} \overline{u_1'^2} & \overline{u_1' u_2'} & \overline{u_1' u_3'} \\ \overline{u_2' u_1'} & \overline{u_2'^2} & \overline{u_2' u_3'} \\ \overline{u_3' u_1'} & \overline{u_3' u_2'} & \overline{u_3'^2} \end{pmatrix}. \tag{18a, b}$$

Here and henceforth, we generally work with Cartesian components; notation, in this case, is always simplified by denoting the averaging operation by an overbar, e.g., $\langle \mathbf{T}_R \rangle = \overline{\mathbf{T}}_R$.

Equations (12) and (13) are viewed as equations for four variables (three for the mean components of the velocity field vector $\vec{u}$ and one for the scalar mean value of the pressure $p$). Constitutive laws must describe all the components of the Reynolds tensor by these variables. The averaging technique produced the averaged turbulent stress tensor $\overline{\mathbf{T}}_R$, which is symmetric

$$\left(\overline{\mathrm{T}}_{\mathrm{R}}\right)_{ij} = \overline{u_i' u_j'} = \overline{u_j' u_i'} = \left(\overline{\mathrm{T}}_{\mathrm{R}}\right)_{ji}, \quad i,j \in \{1,2,3\}. \qquad (19\mathrm{a}-\mathrm{c})$$

Therefore, six unknown components of this tensor must be described by a closure law of turbulence. In early times of turbulence research, empirical and semi-empirical models were applied. This book mainly reviews these approaches and makes an attempt to improve the present situation of zeroth-order turbulence modeling by proposing, in analogy to Newton's shear law for laminar flows, a corresponding nonlocal "constitutive law" for turbulent flows (see Chap. 7).

## References


Egolf, P.W., Hutter, K.: Unified theory of isothermal and non-isothermal turbulent flows in heating, air conditioning and refrigeration. In: Proceeding of XVI European Conference on The Latest Technologies in Air Conditioning and Refrigeration, p. 79, Milano (2015). 12–13th of June

Hutter, K., Jöhnk, K.: Continuum Methods of Physical Modeling. Springer, Berlin (2004). ISBN 13-978-364-205-831-8

Hutter, K., Wang, Y.: Fluid and thermodynamics. In: Basic Fluid Mechanics, vol. 1. Springer, Berlin (2016a). ISBN 978-3-319-33632-9

Hutter, K., Wang, Y.: Fluid and thermodynamics. In: Advanced Fluid Mechanics and Thermodynamic Fundamentals, vol. 2. Springer, Berlin (2016b). ISBN 978-3-319-33635-0

Landau, L.D., Lifshitz, E.M.: Lehrbuch der Theoretischen Physik, Mechanik, vol. 1. Akademie Verlag, Berlin (1979). (in German)

Reynolds, O.: On the dynamical theory of incompressible viscous fluids and the determination of the criterion. Phil. Trans. R. Soc. Lond. A. **186**, 123 (1894)

Sreenivasan, K.R., Antonia, R.A.: The phenomenology of small-scale turbulence. Annu. Rev. Fluid Mech. **29**(1), 435 (1997)

# Chapter 3
# The Closure Problem

Boltzmann and Maxwell, by developing the statistical theory of monatomic gases, and Reynolds, applying these ideas to turbulence, paved the way for the development of higher order turbulence models. One is not at all restricted to equations containing second-order correlations only. One may multiply the momentum equation with a velocity component and apply the averaging technique in the same manner again. This leads to new relations, describing the second-rank tensor elements. This is good per se, but it has the essential disadvantage that in these equations also unknown third-order moments, which are triple correlations, respectively, occur. One may criticize that such a procedure does not bring new physical information. However, this statement needs to be clarified: It is well known that scalar multiplication of the momentum equation with, say, the velocity, yields the balance law of mechanical energy, which for ideal fluids is the Bernoulli equation (Batchelor 1967; Tritton 1988, etc.). Furthermore, it is also well known that for spatially 1-*d* models this equation may replace the momentum equation as an equivalent dynamical equation. If this equation is subjected to a turbulence filter, the resulting equation is independent of the unfiltered original because

$$\langle mom.\ eq. \rangle \cdot \left\langle \vec{u} \right\rangle = 0 \neq \left\langle mom.\ eq. \cdot \vec{u} \right\rangle = 0. \qquad (20a - c)$$

This second-order averaging correlation of the momentum equation with any tensor, formed with the velocity, yields a new and independent relation. In a "Gedanken-Experiment," one may go on in this manner and produce relations with third-, fourth-, and higher order correlations up to infinity. So, in the end, one is led to an infinite set of coupled partial differential equations, which reflects one possibility to demonstrate the high complexity of the mathematical problem of fully solving a turbulence problem under consideration.

To explain this in more detail—in analogy to the theory of probability and statistics—the moments of the average of the correlation tensor elements are introduced (see, e.g., Kraichnan 1961). Such a moment is, e.g.,

---

The original version of this chapter was revised. A correction to this chapter can be found at https://doi.org/10.1007/978-3-030-26033-0_14



P. W. Egolf, K. Hutter, *Nonlinear, Nonlocal and Fractional Turbulence*,
https://doi.org/10.1007/978-3-030-26033-0_3

$$\overline{M}^{(\nu)}_{\nu_1,\nu_2,\ldots,\nu_d} = \overline{u_1^{\nu_1} u_2^{\nu_2} \ldots u_d^{\nu_d}}. \tag{21}$$

Actually, it turns out that the "shifted" moments

$$\overline{M}'^{(\nu)}_{\nu_1,\nu_2,\ldots,\nu_d} = \overline{M}^{(\nu)}_{\nu_1,\nu_2,\ldots,\nu_d} - \overline{u}_1^{\nu_1} \overline{u}_2^{\nu_2} \ldots \overline{u}_d^{\nu_d}, \tag{22}$$

which, if the statistical average [see Eq. (6)] applies, can be written in the form,

$$\overline{M}'^{(\nu)}_{\nu_1,\nu_2,\ldots,\nu_d} = \overline{u'^{\nu_1}_1 u'^{\nu_2}_2 \ldots u'^{\nu_d}_d}. \tag{23}$$

Equation (23) is analytically more convenient than Eq. (21), but mathematically equivalent as these moments span, as a whole, the same velocity space $u_1 \otimes u_2 \otimes \ldots \otimes u_d$ with dimension

$$\nu = \nu_1 + \nu_2 + \ldots + \nu_d, \tag{24}$$

which also is the order of the moment.

As an example,

$$\overline{M}'^{(2)}_{1,1,0} = \overline{u'^1_1 u'^1_2 u'^0_3} = \overline{u'_1 u'_2}, \quad \dim\left(\overline{M}'^{(2)}_{1,1,0}\right) = 2, \tag{25a – c}$$

is an element of a second-rank tensor, one of the independent components of

$$\overline{M}'^{(2)}_{\nu_1,\nu_2,\nu_3} = \overline{u'^{\nu_1}_1 u'^{\nu_2}_2 u'^{\nu_3}_3}. \tag{26}$$

Now, let us define a moment of order $\nu$ in terms of all occurring lower-order moments, namely those of order 1, 2, 3,. . ., $\nu - 2$, $\nu - 1$, viz.,

$$\overline{M}^{(\nu)}_{\nu_1,\nu_2,\ldots,\nu_d} = \Phi\left\langle \left\{\overline{M}^{(\nu-1)}_{\nu_1,\nu_2,\ldots,\nu_d}\right\}, \left\{\overline{M}^{(\nu-2)}_{\nu_1,\nu_2,\ldots,\nu_d}\right\}, \ldots, \left\{\overline{M}^{(1)}_{\nu_1,\nu_2,\ldots,\nu_d}\right\}\right\rangle. \tag{27}$$

The terms in curly brackets denote the complete set of moments of a certain order. The abbreviation $\Phi$ stands for a set of functions or/and operators.

The closure method is now defined by the following procedure; in a first step, if one wants to work with moments up to order $\nu$, one cuts away all the equations with moments of order $\nu$ + 1, and higher (up to infinity). To close the system, it is sufficient to describe the moments of order $\nu$ by equations of the type (27) containing moments of order 1, 2, . . ., $\nu - 1$. If this description is known, the closure is successful and is called a closure of a system of equations of order $\nu - 1$. Because the correct method to determine these relations between moments of different orders is not really known, several methods of developing turbulence closures have been proposed (e.g., Keller and Friedmann 1924).

Following Saad (2015), low-order turbulence closure models are characterized in the three different manners:

1. *Algebraic (zero equation) models*, which are called mixing length models or eddy viscosity models (see Sect. 5.2),
2. *One-equation models*, e.g., $k$-model and $\mu_t$-model (see Rodi 1993),
3. *Two-equation models*, e.g., $k$-$\varepsilon$, $k$-$l$, $k$-$\omega$, low Reynolds number $k$-$\varepsilon$ models, and generalizations, e.g., the $k$-$\varepsilon$-model of the renormalization group (see Launder and Spalding 1974).

"*Zero equation model*" is a confusing denotation, as it is based on the fact that in such turbulence models no partial differential equations (PDE's) describing the transport of the turbulent stresses and fluxes are introduced. They are of the form (turbulent stress) = (parameter) times (mean strain).

Second-order turbulence closure models are:

1. Algebraic stress models,
2. Reynolds stress models, etc.

Algebraic stress models are Reynolds stress parameterizations in terms of deformation measures, which are algebraic relations. Reynolds stress models are derived from a postulated balance law for the Reynolds stress tensor with associated closure statements, e.g., for the third-order stress flux.

Based on scaling and self-similarity arguments, the present authors assume that each moment of order $\nu$ can be modeled by the moments of lower orders $\nu-1, \nu-2$ down to 1. Then it follows that

$$\begin{aligned} \overline{M}^{(\nu)}_{\nu 1,\nu 2,\ldots,\nu d} = \Phi_{\nu-1}\Big(&\Big\{\overline{M}^{(\nu-1)}_{\nu 1,\nu 2,\ldots,\nu d}\Big\}, \\ &\Big\{\overline{M}^{(\nu-2)}_{\nu 1,\nu 2,\ldots,\nu d}\Big\}, \ldots, \Big\{\overline{M}^{(1)}_{\nu 1,\nu 2,\ldots,\nu d}\Big\}\Big), \quad \nu = [2, \ldots, \infty). \end{aligned} \tag{28}$$

This demonstrates that evolution equations are needed for the moments in Eq. (21) from which the latter can be determined. Such equations can be constructed to any order by forming

$$\langle \underbrace{(mom.\ eq.)}_{\substack{\text{vector=tensor} \\ \text{of order 1}}} \otimes \underbrace{\left(\vec{u} \otimes \vec{u} \otimes \ldots \otimes \vec{u}\right)}_{\text{tensor of order } \nu-1} \rangle = 0. \tag{29}$$

Here "*mom eq.*" is the momentum equation stated in the form "time rate of change of momentum minus sum of the forces = 0." It constitutes a tensor expression of order 1; $\otimes$ is the exterior product and $\vec{u}$ is the velocity vector. The bracket, $\langle \ldots \rangle$, denotes the averaging operation. So, as a whole, the left-hand side of Eq. (29) is a tensor expression of order $\nu$. Before averaging, statement (29) is no independent equation, because satisfaction of the momentum equation makes the statement (without the operation of averaging) to be identically satisfied. However, with

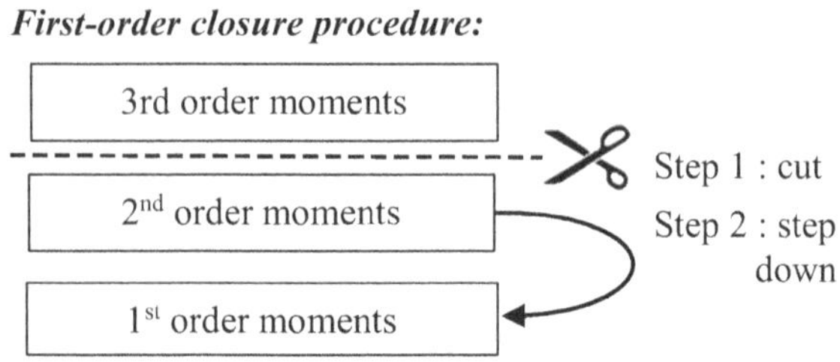


**Fig. 3.1** First-order modeling means cutting away moments of higher order than two and describing second-order by first-order moments

averaging, Eq. (29) is an independent statement owing to the property that $\langle ab \rangle \neq \langle a \rangle \langle b \rangle$, in general. Moreover, it forms an evolution equation for $\overline{M}^{(\nu)}$, and it also contains (e.g., through the divergence of the momentum flux) a moment of order $\nu + 1$. This fact is called a forward coupling.

The system consists of all equations of orders 1 to $\nu$; it constitutes $\nu$ equations for $\overline{M}^{(1)}, \ldots, \overline{M}^{(\nu)}$, of which the last equation also involves $\overline{M}^{(\nu+1)}$, for which a closure postulate must be formulated to make it complete. One can, in principle, look at this system of equations in the limit as $\nu \to \infty$. It is assumed that in this limit, the solution of all these infinitely many equations will converge to the true solution of the (averaged) NSE's (e.g., solved by Direct Numerical Simulations (DNS)).

A direct consequence of applying Eq. (28) successively and, thereby, decreasing step by step the highest occurring order from $\nu$ down to 1 is that $\overline{M}^{(\nu+1)}$ can finally be viewed as a functional relation explicitly expressed in terms of the lowest moment $\overline{M}^{(1)}$. This consequent application of the "scaling idea" is expressed by

$$\overline{M}^{(\nu+1)} = \Phi_\nu\left(\overline{M}^{(1)}\right). \tag{30}$$

This implies that in the limit as $\nu \to \infty$,

$$\lim_{\nu \to \infty} \overline{M}^{(\nu+1)} = \Phi_\infty\left(\overline{M}^{(1)}\right). \tag{31}$$

Therefore, the modeling expectation is that a first-order description exists, which contains full information, so that the establishment of second and higher order modeling becomes unnecessary. An example of a closure of a system containing third-order moments that are cutaway is shown in Fig. 3.1.

## References

Batchelor, G.K.: An Introduction to Fluid Dynamics. Cambridge University Press, Cambridge, UK (1967). ISBN 0-52104118-x

Keller, L., Friedmann, A.: Differentialgleichungen für die turbulente Bewegung einer kompressiblen Flüssigkeit. In: Proceedings of the First International Congress for Applied Mechanics, p. 395, Delft (1924). (in German)

Kraichnan, R.H.: The Closure Problem of Turbulent Theory. Research Report of Office of Naval Research (1961). No. HSN-3, USA, pp. 1–48

Launder, B., Spalding, D.B.: The numerical computation of turbulent flows. Comput. Methods Appl. Mech. Eng. **3**, 269 (1974)

Rodi, W.: Turbulence Models and Their Application in Hydraulics. CRC Press, Boca Raton, FL (1993). ISBN 9789054101505

Saad, T.: Turbulence Modeling for Beginners, Note at University of Tennesse Space Institute. See http://www.cfd-online.com/W/images/3/31/TurbulenceModelingfor_Beginners.pdf (2015)

Tritton, D.J.: Physical Fluid Dynamics, 2nd edn. Oxford Science, Oxford, UK (1988). ISBN 0-19-854489-8

# Chapter 4
# Boussinesq's "Constitutive Law"

Looking at Eqs. (16a, b), one recognizes the identical structure of the two last terms: (1) the divergence of a viscous dissipation tensor and (2) the divergence of the Reynolds stress tensor. Dissipation is related to a momentum transfer by molecular motion. To describe this process, Newton proposed the shear stress law, also called Newton's law of friction (see, e.g., Schlichting 1979), which we prepare for the turbulent case by applying the averaging procedure[1]

$$\bar{t}_{\mathrm{D}_{ij}} = 2\mu \overline{D}_{ij}, \tag{32}$$

with the dynamic viscosity $\mu$ of the fluid and the mean rate of the strain tensor

$$\overline{\mathbf{D}} = \frac{1}{2}\left[\nabla\left\langle\vec{u}\right\rangle + \left(\nabla\left\langle\vec{u}\right\rangle\right)^T\right], \tag{33}$$

respectively, or in Cartesian component form

$$\overline{D}_{ij} = \frac{1}{2}\left(\frac{\partial \bar{u}_i}{\partial x_j} + \frac{\partial \bar{u}_j}{\partial x_i}\right). \tag{34}$$

Such "gradient-type" transport laws have also been applied with moderate success for the transport of heat (Fourier's law) and diffusion of suspended substances (Fick's law).

It was Boussinesq (Boussinesq 1877), based on ideas of Saint-Venant (1843), who described turbulent momentum transport by turbulent eddies, representing fluctuations—and by an analogy to the molecular case—to propose a first closure

[1] $\overline{D}_{ij}$ is the symmetrized mean velocity gradient as $\overline{T}_{Dij}$ is symmetric by definition. As a somewhat sloppy denotation, we nevertheless speak of a gradient law.



P. W. Egolf, K. Hutter, *Nonlinear, Nonlocal and Fractional Turbulence*,
https://doi.org/10.1007/978-3-030-26033-0_4

relation of turbulence, referred to as Boussinesq's "constitutive law"[2] or Boussinesq's approximation. Therefore, in analogy to Eq. (32), he proposed

$$\bar{t}_{\mathrm{R}_{ij}} = 2\mu_{\mathrm{T}} \overline{D}_{ij}, \tag{35}$$

with the scalar-valued turbulent dynamic viscosity $\mu_{\mathrm{T}}$. In the remainder, this Ansatz is discussed and it is demonstrated how it is applied to build algebraic turbulence models (see also Sect. 5.2). *However, it is emphasized that this effective viscosity does not describe real physical properties, but instead is rather a model parameter, which in certain cases depends on a space coordinate or even the dynamics of the fluid.*

The definitions (17) and (18) of the Reynolds stress tensor $\overline{\mathbf{R}}$ (or $\mathbf{R}$) $= \bar{\mathbf{t}}_{\mathrm{R}}$ imply that

$$\operatorname{tr}\left(\overline{\mathbf{R}}\right) = -\rho \operatorname{tr}\left(\left\langle \vec{u}' \otimes \vec{u}' \right\rangle\right) = -\rho \left\langle \vec{u}' \cdot \vec{u}' \right\rangle = -2\rho\, k, \tag{36a – c}$$

where $k$ is the turbulent kinetic energy of the velocity fluctuations per unit mass,

$$k := \frac{1}{2} \left\langle \vec{u}' \cdot \vec{u}' \right\rangle. \tag{37}$$

Please note that in this quantity it is customary to omit the average sign and the density.

Next, let us write

$$\overline{\mathbf{R}} = -\left[\rho\left(\left\langle \vec{u}' \otimes \vec{u}' \right\rangle - \frac{2}{3} k\, \mathbf{I}\right)\right] - \frac{2}{3} \rho\, k\, \mathbf{I}, \tag{38}$$

where $\mathbf{I}$ stands for the $3 \times 3$ unit tensor. The term in brackets is the deviatoric Reynolds stress tensor,

$$\overline{\mathbf{R}}_{\mathrm{D}} = -\rho\left(\left\langle \vec{u}' \otimes \vec{u}' \right\rangle - \frac{2}{3} k \mathbf{I}\right). \tag{39}$$

Because for a density preserving fluid, a Boussinesq fluid or a free convection fluid the strain rate tensor is deviatoric, it is advantageous to request a turbulent closure of the form

$$\overline{\mathbf{R}}_{\mathrm{D}} = \overline{\mathbf{R}}_{\mathrm{D}}\left(\overline{\mathbf{D}}\right). \tag{40}$$

Equations (38–40) then imply

[2] Actually it is not customary to call turbulent closure conditions "constitutive equations" even though the relations fulfill mathematically an analogous role. We shall occasionally deviate from this custom, whereas the signs "..." shall remind us of the slightly unfamiliar use.

$$\overline{\mathbf{R}} = -\frac{2}{3}\rho k \mathbf{I} + \overline{\mathbf{R}}_{\mathrm{D}}(\overline{\mathbf{D}}), \tag{41}$$

or with the Boussinesq closure $\overline{\mathbf{R}}_{\mathrm{D}} = \overline{\mathbf{t}}_{\mathrm{R}}$ (Eq. 35)

$$\overline{\mathbf{R}} = -\frac{2}{3}\rho\, k\, \mathbf{I} + 2\mu_{\mathrm{T}}\overline{\mathbf{D}}, \tag{42}$$

or with the kinematic turbulent viscosity, defined in analogy to the material kinematic viscosity,

$$\nu_{\mathrm{T}} = \frac{\mu_{\mathrm{T}}}{\rho}, \quad \frac{1}{\rho}\overline{\mathbf{R}} = -\frac{2}{3}k\, \mathbf{I} + 2\nu_{\mathrm{T}}\overline{\mathbf{D}}. \tag{43a, b}$$

The quantity $\nu_{\mathrm{T}}$ is frequently also called *eddy viscosity* and denoted by $\varepsilon_{\mathrm{m}}$ (*m* for momentum)

$$\varepsilon_{\mathrm{m}} = \nu_{\mathrm{T}}. \tag{44}$$

Equations (43b) and (44) stand for a general description of Boussinesq's turbulence closure model.

Boussinesq's "constitutive law" is at the basis of practically all turbulence models, presently applied to numerically solve flow problems. This is rather surprising because it is a linear gradient law and it is known that turbulence is a highly complex phenomenon showing memory and nonlocality effects. The shortcomings of the Boussinesq approach are discussed in Sect. 5.3 and some reasons to that end are presented in Sect. 5.4.

Boussinesq's "constitutive law" also constitutes a fundamental part of the most frequently applied turbulence model, the *k*-*ε* model, and other higher order models. Their derivation is theoretically outlined in great detail, e.g., in Launder and Spalding (1974) and Hutter and Wang (2016).

## References

Boussinesq, J.: Essai sur la théorie des eaux courants. Mémoires présentés par divers savants à l'Académie des Sciences. **23**, 1 (1877). (in French)

Hutter, K., Wang, Y.: Fluid and thermodynamics. In: Advanced Fluid Mechanics and Thermodynamic Fundamentals, vol. 2. Springer, Berlin (2016). ISBN 978-3-319-33635-0

Launder, B., Spalding, D.B.: The numerical computation of turbulent flows. Comput. Methods Appl. Mech. Eng. **3**, 269 (1974)

Saint-Venant, A.J.C. (Barré): Note à joindre au mémoire sur la dynamique des fluides. CRAS. **17**, 1240 (1843). (in French)

Schlichting, H.: Boundary-Layer Theory. McGraw-Hill, New York (1979). ISBN 0-07-055-334-3

# Chapter 5
# First Turbulence Models for Shear Flows

## 5.1 Shear Flows and the Works of Prandtl, Taylor, and Contemporaries

In daily life, turbulent motions constitute ubiquitous fluid mechanical elements, which can be observed in various forms, e.g., in wind gusts and surface water flows, in rivers, lakes, and the oceans.

Scientists have investigated turbulent phenomena for hundreds of years. As a prominent early scientist, who studied turbulent flows, Leonardo da Vinci (1452–1519) may be mentioned; he produced several hand drawings showing eddies of various sizes and their interrelations (see Fig. 5.1).

It was, however, not before the nineteenth century that engineers started to investigate theoretically and experimentally fundamental flows, e.g., flows with homogeneous and isotropic turbulence, but also elementary shear flows. Based on ideas of Boussinesq (1877, 1897) and Reynolds (1894), the constitutive law of Boussinesq was applied to shear flows by Prandtl (1925, 1942), Taylor (1932), and contemporaries. This process revealed a number of recognized algebraic turbulence models. Today, turbulent shear flows are categorized as two subclasses, namely "free turbulent shear flows" and "wall turbulent shear flows," which should be called more precisely "near-wall turbulent shear flows." In this text, for traditional reasons, we will use the term "wall turbulence." The two types of shear flows may also be distinguished by the preferential axis of their vorticity distributions. In "wall turbulence" it is parallel to the wall, respectively the main flow direction, whereas in free turbulence it is mainly perpendicular to this direction. Furthermore, the term "free" in free turbulent shear flows means that there is no direct influence of boundaries on this type of flow in the main domain of investigation (see, e.g., Hinze 1975). An indirect influence is only present by the fact that some boundaries are likely involved in the (upstream) generation of the flow. Such could be an obstacle, e.g., a sphere or a vertical cylinder mounted into the flow. Consequently, at downstream positions, one observes a region where there exists a flow with large

---

The original version of this chapter was revised. A correction to this chapter can be found at https://doi.org/10.1007/978-3-030-26033-0_14



P. W. Egolf, K. Hutter, *Nonlinear, Nonlocal and Fractional Turbulence*,
https://doi.org/10.1007/978-3-030-26033-0_5

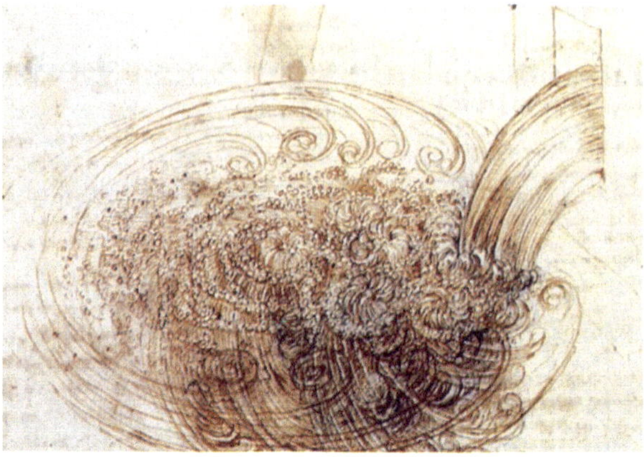

**Fig. 5.1** One of several drawings of turbulent eddies in water motion produced by Leonardo da Vinci (see Da Vinci undated). Leonardo wrote (translation by Ugo Piomelli, University of Maryland) "Observe the motion of the surface of the water, which resembles that of hair, which has two motions, of which one is caused by the weight of the hair, the other by the direction of the curls, thus, the water has eddying motion, one part of which is due to the principal current, the other to the random and reverse motion. Figure text courtesy by Gad-el-Hak (2000). From reference: https://www.bing.com/images/search?q=Draw++ings+Leonardo+da+Vinci+turbulence

spatial directional derivatives (often named "gradients") of the averaged downstream velocity profiles in the transverse direction to the main flow direction. It is the combination of this mean shear with the mean shear stress that locally generates turbulent kinetic energy, which in a model consideration is found in non-isotropic large-scale eddies. In a cascading process (see, e.g., Richardson 1922; Kolmogorov 1941a, b), large eddies decay to smaller ones in a multiple splitting process and smaller eddies merge together to create larger ones. In quantum field theory, the creation and annihilation operators, respectively, describe the creation or reconnection and destruction of elementary particles. This is within a kind of analogy to the equivalent processes of eddy creation (or reconnection) and eddy destruction (decay) in turbulence. One observes that in the average net process there is only a flux of energy stepping down the cascade to eddies of a size of the Kolmogorov dissipation length (see, e.g., Frisch 1995). Eddies with this small diameter are then dissolved and turbulent kinetic energy is transformed to molecular motion, which can be interpreted as heat and leads to a usually small temperature rise of the fluid.

The following subsections review the first momentum and vorticity transfer models, which—to describe the turbulent transport processes—only take one or two eddy sizes of this cascade into consideration. Everybody standing on a bridge and looking over a railing of it, notices in the water wakes behind its piers, the multiple sizes of the occurring eddies and, therefore, immediately realizes a serious lack in these first turbulence models. In Chap. 6, in an attempt to try to beat Boussinesq-type turbulence modeling, first nonlinear and nonlocal closure proposals are thoroughly reviewed. Then, in Chap. 7, a newer nonlinear and nonlocal zero-equation turbulence model—without any of the discussed deficiencies—will be derived, explained and applied to elementary flow problems.

## 5.2 Momentum and Vorticity Transfer Models

### *5.2.1 Prandtl's Mixing Length Model*

The following turbulence models, which are introduced in this and some following subsections, are named phenomenological models. In science, a phenomenological model is created by taking knowledge gained by empirical observations of a phenomenon, possibly modifying it and treating it in a manner that it becomes consistent with fundamental laws, which in our case are basic laws of physics (e.g., balances of mass, momentum, symmetry rules, and Galilean invariance), and dimensional homogeneity (e.g., introduction of a mixing length). In Chap. 4, it has been explained already that there is an analogy between molecular and turbulent transport of momentum. Making use of this analogy is the reason that these turbulence models actually have a certain (even though limited) physical basis.

In the early mixing length models, the transports of the following quantities have been treated with moderate success:

1. Fluid mass (subject of this chapter)
2. Momentum (subject of this chapter)
3. Heat
4. Mass of constituents.

These field variables form the fundamental quantities for the conservation laws of mass, linear momentum, energy, and constituent mass concentrations in continua of the Boltzmann type (without a separate spin balance).

It was Prandtl's idea to link Boussinesq's local gradient-type turbulent transport equation, which is consistent with numerous kinds of diffusion processes, with the real characteristics of turbulence. The kinetic theory of gases (Chapman and Cowling 1970; Truesdell and Muncaster 1980) leads from a microscopic model, describing Brown's motion (see Einstein 1905) to the macroscopic description and a simple analytical formula for the viscosity. This formula contains an exchange length, which is the mean free path of the molecules, describing the mean distance between binary collisions of molecules, and their root-mean-square velocity.

Inspired by this, and having knowledge of the unit of the eddy viscosity (which is the unit of a characteristic length and a velocity (unit: $m^2s^{-1}$), Prandtl assumed for the turbulent momentum exchange a movement of a fluid lump perpendicular to the main motion, which transports portions of conserved momentum over a certain distance (see Fig. 5.2). A "fluid particle" has a characteristic length that is larger than a mathematical differential element and smaller than the Kolmogorov dissipation length scale, whereas a "fluid lump" has a somewhat larger characteristic length

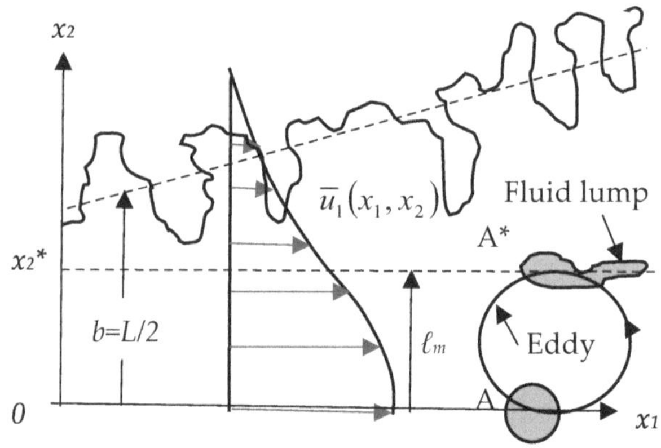


**Fig. 5.2** A fluid lump, that consists of a conglomerate of fluid particles, is transported from location A to A* and during this process conserves its momentum and leads to a momentum exchange over the distance of the mixing length $\ell_{\rm m}$, which is identical to the diameter of the presented type of eddy

than the dissipation length scale, but is smaller than the overall characteristic length of a fluid flow domain. With this concept in mind, Prandtl introduced into the eddy diffusivity a mixing length $\ell_{\rm m}$ and a characteristic velocity, which is the root-mean-square of the fluctuation velocity $\sqrt{\overline{u'^2_2}}$, both in the direction perpendicular to the main flow direction, in which the exchange takes place (see Fig. 5.2), viz.

$$\varepsilon_{\rm m} = c\,\ell_{\rm m}\sqrt{\overline{u'^2_2}}, \tag{45}$$

where $c$ is a dimensionless constant. The lifetime of a fluid lump must be sufficiently large to transport an excess momentum over the distance of the mixing length. The time and length scales, and by this also that for the velocity, are available by approximate considerations limited by the Kolmogorov microscales (see, e.g., Pope 2000).

Then, Prandtl made the following assumptions (Hinze 1975):

1. The mixing length $\ell_{\rm m}$ is small in comparison to the fluid domain, in Fig. 5.2 the width $L$:

$$\ell_{\rm m} \ll L. \tag{46}$$

2. The downstream and transverse components of the fluctuation velocity are approximately the same:

$$\sqrt{\overline{u'^2_1}} \approx \sqrt{\overline{u'^2_2}}. \tag{47}$$

In the remainder, it will become clear that assumption 1 does not hold at all. It is even the key point for further improvement. However, in these early years of first attempts to introduce such closure relations, by the adaptation of empirical constants determined by some experiments, this shortcoming was partly overlooked or counterbalanced. In the remainder, we will see that Prandtl in his later years gave up the first assumption, a fact that directly led him toward more successful models. On the other hand, assumption 2 is approximately true; this will be demonstrated in Sect. 9.2.1 in the validation model of an axisymmetric turbulent jet with experimental data. Nowadays, it is well accepted that fluid particles are rapidly dispersed. Particles, which are initially neighbors, are exponentially separated (in phase space) with a measure given by positive Lyapunov exponents (for more specific information see, e.g., Arrowsmith and Place 1990). This demonstrates the limited validity of the idea of a fluid lump. Yet still, the rough assumption led to predictions of an approximately correct momentum transfer.

There are numerous derivations of different complexity of Prandtl's mixing length model; all of them are easy to be grasped (e.g., Schlichting 1979; Stull 1988; Libby 1996; Stanišić 1988; Lesieur 1990; Pope 2000). Because a more sophisticated derivation will be applied to derive the DQTM (see Sect. 7.2.2) in an analogous manner (see also in Egolf and Weiss 2000), here the simplest and most straightforward method is presented. For some alternative and more sophisticated explanations, see Prandtl (1925), Hinze (1975), Tennekes and Lumley (1972) or Stanišić (1988) and others. In a simple two-dimensional approach, the Reynolds shear stress is modeled by

$$\tau_{21} = -\rho\left(\overline{\mathrm{T}}_{\mathrm{R}}\right)_{21} = -\rho\,\overline{u_2' u_1'} = -\rho\,\tilde{c}\sqrt{\overline{u_2'^2}}\sqrt{\overline{u_1'^2}}, \qquad (48\text{a} - \text{c})$$

with a nonzero correlation coefficient $\tilde{c}$. Furthermore, an ensemble of eddies being displaced from the level $x_2 = 0$ to $x_2^*$ create turbulent velocity fluctuations that fulfill the following root-mean-square (rms) parametrization (see Fig. 5.2):

$$\begin{aligned}\sqrt{\overline{u_2'^2}} &= \hat{c}\left\{\overline{\left[\overline{u}_1(0) + u_1'(0)\right] - \left[\overline{u}_1\left(x_2^*\right) + u_1'\left(x_2^*\right)\right]}\right\} \\ &= \hat{c}\left[\overline{u}_1(0) - \overline{u}_1\left(x_2^*\right)\right] = \hat{c}\,\Delta\overline{u}_1\left(x_2^*\right)\end{aligned} \qquad (49\text{a} - \text{c})$$

with a constant $0 \leq \hat{c} \leq 1$. However, applying a Taylor series expansion of $\Delta\overline{u}_1\left(x_2^*\right)$ Eq. (49c), with $x_2^* = \ell_{\mathrm{m}}$, leads to the following central relation for Prandtl's turbulence closure

$$\sqrt{\overline{u_2'^2}} = \hat{c}\,\ell_{\mathrm{m}}\frac{\partial\overline{u}_1}{\partial x_2}. \qquad (50)$$

Combining Eqs. (47), (48c), and (50), one obtains the following formula for the turbulent shear stress:

$$\tau_{21} = -c\rho\,\ell_{\mathrm{m}}^2 \left(\frac{\partial \overline{u}_1}{\partial x_2}\right)^2, \tag{51}$$

with a new constant

$$c = \widetilde{c}\,\widehat{c}^{\,2}. \tag{52}$$

Prandtl naturally knew that the sign of the Reynolds shear stress must be identical to that of the velocity's directional derivative; that is why he rewrote the mixing length model (51) in its final version as:

$$\tau_{21} = \rho\,\varepsilon_{\mathrm{m}} \frac{\partial \overline{u}_1}{\partial x_2} = c\rho\,\ell_{\mathrm{m}}^2 \left|\frac{\partial \overline{u}_1}{\partial x_2}\right| \frac{\partial \overline{u}_1}{\partial x_2}. \tag{53a, b}$$

By comparison of Eqs. (53a) with (53b), one concludes that the eddy viscosity $\varepsilon_{\mathrm{m}}$ in the mixing length model is

$$\varepsilon_{\mathrm{m}} = c\,\ell_{\mathrm{m}}^2 \left|\frac{\partial \overline{u}_1}{\partial x_2}\right|. \tag{54}$$

It is important to notice that a fluid lump at location A, with a lump diameter somewhat smaller than the mixing length, is smoothly transported by an eddy, with a diameter that is just identical to the size of the mixing length $\ell_{\mathrm{m}}$, at its new location A* (see Fig. 5.2). Here the momentum is increased by the momentum exchange of the fluid lump arriving from a region of higher (specific) momentum.

Prandtl's mixing length model shows numerous shortcomings. For this, we refer to Sect. 5.3 where shortcomings of several zero-order turbulence models are listed and discussed in detail. The observation of some pitfalls led Prandtl to improve his mixing-length model. A first improvement led him to the mean gradient model.

## *5.2.2 von Kármán's Local Model*

In times of their competitive, but fair struggle to obtain more insight to the mechanisms of turbulent momentum exchange, Prandtl and von Kármán debated on the further development of momentum transfer models. Von Kármán stressed the importance of locality, whereas Prandtl in this discussion was more intuitive and creative and was ready to take risks of neglecting usual well-probed and established methods. It was him who started to relate the transport processes in mixing regions of free and "wall turbulent" flows to the overall flow quantities, e.g., instead of applying a small mixing length choosing the entire width of a mixing zone to represent this quantity. By this, he started to pave the way for nonlocal models toward the mean gradient model. However, this is an a posteriori judgment. It is not at all certain whether fluid dynamicists at his time were explicitly aware of the local–nonlocal difference. After all, rational mechanics was then not yet born. At least it is not visible in their early written texts.

On the other hand, von Kármán assumed a clearly local mixing length generated by local flow conditions (von Kármán 1930). He proposed to generate a mixing length by a fraction of two succeeding derivatives,

$$\ell_{\mathrm{m}} = k \frac{\dfrac{\partial \bar{u}_1}{\partial x_2}}{\dfrac{\partial^2 \bar{u}_1}{\partial x_2^2}}, \quad \varepsilon_{\mathrm{m}} = k^2 \frac{\left(\dfrac{\partial \bar{u}_1}{\partial x_2}\right)^2}{\left(\dfrac{\partial^2 \bar{u}_1}{\partial x_2^2}\right)^2} \left| \frac{\partial \bar{u}_1}{\partial x_2} \right|. \tag{55a, b}$$

The dimensionless constant $k$ is called von Kármán constant. With Eqs. (53a) and (55b), his turbulence model for simple shearing shows the following shear–stress formula:

$$\tau_{21} = \rho\, k^2 \frac{\left(\dfrac{\partial \bar{u}_1}{\partial x_2}\right)^2}{\left(\dfrac{\partial^2 \bar{u}_1}{\partial x_2^2}\right)^2} \left| \frac{\partial \bar{u}_1}{\partial x_2} \right| \frac{\partial \bar{u}_1}{\partial x_2}. \tag{56}$$

With this approach, at inflection points of the average downstream velocity profile, the eddy viscosity is infinitely large, a fact that cannot survive any critical debate. A next rather artificial approach was then to take for the square of the mixing length the ratio of the first and third derivative of the average downstream velocity (see, e.g., Prandtl 1942). However, these models are not at all convincing and, therefore, we dispense ourselves from discussing these rather unsuccessful turbulence models in more detail.

### *5.2.3 Reichardt's Inductive Model*

Experimental observations suggest Gaussian distribution functions for the averaged main velocity and constituent concentration profiles of some turbulent mixing zones, e.g., in a free axisymmetric jet for the average downstream velocity (see Sect. 9.2.1). Furthermore, this function is also found in smoothening by molecular diffusion and heat conduction. Assuming a certain analogy between ordinary and turbulent diffusion, Reichardt (1941) proposed his inductive turbulence model.

His ideas are outlined in Hinze (1975). To describe a simple quasi-two-dimensional turbulent quasi-steady flow, neglecting pressure variations and viscosity effects, from Eq. (16b) it follows that

$$u_1 \frac{\partial u_1}{\partial x_1} + u_2 \frac{\partial u_1}{\partial x_2} = 0. \tag{57}$$

The continuity equation for this flow configuration,

$$\frac{\partial u_1}{\partial x_1} + \frac{\partial u_2}{\partial x_2} = 0, \tag{58}$$

is multiplied with $u_1$ and added to Eq. (57), yielding

$$2u_1 \frac{\partial u_1}{\partial x_1} + u_1 \frac{\partial u_2}{\partial x_2} + u_2 \frac{\partial u_1}{\partial x_2} = 0. \tag{59}$$

This equation is identical to

$$\frac{\partial u_1^2}{\partial x_1} + \frac{\partial}{\partial x_2}(u_1 u_2) = 0. \tag{60}$$

Now, to Eq. (60) the averaging operator is applied

$$\frac{\partial \overline{u_1^2}}{\partial x_1} + \frac{\partial}{\partial x_2}(\overline{u_1 u_2}) = 0. \tag{61}$$

"Gaussian diffusion" in one dimension is described by

$$\frac{\partial \overline{u_1^2}}{\partial x_1} - D \frac{\partial^2 \overline{u_1^2}}{\partial x_2^2} = 0. \tag{62}$$

With a coordinate-dependent positive diffusion coefficient, $D(x_1, x_2)$, this equation is a parabolic partial differential equation with a variable coefficient. It can be straightforwardly handled, see, e.g., Courant and Hilbert (1989). Up to this point, Reichardt's derivation is physically sound. However, his next step was of phenomenological nature. He, namely, proposed to compare Eqs. (61) and (62) and to set $D = \ell_\mathrm{m}$ to obtain

$$\overline{u_1 u_2} = -\ell_\mathrm{m} \frac{\partial \overline{u_1^2}}{\partial x_2}. \tag{63}$$

With $D$, also $\ell_\mathrm{m}$ may depend on the $x_1$-coordinate, a choice often made by the early turbulence modelers. Next, to modify the left-hand side of Eq. (63), the following relation is valid:

$$\overline{u_1' u_2'} = -\overline{u_1} \cdot \overline{u_2} + \overline{u_1 u_2}. \tag{64}$$

Substituting it into Eq. (63) leads to the final model equation

$$\tau_{21} = -\rho\, \overline{u_1' u_2'} = \rho \left( \overline{u_1} \cdot \overline{u_2} + \ell_{\mathrm{m}} \frac{\partial \overline{u_1^2}}{\partial x_2} \right). \tag{65a, b}$$

With this result, Reichardt found a diffusive relation connecting the second-order with the first-order correlation terms; and as we have seen in Chap. 3 such a relation defines an algebraic turbulence model. However, as in most of these approaches, a single necessary phenomenological postulate is the critical step, that usually induces problems. In Reichardt's case, the problem is that his relation seriously conflicts with Galilean invariance [see the averaged square of the velocity in Eq. (62)]. As a consequence, we omit a further discussion of this model and refer the critical reader to Reichardt's paper.

### 5.2.4 *Prandtl's Mean Gradient Model*

Prandtl (1942), by comparing measured with algebraically calculated results, that were derived with the mixing length model, found substantial deviations at zero transverse derivative of the average downstream velocity profile. At these points, the mixing length model leads to a vanishing turbulent shear stress, a fact that in some cases does not hold. Furthermore, the obtained average velocity profiles become too pointed at their maxima. Therefore, Prandtl generalized the eddy diffusivity Eq. (54) by adding a curvature-dependent term (the second derivative weighted with a second mixing length, $\ell'_{\mathrm{m}}$),

$$\frac{\partial \overline{u}_1}{\partial x_2} \rightarrow \frac{\partial \overline{u}_1}{\partial x_2} + \ell'_{\mathrm{m}} \frac{\partial^2 \overline{u}_1}{\partial x_2^2} + O\left( \left| \frac{\partial \overline{u}_1}{\partial x_2} \right|^3 \right). \tag{66}$$

Strictly, the replacement of $\partial \overline{u}_1 / \partial x_2$ by the expression in Eq. (66) on the right-hand side may be interpreted as a derivative somewhere between $x_2$ and $x_2 + \ell'_{\mathrm{m}}$. The representation

$$\varepsilon_{\mathrm{m}} = c\, \ell_{\mathrm{m}}^2 \sqrt{\left( \frac{\partial \overline{u}_1}{\partial x_2} + \ell'_{\mathrm{m}} \frac{\partial^2 \overline{u}_1}{\partial x_2^2} \right)^2}, \tag{67}$$

therefore, is a weak nonlocal parameterization. One notices that a model with two mixing lengths and two derivatives of different order, physically describes momentum exchange by eddies of two different sizes. Equation (67) is identical to

$$\varepsilon_{\mathrm{m}} = c\,\ell_{\mathrm{m}}^2 \cdot \sqrt{\left(\frac{\partial \overline{u}_1}{\partial x_2}\right)^2 + \overline{\ell_{\mathrm{m}}'^{\,2}} \left(\frac{\partial^2 \overline{u}_1}{\partial x_2^2}\right)^2}, \tag{68}$$

provided, as Prandtl assumed that the second mixing length is in the negative and positive direction statistically equally distributed. This explains why in Eq. (68) on the average, the cross-terms have dropped out. After a small rearrangement and by applying Eq. (53a), the following algebraic turbulence model is obtained:

$$\tau_{21} = c\,\rho\,\ell_{\mathrm{m}}^2 \cdot \sqrt{\left(\frac{\partial \overline{u}_1}{\partial x_2}\right)^2 + \overline{\ell_{\mathrm{m}}'^{\,2}} \left(\frac{\partial^2 \overline{u}_1}{\partial x_2^2}\right)^2}\,\frac{\partial \overline{u}_1}{\partial x_2} \tag{69}$$

and is called the mean gradient model (see Prandtl 1942).

He noticed an improvement at the critical points where in the mixing length model the eddy diffusivity vanishes, whereas in the mean gradient model it does not. Nonetheless, he also was aware that some improvement of this model stems from the introduction of a second free parameter $\ell_{\mathrm{m}}'$ to fit experiments. Furthermore, he confessed that the application of the mean gradient model usually leads to mathematical complications and that more simplicity would be highly welcome. Further comparison with experiments and the demand for more simplicity in a next step led him to the shear layer model.

### 5.2.5 *Prandtl's Shear Layer Model*

Prandtl experienced that for free turbulent shear flows, better results are obtained by setting the mixing length equal to some overall flow quantity, e.g., the entire width of the turbulent zone. Therefore, in a next very promising model, he proposed an eddy diffusivity constituted by the overall (half) width $b$ of the mixing zone, and for the characteristic velocity the maximum velocity difference across the turbulent region (Prandtl 1942),

$$\varepsilon_{\mathrm{m}} = cb(\overline{u}_{1\,\max} - \overline{u}_{1\,\min}). \tag{70}$$

With this new eddy viscosity formula, the closure model takes the form

$$\tau_{21} = c\rho b(\overline{u}_{1\,\max} - \overline{u}_{1\,\min})\frac{\partial \overline{u}_1}{\partial x_2}. \tag{71}$$

Görtler (1942), a mathematician in Prandtl's group in Göttingen, at that time worked out solutions of free turbulent shear flows with the shear layer model and found good agreement with experimental findings, but critically noticed certain deviations in their boundary regions, because there the eddy diffusivity, Eq. (70), does not vanish

(Görtler 1942). The origin of this problem is tied to the fact that the width $b$ and the averaged velocity difference $(\overline{u}_{1\,\max} - \overline{u}_{1\,\min})$ are treated as constants. In Chap. 7, the DQTM will be developed. In a retrospective view, it will be shown that Prandtl's shear layer model comes closest to this complete model. However, some small imperfections are responsible for the occurrence of formulae that are long and heavy and lack simplicity. Because of this, the shear layer model—which actually is closer to the "right" solution—produces less successful results than the basic mixing length theory, and as a result was less recognized, appreciated and, therefore, also less applied.

### *5.2.6 Taylor's Vorticity Transfer Model*

Turbulence is rotational and, therefore, flows are characterized by high levels of fluctuating vorticity (see Tennekes and Lumley 1972). Especially quasi-two-dimensional flows are known for a strong production rate of vorticity filaments (see Kevlahan and Farge 1997). Vorticity is a feature that distinguishes turbulent motion from other types of irregular motions as, e.g., numerous kinds of nonlinear waves, ocean waves, and atmospheric gravity waves. Furthermore, it plays a crucial role in critical phenomena observed in the solutions derived with the DQTM and presented here. Vorticity rich patterns define the high-order phase, whereas laminar streaks characterize the low-order phase of a phase transition, which occurs and is described by a function of the inverse Reynolds number (for more details see Chap. 11). In Fig. 5.3, iso-surfaces of vorticity, calculated by DNS, are shown for a typical turbulence field. The characteristic data are given in the Fig. 5.3 caption.

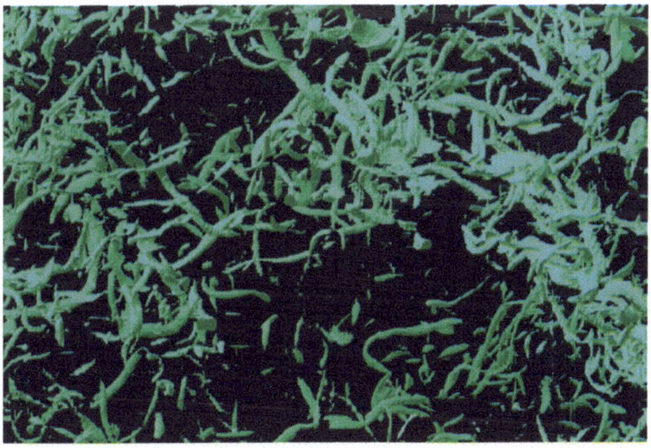

**Fig. 5.3** A zoom into a picture of iso-surfaces of vorticity of a turbulent velocity field with a Taylor Reynolds number, $Re_\lambda = 732$ [for the definition of this number see in Sect. 5.5, Eq. (145)]. Clearly entangled vorticity filaments are seen. From Kaneda and Ishihara (2006), J. Turbulence, reproduced with permission by Taylor and Francis, see: https://www.tandfonline.com/

Vorticity is defined as the curl of the velocity field,

$$\vec{\omega} = \text{curl } \vec{u} = \left( \frac{\partial u_3}{\partial x_2} - \frac{\partial u_2}{\partial x_3}, \frac{\partial u_1}{\partial x_3} - \frac{\partial u_3}{\partial x_1}, \frac{\partial u_2}{\partial x_1} - \frac{\partial u_1}{\partial x_2} \right). \tag{72a, b}$$

We are mainly concerned with quasi-two-dimensional flows for which the vorticity simplifies to

$$\vec{\omega} = (0, 0, \omega_3), \qquad \omega_3 = \frac{\partial u_2}{\partial x_1} - \frac{\partial u_1}{\partial x_2}. \tag{73a, b}$$

To describe quasi-two-dimensional turbulent flows, it is tempting to work only with the third component of the average vorticity

$$\overline{\omega}_3 = \frac{\partial \overline{u}_2}{\partial x_1} - \frac{\partial \overline{u}_1}{\partial x_2}, \qquad \omega_3' = \frac{\partial u_2'}{\partial x_1} - \frac{\partial u_1'}{\partial x_2}. \tag{74a, b}$$

Moreover, it follows with Eq. (4) that

$$\left\langle \vec{\omega} \right\rangle = \left\langle \text{curl } \vec{u} \right\rangle = \text{curl } \left\langle \vec{u} \right\rangle. \tag{75a, b}$$

To develop Taylor's turbulence model, it is not necessary to develop the full vorticity dynamics that starts by applying the curl operator to the NSE. To develop his simple closure relation, Taylor (1932) considered a quasi-two-dimensional flow, for which

$$\frac{\partial \tau_{21}}{\partial x_2} = -\rho \frac{\partial \left( \overline{u_2' u_1'} \right)}{\partial x_2}. \tag{76}$$

In shear flows, there are orders of magnitude differences between quantities in the two main directions. At first, for the main velocities one has the following requirement:

$$|\overline{u}_2| \ll |\overline{u}_1| \tag{77}$$

and then for the gradients

$$\left| \frac{\partial}{\partial x_1} \right| \ll \left| \frac{\partial}{\partial x_2} \right|. \tag{78}$$

From Eq. (74b), we conclude that

$$\frac{\partial u_1'}{\partial x_2} = \frac{\partial u_2'}{\partial x_1} - \omega_3', \tag{79}$$

which, owing to Eqs. (77) and (78), reduces approximately to

$$\omega_3' = -\frac{\partial u_1'}{\partial x_2}. \tag{80}$$

Thus, Eq. (76) is modified to

$$\frac{\partial \tau_{21}}{\partial x_2} = -\rho\left(\overline{u_1'\frac{\partial u_2'}{\partial x_2} + u_2'\frac{\partial u_1'}{\partial x_2}}\right), \tag{81}$$

from which, with Eq. (47), we conclude that approximately,

$$\frac{\partial \tau_{21}}{\partial x_2} = -2\rho\overline{u_2'\frac{\partial u_1'}{\partial x_2}} = 2\rho\overline{u_2'\omega_3'}, \tag{82a, b}$$

where Eq. (80) was substituted to obtain Eq. (82b). This relation will now be compared with Eq. (48b). Furthermore, the treatment between Eqs. (48b) and (54) can be applied to the transport of vorticity instead of linear momentum. Instead of repeating the entire development in analogy, just the key steps of this analysis are presented. Equation (48c) takes the form

$$\frac{\partial \tau_{21}}{\partial x_2} = 2\rho\,\tilde{c}\sqrt{\overline{u_2'^2}}\sqrt{\overline{\omega_3'^2}}. \tag{83}$$

Notice the derivative on the left-hand side in Eq. (83), which is missing in Eq. (48c)! In analogy to Eq. (50), the vorticity exchange can be written as

$$\sqrt{\overline{\omega_3'^2}} = \breve{c}\,\ell_\omega\frac{\partial \overline{\omega}_3}{\partial x_2}. \tag{84}$$

Substituting Eqs. (50) and (84) into Eq. (83) leads to

$$\frac{\partial \tau_{21}}{\partial x_2} = 2c\rho\,\ell_{\mathrm{m}}\ell_\omega\frac{\partial \overline{u}_1}{\partial x_2}\frac{\partial \overline{\omega}_3}{\partial x_2} \tag{85}$$

with the constant given by

$$c = \tilde{c}\,\breve{c}\,\hat{c}. \tag{86}$$

Moreover, from Eqs. (74a) with (77) and (78), the approximation

$$\overline{\omega}_3 = -\frac{\partial \overline{u}_1}{\partial x_2} \tag{87}$$

follows, from which, after substitution into Eq. (85), one deduces

$$\frac{\partial \tau_{21}}{\partial x_2} = -2c\rho \ell_{\mathrm{m}} \ell_{\omega} \frac{\partial \overline{u}_1}{\partial x_2} \frac{\partial^2 \overline{u}_1}{\partial x_2^2}, \tag{88}$$

or after integration and Prandtl's sign adaptation

$$\tau_{21} = \rho \varepsilon_{\mathrm{m}} \frac{\partial \overline{u}_1}{\partial x_2}, \qquad \varepsilon_{\mathrm{m}} = c \ell_{\mathrm{m}} \ell_{\omega} \left| \frac{\partial \overline{u}_1}{\partial x_2} \right|. \tag{89a, b}$$

A possible constant of integration vanishes because $\tau_{21} = 0$ at points of vanishing derivative of the average velocity. Equations (89a,b) agrees with Prandtl's mixing length theory if $\ell_{\mathrm{m}} = \ell_{\omega}$.

It is clear that the physical mechanisms are different in the linear momentum mixing and the vorticity transport model. On the other hand, the identical final formulas of these models make it clear that all the deficiencies of the mixing length model are also inherently those of Taylor's vorticity transport model.

For three-dimensional turbulent flows, Taylor generalized his model and denoted it *the generalized vorticity transport model*. According to Hinze (1975), the expressions to solve turbulent flow problems with this generalized model, became practically intractable. This was already noticed by Taylor, who found a method of small modification to apply the simple model to three-dimensional turbulence and coined this approach *modified vorticity transport theory*. Because these generalizations are of minor importance to our intentions, they are not presented here (for further study, see Taylor 1932).

## 5.3 Overview of Deficiencies of Local Models

Prandtl's mixing length model applies to shear flow problems and, compared with experimental data, shows that the mixing length has nothing to do with the characteristic size of a small fluid lump, but is rather of the overall size of the fluid domain. This fact led Prandtl to develop the shear layer model.

In certain flow problems, e.g., "wall turbulence," the mixing length is not a constant and has to be generalized to become proportional to the transverse coordinate, which, e.g., in "wall turbulence," is the distance from the wall.

According to this model, whenever the average velocity profile reaches its local maximum, the shear stress vanishes. This is experimentally usually not corroborated. This fact led Prandtl to develop the mean gradient model.

*Von Kármán's local model* shows problems at inflection points, where the shear stress becomes infinite, not a very realistic picture of true facts. Furthermore, it turned out that this model leads to more complexity, but no better accuracy than Prandtl's models (Hinze 1975).

*Reichardt's inductive model* was motivated by Gaussian averaged velocity profiles that occur in certain turbulent shear flows. However, working in analogy to linear diffusion it had to fail. The strongest argument against this model is that it does not fulfill Galilean invariance.

*Prandtl's mean gradient model* avoids a vanishing shear stress at zero directional derivative of the averaged velocity profile. However, the second derivative of this second-order Taylor approximation model tremendously complicates the calculus of averaged flow quantities. Furthermore, it is not clear how much improvement is achieved by the fact that two mixing lengths (which are free parameters to fit the algebraic results to experimental data) are responsible for the higher quality of this model.

*Prandtl's shear layer model* was a further step in the right direction, but because of its incompleteness could not help to reveal better results than the more rudimentary models just discussed. This model has a constant eddy viscosity and, therefore, leads to deviations at the boundaries of the turbulent domains, where the eddy diffusivity must vanish.

*Taylor's vorticity transfer model* has the same structure as Prandtl's mixing length model and, therefore, also the same deficiencies. From a point of view of physics, it is more difficult to explain the adaptation of vorticity to its new environment after it has been transported over a distance of the vorticity mixing length $\ell_\omega$ than that of linear momentum over the distance of the usual mixing length $\ell_{\mathrm{m}}$. In any case, viscosity plays a crucial role in the mechanism of embedding higher vorticity fluid lumps into regions with smaller fluid vorticity.

An overview of the deficiencies of the models is listed in Table 5.1, where the most pronounced one is the attempt of earlier turbulence researchers to compensate for the missing nonlocality in their theories by modifying the mixing length, a method that could not lead to full satisfaction.

**Table 5.1** Zero equation turbulence models and their deficiencies

| Model | Deficiencies |
|---|---|
| Prandtl's mixing length model | Mixing length from experiments is of larger size than anticipated |
| | Mixing length is not a constant |
| | Deviations at zero gradient of mean velocity profile |
| von Kármán's local model | Problems at inflection points of mean velocity profile |
| Reichardt's inductive model | Violation of Galilean invariance |
| Prandtl's mean gradient model | Solving flow problems leads to mathematical complications |
| Prandtl's shear layer model | Too stiff eddy viscosity causes deviations at boundaries |
| Taylor's vorticity transfer model | Same deficiencies as Prandtl's mixing length model |

## 5.4 More General Deficiencies and Fallacies

In the last hundred years, when scientists became slowly accustomed to nonlinear dynamics, mathematical approaches to scientific problems of higher complexity still dealt with the development and application of different kinds of linearization techniques. The work of those scientists, who were ahead of the mainstream with their insight into nonlinear phenomena, e.g., Poincaré, seems not to have been able to trigger immediate innovation.

The favoring of linearization had also a negative impact on turbulence research. Boussinesq's linear "gradient" law stands at the origin of a variety of different methods, laws, and turbulence models that show weaknesses because of the application of linearization techniques. Therefore, also the greatest deficiency of zero-equation turbulence models are due to the crude approximation, which these models are based upon. The models, discussed in the previous Sect. 5.2, are listed in Table 5.1. It seems that in the development of all first- and higher order models, the modelers were "trapped" in the Boussinesq-type "gradients" as if this eddy diffusivity concept would represent some physical certainity. In two articles (Egolf 1994, 2009), the author cites works of Corrsin (1974), Hinze (1975), and Bernard and Handler (1990), who were likely the first who stressed that the equation of Boussinesq's turbulence parameterization—being the basis of almost all approaches to turbulence modeling—is a highly problematic and insufficient description for moderate to high Reynolds number flows.

Hutter and Wang (2016a, b) wrote: *Following-up suggestions for the turbulent heat flux, $\vec{q}_t$, and turbulent mass flux, $\vec{j}_t$, were then analogously proposed with gradient-type proposals, 'setting in motion' the victorious advance of the gradient-type closure relations. This apparent gradient-mania found its continuation in the first-order closure scheme in the balance laws of turbulent kinetic energy, k, and the turbulent dissipation rate, ε, and the closure postulates for the flux of turbulent kinetic energy and turbulent dissipation rate. Here too, it seems as if the second and third generations of turbulence modelers would have forgotten already Prandtl's attempts in his papers (Prandtl* 1933 *and Prandtl* 1942*) with the intention to reach better agreement of the theoretical mean velocity distributions with a (partly) gradient-free parameterization of flux terms.*

Moore et al. (1996) investigated the principal directions (as analyzed in elastic media treated in classical mechanics) of the turbulent Reynolds stress tensor at numerous points in a tip leakage vortex flow.

In a $k$-$\varepsilon$ model, the turbulent viscosity is defined as follows (see, e.g., Hutter and Jöhnk 2004):

$$\nu_{\mathrm{T}} = C_{\mu} \frac{k^2}{\varepsilon}, \tag{90}$$

where $C_\mu$ is an empirical constant evaluated from experiments and $k$ is the turbulent kinetic energy defined in Eq. (37). The dissipation rate is given by the formula (see, e.g., Hutter and Jöhnk 2004)

$$\varepsilon = 4\nu\overline{II}_{\mathbf{D}'} \tag{91}$$

(notice the overbar), where the second invariant of the symmetric tensor $\mathbf{A}$ is defined by

$$II_{\mathbf{A}} := \frac{1}{2}\mathrm{tr}\left(\mathbf{A}^2\right) = \frac{1}{2}\mathbf{A}\cdot\mathbf{A} = \frac{1}{2}I_{\mathbf{A}^2}, \qquad (92\mathrm{a}-\mathrm{c})$$

and $\mathbf{D}'$ is the fluctuation of the rate of the strain tensor, defined in analogy to the full deformation tensor $\overline{\mathbf{D}}$ in Eq. (33). Moore et al. (1996) calculated the turbulent kinetic energy $k$ and the dissipation rate $\varepsilon$ from experimental data and state that their evaluation is more an assessment of the potential accuracy of the models based on the Boussinesq hypothesis than of an assessment of any particular turbulence model under consideration. Applying Mohr's circle technique of plane stress analysis (see, e.g., Beer et al. 1981), they found angle differences in the main principal directions of the calculated and measured mean Reynolds shear stress of up to 30° and deviations of the magnitude of the maximum shear stress within a range of 50%. Considering the moderate quality of their results, the main conclusions of their study were rather favorable for the Boussinesq approach, namely that it shows reasonable agreement with the measured Reynolds stresses.

Possibly inspired by the work of Moore et al. (1996), Schmitt and Hirsch (2000) published a remarkable paper on a numerical study of a Boussinesq-type constitutive equation for an axisymmetric complex turbulent flow. They state that the usual linear $k$-$\varepsilon$ model gives reasonable results only for the mean behavior of simple turbulent flows.[1] However, they proved that the $k$-$\varepsilon$ equations have serious deficiencies when considering more complex flows, possessing anisotropies, recirculations, large swirls, etc. They propose to generalize the linear standard $k$-$\varepsilon$ model, a goal which they themselves focused much attention to. As a result, Schmitt and his coauthor found different interesting attempts to the problem of generalizing the closure rule of turbulence (see also in Chap. 6).

As we already did in Chap. 4, Schmitt and Hirsch (2000) associate the Reynolds stress tensor (see Eqs. 17 and 18a,b) with the deviatoric (traceless) Reynolds stress tensor $\overline{\mathbf{R}}_{\mathrm{D}}$ (Eq. 39). Furthermore, they introduce the velocity gradient tensor $\overline{\mathbf{L}}$ (see, e.g., Wallace 2009), with the components

[1]Be aware that this moderate success mainly stems also in these cases from the fact of adjusting free parameters to the problem under consideration.

$$\overline{\mathrm{L}}_{ij} = \frac{\partial \bar{u}_i}{\partial x_j}. \tag{93}$$

It may be decomposed into its symmetric part, the strain rate tensor $\overline{\mathbf{D}}$ (see Eq. 33), and its skew symmetric part $\overline{\mathbf{S}}$, with Cartesian components

$$\overline{\mathrm{S}}_{ij} = \frac{1}{2}\left(\frac{\partial \bar{u}_i}{\partial x_j} - \frac{\partial \bar{u}_j}{\partial x_i}\right), \tag{94}$$

which is known as vorticity tensor (compare with results in Sect. 5.2.6). Obviously,

$$\overline{\mathbf{L}} = \overline{\mathbf{D}} + \overline{\mathbf{S}}. \tag{95}$$

Schmitt and Hirsch's extension of the Reynolds linear stress deviator parameterization of the form

$$\overline{\mathbf{R}}_{\mathrm{D}} \approx \overline{\mathbf{R}}_{\mathrm{DL}} = 2\nu_{\mathrm{T}}\overline{\mathbf{D}}, \tag{96a, b}$$

(index L ~ linear) with constant turbulent viscosity $\nu_{\mathrm{T}}$ consists in making $\nu_{\mathrm{T}}$ functionally dependent on the invariants[2] of $\overline{\mathbf{D}}$, in today's notation,

$$II_{\overline{\mathbf{D}}} = \frac{1}{2}\mathrm{tr}\left(\overline{\mathbf{D}}\,\overline{\mathbf{D}}\right) \quad \text{and} \quad III_{\overline{\mathbf{D}}} = \det \overline{\mathbf{D}}, \tag{97a, b}$$

where

$$I_{\overline{\mathbf{D}}} = 0. \tag{98}$$

Explicitly one has

$$\nu_{\mathrm{T}} = \nu_{\mathrm{T}}\left(II_{\overline{\mathbf{D}}}, III_{\overline{\mathbf{D}}}\right). \tag{99}$$

The authors ignore a dependence on $III_{\overline{\mathbf{D}}}$ [as is also the custom in applications of plasticity and some parts of rheology (polymeric creep)] and rewrite $II_{\overline{\mathbf{D}}}$ in terms of the Euclidean norm of $\overline{\mathbf{D}}$:

$$II_{\overline{\mathbf{D}}} = \frac{1}{2}\left\|\overline{\mathbf{D}}\right\|^2 \Rightarrow \left\|\overline{\mathbf{D}}\right\| \equiv \sqrt{2II_{\overline{\mathbf{D}}}}. \tag{100a, b}$$

Hence, Prandtl's turbulent viscosity (see also Eq. 54)

[2] These are the coefficients of the eigenvalue equation of $\overline{\mathbf{D}}$.

$$\nu_{\mathrm{T}} = c\ell_{\mathrm{m}}^2 \left| \frac{\partial \bar{u}_1}{\partial x_2} \right| \tag{101}$$

is replaced by

$$\nu_{\mathrm{T}} = \nu_{\mathrm{T}}\left(\left\|\overline{\mathbf{D}}\right\|\right) \Leftrightarrow \overline{\mathbf{R}}_{\mathrm{D}} = 2\nu_{\mathrm{T}}\left(\left\|\overline{\mathbf{D}}\right\|\right)\overline{\mathbf{D}}, \tag{102a, b}$$

and by recognizing that in a shear flow $|\partial \bar{u}_1/\partial x_2|$ equals $\left\|\overline{\mathbf{D}}\right\|$; with (100b) this corresponds to

$$\nu_{\mathrm{T}} = c\ell_{\mathrm{m}}^2 \left\|\overline{\mathbf{D}}\right\| = \sqrt{2}c\, \ell_{\mathrm{m}}^2 \left(II_{\overline{\mathbf{D}}}\right)^{1/2}, \tag{103a, b}$$

an expression already anticipated in the 1960s by Goldstein (1965). Furthermore, Schmitt and Hirsch (2000) mentioned that this expression was also given by Smagorinsky (1963) for a subscale grid model of large eddy simulations.

By generalizing Eq. (103a), there is still room for speculations, e.g., for power or polynomial laws,

$$\nu_{\mathrm{T}} = A_n \left\|\overline{\mathbf{D}}\right\|^n, \quad \text{or} \quad \nu_{\mathrm{T}} = \sum_{j=1}^{N} A_j \left\|\overline{\mathbf{D}}\right\|^{n_j} \tag{104a, b}$$

in which $A_j$ are constants and

$$n, n_j, \quad j = 1, 2, \ldots, N \tag{105}$$

are assignable exponents. Or one may invert Eq. (102b) as (106b)

$$\overline{\mathbf{D}} = \frac{1}{2} f_{\mathrm{T}}\left(II_{\overline{\mathbf{R}}_{\mathrm{D}}}\right)\overline{\mathbf{R}}_{\mathrm{D}}, \quad f_{\mathrm{T}}\left(II_{\overline{\mathbf{R}}_{\mathrm{D}}}\right) = \frac{1}{\nu_{\mathrm{T}}}, \tag{106a, b}$$

where $f_{\mathrm{T}}$ may be called *turbulent fluidity*, which is a function of the second invariant of the stress deviator $II_{\overline{\mathbf{R}}_{\mathrm{D}}}$. The inversions from Eqs. (102b) to (106a) or vice versa exist uniquely for monotonic $\nu_{\mathrm{T}}$ or $f_{\mathrm{T}}$. Analogous specifications as for $\nu_{\mathrm{T}}$ in Eqs. (104a,b) can also be written down for the fluidity (106b). However, the transformation from viscosity to a fluidity formulation and vice versa is for non-power laws generally a difficult task.

Before proceeding further, let us emphasize that the above closure schemes transpire locality and gradient-type structure, e.g., Prandtl's spirit in the strongest possible sense. The proposed closure class also maintains affinity (coaxiality) of $\overline{\mathbf{R}}_{\mathrm{D}}$ to $\overline{\mathbf{D}}$, which Prandtl never gave up.

Schmitt and Hirsch (2000) do not proceed in this way, but their approach corresponds to it. For a particular selection of $\nu_{\mathrm{T}}$ or $f_{\mathrm{T}}$, the corresponding turbulent closure conditions according to Eqs. (103a,b) or Eq. (106b) are nonlinear relations.

With Eqs. (104a,b) a linear $\overline{\mathbf{R}}_{\mathrm{DL}}$ can be replaced by a nonlinear equivalent $\overline{\mathbf{R}}_{\mathrm{DN}}$. Schmitt and Hirsch (2000) apply a quadratic development of $\overline{\mathbf{R}}_{\mathrm{DN}}$ as a polynomial series of four basic symmetric and traceless stress tensors. They state that the scalar coefficients are invariants of the flow under consideration (for more technical details see Chap. 6, Hutter and Jöhnk 2004 or Schmitt and Hirsch 2000). With this procedure it is possible to estimate a ratio $r$ of nonlinear to linear (deviatoric) turbulent stress, which is an approximate quantitative measure for the validity of the Boussinesq-type closure

$$r = \frac{\left\|\overline{\mathbf{R}}_{\mathrm{DN}} - \overline{\mathbf{R}}_{\mathrm{DL}}\right\|}{\left\|\overline{\mathbf{R}}_{\mathrm{DL}}\right\|}. \tag{107}$$

Experimental evaluation of a turbulent flow in a nozzle region of a burner chamber shows large ratios $r$ [e.g., of more than thirty ($r > 30$)], this also in regions where the denominator of Eq. (107) does not show small values. So Schmitt and Hirsch (2000), with their critical analysis of the derived results, performed an important quantitative proof of the invalidity of Boussinesq's hypothesis for turbulence modeling. In their concluding section, they cite some papers of authors who criticize local or one-point turbulence models and mention three possible types of new "constitutive equations" of turbulence that could help to overcome the unsatisfactory situation (see Chap. 6).

Schmitt et al. (2003) continued their former work of testing $k$-$\varepsilon$ and $k$-$\omega$ models with the underlying one-point Boussinesq gradient law by applying also large eddy simulation data. The $k$-$\varepsilon$ model transport equations are derived for the turbulent kinetic energy $k$ and the dissipation rate $\varepsilon$. In this procedure the same idea, as we have used in Chap. 3 to produce higher-order momentum equations, is applied—but only a single step is conducted.

They start their theory with the $k$-transport equation only, which is derived from the Reynolds averaged Navier–Stokes equations (16a,b)

$$\frac{\mathrm{D}k}{\mathrm{D}t} = \frac{\partial k}{\partial t} + \left\langle \vec{v} \right\rangle \cdot \vec{\nabla} k = \mathit{\Pi} - \varepsilon + \vec{\nabla} \cdot \left\langle \vec{T} \right\rangle + \mathrm{D}_{\mathrm{m}}. \tag{108}$$

Here $\mathit{\Pi} = \overline{\mathbf{R}}_{\mathrm{D}}\overline{\mathbf{D}}$ denotes the (averaged) production rate of turbulent kinetic energy; $\varepsilon$ is the (averaged) turbulent dissipation rate; and $\mathrm{D}_{\mathrm{m}}$ the molecular diffusion term. An important term is the turbulent diffusion term $\nabla \cdot \left\langle \vec{T} \right\rangle$ with the averaged turbulent enthalpy flux (Speziale 1998)

$$\left\langle \vec{T} \right\rangle = \left\langle \vec{T}_k \right\rangle + \left\langle \vec{T}_p \right\rangle. \tag{109}$$

The additive two terms are derived to be

$$\left\langle \vec{T}_k \right\rangle = -\left\langle \left( \vec{u}' \cdot \vec{u}' \right) \vec{u}' \right\rangle, \quad \left\langle \vec{T}_p \right\rangle = -\frac{1}{\rho} \left\langle p' \vec{u}' \right\rangle, \tag{110a, b}$$

where the second term $p'$ is the pressure fluctuation. In this theory, the closure is performed with a gradient of the averaged turbulent kinetic energy

$$\left\langle \vec{T} \right\rangle = \gamma_{\mathrm{T}} \vec{\nabla} k. \tag{111}$$

With the scalar turbulent kinetic energy diffusivity, $\gamma_{\mathrm{T}}$, the turbulent Prandtl number is defined as the ratio of the turbulent viscosity to the turbulent diffusivity of kinetic energy,

$$\mathrm{Pr}_{\mathrm{T}} = \frac{\nu_{\mathrm{T}}}{\gamma_{\mathrm{T}}}. \tag{112}$$

Then, Schmitt et al. (2003) generalize Eq. (111) by replacing the scalar $\gamma_{\mathrm{T}}$ by the second-rank tensor $\mathbf{\Gamma}_{\mathrm{T}}$,

$$\left\langle \vec{T} \right\rangle = \mathbf{\Gamma}_{\mathrm{T}} \vec{\nabla} k, \tag{113}$$

where $\mathbf{\Gamma}_{\mathrm{T}}$ was already proposed in the late 1940s by Batchelor (1949). It will therefore, be referred to as Batchelor's $k$-diffusivity tensor; it suggests the definition of "Batchelor's Prandtl number" by

$$\mathrm{Pr}_{\mathrm{T}}^{(\mathrm{B})} = \frac{\nu_{\mathrm{T}}}{\|\mathbf{\Gamma}_{\mathrm{T}}\|}. \tag{114}$$

Now, a small calculus and applying Eq. (96b) leads to the following chain of "identities":

$$1 \equiv \frac{2\nu_{\mathrm{T}} \operatorname{tr}\left(\overline{\mathbf{D}}\, \overline{\mathbf{D}}\right)}{2\nu_{\mathrm{T}} \operatorname{tr}\left(\overline{\mathbf{D}}^2\right)} \overset{(96\mathrm{b})}{=} \frac{1}{2} \frac{\operatorname{tr}\left(\overline{\mathbf{R}}_{\mathrm{D}} \overline{\mathbf{D}}\right)}{\nu_{\mathrm{T}} \operatorname{tr}\left(\overline{\mathbf{D}}^2\right)}, \tag{115a, b}$$

where the trace of the tensor products was introduced.[3] Equations (115a,b) show, in view of the abbreviation $\Pi = (1/2) \operatorname{tr}\left(\overline{\mathbf{R}}_{\mathrm{D}} \overline{\mathbf{D}}\right)$, that

[3]Note that Eq. (115a) is an identity, but Eq. (115b) is an equation by replacing $\overline{\mathbf{D}}$ by the Reynolds stress constitutive postulate (Eq. 96b).

$$\nu_{\mathrm{T}} = \frac{1}{2}\frac{\operatorname{tr}\left(\overline{\mathbf{R}}_{\mathrm{D}}\overline{\mathbf{D}}\right)}{\operatorname{tr}\left(\overline{\mathbf{D}}^{2}\right)} = \frac{\Pi}{\operatorname{tr}\left(\overline{\mathbf{D}}^{2}\right)}. \tag{116a, b}$$

Furthermore, with Eq. (113), one concludes that the following relation holds:

$$\frac{\vec{\nabla}k \cdot \left\langle \vec{T} \right\rangle}{\left(\vec{\nabla}k\right)^{2}} = \frac{\vec{\nabla}k \cdot \mathbf{\Gamma}_{\mathrm{T}}\vec{\nabla}k}{\left(\vec{\nabla}k\right)^{2}} = \|\mathbf{\Gamma}_{\mathrm{T}}\|, \tag{117a, b}$$

and from this

$$\|\mathbf{\Gamma}_{\mathrm{T}}\| = \frac{\vec{\nabla}k \cdot \left\langle \vec{T} \right\rangle}{\left(\vec{\nabla}k\right)^{2}}. \tag{118}$$

By inserting Eqs. (116b) and (118) into Eq. (114), the Batchelor Prandtl number takes the form

$$\mathrm{Pr}_{\mathrm{T}}^{(\mathrm{B})} = \frac{\Pi\left(\vec{\nabla}k\right)^{2}}{\operatorname{tr}\left(\overline{\mathbf{D}}^{2}\right)\vec{\nabla}k \cdot \left\langle \vec{T} \right\rangle}. \tag{119}$$

If Bossinesq's approximation withstands experimental scrutiny, calculated values on the right-hand side of Eq. (119) will yield values close to unity. Inferences revealed by numerical simulations are shown by the distribution of $\mathrm{Pr}_{\mathrm{T}}^{(\mathrm{B})}$, which is plotted in Fig. 5.4. Because the formula involves the ratio of numerically evaluated values, it is in domains with small denominators rather noisy. These regions are left blank (white). *Nevertheless, this figure shows that the generalized turbulent Prandtl number is highly variable and its constancy as assumed in the k-ε and k-ω models is not justified.*

Another measure to judge the performance of the model is by the formula

$$\rho_k := \frac{\left|\left\langle \vec{T} \right\rangle \cdot \vec{\nabla}k\right|}{\left|\left\langle \vec{T} \right\rangle\right| \left|\vec{\nabla}k\right|} = |\cos\theta|. \tag{120a, b}$$

It measures the alignment of the mean flux $\left\langle \vec{T} \right\rangle$ with the gradient of $k$. The angle $\theta$ denotes the deviation between the flux vector and the directional derivative ("gradient") of the turbulent kinetic energy $k$. Schmitt et al. (2003) call this their main test of the $k$-transport equation. We are not discussing the influence of the pressure fluctuation flux, which these authors find to be larger than assumed. However, one could

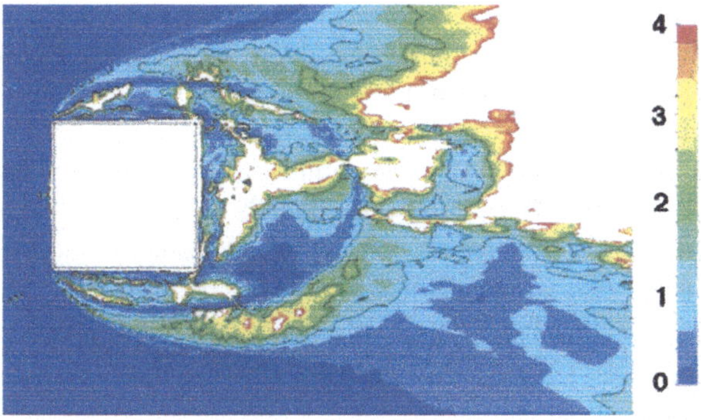


**Fig. 5.4** The generalized turbulent Prandtl number (Eq. 119) is evaluated from large eddy-simulation data. The isoline with value "1" corresponds to the line where the $k$-$\varepsilon$ model is valid, the isoline with value "2" marks the location where the $k$-$\omega$ model applies correctly, but in the other regions both models mainly fail. From Schmitt et al., (2003), J. Turbulence, reproduced with permission by Taylor and Francis, see: https://www.tandfonline.com/

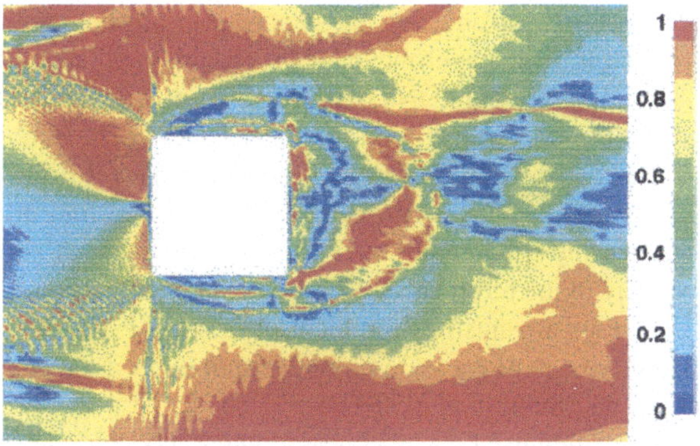


**Fig. 5.5** The validity of the classical $k$-transport equation is visualized. The quantitative measure (see Eq. 120b) is the absolute value of the cosine of the angle between the total flux and the gradient of $k$. Low values, occurring in the main turbulent wake domain behind the object, signalize an invalidation of Boussinesq's hypothesis. From Schmitt et al., (2003), J. Turbulence, reproduced with permission by Taylor and Francis, see: https://www.tandfonline.com/

imagine that an alignment between the total flux and the gradient of $k$ could be observed. That this is not the case is demonstrated in Fig. 5.5. The validity of the $k$-$\varepsilon$ model corresponds to $\rho_k = 1$. For values of $|\nabla k|$ smaller than 0.04 the value of $\rho_k$ was not computed. It can easily be verified that in regions of high turbulent shearing the gradient diffusion hypothesis (Eq. 111) fails. Results of Schmitt et al. (2003) show that one of the basic assumptions does not rest on solid grounds and must be replaced by some generalization.

The main results of these authors are the following: (1) the pressure flux (Eq. 110b) is not negligible as compared to the kinetic energy flux (because this is of minor importance for our purposes, it is not explained in detail in this review); (2) the gradient diffusion hypothesis is not confirmed, especially in complex flow regions, where there is no alignment between the total flux and the gradient of $k$; and (3) the turbulent Prandtl number, $\mathrm{Pr_T}$, which in the strict $k$-$\varepsilon$ model equals "1" (see Launder and Spalding 1974) and in the $k$-$\omega$ model equals "2" (see Wilcox 1998), here shows a huge variability (see Fig. 5.5).

In a further paper, Schmitt (2007a) complements the main measures for the quality of down-gradient-flux turbulence models by introducing the following expression:

$$\rho_{\mathrm{RD}} := \frac{\left|\overline{\mathbf{R}}_{\mathrm{D}} \cdot \overline{\mathbf{D}}\right|}{\left\|\overline{\mathbf{R}}_{\mathrm{D}}\right\| \left\|\overline{\mathbf{D}}\right\|}. \tag{121}$$

This measure is based on a tensor projection procedure proposed by Jongen and Gatski (1998) in the context of algebraic stress models and is applied to linear and nonlinear eddy viscosity models. It consists of projecting—by an inner product of tensors—the constitutive equation onto a tensor basis.

Just as for Eqs. (120a,b), this measure is bounded between zero and one. Zero means that Boussinesq's hypothesis fails, whereas if the value "1" is reached, it is correct. To calculate this measure, one needs access to all the components of the Reynolds stress deviator. With experimental data sets, this is not always the case. On the other hand, Schmitt's bases are DNS data sets that are complete and lead to most accurate evaluations. For his sophisticated study, Schmitt (2007a, b) chose DNS data sets of simple wall shear flows (boundary-layer flows, channel flows, and Couette flows) with Reynolds numbers from $10^3$ to $10^5$. The results also clearly consolidate the invalidity of Boussinesq's hypothesis, even for a treatment of simple turbulent shear flows, especially in the turbulent fluid regions not close to the wall. It is further concluded that in regions of vanishing turbulent shear stress even more sophisticated polynomial constitutive equations and nonlinear $k$-$\varepsilon$ models are incapable of delivering reasonable results (Schmitt et al. 2003).

Another argument, often used against linear gradient models, is the statement that these models are unable to display anisotropic (non-coaxial) Reynolds stress; in other words, they do not correctly express normal stress effects; this argument is not true. For steady one-dimensional shear flows within the $(x,y)$ plane Eq. (39) implies

$$\overline{\mathbf{T}}_{\overline{\mathbf{D}}'} = \frac{\overline{\mathbf{R}}_{\mathrm{D}}}{\rho} = \begin{pmatrix} \frac{2}{3}k - \overline{u_1'^2} & \tau & 0 \\ \tau & \frac{2}{3}k - \overline{u_2'^2} & 0 \\ 0 & 0 & \frac{2}{3}k - \overline{u_3'^2} \end{pmatrix}, \tag{122a, b}$$

while $\overline{\mathbf{D}}$ takes the form

$$\overline{\mathbf{D}} = \frac{c}{2}\begin{pmatrix} 0 & 1 & 0 \\ 1 & 0 & 0 \\ 0 & 0 & 0 \end{pmatrix}, \tag{123}$$

where $c$ is the averaged shear–strain rate. Now, $\mathbf{T}_{\overline{\mathbf{D}}'}$ is coaxial (affine) to $\overline{\mathbf{D}}$ (a deviator) in linear "gradient" models. In numerous presentations at this point, it is reasoned that the diagonal elements of (Eq. 122b) must vanish leading to the unique solution

$$\overline{u'^2_1} = \overline{u'^2_2} = \overline{u'^2_3}, \tag{124}$$

which is reminiscent of Reynolds stress isotropy. Correct is that only the trace of (Eq. 122b) vanishes

$$2k - \left(\overline{u'^2_1} + \overline{u'^2_2} + \overline{u'^2_3}\right) = 2k - 2k = 0, \tag{125a, b}$$

which, by an application of Eq. (37), shows that this condition is identically satisfied for any values of $\overline{u'^2_i}$, $i \in \{1, 2, 3\}$ (see also Spalart 2015 or Spalart and Allmaras 1994).

What inference can be concluded from this result? We either accept the linear gradient model, $\overline{\mathbf{R}}_{\mathrm{D}} = 2\nu_{\mathrm{T}}\overline{\mathbf{D}}$, and take the position that simple turbulent shear flows do exist (consistency of Eqs. (122b) and (123) then implies that Eq. (124) is valid, which is not supported experimentally (see Sect. 9.2.1)) or else, we accept the representation (Eqs. 122a,b), paired with anisotropic turbulent normal stress, and deny pure turbulent simple shearing. In both cases, we come to the final conclusion that the turbulent stress parameterization $\overline{\mathbf{R}}_{\mathrm{D}} = 2\nu_{\mathrm{T}}\overline{\mathbf{D}}$ is inacceptable.

In the rheological literature it is well known that differing normal stress components can be obtained with a nonlinear turbulent stress parameterization (see Chap. 6) of, e.g., Reiner-Riwlin type (see Hutter and Wang 2018)

$$\mathbf{T}_{\overline{\mathbf{D}}'} = \frac{2}{3}k\,\mathbf{I} - 2\nu_{\mathrm{T}}\overline{\mathbf{D}} + \alpha\overline{\mathbf{D}}^2. \tag{126}$$

By applying this representation to a simple shear flow, Eq. (123), one obtains the following result, in which $c$ is the average shear strain rate:

$$\mathbf{T}_{\overline{\mathbf{D}}'} = \frac{2}{3}k\begin{pmatrix} 1 & 0 & 0 \\ 0 & 1 & 0 \\ 0 & 0 & 1 \end{pmatrix} - \nu_{\mathrm{T}}c\begin{pmatrix} 0 & 1 & 0 \\ 1 & 0 & 0 \\ 0 & 0 & 0 \end{pmatrix} + \frac{1}{4}\alpha c^2\begin{pmatrix} 1 & 0 & 0 \\ 0 & 1 & 0 \\ 0 & 0 & 0 \end{pmatrix}. \tag{127}$$

Thus, with $\alpha \neq 0$, the first two normal turbulent stress components $\overline{u'^2_1} = \overline{u'^2_2}$ are non-zero, and differ from the third normal stress component $\overline{u'^2_3}$, see Eq. (18b).

A most convincing example to demonstrate by simple means that the Boussinesq approximation fails is plane turbulent Couette flow. In this flow configuration, one finds that for flow regions distant from the wall the correlation $\overline{u'_2 u'_1}$ is non-zero,

$$\overline{u'_2 u'_1} = u^{*2}, \qquad \forall \ \mathrm{Re} > \mathrm{Re}_c, \tag{128}$$

in which the shear velocity $u^*$ is defined with the help of the constant boundary shear stress $u^* = \tau_0/\rho$.

Furthermore, by definition, the turbulent shear stress is symmetric, and because it is non-zero for finite Reynolds numbers above its critical value (see Fig. 9.31), it is a direct consequence that for flow regions distant from the wall

$$\overline{u'_2 u'_1} \neq 0, \qquad \forall \ \mathrm{Re} > \mathrm{Re}_c, \tag{129}$$

which is in agreement with Eq. (128). This behavior is also proven by numerical experiments (see Fig. 9.32). On the other hand, the average downstream velocity profile, because of the anti-parallel movement of the two plates relative to its central axis, is anti-symmetric, with a vanishing derivative for increasing Reynolds numbers. Therefore, the equation

$$\frac{\mathrm{d}\bar{u}_1}{\mathrm{d}x_2} = 0, \qquad \mathrm{Re} \to \infty \tag{130}$$

holds (see Fig. 9.29). Combining the two relations (129) and (130) leads to

$$\overline{u'_2 u'_1} \neq \propto \frac{\mathrm{d}\bar{u}_1}{\mathrm{d}x_2}, \qquad \alpha < \infty, \qquad \mathrm{Re} \to \infty \tag{131}$$

which is a contradiction with Boussinesq's approximation applied to infinite Reynolds number plane Couette flows for constant and any space and velocity dependent $\alpha$'s. Equation (131) is valid for this type of wall-bounded shear flows and directly disproves the entire class of models based on the Boussinesq approximation (see, e.g., Eqs. 53a,b, 69, 71, and 89a,b). *It immediately becomes clear that this behavior can only be described by a model that shows nonlocality effects. The higher the Reynolds number is, the more distant will be the influence of the average velocity field on the Reynolds shear stress on the axis of the flow. In the infinite Reynolds number limit, it is only the location at the wall that is of any influence.*

Summarizing, all these different results imply a disillusion for the conventional theories and numerical modeling and simulation efforts based on Boussinesq's turbulence closure. It becomes clear now that other turbulence models must be developed, based on new and alternative concepts to accurately describe elementary turbulent shear flows (see Chaps. 6 and 7).

Our critical viewpoint concerning several aspects of present turbulence modeling is also supported by Tsinober (2009, 2014), who reports on key misconceptions and outdated paradigms (distinct sets of concepts, theories, models, research methods, postulates, and standards for legitimate contributions) of turbulence. He highly criticizes the excessive belief in local properties of turbulence and the belief in simple "structures," respectively, "objects" and their weak interactions. Small (unresolved) scales depend nonlinearly, temporally, and spatially nonlocally on the large (resolved) scales. Furthermore, he draws a rather pessimistic image concerning a success of surpassing the closure problem, which he does not see to happen in the near future if at all. When discussing intermittency in space and time, he draws a clear picture of two separated flow regions, one "vorticity-rich region" and another, which he calls "voids of lowest intensity," however, without relating (yet) his twofold image to the two phases of a critical phenomenon and phase transition concept, respectively (see Chaps. 10 and 11).

Similar critical articles were published by Bradshaw (1994) and Spalart (2015), later with the title, *Philosophies and Fallacies in Turbulence Modeling*. In this treatise Spalart addresses a set of controversial positions concerning RANS modeling, namely:

1. The fundamental paradoxon
2. The principle of receding influence and
3. The fallacies of turbulence modeling.

The fundamental paradoxon is the local nature of the Navier–Stokes system of equations, which involves the RANS equations and the obviously nonlocal nature of turbulence as a physical phenomenon.

The principle of receding influence addresses the role of higher-order moments. If a vanishing influence of higher-order moments would be observed, then higher-order turbulence modeling would be superior to lower-order methods. The reasoning is that simplified or even slightly erroneous assumptions, applied to higher-order moment closures, would be of less influence and by this also less harmful. But according to Spalart's experience this is not at all the case.

Let us now discuss the various items in Table 5.2:

- A *hard fallacy* occurs if by a mathematical argument or (small) calculus it is demonstrated that a model is meaningless. A further example is a violation of a physical law or realizability condition, e.g., frame indifference and Galilean invariance. Therefore, they are unquestionable fallacies.
- Distinguishing a *medium fallacy* from a *soft fallacy* is not so easy. Therefore, we give one joint explanation for the two groups. For these fallacies, approximate physical arguments can be used for their justification, but more intense discussions are overdue, because the underlying problem in discussion contains wishful thinking that, during the past decades, became well-accepted standard knowledge in numerous cases. An example of such a fallacy is the logarithmic mean velocity law for shear flows (see Sects. 5.5 and 5.6), which is not the final solution to the wall problem and by which all discussions on the universality of the von Kármán constant become superfluous.

In this chapter, we carefully sort out all the fallacies which could be related to our work, and we discuss them in our context (see Table 5.2). On the other hand, a large number of fallacies concern higher-order modeling and specific numerically motivated turbulence modeling methods (e.g., wall functions), which are not so relevant to the purposes of this book. Finally, in the conclusions, we will come back to the three main themes of Spalart and discuss them in the light of the present work.

Spalart mentions that the application of algebraic nonlocal approaches in the 1960's and 1970's, because of the ambition to treat complex geometries, was mainly

**Table 5.2** A selection of themes and fallacies of turbulence modeling and their importance in the context of the present work (derived from Tsinober 2009, 2014 and Spalart 2015)

| Theme/misconception/fallacy | Importance |
|---|---|
| 1. *Themes and misconceptions* | |
| 1.1 Fundamental paradoxon (locality) | 7.3 |
| 1.2 Principle of receding influence | 3 |
| 2. *Fallacies* | |
| 2.1 *Hard fallacies* | |
| 2.1.1 Isotropy of normal Rey. stress | 5.4/9.3 |
| 2.1.2 Isotr. of lin. eddy viscos. mod. | ● |
| 2.1.3 Velocity is an entry in the model[a] | 7 |
| 2.1.4 Acceleration is an entry in the model[a] | ● |
| 2.1.5 Unsteady flows are difficult | o |
| 2.1.6 Wall functions are useful | ● |
| 2.2 *Intermediate fallacies* | |
| 2.2.1 von Kármán constant is universal | ● |
| 2.2.2 Realizability (stress tens. pos. def.) | 5.4 |
| 2.2.3 "Equilibrium" turb. state exists | o |
| 2.2.4 Artificial turbulence as test case | o |
| 2.2.5 Reynolds stress of order $y^{+3}$ at wall | 9.5 |
| 2.2.6 Correct $k$ and $\varepsilon$ is important | ● |
| 2.2.7 Const. total shear stress at wall | 9.5 |
| 2.2.8 One-eq. models are not complete | 3 |
| 2.2.9 Extra strains for curvature | 7.3 |
| 2.3 *Soft fallacies* | |
| 2.3.1 Hom. and isotr. turbulence as test | ● |
| 2.3.2 Algeb. Reyolds stress models superior | ● |
| 2.3.3 Wall distance should not be used | 9.5 |

If a theme or fallacy is important to us and is discussed in this chapter, the section number is given, otherwise just the symbol 'o' is used. The bullet ● denotes that the corresponding fallacy is less important to us, mainly because of new insights, or by its own terms

[a]Galilean invariance demands velocity derivatives or differences

replaced by the development of models based on transport equations, but by retaining local formulations. In this context, a series of publications support this idea and in our opinion justifies them with incorrect arguments, e.g., with the statement that diffusion-like processes lead to an influence of flow properties far apart from the streamline under consideration. It is summarized that the equations are local, but the mature flow field solution is not. Obviously, this is a misunderstanding of the nonlocality concept (compare with explanations in Chap. 6).

The use of the wall distance or boundary layer thickness concepts (see Sect. 5.5) in models (e.g., model of Cebeci and Smith, see Smith and Cebeci 1967 or/and Wilcox 1998) is essentially the same concept as Prandtl's introduction of the width of the overall turbulent domain in his shear layer model for free turbulent shear flows (Sect. 5.2.5). This method is a first step toward the introduction of nonlocality, but in the end also these models were criticized for poor agreement with experimental data, etc. (see Hinze 1975).

The Spalart Allmaras Model (SA model) (Spalart and Allmaras 1994) with different modifications [e.g., SARC (SA with Rotation and Curvature)] introduce in the Reynolds stresses a higher-order term by products of first-order velocity "gradients." Generally, use of higher-order velocity derivatives with respect to time and space are also a step toward a description of history and nonlocal spatial effects. Also, this method was criticized, e.g., with the argument that viscous stresses neither do contain such derivatives. With such and similar putative physical arguments these methods were set in an unfavorable light. Notice that it is evident that exactly the incorporation of higher derivatives in closure statements leads to a welcome flexibility of adequately describing curvature (vortices) of streamlines. Spalart writes that it would be preferable to eliminate such derivatives, but no satisfactory alternative that is properly invariant has been found to sensitize eddy-viscosity turbulence models to streamline curvature. Nevertheless, such models are, in contrast to simpler models, able to describe corner vortices.

Most problems related to the fallacies, listed in Table 5.2, are clear. Concerning further details, we refer to the publication of Spalart (2015). Nevertheless, still a few additional clarifying remarks are made to these fallacies:

2.1 *Hard fallacies:* (2.1.1) The isotropy of the Reynolds stress and a possible solution has been discussed in this section. Spalart states that the isotropy is only an artifact of the special choice of coordinates in a shear flow. (2.1.2) Heavily criticizing linear eddy viscosity models and showing an alternative path to turbulence modeling, their isotropic, respectively, scalar nature is of minor importance to us. (2.1.3) Furthermore, the entry of velocities plays a crucial role in the presented zero-equation turbulence model—the DQTM (see Chap. 7)—whereas (2.1.4) in our considerations acceleration plays no role. These last two topics are directly coupled with Galilean invariance that a model is required to fulfill. (2.1.5) Discussing whether unsteady flows are more difficult than steady flows is at this initial state of the development of our ideas outside our scope as we, in this initial phase, are only concerned with quasi-steady flows. (2.1.6) The implementation of wall functions is a numerical artifact, not of central importance in a book mainly focusing on a new zero-equation turbulence model.

2.2 *Intermediate fallacies:* (2.2.1) In Sect. 9.5 we will give evidence that the logarithmic law finally must be replaced by a deficit power law. If this is accepted, the discussion about the universality property of the "von Kármán constant" proves to be vacuous. (2.2.2) It is emphasized that linear Boussinesq-type turbulence models do not fulfill realizability. Spalart relates this condition mainly to the occurrence of three positive eigenvalues demanded for the stress tensor. (2.2.3) Turbulent equilibrium flows have equal production and dissipation rates for the turbulent kinetic energy. Spalart doubts that eddy viscosity models refer to non-equilibrium flows as decaying turbulence. (2.2.4) DNS can produce turbulent fields (e.g., with periodic shearing in the continuum) that are not realized in nature nor realizable in technical devices. Such turbulence problems need not be taken into consideration to test a turbulence model. (2.2.5) The Reynolds stress shows toward the wall a third-order power dependence. This behavior also depends on transition and damping functions; therefore, Spalart does not give it a major importance in the evaluation of a turbulence model. (2.2.6) This topic is only important for turbulence modelers applying transport equations; these, dealing only with zero-equation turbulence models, miss these quantities in their models. (2.2.7) This fallacy property is directly related to the log law and therefore of no significance for our purposes. (2.2.8) Models with fewer than two transport equations cannot be complete. This agrees well with our presentation on closure demonstrated in Chap. 3. (2.2.9) In transport equations extra strains for curvature are required.

2.3 *Soft fallacies:* (2.3.1) According to Spalart there are problems in accepting the decay of turbulent kinetic energy in homogenous and isotropic turbulence. This is predicted by Kolmogorov and some extended multi-fractal models as test cases for turbulence models (for details see Spalart 2015). (2.3.2) Newer algebraic stress models (ARSM) introduce nonlinear extensions of eddy viscosity models, by combining Reynolds stress transport models (RST) with a "weak nonequilibrium hypothesis" set-up by Rodi (1976a, b). In these models, one postulates that a dimensionless Reynolds stress anisotropy varies less than the turbulent kinetic energy. Spalart suspects in the method of evaluation of this hypothesis a fallacy (see Spalart 2015). (2.3.3) He sees the appearance of the wall distance and the wall-normal vector not at all as flaws, but as ideal candidates to improve "constitutive equations" of turbulence. These two quantities are of nonlocal character and correspond with important quantities occurring in the DQTM, presented in Chap. 7.

George (2013) *addresses the importance of turbulence research to society and the costs of our ignorance of the physics of turbulence and states that it is difficult to place a price tag on the costs of our limited understanding of* this *phenomenon. He is convinced that it must be enormous.* It includes inadequate weather forecasting (e.g., for ships and planes), fluid-thermal systems with a need for abundant safety factors and by this occurring smaller system's efficiency and higher energy consumption. He also mentions a never-ending "numerical code" validation and calibration for all flow configurations by experimentation that would be just a small percentage of the present, if the "right" physical laws and models would be known.

## 5.5 Questioning the Logarithmic Law

In the last section, numerous insufficiencies of the early momentum and vorticity mixing length models were discussed; they will subsequently further become transparent. However, it is rather futile to spend also space for solutions of shear flows, calculated with incomplete models. To still satisfy this demand a little, here at least a single exemplary and important case—the "wall-turbulent" shear flow—is treated and presented in detail by an application of Prandtl's mixing length model. The standard solution of this important problem demonstrates how an incomplete solution was obtained and, by adjusting free parameters, how it was adapted to become quite useful and finally even generally accepted.

In the following, we are inspired by Schlichting's (1979) book. A plane turbulent stream along a smooth and flat plate with vanishing derivative of the pressure in the downstream direction is studied.

Extrapolating experimental data from pipe flows to turbulent flows along a wall and some simple theoretical considerations led Prandtl already in 1913 (Davidson et al. 2011) to the $x_2^{1/7}$–power law for the average downstream velocity profile, where $x_2$ is the distance from the wall. A more detailed analysis revealed a slight dependence of the exponent of the power law on the Reynolds number (Wieghardt 1946). In pipe flows, for high Reynolds numbers the power 1/6 decreases to 1/10 (compare with results in Fig. 5.7).

In solving this problem, Prandtl also made a first step of adapting his mixing length model to experimental results by stating that the mixing length is not constant, but proportional to the distance from the wall,

$$\ell_{\mathrm{m}} = \kappa\, x_2. \tag{132}$$

We have learnt that the mixing length is experienced to be rather the size of the transporting eddy than the diameter of the fluid lump. By the hindrance effect of the wall, assumption (132) seems to be very reasonable. From Eq. (54), for the eddy viscosity, it now follows that

$$\varepsilon_{\mathrm{m}} = \kappa^2 x_2^2 \left|\frac{\partial \overline{u}_1}{\partial x_2}\right|. \tag{133}$$

We have set $c = 1$ because this arbitrariness can be absorbed by the new constant $\kappa$. Following Eq. (53a) and applying the eddy viscosity (Eq. 133), for a monotonically increasing averaged velocity profile at some distance from the wall, the following turbulent shear stress formula is obtained

$$\tau_{21} = \rho\, \kappa^2 x_2^2 \left(\frac{\partial \overline{u}_1}{\partial x_2}\right)^2. \tag{134}$$

A second assumption by Prandtl is that the turbulent shear stress in the immediate vicinity of the wall remains constant, i.e.,

$$\tau_{21} = \tau_0, \tag{135}$$

where $\tau_0$ is the shear stress at the wall ($x_2 = 0$). For reasons of non-dimensionalization the equally constant friction velocity

$$u^* = \sqrt{\frac{\tau_0}{\rho}} \tag{136}$$

is introduced. By combining Eqs. (134) and (135) with Eq. (136), one obtains

$$u^* = \kappa\, x_2 \frac{\partial \bar{u}_1}{\partial x_2} = \text{const.} \tag{137a, b}$$

With the dimensionless inner variables

$$y^+ = \frac{u^* x_2}{\nu}, \qquad u^+ = \frac{\bar{u}_1}{u^*}, \tag{138a, b}$$

Equation (137a) transforms to

$$\kappa\, y^+ \frac{du^+}{dy^+} = 1, \tag{139}$$

revealing as its solution (Prandtl 1949) the universal logarithmic law of the wall

$$u^+ = A \log_{\mathrm{e}} y^+ + B, \quad A = \frac{1}{\kappa}. \tag{140a, b}$$

After following this or a similar presentation of the model, in practically all articles and textbooks it is then stated that comparison with experiments determines the two constants to be

$$A = 2.5 \pm 0.4, \quad B = 5.0 \pm 1.3. \tag{141a, b}$$

The experimental values published over the years spread a little. Here, they are taken from Reynolds (1974). At first, Prandtl, who believed very strongly in the adequacy of the 1/7-power law, was disappointed by this result and struggled further to derive with some modified assumptions the power law. However, influenced by von Kármán, who strongly supported the logarithmic law, Prandtl also got slowly acquainted to this alternative law. Eisner (1932) presented the logarithmic law as the best fit to experimental data and, thus, terminated the "interregnum of power laws." In Sect. 9.5, it will be shown that in the recent physical and mathematical fluid mechanics literature the discussion on the logarithmic law or power law has reappeared.

In the following sections, it is demonstrated that the classical derivation of the logarithmic law with Prandtl's mixing length formulation constitutes a serious problem. To see this, an extended model, due to Obermeier (1996) (personal communication), is presented.

For a turbulent flow at a wall, it makes sense to write the full equation as

$$\tau_0 = \mu \frac{\partial \overline{u}_1}{\partial x_2} - \rho\, \overline{u_2' u_1'}, \tag{142}$$

where the viscous dissipation term is added; it plays a crucial role at moderate Reynolds numbers

$$\mathrm{Re} = \frac{\overline{u}_1 L}{\nu}, \tag{143}$$

and even at high Reynolds numbers, but then only very close to the wall. The Reynolds number is the dimensionless stress parameter of the fluid dynamic system and contains, with the characteristic velocity, the main driving parameter. One can see that a high value of the characteristic overall length $L$ of the fluid domain also favors the creation of turbulence. On the other hand, a very large kinematic viscosity $\nu$ of the fluid leads to a damping of perturbations and brings the fluid system back into a stable state, which in this case is the laminar flow configuration.

To investigate the internal structure of the turbulence, it makes sense to replace the overall characteristic length $L$ by an internal characteristic length scale $\lambda$, that describes the internal properties of the turbulence under investigation, and the average velocity by its rms-fluctuation velocity. A definition of such a characteristic length $\lambda$ was proposed by Taylor (see, e.g., Tennekes and Lumley 1972):

$$\overline{\left(\frac{\partial u_1'}{\partial x_1}\right)^2} \equiv \frac{\overline{u_1'^2}}{\lambda^2}. \tag{144}$$

This length scale is also related to the curvature of the spatial velocity autocorrelation (see Hinze 1975). In Eq. (143), by replacing $L$ by $\lambda$ and the average velocity by the rms-fluctuation velocity, the Taylor-Reynolds number is obtained as

$$\mathrm{Re}_\lambda = \frac{\sqrt{\overline{u_1'^2}}\,\lambda}{\nu}. \tag{145}$$

Returning to Eq. (143), the quantity $L$ denotes an overall characteristic length of the problem. In a description of "wall turbulence" this length scale is, e.g., the entire length of the wall.

Introducing Prandtl's mixing length model Eq. (53a), with the eddy viscosity (Eq. 133), into Eq. (142) and substituting again dimensionless variables leads to a nonlinear generalization of Eq. (139),

$$\frac{du^+}{dy^+} + \kappa^2 y^{+2} \left( \frac{du^+}{dy^+} \right)^2 = 1, \tag{146}$$

where, in analogy to Eq. (43a), the kinematic viscosity $\nu = \mu/\rho$ was introduced. The no-slip boundary condition states that

$$u^+(y^+)|_{y^+=0} = 0 \tag{147}$$

In order to solve the nonlinear differential equation (146), the following substitution is applied:

$$p = \frac{du^+}{dy^+}. \tag{148}$$

This transforms Eq. (146) to the quadratic equation

$$p^2 + \frac{1}{\kappa^2 y^{+2}} p - \frac{1}{\kappa^2 y^{+2}} = 0 \tag{149}$$

with the two solutions (see also Obermeier 2006)

$$p^{\pm} = -\frac{1}{2\kappa^2 y^{+2}} \left( 1 \mp \sqrt{1 + 4\kappa^2 y^{+2}} \right). \tag{150a, b}$$

Substitution of Eq. (148) and a subsequent integration of the positive branch $p^+$, yields the general solution subject to the boundary condition (147)

$$u^+ = \frac{1}{2\kappa^2 y^+} \left( 1 - \sqrt{1 + 4\kappa^2 y^{+2}} \right) + \frac{1}{\kappa} \log_e \left( 2\kappa y^+ + \sqrt{1 + 4\kappa^2 y^{+2}} \right). \tag{151}$$

We notice that two terms occur, namely a hyperbolic term and again a logarithmic term with the inverse von Kármán constant $1/\kappa$ as prefactor.

It is meaningful to determine two limits, the first is $y^+ \to 0$. The root is approximated by a first-order Taylor series approximation (to second order in $y^+$)

$$\sqrt{1 + 4\kappa^2 y^{+2}} = 1 + 2\kappa^2 y^{+2} + O(y^{+4}). \tag{152}$$

In this limit, the first term in Eq. (151) vanishes as $\sim(-y^+)$. So, in view of the wrong sign of the resulting expression, we expect also a contribution of the logarithmic term to occur. Applying then the result (Eq. 152) in the second part of Eq. (151) and writing a Taylor series expansion leads to

$$\log_{\mathrm{e}}\left(1+2\kappa y^{+}+2\kappa^{2}y^{+2}\right)\approx 2\kappa y^{+}+O\left(y^{+2}\right). \tag{153}$$

Thus, as the quantity $y^{+}\rightarrow 0$, Eq. (151) reduces to $u^{+}\rightarrow y^{+}+[O(y^{+2})]$. For small values of $y^{+}$, we thus have the limit behavior

$$y^{+}\rightarrow 0:\quad u^{+}=y^{+}\;\Rightarrow\; u^{+}(0)=0. \tag{154a, b}$$

This is the linear solution for the average velocity profile, valid in a small sublayer with decreasing thickness as a function of increasing Reynolds number. This layer is the so-called *laminar viscous sublayer.*

The second limit of Eq. (151) is that for very large distances from the wall. Then a similar asymptotic argument, now for $y^{+}\rightarrow\infty$ implies

$$\lim_{y^{+}\rightarrow\infty}\sqrt{1+4\kappa^{2}y^{+2}}\rightarrow 2\kappa y^{+} \tag{155a, b}$$

and

$$\lim_{y^{+}\rightarrow\infty}\log_{\mathrm{e}}\left(2\kappa y^{+}+\sqrt{1+4\kappa^{2}y^{+2}}\right)\rightarrow\log_{\mathrm{e}}(4\kappa y^{+}). \tag{156a, b}$$

Therefore, it follows

$$y^{+}\rightarrow\infty:\Rightarrow\; u^{+}=\frac{1}{\kappa}[\log_{\mathrm{e}}(4\kappa y^{+})-1]=\frac{1}{\kappa}\log_{\mathrm{e}}(y^{+})+\frac{1}{\kappa}[\log_{\mathrm{e}}(4\kappa)-1]. \tag{157a, b}$$

Comparison of Eqs. (140a,b) with (157b) leads to a mathematical relation between the constants $\kappa$ and $A$ and $B$, respectively,

$$B=A\left[\log_{e}\left(\frac{4}{A}\right)-1\right],\quad A=\frac{1}{\kappa}. \tag{158a, b}$$

If we now assume that $A=2.5$ as extracted from experimental data, with Eq. (158a), it follows that

$$B=-1.33 \tag{159}$$

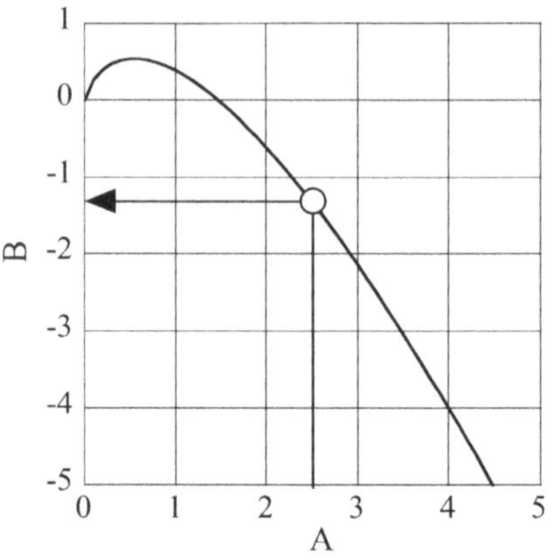


**Fig. 5.6** The derivation of the logarithmic law leads to an unrealistic relation between the multiplicative constant $A$ and additive constant $B$. If the constant $A$ is taken from experiment to be 2.5, $B$ is $-1.33$. The experimental value $B = 5.0$ does not even arise in the range of this graph

instead of the experimental value $B = 5.0$ in Eq. (141b). This is off by an absolute error of 6.33![4] The relation between $A$ and $B$ is shown in Fig. 5.6. This inacceptable result seems to the authors to be corroboration that the logarithmic law—at least in its version that is developed with Prandtl's mixing length model—is neither a realistic nor trustworthy solution for the "wall turbulent" flow problem.

It is also symptomatic that, because of a lack of good basic physical models, phenomenological, semi-empirical and fully empirical models were constructed in processes without committing consultancy of experimental data. Some of these solutions present today a performance in much better agreement with experimental data than would be expected on the basis of their boldness. An argument often heard is that "all zero-equation models, for elementary turbulent flow problems, lead to good results." This statement is not true if one is strict in assessing a solution as we have just done with the logarithmic "law of the wall" in this subsection.

Doering and Constantin (1994) applied the same closure to plane turbulent Couette flows and also derived a logarithmic law of a similar form as Obermeier did in his calculus presented above.

## 5.6 Logarithmic Versus (Deficit) Power Law

Over the last hundred years, there was for the averaged downstream velocity profile in the overlap region of zero-pressure-gradient boundary layers, smooth channels, and pipes an alternating believe in the power and logarithmic law as the best solution (see Table 5.3). Today, the consensus is that the solution is a logarithmic function. However, in this chapter it is stated that the right solution is a deficit, respectively, a defect power law. Such a law describes the difference of the maximum free stream velocity and the actual average velocity by a power law. The main historical

[4]This is the end of the calculation by Frank Obermeier.

**Table 5.3** The preference for a logarithmic and a power law was alternating over the last hundred years

| Year of discovery/ contributor(s) | Contribution (preferred law by authors) |
|---|---|
| *~1913[a]/1921*<br>Prandtl (1913, 1921a)<br>von Kármán (1921) | 1/7th, later $1/n(\mathrm{Re})$ power law (see additional comments in Bodenschatz and Eckert 2011) |
| *1925/1931*<br>Prandtl (1925)<br>von Kármán (1931) | Logarithmic law with experimentally determined coefficients was considered to be the universal final solution |
| *1938*<br>Millikan (1938) | Similarity hypothesis favors the logarithmic law |
| *1993*<br>Barenblatt (1993a, b) | Supported by measurements (super pipe) the power law receives attention again |
| *1997*<br>Zagarola et al. (1997) | These researchers defend the logarithmic law for high Reynolds numbers |
| *2000*<br>Barenblatt et al. (2000) | Postulated power law of the wall based on dimensional analysis |
| *2001*<br>Afzal (2001) | Shows that similarity consideration of Millikan can also be applied to confirm a power law |
| *2001*<br>Oberlack (2001),... | ...applying Lie group methods, shows that for turbulent shear flows any function and, thus, any scaling law is admissible |
| *2007*<br>George (2007)... | ...shows that, derived from first principles and using equilibrium similarity analysis and near-asymptotic methods, a power law theory is the consequence |
| *2009*<br>Egolf (2009) | With DQTM theoretically a deficit power law with Reynolds number dependent coefficients (incomplete similarity) is obtained. |

Today, based on this chapter, the authors show a preference for a deficit, respectively, defect power law

[a]Not confirmed (Sect. 9.5)

developments concerning this very important scientific and engineering theme shall be outlined in this section.

Prandtl (1904) introduced the important boundary layer concept. At the beginning of the 1920s, von Kármán (1921) presented Prandtl a theory for the averaged velocity profile in the overlap region (see Sect. 9.5), but Prandtl responded that he knew this law, which is a power law with the exponent 1/7, for years, say since 1913 (see Davidson et al. 2011 and also Table 5.3). Indeed, a presentation of a resistance coefficient (Prandtl 1921b) could not have been achieved by Prandtl and his collaborators without having any knowledge of this first discovered power law of the wall, which describes the average downstream velocity in the overlap region, located at some distance from the wall,

$$\frac{\overline{u}_1}{U} = \left(\frac{x_2}{R}\right)^{1/7}, \tag{160}$$

where in this case $R$ denotes the radius of the pipe and $U$ the velocity on its centerline $U = \overline{u}_1(0)$. Toward the wall, approximately the same laws are valid for plane (plate or channel) and cylindrical (pipe) geometries.

Experiments of Jakob and Erk (1924) and Nicuradse (1926) confirmed the 1/7th power law, but raised the question whether this rather static law was good enough for the entire Reynolds number range. Occurring discrepancies were corrected by formulating an empirical power law with a Reynolds-dependent exponent

$$\frac{\overline{u}_1}{U} = \left(\frac{x_2}{R}\right)^{1/n}. \tag{161}$$

In this law, the exponent varies between six and ten (see Wieghardt 1946 and Fig. 5.7).

As is shown in Sect. 5.5, the mixing length model with a mixing length depending on the wall distance led Prandtl (1925) to obtain the logarithmic law that he disliked very much. He took strong efforts to derive the 1/7th power law with his mixing length model, but did not succeed. In his 1931 Stockholm conference contribution, von Kármán (1931) compares the logarithmic law with Nicuradse's (1926) experiments and brings up the idea that in the logarithmic law the multiplicative constant $\kappa$ could be universal. Furthermore, he finally provides his full conviction for the logarithmic law to be correct.

Some years later Millikan and Clauser (see Millikan 1938) introduced a singular matched asymptotic perturbation approach, based on scaling and matching inner and outer variables, that leads to an even more convincing argument toward the logarithmic law than the former developments. They called the region where inner and outer solutions agree with one another except for exponentially small terms the overlap region (see, e.g., van Dyke 1964 or Kevorkian and Cole 2010). As a

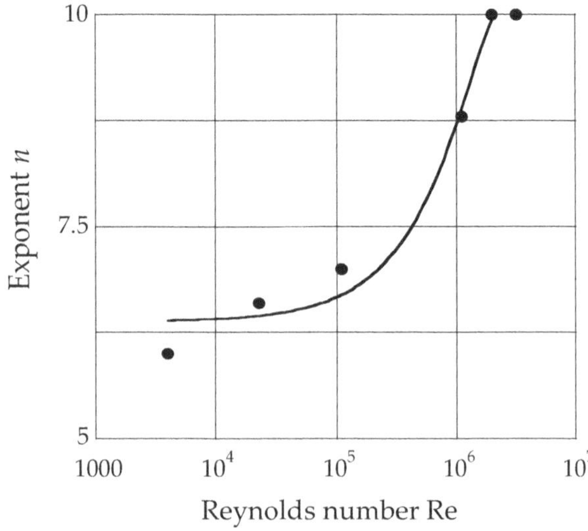


**Fig. 5.7** The variation of the exponent $n$ of the power law with Reynolds number Re is shown and varies in the presented domain from 6 to 10 (redesigned from data in Schlichting 1979)

consequence, it looked as if the power law then definitively had lost all chance to become the established "law of the wall." However, George and Castillo (1997) give three characteristic results of the Millikan/Clauser theory that are questionable. The main critic is that the asymptotic averaged velocity profile for infinite Reynolds number tends to a Heaviside distribution and does not reveal an asymptotic curved profile as it might be expected:

$$\lim_{\mathrm{Re}\to\infty} \frac{\overline{u}_1(\mathrm{Re}, x_2)}{U} = 1. \tag{162}$$

However then, beginning in the 1990s, Barenblatt (1993a, b) and Barenblatt et al. (2000) renewed the discussion whether there is a favor for a power law or the logarithmic law and gave preference to the power law, that was also applied in early years by engineers,

$$u^+ = C(y^+)^\alpha; \tag{163}$$

he states that the multiplicative factor $C$ and the exponent $\alpha$ depend on the logarithm of the Reynolds number. A logarithmic function is a kind of envelope to all the different power laws for different Reynolds numbers. More precisely, the power laws actually slightly overshoot the logarithmic envelope a little at the high Reynolds number end (see Fig. 5.8). Numerous arguments were supported by precise measurements that were published and performed in a super pipe at Princeton University, which produced a flow of the highest Reynolds number ever achieved in a laboratory (of the order of $\mathrm{Re} = 35$ millions) (see Zagarola et al. 2015). In a publication of Cipra (1996) in "Science" Chorin is cited: *"The "law of the wall" was viewed as one of the few certainties in the difficult field of turbulence, and now it has been dethroned."* He further continues by writing *"Generations of engineers, who learned the law, will have to abandon it."*

On the other hand, Zagarola et al. (1997) published an article in the spirit of matched asymptotics and showed again a slight preference for the logarithmic law for high Reynolds number flows, whereas they relate a power law to a second overlap region. Finally, they give preference to a defect law and pronounce, that if it is independent of the Reynolds number, complete similarity occurs and if not, the law shows incomplete similarity.

Then, at the very beginning of the new millennium Afzal (2001) demonstrated that the Millikan/Clauser idea can also be applied to prove consistency with a power law as averaged velocity profile of the overlap region. This work then immediately removed a believed theoretical obstacle for acceptance of a power law.

Almost at the same time, Oberlack (1999) (see also Oberlack 2001) came to a similar conclusion by other means. By a completely new approach to turbulent shear

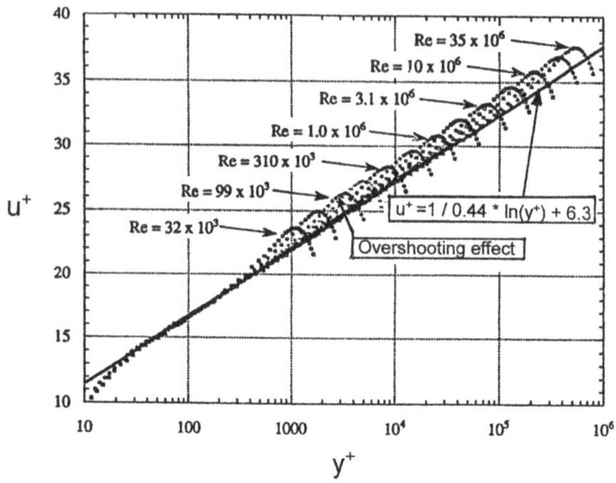


**Fig. 5.8** The super pipe data for different Reynolds numbers and the universal logarithmic law. The power laws follow with good accuracy the experimental curves (Reproduced with changes from Zagarola et al. 1996). For more information see the main text of this section

flows, using Lie group analysis, he demonstrated that any function is fully admissible.

Seven years later, George (2007) finds that the original idea of justifying a logarithmic law on the basis of a constant Reynolds stress layer argument is deficient. Furthermore, he finds that for a boundary layer along a flat plate, a power law is consistent and in good agreement with experimental data.

A slightly different finding was published ten years ago by Egolf (2009) (briefly in Sect. 9.5). By applying the new turbulence model DQTM, a deficit power law was discovered, viz.,

$$u^{+} = \frac{U}{u*}\left[1 - C\frac{1}{(y^{+})^{m}}\right]. \tag{164}$$

Such a law does not describe the average velocity profile directly, but instead is a description of the difference between the maximum free stream velocity and the averaged velocity profile. This law is in excellent agreement with the superpipe data set. It predicts, e.g., for an infinite Reynolds number flow the following exponent in Eq. (164) (Egolf 2009):

$$m(\mathrm{Re} = \infty) = \frac{\sqrt{17} - 3}{4} = 0.2808. \tag{165a, b}$$

Equation (165a) is a quadratic irrationality as the golden mean (Penrose 2007)! Furthermore, the highest Reynolds number experiment leads to

$$m\left(\mathrm{Re} = 3.5 \cdot 10^{7}\right) = 0.2808 \pm 0.0001. \tag{166}$$

The relative error between the experimental (see Eq. 166) and the theoretical result (see Eq. 165) is less than 0.1%. To the authors' best knowledge up-to-present no other closure showed such predictive accuracy.

In meteorology and especially in air contamination studies, to model the turbulent planetary boundary layer, turbulent diffusion is also described by local Boussinesq-type down-gradient transport. However, it is well known that scalar quantities are mixed by large eddies in the counter direction to the local gradient (see, e.g., Deardoff and Willis 1975; Zilitinkevich et al. 1999; Vilhena et al. 2008). Too simple descriptions only assume a net movement of a contaminant down the gradient of material concentration at a rate which is directly proportional to the magnitude of the gradient

$$\overline{w'c'} = -K_z \frac{\partial c}{\partial z}, \tag{167}$$

where $c$ is the mass concentration of an air pollutant, $z$ the vertical coordinate, $w'$ the vertical velocity fluctuation, and $K_z$ the vertical eddy diffusivity. Clouds may grow and larger eddies will come into play, which alter the diffusion process observed at a local dimension by nonlocal (far-distant) behavior. *Counter-gradient fluxes are a result of large boundary layers in opposition to local configurations and may comprise in successful descriptions of the turbulent flow phenomena, "ruled" by self-similarity and fractality, only dependence on the largest scales.* Despite the bold statement in Chap. 3, that higher order modeling is not necessary, even on intermediate and eventually also low levels, small-scale eddy transport must be complemented (or advantageously replaced) by nonlocal fluxes (see Buske et al. 2007). In a first attempt to solve this additional complexity, Ertel (1942) and Deardoff (1966, 1972) made a proposal of complementary nature

$$\overline{w'c'} = -K_z \left( \frac{\partial c}{\partial z} - \gamma \right), \tag{168}$$

where $\gamma$ denotes the counter-gradient term. Recently further progress was made with nonlocal models (see Chap. 6). Vilhena et al. (2008) wrote that the phenomenon of large-scale counter-gradient pollutant flux is proportional to the mean particle concentration gradient, an observation that is in agreement with the DQTM (see Chap. 7). However, such counter-gradient contributions also raise questions of thermodynamic stability and/or violation of the second law of thermodynamics (see, e.g., Hutter and Jöhnk 2004 and ref. therein).

In astrophysics, Canuto (1996) takes a one-eddy Mixing Length Theory (MLT) that underestimates and overestimates the convective flux in the high- and low-efficiency $k$-number spectral regime, respectively. He proposed a generalized model adopted to the Kolmogorov spectrum, which led to improved results.

It is now time to move from the old ideas concerning the persuasiveness using gradient laws and the application of strict linearization and localization to turbulence phenomena, which actually are inherently nonlinear and nonlocal. In this book, in a preliminary step, we have the intention to demonstrate the need of the new approach only in zero-equation turbulence modeling. Consequently, then, one may also ask the question whether these new ideas could be adjusted and implemented into numerical algorithms to solve more complex flow configurations, with higher complexity of the geometry (boundaries), the fluid properties, non-isothermal flows, or flows with constituents or even exhibiting additional chemical reactions, etc. Despite the bold statement in Chap. 3 that higher-order modeling is not necessary; indeed, for more accurate and successful developments these new ideas could also be taken to improve some standard turbulence models, such as $k$-$\varepsilon$, $k$-$l$, $k$-$\omega$, and *SST* (shear stress transport) models.

The second part of this review, therefore, radically opts out of linearization and "gradient"-type turbulence modeling. This proposed alternative is strongly supported by a basic example, which is strictly derived from Newton's second axiom. Such a procedure seems to be preferable, because the essential features become quite clear this way. Generalizations of following ideas to higher-order modeling would not be false, but probably superfluous.

## References

Afzal, N.: Power law and log law velocity profiles in turbulent boundary-layer flow: equivalent relations at large Reynolds numbers. Acta Mech. **151**, 195 (2001)

Arrowsmith, D.K., Place, C.M.: Dynamische Systeme. Academic Edition Spectrum, Heidelberg (1990). ISBN 3-86025-308-5 (in German)

Barenblatt, G.I.: Scaling laws for fully developed turbulent shear flows. Part 1. Basic hypotheses and analysis. J. Fluid Mech. **248**, 513 (1993a)

Barenblatt, G.I.: Intermediate asymptotics, scaling laws and renormalization group in continuum mechanics. Meccanica. **28**(3), 177 (1993b)

Barenblatt, G.I., Chorin, A.J., Prostokishin, V.M.: Self-similar intermediate structures in turbulent boundary layers at large Reynolds numbers. J. Fluid Mech. **410**, 263 (2000)

Batchelor, G.K.: Diffusion in a field of homogeneous turbulence. Aust. J. Sci. Res. A. **2**, 437 (1949)

Beer, F.P., Johnston Jr., E.R., DeWolf, J.T.: Mechanics of Materials. McGraw-Hill, New York (1981). ISBN 13-978-0073398235

Bernard, P.S., Handler, R.A.: Reynolds stress and the physics of turbulent momentum transport. J. Fluid Mech. **220**, 99 (1990)

Bodenschatz, E., Eckert, M.: "Prandtl and the Göttingen School", in Davidson et al. (see in this reference list and notice the remarks: *Bodenschatz and Eckert report, on the basis of correspondences of Prandtl with a number of involved scientists, that the 1/7th law and the struggle on it had been around between 1913 and 1921, say. In a letter of Prandtl to von Kármán in 1921 Prandtl stated that he already knew that the velocity distribution was proportional to* $y^{1/7}$*, where y is the distance from the wall. Prandtl stated in this letter that he had known this "already for a pretty long time, say since 1913". It appears (see Bodenschatz and Eckert, 2011) that Prandtl was too much overloaded in that period with other work, to formally publish his derivation before the 1920s. However, there is evidence that Prandtl's statement is correct (more details in Bodenschatz and Eckerts)* (2011))

Boussinesq, J.: Essai sur la théorie des eaux courants. Mémoires présentés par divers savants à l'Académie des Sciences. **23**(1), (1877). (in French)

Boussinesq, J.: Théorie de l'écoulement tourbillonnant et tumulteux des liquids. Gaulthier-Villars et fils, Paris (1897)

Bradshaw, P.: Turbulence: the chief outstanding difficulty of our subject. Exp. Fluids. **16**, 203 (1994)

Buske, D., Vilhena, M.T., Moreira, D.M., Tirabassi, T.: An analytical solution of the advection-diffusion equation considering non-local turbulence closure. Environ. Fluid Mech. **7**, 43 (2007)

Canuto, V.M.: Turbulent convection: old and new models. Astrophys. J. **467**, 385 (1996). https://doi.org/10.1086/177613

Chapman, S., Cowling, T.G.: The Mathematical Theory of Non-Uniform Gases: An Account of the Kinetic Theory of Viscosity, Thermal Conduction and Diffusion in Gases. Cambridge University Press, Cambridge, UK (1970). ISBN 10-0521075777

Cipra, B.: A new theory of turbulence causes a stir among experts. Science. **272**(5264), 951 (1996). https://doi.org/10.1126/science.272.5264.951

Corrsin, S.: Transport model limitations in turbulence. Adv. Geophys. **18A**, 25 (1974)

Courant, R., Hilbert, D.: Methods of Mathematical Physics, Differential Equations, vol. I. Wiley, New York (1989). ISBN 978-0-471-50439-9

Davidson, P.A., Kaneda, Y., Moffat, K., Sreenivasan, K.R.: A Voyage Through Turbulence. Cambridge University Press, Cambridge, UK (2011). ISBN 978-0-521-19868-4

Deardoff, J.W.: The counter-gradient heat flux in the lower atmosphere and in the laboratory. J. Atmos. Sci. **23**, 503 (1966)

Deardoff, J.W.: Numerical investigation of neutral and unstable planetary boundary layer. J. Atmos. Sci. **29**, 91 (1972)

Deardoff, J.W., Willis, G.E.: A parameterization of diffusion into the mixed layer. J. Appl. Meteorol. **14**, 1451 (1975)

Doering, C.R., Constantin, P.: Variational bounds on energy dissipation in incompressible flows: shear flow. Phys. Rev. E. **49**(5), 4087 (1994)

Egolf, P.W.: Difference-quotient turbulence model: a generalization of Prandtl's mixing-length theory. Phys. Rev. E. **49**(2), 1260 (1994)

Egolf, P.W.: Lévy statistics and beta model: a new solution of "wall" turbulence with a critical phenomenon. Int. J. Refrig. **32**, 1815 (2009)

Egolf, P.W., Weiss, D.A.: Difference-quotient turbulence model: analytical solutions for the core region of plane Poiseuille flow. Phys Rev E. **62**(1-A), 553 (2000)

Einstein, A.: Über die von der molekularkinetischen Theorie der Wärme geforderte Bewegung von in ruhenden Flüssigkeiten suspendierten Teilchen. Ann. Phys. **17**, 549 (1905). (in German)

Eisner, F.: Reibungswiderstand. Werft, Reederei, Hafen. **13**, 207 (1932). (in German)

Ertel, H.: Der vertikale Turbulenz-Wärmestrom in der Atmosphäre. Meteorol. Z. **59**, 250 (1942). (in German)

Frisch, U.: Turbulence – The Legacy of A.N. Kolmogorov, 1st edn. Cambridge University Press, Cambridge, UK (1995). ISBN 0-85-403-441-2

Gad-el-Hak, M.: Flow Control: Passive, Active and Research Flow Management. Cambridge University Press, Cambridge, UK (2000). ISBN 13-978-0521036719

George, W.K.: Is there a universal log law for turbulent wall-bounded flows? Phil. Trans. R. Soc. A. **365**, 789 (2007)

George, W.K.: Lectures in Turbulence for the 21st Century. www.turbulence-online.com (2013)

George, W.K., Castillo, L.: Zero-pressure-gradient turbulent boundary layer. Appl. Mech. Rev. **50** (11), 689 (1997)

Goldstein, S.: Modern Developments in Fluid Dynamics. Dover Press, New York (1965). ISBN 13-978-0486613574

Görtler, H.: Berechnung von Aufgaben der freien Turbulenz auf Grund eines neuen Näherungsansatzes. ZAMM. **22**, 241 (1942). (in German)

Hinze, J.O.: Turbulence, 2nd edn. McGraw-Hill, New York (1975). ISBN 0-07-029037-7

Hutter, K., Jöhnk, K.: Continuum Methods of Physical Modeling. Springer, Berlin (2004). ISBN 13-978-364-205-831-8

Hutter, K., Wang, Y.: Fluid and thermodynamics. In: Basic Fluid Mechanics, vol. 1. Springer, Berlin (2016a). ISBN 978-3-319-33632-9
Hutter, K., Wang, Y.: Fluid and thermodynamics. In: Advanced Fluid Mechanics and Thermodynamic Fundamentals, vol. 2. Springer, Berlin (2016b). ISBN 978-3-319-33635-0
Hutter, K., Wang, Y.: Fluid and thermodynamics. In: Structured and Multiphase Fluids, vol. 3. Springer, Berlin (2018). ISBN 978-3-319-77745-0
Jakob, M., Erk, S.: Der Druckabfall in glatten Rohren und die Durchflussziffer von Normaldüsen. In: Forschungsarbeiten auf dem Gebiet des Ingenieurwesens. Verein der Deutschen Ingenieure (V.D.I.), Berlin (1924). (in German)
Jongen, T., Gatski, T.B.: General explicit algebraic stress relations and best approximation for three-dimensional flows. Int. J. Eng. Sci. **36**, 739 (1998)
Kaneda, Y., Ishihara, T.: High-resolution direct numerical simulation of turbulence. J. Turbul. **7**, 20 (2006)
Kevlahan, N.K.R., Farge, M.: Vorticity filaments in two-dimensional turbulence: creation, stability and effect. J. Fluid Mech. **346**, 49 (1997)
Kevorkian, J., Cole, J.D.: Perturbation Methods in Applied Mathematics. Springer, New York (2010). ISBN 978-1-4419-2812-2
Kolmogorov, A.N.: The local structure of turbulence in incompressible viscous fluid with very large Reynolds numbers. Dokl. Akad. Nauk. SSSR Seria fizichka. **30**, 301 (1941a). (in Russian). English translation in Proc. R. Soc. Lond. A **434,** 9
Kolmogorov, A.N.: Dissipation of energy in the locally isotropic turbulence. Dokl. Akad. Nauk SSSR. **32**, 16 (1941b). (in Russian). English translation in Proc. R. Soc. Lond. A **434,** 15
Launder, B., Spalding, D.B.: The numerical computation of turbulent flows. Comput. Methods Appl. Mech. Eng. **3**, 269 (1974)
Lesieur, M.: Turbulence in Fluids: Stochastic and Numerical Modeling, 2nd edn. Kluwer Academic, Dordrecht (1990). ISBN 0-7923-0645-7-00
Libby, P.A.: Introduction to Turbulence Combustion, An International Series. Taylor & Francis, Abingdon, UK (1996). ISBN 1-56032-100-8
Millikan, C.B.: A critical discussion of turbulent flows in channels and circular tubes. In: Den Hartog, J.P., Peters, H. (eds.) Proceedings of the Fifth International Congress for Applied Mechanics, Massachusetts (1938)
Moore, J.G., Schorn, S.A., Moore, J.: Methods of classical mechanics applied to turbulence stresses in a tip leakage vortex. ASME J. Turbomach. **118**(4), 622 (1996)
Nicuradse, J.: Untersuchungen über die Geschwindigkeitsverteilung in turbulenten Strömungen, PhD thesis Göttingen, VDI Res. J. 281 (1926)
Oberlack, M.: Symmetries and scaling-laws in turbulence. ZAMM. **79**, 123 (1999)
Oberlack, M.: A unified approach for symmetries in plane parallel turbulent shear flows. J. Fluid Mech. **427**, 299 (2001)
Obermeier, F.: Letter of Frank Obermeier (Technical University, Bergakademie Freiberg, Germany) to Daniel Weiss with a copy to the first author on 22nd of January (1996)
Obermeier, F.: Prandtl's mixing length model – revisited, PAMM. Proc. Appl. Math. Mech. **6**, 577 (2006)
Penrose, R.: The Road to Reality. Vintage Books, New York (2007). ISBN 978-0-679-77631-4
Pope, S.B.: Turbulent Flows. Cambridge University Press, Cambridge, UK (2000). ISBN 978-0-521-59125-6
Prandtl, L.: Über Flüssigkeitsbewegung bei sehr kleiner Reibung. In: Proceeding of Third International Mathematical Congress, Heidelberg (1904). (in German), 484 (Engl. transl. in NACA Tech. Memo. 452)
Prandtl, L.: Approximative time of discovery of 1/7-power law of the wall by Prandtl (see also Davidson et al., (2011), p. 53) (1913)
Prandtl, L.: Letter of Prandtl to von Kármán on 16 February 1921, MPGA, Abt. III, rep. 61, no. 792 (see also in Davidson et al. (2011)), p. 53 (1921a)

Prandtl, L.: Ergebnisse der Aerodynamischen Versuchsanstalt zu Göttingen. Bericht, Oldenbourg (1921b). (in German)
Prandtl, L.: Bericht über Untersuchungen zur ausgebildeten Turbulenz. ZAMM. **5**(2), 136 (1925). (in German)
Prandtl, L.: Neuere Ergebnisse der Turbulenzforschung. Zeitschr. VDI. **77**, 105 (1933). (in German)
Prandtl, L.: Bemerkungen zur Theorie der freien Turbulenz. ZAMM. **22**(5), 241 (1942). (in German)
Prandtl, L.: Führer durch die Strömungslehre, 3rd edn. Friedr. Vieweg & Sohn, Braunschweig (1949). (in German)
Reichardt, H.: Über eine neue Theorie der freien Turbulenz. ZAMM. **21**, 257 (1941). (in German)
Reynolds, O.: On the dynamical theory of incompressible viscous fluids and the determination of the criterion. Phil. Trans. R. Soc. Lond. A. **186**, 123 (1894)
Reynolds, A.J.: Turbulent Flows in Engineering. Wiley, London (1974). ISBN 10-0471717827
Richardson, L.F.: Weather Prediction by Numerical Methods. Cambridge University Press, Cambridge, UK (1922)
Rodi, W.: In: Launder, B.E. (ed.) A Review of Experimental Data of Uniform Density Free Turbulent Boundary Layers, Studies in Convection, vol. 1. Academic, London (1976a)
Rodi, W.: A new algebraic relation for calculating the Reynolds stresses. ZAMM. **56**, T 219 (1976b)
Schlichting, H.: Boundary-Layer Theory. McGraw-Hill, New York (1979). ISBN 0-07-055-334-3
Schmitt, F.G.: Direct test of a nonlinear constitutive equation for simple turbulent shear flows using DNS data. Commun. Nonlinear Sci. Numer. Simul. **12**, 1251 (2007a)
Schmitt, F.G.: About Boussinesq's turbulent viscosity hypothesis: historical remarks and a direct evaluation of its validity. C.R. Mécanique. **335**, 617 (2007b)
Schmitt, F.G., Hirsch, C.: Experimental study of the constitutive equation for an axisymmetric complex turbulent flow. ZAMM. **80**, 815 (2000)
Schmitt, F.G., Merci, B., Dick, E., Hirsch, C.: Direct investigation of the K-transport equation for a complex turbulent flow. J. Turbul. **4**, 21 (2003)
Smagorinski, J.: General circulation model of the atmosphere. Mon. Weather Rev. **91**, 99 (1963)
Smith, A.M.O., Cebeci, T.: Numerical Solution of the Turbulent Boundary Layer Equations. Douglas Aircraft Division Report DAC 33735, USA (1967)
Spalart, P.R.: Philosophies and fallacies in turbulence modeling. Prog. Aerosp. Sci. **74**, 1 (2015)
Spalart, P.R., Allmaras, S.R.: A one-equation turbulence model for aerodynamic flows. Recherche Aero-spatiale. **1**, 5–21 (1994)
Speziale, C.G.: A consistency condition for non-linear algebraic Reynolds stress models in turbulence. Int. J. Non-Linear Mech. **33**, 579 (1998)
Stanišić, M.M.: The Mathematical Theory of Turbulence, 2nd edn. Springer, Berlin (1988). ISBN 0-387-96685-4
Stull, R.B.: An Introduction to Boundary Layer Meteorology, Atmospheric Sciences Library. Kluwer Academic, Dordrecht (1988). ISBN 90-277-2769-4
Taylor, G.I.: The transport of vorticity and heat through fluids in turbulent motion. Proc. R. Soc. Lond. **135A**, 685 (1932)
Tennekes, H., Lumley, J.L.: A First Course in Turbulence. MIT Press, Cambridge, MA (1972). ISBN 0-262-200-19-8
Truesdell, C.A., Muncaster, R.G.: Fundamentals of Maxwell's Kinetic Theory of a Simple Monoatomic Gas: Treated as a Branch of Rational Mechanics. Academic, Cambridge, MA (1980). ISBN 0-12-701350-4
Tsinober, A.: An Informal Conceptual Introduction to Turbulence. Springer, Cham (2009). ISBN 978-90-481-3174-7
Tsinober, A.: The Essence of Turbulence as a Physical Phenomenon with Emphasis on Issues of Paradigmatic Nature. Springer, Dordrecht (2014). ISBN 978-94-007-7179-6
van Dyke, M.: Higher approximations in boundary-layer theory part 3. Parabola in uniform stream. J. Fluid Mech. **19**, 145 (1964)

Vilhena, M.T., Cost, C.P., Moreira, D.M., Tirabassi, T.: A semi-analytical solution for the three-dimensional advection–diffusion equation considering non-local turbulence closure. Atmos. Res. **90**, 63 (2008)

von Kármán, T.: Über laminare und turbulente Reibung. ZAMM. **1**(4), 233 (1921). (in German)

von Kármán, T.: Mechanische Ähnlichkeit und Turbulenz. Nachr. Akad. Wiss. Göttingen, Math-Phys. Kl., 58 (1930). (in German). Proc. 3. Intern. Congr. Appl. Mech. Stockholm 1930. Also in: Collected Works, vol. 2, pp. 322–346. London, Butterworth (1956)

von Kármán, T.: Mechanical similitude and turbulence. In: Proceedings of the Third International Congress of Applied Mechanics, vol. 1, p. 85, Stockholm (1931)

Wallace, J.M.: Twenty years of experimental and direct numerical simulation access to the velocity gradient tensor: What have we learned about turbulence? Phys. Fluids. **21**(2), 021301 (2009)

Wieghardt, K.: Turbulente Grenzschichten. Göttinger Monographie, Part B, 5 Germany (in German) (1946)

Wilcox, D.C.: Turbulent Modeling for CFD, 2nd edn. DCW Industries, La Canada (1998)

Zagarola, M.V., Smits, A.J., Orszag, S.A., Yakot, V. : Experiments in high Reynolds number turbulent pipe flow. AIAA paper No. 96-0654, Reno Nevada (1996)

Zagarola, M.V., Perry, A.E., Smits, A.J.: Log laws or power laws: the scaling in the overlap region. Phys. Fluids. **9**(7), 2094 (1997)

Zagarola, M.V., Perry, A.E., Smits, A.J.: http://www.princeton.edu/~gasdyn/#superpipe_data (2015)

Zilitinkevich, S., Gryanik, V.M., Lykossov, V.N., Mironov, D.V.: Third-order transport and nonlocal turbulence closure for convective boundary layers. J. Atmos. Sci. **56**, 3463 (1999)

# Chapter 6
# Review of Nonlinear and Nonlocal Models

## 6.1 Nonlocality in Phase Space

With the development of nonlinear dynamics in the last 50 years (see, e.g., Guckenheimer and Holmes 1983; Schuster 1988; Arrowsmith and Place 1990; Strogatz 1994) much has been learned about nonlinear systems. It became clear that instabilities and bifurcations (see, e.g., Mickens 1981; Guckenheimer and Holmes 1983; Collet and Eckmann 1990) are important mechanisms for the creation of complex behavior as it is observed in many low-, medium-, and high-dimensional chaotic and turbulent physical, biological, chemical, medical, economic, etc. systems. Only nonlinear systems can reveal self-organization (see, e.g., Haken 1983, 1987; Takayama 1989; Schroeder 1991) in time and space, which involve impressive pattern creations (see, e.g., Ruelle 1992; Broer et al. 1991). Fascination evokes by experiencing that a nonlinear system with a constant energy influx, by a cooperative phenomenon occurring in its microstructure, can create overall periodic, quasi-periodic, chaotic, and, thus, pulsating and turbulent behaviors.

Because the convection term of the Navier–Stokes equations (NSE) shows a quadratic dependence on the velocity field, it is quite natural to equally suppose closure relations for the Reynolds stress tensor as nonlinear functionals of the independent constitutive variables. The DQTM (see Chap. 7) adds a linear and also a second quadratic term of the velocity to the NSEs (see Chap. 9).

As theoretical tools to study nonlinear system concepts, like phase spaces, Poincaré maps, Lyapunov functions, stability analyses, blowing-up techniques, etc. have been developed and applied. Ott et al. (1990) introduced the Ott-Grebogi-Yorke (OGY) method to control chaos. Today, all these methods are fairly well understood by mathematicians, natural scientists, and engineers. Because of this, no further space will be reserved for a deeper discussion of nonlinearity.

It should also be emphasized that the forward coupling of the statistical moments of the NSEs of all orders up to infinity is a direct consequence of the quadratic nonlinearity of the convective acceleration terms of these equations. The hierarchy

The original version of this chapter was revised. A correction to this chapter can be found at https://doi.org/10.1007/978-3-030-26033-0_14



P. W. Egolf, K. Hutter, *Nonlinear, Nonlocal and Fractional Turbulence*,
https://doi.org/10.1007/978-3-030-26033-0_6

of these equations yields the stress closure iteration, which is equivalent to the joint probability velocity distribution and, thus, directly points at the ultimate source of the complex behavior of the turbulent processes (see also Kraichnan 1961).

On the other hand, as already discussed in Sect. 5.4, nonlocalities are rather poorly known and understood. For example, subject indices of standard books on fluid dynamics and turbulence phenomena practically never contain the keyword "nonlocal." This is even more astonishing as there is a wide consensus that in turbulence "nonlinearity," as well as "nonlocality" plays a crucial role. Therefore, this concept will be more thoroughly introduced and discussed in the next few paragraphs.

A general review of nonlocality brings us also to nonlocality in quantum mechanics, one of the most intriguing topics of modern physics (see, e.g., Zeilinger 2010). However, classical nonlocality is easier to be understood; in the next few sections it will be discussed on a very conceptual basis.

In Sect. 5.4, we saw that classical nonlocality in time and space is also not so easy to be grasped and explained. A definition by Eringen (2002) is, however, very clear: "*A material point of the body is considered to be attracted* [influenced] *by all points of the body at all past times*" (Principle of causality).

As many aspects in physics, especially of nonlinear dynamics, can be better understood in a phase space representation, we explain "nonlocality" also in this way. To arrive at this goal, we give a two-stage explanation (see Fig. 6.1a–d).

Assume that a particle moving through three-dimensional space is described by three space coordinates as functions of time: $x_1(t)$, $x_2(t)$, $x_3(t)$. If two particles are present, in real space their motion can be transformed to a virtual space with six coordinates: $x_1(t), \ldots, x_6(t)$. The movement of a fluid system, with a very high number of particles $N$, in the same spirit, is then described by a $3N$-dimensional Euclidean space with the real-space coordinates $x_1(t), \ldots, x_{3N}(t)$. Notice that the movement of all these numerous particles in this high-dimensional space leads to only a single trajectory. By taking the directional derivative of the trajectory at a defined time, a $3N$-dimensional velocity vector is obtained, a single element in a $3N$-dimensional velocity space.

Now let us discuss the simplest case of a one-dimensional movement of a single particle (or fluid lump) forth and back on the $x_1 = x$ axis as shown in Fig. 6.1a). This particle, with trajectory $x_A(t^-)$, arrives at $t = t_0 = 0$ at the location $x_0$. The trajectory $x_A(t^-)$ describes the movement that the particle performed in the past, and $x_A(t^+)$ that predicted by a differential equation, e.g., a first-order evolution equation (the NSE for a fluid element) into the future. In a fluid this path can, on the one hand, be the trajectory of the real particle, contributing to fluctuations and intermittency, or, on the other hand, it could also just represent the ensemble-averaged movement of the fluid lump "*A*." Now, let us assume that another particle "*B*," at another time $t_1$, arrives at the same location $x_0$. In a one-dimensional situation, this occurrence shows a high probability. Without any consequences for our example, we normalize the arrival of the second particle "*B*" at location $x_0$ by a time shift $t \to t - t_1$ and also set it equal to $t_0 = 0$. However, in the mathematical methodology of physics, it is more customary to look at the movements of these two particles as two different

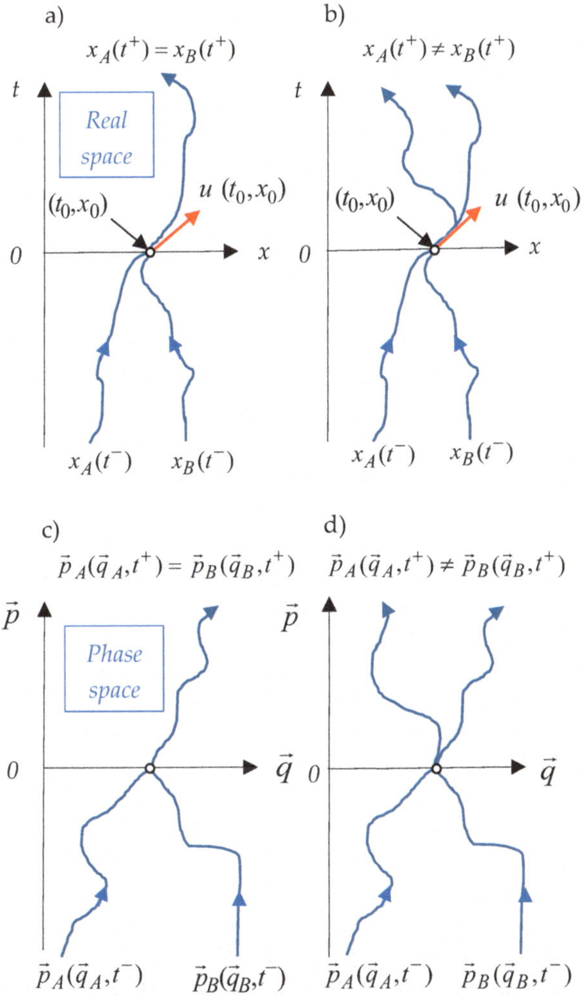


**Fig. 6.1** A new manner to explain local and nonlocal system's behavior with the help of phase space considerations. Real spaces of single particle movements are shown in panels (**a**) and (**b**). The trajectories of a local particle flight (**a**) and of a nonlocal particle flight (**b**) cross the same time-space point $(t_0, x_0) = (0, x_0)$ with exactly the same velocity (tangent) $u_0 = u(t_0, x_0)$. The result in the local system is always an identical continuation to positive times $t^+$, whereas this is usually not the case in a nonlocal system. Panels (**c**) and (**d**) present a movement of a multitude of fluid particles in a phase space representation. However, the statement about locality and nonlocality in (**a**,**b**) remains the same in the lower panels (**c**,**d**)

realizations of two succeeding movements of a single particle, e.g., "$A$" (in the context of ensemble consideration and ergodic hypothesis). We further demand that the derivatives (velocities) of the two realizations at $x_0$ are additionally identical: $u(t_0, x_0) = u_A(t_0, x_0) = u_B(t_0, x_0)$ (see Fig. 6.1a). From a classical deterministic mechanics point-of-view, it is known that the basic system of differential equations, together with the initial conditions (location and velocity), determine the entire future movement. Because both are identical for the two particles (or realizations), in the local systems, the future trajectories must also be identical: $x_A(t^+) = x_B(t^+)$ for any $t^+ > 0$ (this can be seen in Fig. 6.1a).

On the other hand, in a nonlocal system the future trajectory of particle "$A$" is determined not only by the condition (location and velocity) at time $t_0 = 0$, but it depends on the (entire) past trajectory $x_A(t^-)$. The same is the case for particle "$B$"

with a different history $x_B(t^-)$. Therefore, it is a very seldom occurrence that their trajectories in the future coincide, and usually one has $x_A(t^+) \neq x_B(t^+)$ (see Fig. 6.1b). In real systems, the influence on distance in space and backward time fades. In physics and engineering in linear and local models, there is a high performance of exponential decays. However, we will see in Sect. 7.3 that fading in turbulent flows is not as strong as assumed in the mathematical, physical, and engineering literature.

In a high-dimensional system, the variables $x$ and $u$ must be replaced by $\vec{x} = (x_1, x_2 \ldots, x_{3N})$ and $\vec{u} = (u_1, u_2 \ldots, u_{3N})$. In the terminology of mathematical physics the variables are renamed; for the space location one writes the 3$N$-dimensional vector as $\vec{q} = (q_1, q_2 \ldots, q_{3N})$, and instead of the velocity the momentum vector $\vec{p} = (p_1, p_2 \ldots, p_{3N})$ is introduced, in which the single momentum components are: $p_j = \rho_j u_j$. The pair $\left(\vec{p}, \vec{q}\right)$ then represents a point in phase space of dimension 6$N$. In this complete phase-space description, the information of the velocity components is contained in the additional 3$N$ momentum coordinates. Therefore, in this more general representation, velocity vectors at $t = t_0 = 0$, with identical tangents, are no longer required. As long as the fluid dynamic system is local (Fig. 6.1c), crossing the same point in phase space is sufficient to guarantee an identical future development of two different realizations with different trajectories in the past. Be aware, however, in such high-dimensional systems the crossing of the same point is extremely improbable and ought to be regarded as a "Gedanken-Experiment." On the other hand, in a nonlocal system in analogy to the situation in Fig. 6.1b here usually also differing developments to the future are highly probable (see Fig. 6.1d).

As we have taken a microscopic particle picture to explain nonlocality in detail, these ideas must now be transformed into a continuous field notion, where, instead of particles, material points of a body are considered, which are continuously related. To these material points independent objects, called variables, are assigned, e.g., mass, charge, electric field, and magnetic field. After Eringen (2002), the state of a body (in our case the fluid) at a material point is described by relations of the response objects that constitute a second class (e.g., stress, internal energy, and heat) as functions or functionals of independent objects. These relations are called constitutive equations. Now, it becomes clear that a closure functional, describing the Reynolds shear stress rather than the viscous stress, exhibits features analogous to a constitutive equation (see Chap. 4). It also becomes clear now that gradient theories are theories of lacking nonlocality and that with the knowledge of further higher-order derivatives this limitation decreases and becomes unfounded once all the higher-order derivatives up to infinity are accounted for, provided, of course, these objects represent a function or functional.

## 6.2 Atomic and Continuum Theories

Continuum theories are based on the assumption that the space is continuously filled with matter and that an overall characteristic length $L$ of the system, called also a global length, is much larger than an internal length $\lambda$. Correspondingly, overall characteristic times $T$ of time-dependent processes are large as compared to typical internal times $\tau$. Therefore, a classical field theory is valid when (see Eringen 2002)

$$\frac{\lambda}{L} \ll 1, \frac{\tau}{T} \ll 1 : \begin{cases} \text{Continuum theory,} \\ \text{Local model,} \end{cases} \qquad (169\text{a, b})$$

where it is assumed that length and time scales are related by a transmission velocity, which transfers information from one atom (molecule) to the next. When internal and external length and time scales are approximately equal (see Eringen 2002),

$$\frac{\lambda}{L} \approx 1, \frac{\tau}{T} \approx 1 : \quad \begin{cases} \text{Atomic theory,} \\ \text{Nonlocal model,} \end{cases} \qquad (170\text{a, b})$$

a continuum theory fails and one is forced to apply a microscopic or atomic (molecular) model. However, there is an alternative possibility how the validity of a continuum theory may be (slightly) extended. The recipe is to take long-range interatomic forces into consideration and to macroscopically model them. This leads to a nonlocal generalization of the constitutive equations which then, just as its local predecessor, also is combined with the basic continuum field theoretical equations.

It was Prandtl who first applied ideas describing momentum transport of atoms (molecules) in the kinetic theory of gases to turning eddies in turbulent flow fields (see Sects. 7.2.1 and 7.2.2). For this purpose, he replaced the mean free path length between atoms (molecules), corresponding to the interatomic distance in solids, $\lambda$, by an eddy diameter $l$ and the characteristic time $\tau$ by the eddy turnover time $t$. He stayed within the linear and local limit, which means that he, in his analogy, realized inequalities (169a,b) as

$$\frac{l}{L} \ll 1, \frac{t}{T} \ll 1 : \begin{cases} \text{Continuum theory,} \\ \text{Local model.} \end{cases} \qquad (171\text{a, b})$$

Yet, isn't it evident that the largest eddy size $l = l_0$ is dictated by the characteristic length of the overall flow field $L$ and its related turnover time $T$? Therefore, one has also here a transition from local to nonlocal properties as in the atomic theory from (169a,b) to (170a,b), but now for the highly nonlocal fluid dynamic system going from inequalities (171a,b) to the following relations:

$$\frac{l}{L} \approx 1, \quad \frac{t}{T} \approx 1 : \quad \begin{cases} \text{Eddy theory,} \\ \text{Nonlocal model.} \end{cases} \qquad (172\text{a, b})$$

Let us in analogy now change the sentence in Eringen's book: "*Here the material is considered to consist of discrete atoms connected by distant forces* [also] *from other* [than] *neighbouring atoms*" to: "*Here the fluid is considered to consist of discrete eddies related by distant momentum exchanges* [also] *from other* [than] *neighbouring eddies.*" We have made the following evident substitutions listed in Table 6.1:

**Table 6.1** Analogous terms used in molecular and fluid dynamics

| | | |
|---|---|---|
| 1. | Material | Fluid |
| 2. | Atom | Eddy |
| 3. | Connected | Related |
| 4. | Force | Momentum exchange |

Furthermore, we added two words in brackets, [...], to both sentences, which guarantee consistency with Non-Markovian processes. The substitution of "forces" by "momentum exchanges" is in agreement with Newton's second law.

Relations (172a,b) tell us that for the development of a "constitutive equation" or "closure scheme" for turbulence, there are two promising possibilities, namely, first, to create an eddy model and, second, to establish a nonlocal constitutive equation. It is quite evident that the first can also be the basis to develop the second type of physical model (see Sect. 7.2.2).

## 6.3 Stress as an Objective Polynomial Function of the Mean Rate of Strain Tensor

A first class of nonlocal models is based on polynomial series of the mean rate of strain tensor, a development starting with work by Spencer and Rivlin (1959, 1960) and continued by Lumley (1970) and Pope (1975). This idea was supported by an observation by Bradshaw (1973) that the Boussinesq hypothesis fails over curved surfaces; he inferred that this failure is due to the form of the stress–strain rate relation.

Based on such considerations the representation,

$$\frac{1}{\rho}\overline{\mathbf{R}} = -\overline{\mathbf{T}}_R = - < \vec{u}' \otimes \vec{u}' > = a\mathbf{I} + b\overline{\mathbf{D}} + c\overline{\mathbf{D}}^2, \qquad (173a-c)$$

was proposed; it fulfills turbulence objectivity (see, e.g., Eringen 2002; Hutter and Jöhnk 2004) and is of Reiner–Riwlin structure. Any nonlinear isotropic expression $\overline{\mathbf{T}}_R = \overline{\mathbf{T}}_R(\overline{\mathbf{D}})$ is of the form (173c). Neglecting the quadratic dependence ($c = 0$) and requesting that $\mathrm{tr}(\overline{\mathbf{R}}/\rho) = -2k$ leads to

$$\frac{1}{\rho}\overline{\mathbf{R}} = -\frac{2}{3}k\mathbf{I} + 2\nu_{\mathrm{T}}\overline{\mathbf{D}}, \qquad \nu_{\mathrm{T}} = \nu_{\mathrm{T}}(k, \varepsilon), \qquad (174a, b)$$

where the explicit form of $\nu_{\mathrm{T}}$ is obtained by arguments of dimensional analysis. This leads to Eq. (90): $\nu_{\mathrm{T}} = C_\mu k^2/\varepsilon$. The constant coefficient $C_\mu$ in this equation is regarded as universal. Real case applications show differently that the constant has to be adjusted in a wide range to different flow cases under consideration. In this lowest-order approximation, we recognize relations of the $k$-$\varepsilon$ model with the underlying Boussinesq hypothesis.

The realization that the Euclidian invariance is a necessity for turbulence modeling goes back to Pope (1975). Again, a tensor series development is considered;

however, now only for the deviatoric or traceless part $\overline{\mathbf{R}}_{\mathrm{D}}$ of the Reynolds stress tensor $\overline{\mathbf{R}}$,

$$\frac{1}{\rho}\overline{\mathbf{R}}_{\mathrm{D}} = \sum_{k=1}^{N} a_k \overline{\mathbf{T}}_k. \tag{175}$$

Here, the $\overline{\mathbf{T}}_k$'s are taken from a tensorial basis of the tensor space of traceless tensors $\overline{\mathbf{T}}_k$. For $N = 2$, e.g., the Reiner–Riwlin model is obtained, see Sect. 5.4. According to Pope (2000), this number of basic tensors is sufficient for quasi-two-dimensional turbulent flows. Schmitt (2007) introduces a development up to order $N = 3$ (see also Mompean et al. 1998, 2011; Qiu et al. 2011) and presents the symmetric traceless tensor basis; the coefficients $a_1 = \nu_{\mathrm{T}}$, $a_2$ and $a_3$ correspond to invariants of the flow. Non-Newtonian material functions were introduced to derive the three parameters. Therefore, in this model no evolution equations for the kinetic energy $k$ and the turbulent dissipation rate $\varepsilon$ are required, e.g., as in the classical $k$-$\varepsilon$ model approach. Citing Pope (1975), Schmitt remarks that to model three-dimensional turbulence actually ten such basic tensors with five invariants are necessary to obtain a complete representation. Because this method is not advised as best route to reach a satisfying final description of turbulent closure, only a low-order development is further discussed here. The deviatoric Reynolds stress tensor is written by a (linear) series up to three independent traceless tensors formed with $\overline{\mathbf{D}}$ and $\overline{\mathbf{S}}$ (see Eqs. (34) and (94)), which are the stretching or strain rate and the vorticity tensors,

$$\frac{1}{\rho}\overline{\mathbf{R}}_{\mathrm{D}} = 2\nu_{\mathrm{T}}\overline{\mathbf{T}}_1 - \beta\overline{\mathbf{T}}_2 - \gamma\overline{\mathbf{T}}_3. \tag{176}$$

Such tensors are

$$\overline{\mathbf{T}}_1 = \overline{\mathbf{D}}, \overline{\mathbf{T}}_2 = \overline{\mathbf{D}}\,\overline{\mathbf{S}} - \overline{\mathbf{S}}\,\overline{\mathbf{D}}, \quad \overline{\mathbf{T}}_3 = \overline{\mathbf{D}}^2 - \frac{1}{3}\mathrm{tr}\left(\overline{\mathbf{D}}^2\right)\mathbf{I}, \tag{177a – c}$$

which are symmetric and Euklidian objective. Moreover, the scalar coefficients $\nu_{\mathrm{T}}$, $\beta$, $\gamma$, are functions of the five (not six!) invariants of $\overline{\mathbf{T}}_k$, $k = 1,2,3$, and can be expressed as follows (see also Schmitt 2007; Mompean et al. 2011):

$$\begin{aligned}
\nu_{\mathrm{T}} &= -\frac{\overline{u'_2 u'_1}}{a},\\
\beta &= \frac{\overline{u'^2_1} - \overline{u'^2_2}}{a^2},\\
\gamma &= \frac{6}{a^2}\left(\frac{2}{3}k - \overline{u'^2_3}\right),
\end{aligned} \tag{178a – c}$$

where $\overline{u_2'u_1'}$ is the shear stress component, $\overline{u_k'^2}$, $k = 1{,}2{,}3$ are the three normal stress components and $k = 1/2\left(\overline{u_1'^2} + \overline{u_2'^2} + \overline{u_3'^2}\right)$ is the turbulent kinetic energy. Moreover, $a = \partial \overline{u}_1 / \partial x_2$ characterizes simple shearing (Mompean et al. 2011; Qiu et al. 2011).

Also, nonlinear $k$-$\varepsilon$ and $k$-$\omega$ models are based on these ideas, but are not further presented (see also the remarks in the next section). For example, Speziale (1987) extended the Boussinesq hypothesis by adding in the summation (175) the Euklidian objective second Rivlin-Ericksen tensor (see Hutter and Jöhnk 2004)

$$\frac{1}{2}\overline{\mathbf{A}}_2 = \dot{\overline{\mathbf{D}}} + \left(\overline{\mathbf{L}}^{\mathrm{T}}\overline{\mathbf{D}} + \overline{\mathbf{D}}\,\overline{\mathbf{L}}\right) \tag{179}$$

and reduced it to its deviator

$$\frac{1}{2}\overline{\mathbf{A}}_{2,\mathrm{D}} = \frac{\overline{\mathbf{A}}_2}{2} - \frac{1}{6}\mathrm{tr}\left(\overline{\mathbf{A}}_2\right)\mathbf{I}. \tag{180}$$

So, he introduced a history dependence and thereby brought in convective nonlinearities by involving $\overline{\mathbf{L}}^{\mathrm{T}}\overline{\mathbf{D}} + \overline{\mathbf{D}}\,\overline{\mathbf{L}}$ (see also Nieckele et al. (2016), researchers who themselves proposed six extended models with tensorial decompositions and orthogonal projections to relate anisotropy effects to the Reynold stress tensor). Summarizing, all these proposals guarantee improvement compared to Boussinesq-type modeling, but as tensor series developments of order $n$ ($n \geq 2$) they lack the elegance and precision that an adequate and complete physical theory would offer.

## 6.4 Modified Diffusivity Models

A second class of models generalizes turbulence modeling by modifying the eddy diffusivity.

Following work by Corrsin (1974) and Monin and Yaglom (1965, 1971) and Egolf and Hutter (2016), the mean flux $\overline{F}(\vec{x},t)$ of a scalar conservative and transported quantity $\overline{\Gamma}(\vec{x},t)$ is often called mean "concentration." Momentum transport is given by $\overline{\Gamma} = \overline{u}_i$; if it is sensible heat, the transport quantity is the mean absolute temperature $\overline{\Gamma} = \overline{T}$, and for particles or seeds it is their mean specific concentration $\overline{\Gamma} = \overline{c}$. In differential form, the basic conservation law for this situation is

$$\overline{\Gamma}_t = -\overline{F}_x, \tag{181}$$

in which a production term has been ignored.

If the transporting mechanism is one-dimensional and fairly local in space and time, the mean flux $\overline{F}$ can be described as a function of the mean conserved quantity $\overline{\Gamma}$ and some of its derivatives

$$\overline{F} = f\left(\overline{\Gamma}, \overline{\Gamma}_x, \overline{\Gamma}_t, \overline{\Gamma}_{xx}, \overline{\Gamma}_{tt}, \overline{\Gamma}_{xt}, \ldots, x, t\right), \tag{182}$$

where the indices denote partial derivatives, and where the spatial dependence is written only for the one-dimensional case; the three-dimensional extension is evident. The addition of the indices $x$ and $t$ in Eq. (182) reflect a possible dependence of the mean flux on the mean "concentration," $\overline{\Gamma}$, of the fluctuation field and space and time derivatives of it. Usually, $f$ does not depend on $\overline{\Gamma}$ itself. If it is assumed that the diffusion mechanism is homogenous and linear, the following linear additive development is suggested:

$$\overline{F} = K_1\overline{\Gamma}_x + K_2\overline{\Gamma}_t + K_3\overline{\Gamma}_{xx} + K_4\overline{\Gamma}_{xt} + K_5\overline{\Gamma}_{tt} + \ldots \tag{183}$$

Furthermore, it can be assumed that the process is quasi-stationary (i.e., the random transporting process has a sufficiently high transporting speed compared to the overall mean field variations) and is symmetric in space.[1] Then, Eq. (183) simplifies to

$$\overline{F} = K_1\overline{\Gamma}_x + K_3\overline{\Gamma}_{xx} + \ldots \tag{184}$$

If higher order terms are neglected, the transport is of Fourier/Fick type and for turbulence modelling of Boussinesq type,

$$\overline{F} = -K\overline{\Gamma}_x, \qquad K > 0, \tag{185a, b}$$

where for the constant of diffusivity $K = -K_1$ was introduced to obtain a formula of gradient transport that is stable in forward time processes. Because Eq. (185a) is a result of random walk (Brownian motion) and diffusion theory (see Einstein 1905, 1926), and since we know that turbulence is more accurately described by Lévy flights and related anomalous diffusion than Fick-type diffusion, it is clear that the simplifications for an adequate description of turbulence has gone too far (compare also with information in Table 7.1). If Eq. (185a) is inserted into Eq. (181), it follows that

$$\overline{\Gamma}_t = K\overline{F}_{xx}, \qquad K = \text{const} > 0. \tag{186a, b}$$

In Chap. 5, it was explained that the turbulent diffusivity,

[1]The mean flow does not need to be symmetric, but parity invariance (invariance to spatial reflections) is required.

$$K = \varepsilon_{\mathrm{M}} = \nu_{\mathrm{T}}, \tag{187a, b}$$

was introduced as a function of space and in some cases also of velocity. Therefore, early turbulence modellers, e.g., made the generalization

$$\overline{F} = -K(x,t)\,\overline{\Gamma}_x. \tag{188}$$

Corrsin, in stepping back in the context of development of such models, raises a very important point, namely that Eq. (186a) could be very problematic. To demonstrate this, we insert Eq. (188) into Eq. (181)[2] and obtain

$$\overline{\Gamma}_t = \left[K(x,t)\overline{\Gamma}_x\right]_x = K\,\overline{\Gamma}_{xx} + K_x\overline{\Gamma}_x. \tag{189a, b}$$

The condition that the assumption made in early zero-equation turbulence models is valid, is rooted in Eqs. (188) and (189b). For Eqs. (186a) and (189b) to be nearly equivalent the second term on the right-hand side of Eq. (189b) must be small in comparison with the first term. So, early zero-equation turbulence models are only valid, if

$$\left|\frac{K_x}{K}\right| \ll \left|\frac{(\overline{\Gamma}_x)_x}{\overline{\Gamma}_x}\right|. \tag{190}$$

This is tantamount to a $K$ that is nearly constant.

In turbulence modelling, for a turbulent shear flow with the mean downstream velocity $\overline{u}_1$ and shear in the transverse $x_2$-direction, the above result and notation (187a) correspond to

$$\left|\frac{\dfrac{\partial \varepsilon_{\mathrm{m}}}{\partial x_2}}{\varepsilon_{\mathrm{m}}}\right| \ll \left|\frac{\dfrac{\partial}{\partial x_2}\left(\dfrac{\partial \overline{u}_1}{\partial x_2}\right)}{\dfrac{\partial \overline{u}_1}{\partial x_2}}\right|. \tag{191}$$

After multiplication of both sides of this equation with $dx_2$ and introducing the differences $\Delta(\ldots)$, one obtains a condition of local gradient turbulence modeling:

$$\left|\frac{\Delta \varepsilon_{\mathrm{m}}}{\varepsilon_{\mathrm{m}}}\right| \ll \left|\frac{\Delta\left(\dfrac{\partial \overline{u}_1}{\partial x_2}\right)}{\dfrac{\partial \overline{u}_1}{\partial x_2}}\right|, \tag{192}$$

which may be formulated and highlighted as

---

[2]Note that the second spatial derivative of $\overline{F}$ occurs by inserting a gradient turbulence model into the NSEs.

**Theorem 1**

*Eddy diffusivity turbulence models are physically only justified if the relative variation of the turbulent diffusivity in space is at least one order smaller than the relative variation of the mean velocity "gradient."*

Based on criterion (191), the different eddy diffusivity models, which contain a "gradient" may be thoroughly tested. It is our next task to do this for the zero-equation turbulence models of Sect. 5.2.

1. *Prandtl's mixing-length model*

By inserting the eddy diffusivity of Prandtl's mixing length model, given by Eq. (54), into Eq. (191), the condition

$$\left|\frac{\frac{\partial^2 \overline{u}_1}{\partial x_2^2}}{\frac{\partial \overline{u}_1}{\partial x_2}}\right| \ll \left|\frac{\frac{\partial^2 \overline{u}_1}{\partial x_2^2}}{\frac{\partial \overline{u}_1}{\partial x_2}}\right| \quad \Rightarrow \quad 1 \ll 1 \tag{193a, b}$$

is obtained. The conclusion is that Prandtl's mixing length model is unphysical, because Eq. (193a,b) cannot be fulfilled. Furthermore, this condition shows no spatial variation of a quality measure of the model as it was presented in Sect. 5.4 by Schmitt et al. for the Boussinesq law, by applying LES and DNS calculations. Equation (193b) states that this model fails, and that it fails throughout the entire turbulent domain. In Table 5.1, it may be found that this turbulence model is especially inaccurate at points of vanishing gradient of the mean velocity. However, it is concluded that it is not condition (193a,b) and the physical ideas that led to this conclusion, which explicitly explain the breakdown at the critical points of the averaged velocity profile.

2. *Von Kármán's model*

By inserting the eddy diffusivity (Eq. 55b), into Eq. (191), a sufficient, but not a necessary condition,

$$\left|\frac{\frac{\partial \overline{u}_1}{\partial x_2}\frac{\partial^3 \overline{u}_1}{\partial x_2^3}}{\left(\frac{\partial^2 \overline{u}_1}{\partial x_2^2}\right)^2}\right| \ll -1, \tag{194}$$

is obtained. Note that this condition has no solution. This is again a consequence of the considerable deficiency of the von Kármán model, which is based on the problematic Bossinesq approximation.

3. *Reichardt's inductive model*

This model is not a generalized diffusivity model and, therefore, cannot be analyzed by the evaluation method of this section.

4. *Prandtl's mean-gradient model*

If the eddy diffusivity of Prandtl's mean-gradient model (Eq. 68) is inserted into Eq. (191), condition

$$\left| \frac{\left(\frac{\partial \overline{u}_1}{\partial x_2}\right)^2 + l'^2_{\mathrm{m}} \frac{\partial \overline{u}_1}{\partial x_2} \frac{\partial^3 \overline{u}_1}{\partial x_2^3}}{\left(\frac{\partial \overline{u}_1}{\partial x_2}\right)^2 + l'^2_{\mathrm{m}} \left(\frac{\partial^2 \overline{u}_1}{\partial x_2^2}\right)^2} \right| \ll 1 \tag{195}$$

is obtained. This condition is difficult to interpret, but it could be used in a special application to calculate it analytically or numerically to see whether it is fulfilled. It is noted that, if the second mixing length $l'_m$ is zero, this condition corresponds to the one of Prandtl's mixing length model (Eq. 193b).

5. *Prandtl's shear-layer model*

If the constant eddy diffusivity of Prandtl's shear-layer model (Eq. 70) is inserted into (Eq. 191), the condition

$$0 \ll \left| \frac{\frac{\partial^2 \overline{u}_1}{\partial x_2^2}}{\frac{\partial \overline{u}_1}{\partial x_2}} \right| \tag{196}$$

is obtained. This condition states that the expression with the absolute value should be much larger (here not by a factor ten) than the term on the left-hand side, which is zero. However, if the second derivative in the numerator tends faster to zero than the gradient in the denominator, this condition is violated, because $0 \ll 0+$ is a questionable result and at least a weak contradiction. This is in agreement with the observation of the early turbulence modelers that, because of the stiffness of the eddy diffusivity, the theoretical results deviate from experiments (see Table 5.1).

6. *Taylor's vorticity model*

The conclusions for Prandtl's mixing-length model also apply to Taylor's vorticity model, because its eddy diffusivities are algebraically the same as those of Prandtl [compare Eq. (54) with Eq. (89b)].

To summarize the above brief comparative study, it is concluded that Prandtl's shear-layer model exhibits the best fulfillment of condition (191). Furthermore, we tentatively conclude in agreement with Corrsin (1974) that turbulence modeling with generalized diffusion "coefficients" may not be self-consistent and make us less sanguine about the common practice of using simple, variable-diffusivity turbulence models. In this section, it was demonstrated that all early gradient-type turbulence models fail, whether in the entire region or at least partially in regions of special forms of the averaged downstream velocity profile (zero derivative, zero curvature, etc.).

A somewhat more sophisticated proposal of this nature was briefly introduced in Sect. 5.4, namely the generalization of the eddy diffusivity from a scalar to Batchelor's diffusivity tensor (see Eq. (113), Batchelor 1949; Schmitt et al. 2003). This forms a direct continuation of the considerations of this section.

The three-dimensional generalization of Eq. (185a), with the scalar diffusivity $K$, is

$$\left\langle\vec{F}\right\rangle(\vec{x},t) = -K\nabla\overline{\Gamma}(\overline{x},t). \tag{197}$$

If this isotropic transport model is generalized to an anisotropic variant, the scalar or functional diffusivity is replaced by a rank-three tensor, so that

$$\left\langle\vec{F}\right\rangle(\vec{x},t) = -\left(\mathbf{K}\cdot\nabla\overline{\Gamma}\right)(\vec{x},t). \tag{198}$$

**K** is referred to as Batchelor's diffusivity tensor. It was Richardson (1920) who nearly hundred years ago had realized that turbulent diffusivity of heat in the atmosphere has different values in different directions. Yet, it must be remarked that these first discussions were focused on a second-rank tensor with unequal diagonal elements and zero off-diagonal elements. However, Lettau (1952) pointed out that in turbulent shear flows there is no a priori reason to assume that the non-diagonal terms must be zero.

In a turbulent flow, in agreement with Eq. (1), any evolutionary scalar physical quantity is split into two parts

$$\Gamma = \overline{\Gamma} + \Gamma', \tag{199}$$

and according to Kampé de Fériet (1939) its turbulent flux is

$$\overline{F}_i = \overline{u_i'\Gamma'} = -K_{ij}\frac{\partial\overline{\Gamma}}{\partial x_j}, \tag{200a, b}$$

in which Eq. (200b) is a "gradient"-type closure postulate.

As already noticed by Hinze (1975), for turbulent momentum transport, the "flux" of momentum is a second-rank tensor and, therefore, the kinematic diffusivity tensor is of rank four, so that

$$\overline{F}_{ij} = \overline{u_i'u_j'} = -\nu_{ijkl}\left(\frac{\partial \overline{u}_k}{\partial x_l} + \frac{\partial \overline{u}_l}{\partial x_k}\right). \tag{201a, b}$$

It is the most general affine connection between a pair of tensors of rank two. Another tensorial parameterization was proposed by Daly and Harlow (1970). In their scheme, the eddy diffusivity tensor is modeled slightly differently from Eq. (90), namely by

$$\varepsilon_{\mathrm{m}} = \mathrm{tr}\left[(\nu_{\mathrm{T}})_{ij}\right] \propto \frac{k}{\varepsilon}\mathrm{tr}\left(\overline{u_i'u_j'}\right). \tag{202a, b}$$

These authors introduce an anisotropy effect by replacing in Eq. (90) one of the two multiplicative coefficients $k$ by the second-order velocity fluctuation correlations, which also can be of antisymmetric nature. Convenient is the fact that, if one chooses (see also Hamba 2013), namely

$$\overline{u_i'u_j'} = \frac{2}{3}k\delta_{ij}, \tag{203}$$

Equation (202b) reduces to the scalar isotropic eddy diffusivity presented as Eq. (90). Else, proposal (202a,b) cannot objectively be combined with $D_{ij}$.

This class of closure proposals will not further be considered here, because it is known that turbulence is a nonlocal phenomenon; therefore, it is doubtful that a point-like gradient driving mechanism is the right description for complex turbulent flows, even if the eddy diffusivity is generalized in a way as presented above. Another argument against a substantial part of such alternative model approaches is that they mainly leave unconsidered all the older developments of turbulent momentum transfer by turbulent fluid lumps. One still must be aware of the fact that, even in their linear appearance, these models enjoy some rudimentary physical interpretation. Therefore, in a new formulation these early considerations should be somehow conserved, even though new models may contain more sophisticated and generalized concepts to describe complex turbulent motions. Furthermore, if one assumes that the DQTM points into the right direction, then simply nonlocal difference quotients of averaged velocities lie in the focus rather than sophisticated parameterizations of the Reynolds stress tensor in terms of strain-rate measures.

## 6.5 Truly History Dependent and Nonlocal Models

A third class of models contains the most promising ideas and concepts; thus, we shall reserve more space for a detailed discussion of this last class of relations. Models of this class are the history-dependent and nonlocal parameterizations of the Reynolds stress by integral functionals containing a delay-time and a nonlocal kernel.

To achieve acquaintance with nonlocality, the following function is introduced

$$g(x;x') = f(x - x'), \tag{204}$$

where $x$ is the genuine variable and $x'$ a free parameter.[3] The difference $x - x'$ measures the distance from point $x$ to the negative and positive (left and right) sides of $x$. If $x' = ct$, where the constant $c$ denotes a propagation velocity, then $x-ct$ measures the distance of $x'$ from $x$ at time $t$ and $f(x-ct)$ describes a linearly propagating wave of elevation $f$.

Next, assume a dependence on two shifted points

$$g(x;x',x'') = p'f(x - x') + p''f(x - x''), \tag{205}$$

where $p'$ denotes an impact factor of the shifted function by $x'$, and $p''$ is the impact strength of the function shifted by $x''$. These impact factors may sometimes be interpreted as probabilities; then, a normalization leads to the condition

$$p' + p'' = 1. \tag{206}$$

In the analogous continuous case, the probability density function or the weighting function $\omega(x')$, respectively, is considered. In this generalization an impact from a continuous region around the actual position $x$ can be described by the convolution-type representation

$$g(x) = \int_{-\infty}^{+\infty} \omega(x')f(x - x')\mathrm{d}x'. \tag{207}$$

Because $x'$ is ranging from $-\infty$ to $+\infty$, these limits do not need to be explicitly specified in the argument of the functional $g$. In analogy to the discrete case, the normalization condition takes the form

$$\int_{-\infty}^{+\infty} \omega(x')\mathrm{d}x' = 1. \tag{208}$$

The same concept is valid for a time dependence and a temporal shift, respectively. In this case, one must observe determinism (see Sect. 7.3). If history and nonlocality occur simultaneously, a double integral description is needed.

[3]Because $x'$ can be negative, zero or positive one could also write a positive sign, $x + x'$, instead.

For exclusive hereditary effects, one has

$$g(t) = \int_{-\infty}^{0} \omega(t')f(t - t')\mathrm{d}t', \qquad \int_{-\infty}^{0} \omega(t')\mathrm{d}t' = 1, \tag{209a, b}$$

and when spatial nonlocality effects and hereditary properties operate together, double integrals such as

$$\begin{aligned} g(x,t) &= \int_{-\infty}^{0} \int_{-\infty}^{+\infty} \omega(x',t')f(x - x', t - t')\mathrm{d}x'\mathrm{d}t', \\ \int_{-\infty}^{0} \int_{-\infty}^{+\infty} & \omega(x',t')\mathrm{d}x'\mathrm{d}t' = 1, \end{aligned} \tag{210a, b}$$

yield the correct description.

What do these memory and nonlocality concepts have to do with our fluid dynamical problems? An answer to this question can be obtained by studying the velocity-gradient theory of turbulence by Wilczek and Meneveau (2014). These authors introduce the velocity-gradient tensor $\mathbf{L}$ in agreement with Eq. (93), defined for the averaged velocity,

$$\mathbf{L} = \nabla\vec{u}, \qquad L_{ij} = \frac{\partial u_i}{\partial x_j}. \tag{211a, b}$$

Because $\vec{u}$ is solenoidal (see Eq. 15), $\mathbf{L}$ is traceless,

$$\mathrm{tr}(\mathbf{L}) = 0. \tag{212}$$

Taking the gradient of the Navier–Stokes momentum equation leads to

$$\begin{aligned} &\frac{\partial}{\partial t}\nabla\vec{u} + \vec{u}\cdot\nabla\nabla\vec{u} + \nabla\vec{u}\cdot\nabla\vec{u} + \nabla\nabla p \\ &-\nu\Delta(\nabla\vec{u}) - \nabla\vec{f} = 0, \qquad \text{if } \nu = \text{const.} \end{aligned} \tag{213a, b}$$

By inserting Eq. (211a) and applying

$$\mathbf{H} = \nabla\nabla p, \qquad H_{ij} = \frac{\partial^2 p}{\partial x_i \partial x_j}, \tag{214a, b}$$

which is the Hessian tensor of the pressure field (see, e.g., Lawson and Dawson 2015) as well as

$$\mathbf{F} = \nabla\vec{f}, \qquad F_{ij} = \frac{\partial f_i}{\partial x_j}, \qquad \frac{\partial F_{ij}}{\partial x_j} = 0, \tag{215a – c}$$

which is an optionally added solenoidal (Eq. 215c) large-scale forcing of the fluid dynamic system,[4] it follows from Eq. (213a) that

$$\frac{\partial}{\partial t}\mathbf{L} + \vec{v} \cdot \nabla\mathbf{L} + \mathbf{L}^2 + \mathbf{H} - \nu\Delta\mathbf{L} - \mathbf{F} = \mathbf{0}. \tag{216}$$

Of all terms in Eq. (216), $\mathbf{L}^2$ is the term with the lowest order of derivatives. For this reason, it is sometimes called the local term of Eq. (216). It describes nonlinear self-amplification of the velocity gradient and, according to Wilczek and Meneveau (2014), has been extensively investigated for the Euler equation by Cantwell (1992).

It is clear from Eq. (214a) that

$$\mathrm{tr}(\mathbf{H}) = \Delta p. \tag{217}$$

Taking the trace of Eqs. (214a) and (216) leads to

$$\Delta p = -\mathrm{tr}\left(\mathbf{L}^2\right) \tag{218}$$

a Poisson equation, relating the pressure with the velocity (gradient) field, justifying the attribute “local” to $\mathbf{L}^2$.

The Hessian tensor can be decomposed into a local isotropic and a nonlocal second part

$$\mathbf{H} = -\frac{1}{3}\mathrm{tr}\left(\mathbf{L}^2\right)\mathbf{I} + \widetilde{\mathbf{H}}, \tag{219}$$

where $\widetilde{\mathbf{H}}$ denotes the traceless symmetric Hessian tensor. For an unbounded flow the solution for this tensor is a principal value integral of Biot-Savart type (see Novikov 1967; Ohkitani and Kishiba 1995; Tsinober 2009), a method that can also be applied to solve the vorticity equation, see, e.g., Chap. 10,

[4]The forcing of the fluid dynamic system can, e.g., be due to a moving boundary and then this term is not demanded.

$$\widetilde{\mathrm{H}}_{ij}\left(\vec{x},t\right) = -\frac{1}{4\pi}\int_{\Gamma}\left[\frac{\delta_{ij}}{\left|\vec{x}-\vec{x}'\right|^{3}} - 3\frac{\left(\vec{x}-\vec{x}'\right)_i\left(\vec{x}-\vec{x}'\right)_j}{\left|\vec{x}-\vec{x}'\right|^{5}}\right] \times \mathrm{tr}\left[\mathbf{L}\left(\vec{x},t\right)^{2}\right]\mathrm{d}\vec{x}', \tag{220}$$

in which $\Gamma$ denotes the space occupied by the fluid material. The deviatoric part of **H** is obviously nonlocally related with moderately distant points in the fluid domain.[5] Note the formal analogy with the one-dimensional version given by Eq. (207). Therefore, Eq. (220) demonstrates that nonlocal effects do occur in solutions of the NSEs and are inherently acting in fluid dynamical systems involving turbulent flows.

If the pressure field is nonlocally connected with the velocity field, as surmised by Eq. (216), it may be assumed that the same is valid for the Reynolds shear stress. An analogous statistical treatment demonstrating this does not exist yet but would be very instructive.

In Chap. 3, it was stipulated that turbulence closure methods higher than of zeroth order do not seem to be very fruitful. We, however, found that this risky statement is supported by some important theoretical findings by Speziale (1979). At first, he states that turbulence correlations of any degree are frame independent. Then, the conclusion is that good turbulence modeling must conserve this basic property. Continuing he studies the frame indifference of the Reynolds (stress) transport equations, which also contain third-order moments. The conclusion of his calculation (see his article, Speziale 1979) is that these equations are proven to be frame dependent!

Speziale comments this important finding by the following remarks, which are so much in agreement with our considerations that we cite them here explicitly: *What are we to conclude all of this* [his calculus]? *If we model the higher turbulence correlations* [the Reynolds stress transport equations] *in terms of the Reynolds stresses and the mean velocity field..., a set of equations results ... which does not exhibit the frame independence that we have proven them to have. Hence, at best, such* [higher-order] *models can hold for a restricted class of turbulent flows, i.e.* [physically correct] *closure cannot be achieved at this level.* Speziale then continues by writing: *Hence, in our minds, tying the higher turbulence correlations to the local Reynolds stresses and mean velocity field is a highly questionable procedure. Without this ability, the method of moments would completely lose its utility since we would be introducing higher turbulence correlations for which we would have less a priori knowledge than about the Reynolds stresses themselves. So, why not mathematically model the Reynolds stresses directly? At the core of second moment*

[5] The operations used to reach Eq. (220) are second-order differentiations which reflect still a strong degree of locality.

*closure as well as the older phenomenological theories of turbulence is the intrinsic belief that the Reynolds stresses are quantities that are determined by the mean velocity field. In mathematical terms this simply means that the Reynolds stresses are functionals of the history of the mean velocity field of all points* $x'$ *in the fluid volume* $\Gamma$, *i.e.*

$$\tau_{ij}\left(\vec{x},t\right) = \tau_{ij}\left[\vec{u}\left(\vec{x}',t'\right);x,t\right], \quad \vec{x}' \in \Gamma, t' \in (-\infty,t). \tag{221}$$

*Explicit forms can be obtained, subject to frame indifference, by utilizing integral representation theorems from functional analysis. Furthermore, such forms can take into account the nonlocality of turbulence which all current models seem to neglect... Alternative approaches must be pursued if an ultimate solution of turbulence is to be obtained.*

It is concluded that the above rigorous analysis by Wilczek and Meneveau (2014) fully corresponds to Speziale's proposal for Reynolds shear stress modeling.

One should not be too much concerned about the velocity-gradient dependence in Eqs. (219) and, on the other hand, the alternative velocity dependence in the proposal (Eq. 221). In this respect, Eqs. (219) and (221) are not in conflict with one another. Equation (221) is to be interpreted as a general functional representation of the velocity field, while Eq. (220) shows how such a functional relation looks like for $\widetilde{\mathbf{H}}\left(\vec{x},t\right)$. Having Prandtl's momentum exchange ideas in mind, a presentation taking account of the velocity gradients is the one that directly copes with these early ideas (see Sect. 7.3).

It is highly attractive that Kraichnan's (1964) *D*irect *I*nteraction *A*pproximation (DIA) for the transport of a scalar quantity $\theta$ also leads to a nonlocal expression containing the gradient of the transported quantity,

$$\overline{u_i'\theta'}\left(\vec{x},t\right) = -\int_{\Gamma} \mathrm{d}\vec{x}' \int_{0}^{t} \mathrm{d}t' \nu_{ij}\left(\vec{x},t;\vec{x}',t'\right) \frac{\partial}{\partial x_j'}\overline{\theta}\left(\vec{x}',t'\right). \tag{222}$$

To acquire a deeper knowledge of this topic see Kraichnan (1959) and Roberts (1961). In the integral, $\theta$ denotes a velocity component, temperature, concentration of a dispersed additive to the fluid, etc. and $\theta'$ is its fluctuation quantity. In this expression, $\nu_{ij}$ is the nonlocal kernel related to the nonlocal tensorial eddy diffusivity in space and time. It is not so surprising that Prandtl's mixing-length model appears again intrinsically in this nonlocal integral representation (compare with Sect. 7.3). Therefore, Kraichnan's proposal is a perfect combination of Prandtl's idea of a mixing-length concept, (Eq. 53b), with Speziale's requirement.

In Sect. 7.3, Eq. (222) will suffice for our purposes. Nonetheless, a few important relations can be derived, which are based on a profound physical basis. It would be a pity not to present them, because they could become important in future developments of better (nonlocal) turbulence models. Kraichnan's nonlocal kernel satisfies the relation,

$$\nu_{ij}\left(\vec{x},t;\vec{x}',t'\right)=\overline{u_i'\left(\vec{x},t\right)g_K\left(\vec{x},t;\vec{x}',t'\right)u_j'\left(\vec{x}',t'\right)}, \tag{223}$$

in which $g_K$ is Kraichnan's response function (index $K$). It contains the double space and double time correlations. We shall not show Kraichnan's entire field–theoretic development of this equation, because in Sect. 7.3 a modern straightforward development, based on fractional calculus, will lead us directly to relation (222). Furthermore, a development below by Hamba (2004, 2005, 2013) makes Eq. (223) plausible. The response function satisfies the following differential equation:

$$\frac{\partial g_K}{\partial t}+u_i'\frac{\partial g_K}{\partial x_i}-\kappa_{\mathrm{m}}\frac{\partial^2 g_K}{\partial x_i\partial x_i}=\delta\left(\vec{x}-\vec{x}'\right)\delta(t-t'). \tag{224}$$

In this equation, $\kappa_{\mathrm{m}}$ denotes the molecular diffusion coefficient. One can demonstrate that the double velocity correlation is replaced by the Reynolds stress, which is a one-point space and one-point-time correlation and the integral of the response function is approximated by $k/\varepsilon$, whereas the local diffusivity tensor $\nu_{\mathrm{T}_{ij}}$ occurring in Eq. (202b) is obtained as a special case (see Hamba 2013).

Romanof (1989) and Hamba (2013) introduced a generalization to the theory of Kraichnan. Starting with the NSEs, these authors develop a transport equation for the fluctuation of a scalar quantity $\theta'$, viz.,

$$\frac{\mathrm{D}\theta'}{\mathrm{D}t}+\frac{\partial}{\partial x_i}\left(u_i'\theta'-\overline{u_i\theta'}\right)-\kappa_{\mathrm{m}}\frac{\partial^2\theta'}{\partial x_i\partial x_i}=-u_i'\frac{\partial\overline{\theta}}{\partial x_i}, \tag{225}$$

where the D/D$t$ is the substantial time derivative. The term on the right-hand side does not contain the fluctuation quantity $\theta'$. Inspired by this, Hamba formally interprets this expression to be a source term for the fluctuation quantity. Hence, he proposes the modified nonlocal response function

$$g_H=g_H\left(\vec{x},t;\vec{x}',t'\right), \tag{226}$$

where the index $H$ (Hamba) is used. This new response function satisfies (compare with Eq. (224) and replace $\theta'$ by $g_H$) the differential equation

$$\frac{\mathrm{D}g_H}{\mathrm{D}t}+\frac{\partial}{\partial x_i}\left(u_i'g_H-\overline{u_i'g_H}\right)-\kappa_{\mathrm{m}}\frac{\partial^2 g_H}{\partial x_i\partial x_i}=\delta\left(\vec{x}-\vec{x}'\right)\delta(t-t'). \tag{227}$$

With the continuity equation for the fluctuation quantities of the velocity field $\partial u_i'/\partial x_i=0$, Eq. (227) reaches its final form

$$\frac{Dg_H}{Dt} + u_i' \frac{\partial g_H}{\partial x_i} - \frac{\partial}{\partial x_i} \overline{u_i' g_H} - \kappa_m \frac{\partial^2 g_H}{\partial x_i \partial x_i} = \delta\left(\vec{x} - \vec{x}'\right) \delta(t - t'). \tag{228}$$

When compared with Kraichnan's transport Eq. (224), it is immediately recognized that Romanof and Hamba's analogous Eq. (228) contains an additional derivative of the correlation term $\overline{u_i' g_H}$. Using the new response function $g_H$, the formal solution for the scalar fluctuation follows by Green's function associated to Eq. (228) (see also Romanof 1989)

$$\theta'\left(\vec{x},\ t\right) = -\int_{\Gamma} d\vec{x}' \int_0^t dt' g_H\left(\vec{x}, t; \vec{x}', t'\right) u_i'\left(\vec{x}', t'\right) \frac{\partial}{\partial x_j} \bar{\theta}\left(\vec{x}', t'\right). \tag{229}$$

Initial value terms have been neglected, because in turbulence models for mean quantities they do not play a role. With the nonlocal eddy diffusivity

$$\nu_{ij}\left(\vec{x}, t; \vec{x}', t'\right) = \overline{u_i'\left(\vec{x}, t\right) \, g_H\left(\vec{x}, t; \vec{x}', t'\right) \, u_j'\left(\vec{x}, t\right)}, \tag{230}$$

Kraichnan's nonlocal model Eq. (223) is again obtained, the only difference being the slightly differing types of response functions $g_K$ and $g_H$.

If the nonlocal eddy diffusivity tensor is written in a simpler form, namely as

$$\nu_{ij}\left(\vec{x}, t; \vec{x}', t'\right) = K_{ij}\left(\vec{x}, t\right) \, \delta(x - x') \, \delta(t - t'), \tag{231}$$

then Eq. (222) leads to the local representation.

Between the lowest-quality approach (Eq. 231) and the most sophisticated one, (Eq. 230), Romanof (1989) found, listed and cited some intermediate eddy diffusivity expressions, e.g.,

$$\nu_{ij}\left(\vec{x}, t; \vec{x}', t'\right) = D_{ij}^{(s)}\left(\vec{x} - \vec{x}'\right) \delta(t - t'), \tag{232}$$

where $D_{ij}^{(s)}\left(\vec{x} - \vec{x}'\right)$ is the transfer function of turbulent diffusion in the $K$-spectral theory (see Berkowicz and Prahm 1979, 1980) and describes only nonlocal behavior in space, but not in time (no history effect). A further formula for the eddy diffusivity is given by

$$\nu_{ij}\left(\vec{x}, t; \vec{x}', t'\right) = B_{ij}(t - t') \, \delta\left(\vec{x} - \vec{x}'\right), \tag{233}$$

where

$$B_{ij}(t - t') = \overline{u_i'\left(\vec{x}, t\right) u_j'\left(\vec{x}, t'\right)}, \tag{234}$$

is the Eulerian time covariance. This approach describes history effects but neglects nonlocal behavior in space. Equation (233) is named Saffman equation after its founder (see Saffman 1969).

A third approach combines nonlocality and heredity

$$\nu_{ij}\left(\vec{x}, t; \vec{x}', t'\right) = B_{ij}\left(\vec{x} - \vec{x}', t - t'\right) P\left(\vec{x} - \vec{x}', t - t'\right), \tag{235}$$

where

$$B_{ij}\left(\vec{x} - \vec{x}', t - t'\right) = \overline{u_i'\left(\vec{x}, t\right) u_j'\left(\vec{x}', t'\right)} \tag{236}$$

is the Eulerian space–time covariance of the velocity. Furthermore, $P\left(\vec{x} - \vec{x}', t - t'\right)$ is the transition probability (rate) for a fluid lump being at time $t'$ situated at $\vec{x}'$ to be at a later time $t$ at location $\vec{x}$ (see Roberts 1961). This last approach corresponds to Corrsin's (1974) proposal with results and interpretations given also by him.

The diffusion coefficients $K_{ij}$ are related to the generalized diffusion tensor by applying the double integration (compare with Eq. 210a)

$$K_{ij}\left(\vec{x}, t\right) = \int_{\Gamma} \mathrm{d}\vec{x}' \int_{0}^{t} \mathrm{d}t' \nu_{ij}\left(\vec{x}, t; \vec{x}', t'\right). \tag{237}$$

The turbulent diffusivity model (Eq. 222) is based on the fact that a turbulent flux depends on the averaged gradient in space–time not just at the single point under consideration, but rather in a surrounding domain of this single point. This is a manifestation of the space–time (history) dependent statistics of turbulent motion and diffusion, respectively, in contrast to a molecular momentum transport and diffusion process, respectively, determined by the independently moving molecules (Romanof 1989).

Hamba (2004) had the intention to validate the nonlocal eddy diffusivity expression by DNS computation of the scalar transport in a turbulent channel flow. Because of its dual character in space and time, the calculations were very time consuming. This problematic situation led him back into applications of the Taylor expansion technique to the scalar mean gradient and back to derivations of usual-type linear and local special cases.

Romanof's and Hamba's works include first-order transport equations modeling, which in Chap. 3 was judged to be unnecessary. However, then, more concrete knowledge is demanded on the kernels: in Kraichnan's case $\nu_{ij}$ and $g_K$, respectively, in Hamba's theory $g_H$, and finally in Romanof's physical model $K_{ij}$, $D_{ij}$, $B_{ij}$, etc., depending on the complexity of the fluid dynamic problem under consideration. Therefore, an alternative simpler concept would be much welcomed.

Works of Rivlin (1957), Liepmann (1961), Lumley (1970), Hinze et al. (1974) [see also in Hinze's standard textbook on turbulence: Hinze (1975)], Picu (2002) and Huang (2004) indicate that a fluid in turbulent motion behaves like a classical nonlinear non-Newtonian medium describing the motion of viscoelastic "fluids." The equations of motion of materials, such as taffy, hot plastics, and yoghurt, fall outside traditional analytical fluid dynamics (see West 2014), because the material properties are neither those of solids nor of fluids (Rabotnov 1977). The constitutive equations of such materials are of integro-differential form that has an equivalent interpretation in fractional calculus (Scott et al. 1947; Glöckle and Nonnenmacher 1991).

Lumley (1970) suggests that large eddy structures of turbulent shear flows should be explained by the validity of some form of non-Newtonian constitutive relations of integral form, as already introduced by Kraichnan's formulation, Eq. (222). Nonlocal integral formulations were also proposed and discussed by Crow (1968), Mompean et al. (2011), Zhu et al. (2013), and Fiedler (1984) in a study of the atmosphere. The authors of this book drew attention to a tight relation of such nonlocal integro-type approaches for turbulent flows with an analogous use of "fractional calculus" (see Egolf and Hutter 2017). This is presented in more detail in Sect. 7.3, and is a natural and new continuation of the nonlocality discussed in this section.

Beyond the scope of Prandtl's theory, the focus to derive better Reynolds stress models and constitutive equations of turbulence was more on mathematical constructions than fundamental physical considerations, where invariance properties are applied to assign to the derived results a better appeal. Coherent structures with bursts and sweeps are intrinsically coupled to momentum transfer considerations and must be taken as "serious candidates" for alternative closure model developments. Bernard and Handler (1990), following an early direction of development by Taylor (1932) on transport of vorticity, by evaluating Lagrangean particle paths in turbulent fields and in turbulent flows near a wall, found large-scale displacements of momentum in the direction perpendicular to the wall in wall-bounded shear flows.

The basic successful question of Bernard et al. (1989a, b) and Bernard and Handler (1990) was, whether studying the Reynolds shear stress in a Lagrangean frame of coordinates could reveal some new insight. This idea is very intriguing, because it is open for a direct validation by DNS results, a method which was chosen by Bernard and his coauthors already in the early years of its application. The Lagrangean statistical method had already been used by Rhines and Holland (1979), and Handler et al. (1992) for a study of the importance of vorticity transport in turbulent closures.

It is valuable to study the correlation $\overline{u_2' u_1'}$ at a particular point $a$ in Euclidian space at time $t_0$ (see Fig. 6.2). Then, a fluid particle with trajectory $x(a,t)$ satisfies the condition that it arrives at time $t_0$ at location $a$: $x(a, t_0) = a$. The position of this particle at an earlier time $t - \tau$, $\tau > 0$ is denoted by $b$: $x(a, t_0 - \tau) = b$. In a highly fluctuating turbulent flow field, point $b$ varies chaotically (statistically treated like randomly) depending on its realization in the process of determining the ensemble average movement of the fluid particles.

By definition, in the Lagrangean framework the convective terms of the NSEs (see Eq. 16b) do not arise,

$$\frac{\partial \widehat{u}_1}{\partial \widehat{t}} = -\frac{\partial \widehat{p}}{\partial \widehat{x}} + \frac{1}{\mathrm{Re}} \Delta \widehat{u}_1, \tag{238}$$

whereas the physical quantities (with hats which in the following are dropped) are dimensionless (normalized) and the Reynolds number Re has already been defined (see Eq. 143).

By integrating Eq. (238), it follows that

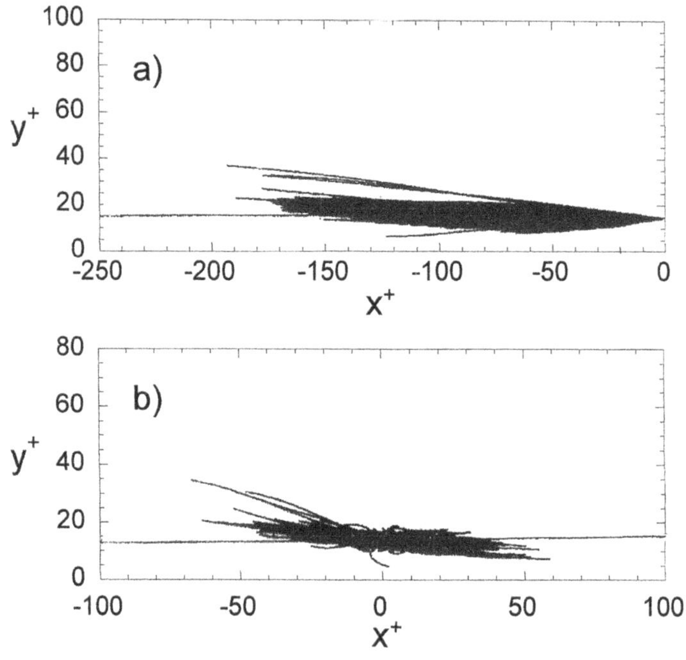


**Fig. 6.2** DNS results of backward tracings of the trajectories of particles in a channel flow arriving at time $t_0$ at location $a$ (wall distance: $y^+ = 15.8$) for a time interval $|\tau|$. The result for the Eulerian frame is shown in panel (**a**) and of the Lagrangean in panel (**b**). For Lagrangean observation the distribution of the starting point's $b$ is more symmetric than that for the Eulerian. From Bernard et al. (1989b), reproduced with the permission of AIP publishing

$$\widehat{u}_{1a} - \widehat{u}_{1b} = -\int_{\tau}^{0} \widehat{\frac{\partial p}{\partial \widehat{x}}}(s)\,\mathrm{d}s + \frac{1}{\mathrm{Re}} \int_{\tau}^{0} \Delta \widehat{u}_1 \mathrm{d}s, \tag{239}$$

where $t_0 = 0$ and $\tau < 0$ is assumed to hold. The momentum of a fluid lump changes by the pressure and viscous forces. Splitting velocities into mean and fluctuation quantities yields the non-normalized equation

$$u_{1a} - u_{1b} = (\overline{u}_{1a} - \overline{u}_{1b}) + \left(u'_{1a} - u'_{1b}\right). \tag{240}$$

This equation is multiplied by the fluctuation quantity $u'_{2a}$, leading to

$$(u_{1a} - u_{1b})\, u'_{2a} = u'_{1a}u'_{2a} - u'_{1b}u'_{2a} + u'_{2a}(\overline{u}_{1a} - \overline{u}_{1b}). \tag{241}$$

Now, also Eq. (239) is multiplied with $u'_{2a}$, averaged and combined with the averaged Eq. (241). This yields

$$\overline{u'_{1a}u'_{2a}} = \overline{u'_{1b}u'_{2a}} + \overline{u'_{2a}(\overline{u}_{1b} - \overline{u}_{1a})} - \frac{1}{\rho} \int_{\tau}^{0} \overline{u'_2(0)\frac{\partial p}{\partial x}(s)}\,\mathrm{d}s + \nu \int_{\tau}^{0} \overline{u'_2(0)\Delta u_1}\mathrm{d}s, \tag{242}$$

where $u'_{2a} = u'_2(0)$ was introduced. Here, $u_1$ is the wall parallel and $u_2$ the wall normal fluid velocity. Note that the overbar denotes ensemble averaging. The term on the left-hand side of Eq. (242) corresponds to the Reynolds stress, and the first term on the right-hand side is a spatial correlation of velocity fluctuations over a usually large distance, which is described by the Lagrangean correlation function

$$R_{u_1 u_2}(\tau) = \overline{u'_{1b}u'_{2a}}. \tag{243}$$

The second term on the right-hand side of Eq. (242) is the term describing large-scale momentum transport by fluid eddies.[6] Furthermore, Bernard and Handler (1990) define

$$\Phi_{\mathrm{P}} = -\frac{1}{\rho} \int_{\tau}^{0} \overline{u'_2(0)\frac{\partial p}{\partial x}(s)}\,\mathrm{d}s, \quad \Phi_{\mathrm{D}} = \nu \int_{\tau}^{0} \overline{u'_2(0)\Delta u_1}\,\mathrm{d}s. \tag{244a, b}$$

Then Eq. (242), with the aid of Eq. (243) and Eqs. (244a,b), transforms to

[6] In this Lagrangean dynamics ensemble, the averaging rules Eqs. (3) and (6) do not hold.

$$\overline{u'_{1a}u'_{2a}} = R_{u_1u_2}(\tau) + \overline{u'_{2a}(\bar{u}_{1b} - \bar{u}_{1a})} + \Phi_P + \Phi_D. \tag{245}$$

The quantities $\Phi_P$ and $\Phi_D$ are important sources for the Reynolds stress, the first is responsible for acceleration and deceleration of fluid particles by pressure forces and the second is due to viscous forces. It is intuitively clear that the correlation term $R_{u_1u_2}(\tau)$ for large $|\tau|$ decays to zero. This was also demonstrated by Bernard et al. (1989a, b). An evident relation is the identity

$$\overline{u'_{1a}u'_{2a}} = \overline{u'_{1b}u'_{2a}} + \overline{u'_{2a}\left(\bar{u}_{1b} - \bar{u}_{1a}\right)} + \overline{u'_{2a}\left(u_{1a} - u_{1b}\right)}. \tag{246}$$

By comparing Eq. (242) with Eq. (246) and taking Eqs. (244a,b) into consideration, one concludes that

$$\overline{u'_{2a}(u_{1a} - u_{1b})} = \Phi_P + \Phi_D. \tag{247}$$

In Eq. (246), for large times $|\tau|$, the first term on the right-hand side can be neglected, because $\overline{u'_{1b}}$ fades with growing $|\tau|$. Then, when also omitting the index $a$, Eq. (246) becomes

$$\overline{u'_2u'_1} = \overline{u'_2(\bar{u}_{1b} - \bar{u}_1)} + \overline{u'_2(u_1 - u_{1b})}, \tag{248}$$

where the velocities with an index $b$ are the velocities at a far-distant location. Naturally this relation is trivial, but the important conclusion is that turbulent momentum transport by momentum advection (second term on the right of Eq. (242)) and acceleration and deceleration of fluid lumps by pressure and dissipation effects (third and fourth term)—the three main physical effects, responsible for the Reynolds stress—can entirely be described by velocity differences of points that lie largely apart from one another! These results were confirmed by several direct numerical simulation (DNS) results (see Bernard et al. 1989a, b).

Because these rather general considerations contain also the momentum transfer by eddying motion, it is possible to search in this formalism for the Boussinesq approximation. To this end, a Taylor series expansion of $\bar{u}_{1b}$ up to second order is introduced,

$$\bar{u}_{1b} = \bar{u}_{1a} - \int_{\tau}^{0} u'_2(s)\mathrm{d}s\frac{\mathrm{d}\bar{u}_1}{\mathrm{d}x_2}(a) + \frac{1}{2}\int_{\tau}^{0}\int_{\tau}^{0} u'_2(s)u'_2(r)\mathrm{d}s\mathrm{d}r\frac{\mathrm{d}^2\bar{u}_1}{\mathrm{d}x_2^2}(\theta), \tag{249}$$

where, because of the backward in time integration the odd (even) derivative terms have a negative (positive) sign. Furthermore, the quantity $\theta$ is located somewhere between the integration limits: $\tau \leq \theta \leq 0$.

This expression is multiplied with $u'_{2a}$ and the first term on the right-hand side is transferred to the left-hand side and finally, the full equation is averaged; this process yields

$$\overline{u'_{2a}(\overline{u}_{1b} - \overline{u}_{1a})} = -\int_{\tau}^{0} \overline{u'_2(0)\, u'_2(s)}\, \mathrm{d}s \frac{\mathrm{d}\overline{u}_1}{\mathrm{d}x_2} + \frac{1}{2}\int_{\tau}^{0}\int_{\tau}^{0} \overline{u'_2(0)\, u'_2(s)\, u'_2(r)\, \mathrm{d}s\, \mathrm{d}r} \frac{\mathrm{d}^2\overline{u}_1}{\mathrm{d}x_2^2}\,(\theta). \tag{250}$$

With the definitions of a local,

$$\Phi_{\mathrm{L}} = -\int_{\tau}^{0} \overline{u'_2(0)\, u'_2(s)}\, \mathrm{d}s \frac{\mathrm{d}\overline{u}_1}{\mathrm{d}x_2}, \tag{251}$$

and nonlocal contribution,

$$\Phi_{\mathrm{NL}} = \overline{\frac{1}{2}\int_{\tau}^{0}\int_{\tau}^{0} u'_2(0)\, u'_2(s)\, u'_2(r)\, \mathrm{d}s\, \mathrm{d}r \frac{\mathrm{d}^2\overline{u}_1}{\mathrm{d}x_2^2}} \tag{252}$$

Equations (247)–(250) can be combined, which then leads to

$$\overline{u'_1 u'_2} = \Phi_{\mathrm{L}} + \Phi_{\mathrm{NL}} + \Phi_{\mathrm{P}} + \Phi_{\mathrm{D}}, \tag{253}$$

in which $\Phi_{\mathrm{L}}$ is defined by Eq. (251), $\Phi_{\mathrm{NL}}$ by Eq. (252), $\Phi_{\mathrm{P}}$ by Eq. (244a) and $\Phi_{\mathrm{D}}$ by Eq. (244b).

By only considering the first term on the right-hand side of Eq. (253) and by introducing Eq. (251), the local Boussinesq approximation is obtained

$$\overline{u'_2 u'_1} = -\nu_{\mathrm{T}} \frac{\mathrm{d}\overline{u}_1}{\mathrm{d}x_2}, \tag{254}$$

with a new expression for the turbulent diffusivity

$$\nu_{\mathrm{T}} = \int_{\tau}^{0} \overline{u'_2(0)\, u'_2(s)}\, \mathrm{d}s. \tag{255}$$

Bernard and Handler (1990) found by their theoretical considerations and by DNS that mainly two mechanisms are responsible for the occurrence of the Reynolds shear stresses. The first and more significant entails an eddy transport, as introduced by Prandtl in which momentum is carried unchanged from one point to another by the turbulent displacement of the fluid lumps. However, of similar importance is their finding that one-point gradient laws are inherently unsuitable to describe this phenomenon. It is inappropriate to view the Reynolds shear stress as arising from a gradient mechanism, nor by supplementing it by non-gradient corrections. These authors state: "*However, a potentially useful non-local approximation to displacement transport, depending on the global distribution of the mean velocity gradient, may be developed as a natural consequence of its definition.*" The second less pronounced mechanism relates to acceleration and deceleration and sweep-type motion, especially occurring in near-wall regions, and it is related to the creation, stretching, folding, and dissipation of vortex structures.

Hence, the following two main results are drawn:

R1) A new useful closure scheme should be developed from a nonlocal integral approach (Kraichnan, Hamba, etc.).

R2) A useful closure scheme contains (large-scale) nonlocal contributions, e.g., mean velocity differences as they occur in the DQTM.

To summarize: The first part of this book, especially Chaps. 1–4, presents some basic fluid dynamics and turbulence problems and numerous techniques of low-order turbulence modeling. Especially the modeling with zero-order turbulence models was discussed. In this context, the linear and local Boussinesq approach, and the models based on it were criticized and numerous arguments and algebraic and physical demonstrations of its failures were presented in detail.

In the following second part of this book, a new turbulence model, the nonlinear and nonlocal Difference-Quotient Turbulence Model (DQTM), will be introduced and presented, in which the local strain rate (~local mean velocity gradient) is replaced by a difference quotient. We will motivate the introduction and consideration of this model, and will demonstrate that for quasi-two-dimensional elementary turbulent flow configurations this model performs better than other zero-equation turbulence models. Furthermore, in combination with the NSEs, it reveals in a natural manner a critical phenomenon (continuous phase change). Owing to this property the DQTM shares properties occurring inherently also in other general physical theories, expressed e.g., as critical phenomena, renormalization structure, non-equilibrium statistical phenomena, fractional mathematical and physical properties, etc.

## References

Arrowsmith, D.K., Place, C.M.: Dynamische Systeme. Academic Edition Spectrum, Heidelberg (1990). ISBN 3-86025-308-5 (in German)

Batchelor, G.K.: Diffusion in a field of homogeneous turbulence. I. Eulerian analysis. Aust. J. Sci. Res. A. **2**, 437 (1949)

Berkovicz, R., Prahm, L.P.: Generalization of K-theory for turbulent diffusion. Part I. Spectral turbulent diffusivity concept. J. Appl. Meteorol. **18**, 266 (1979)

Berkovicz, R., Prahm, L.P.: On the spectral turbulent diffusivity theory for homogeneous turbulence. J. Fluid Mech. **100**(2), 433 (1980)

Bernard, P.S., Ashmawey, M.F., Handler, R.A.: Evaluation of the gradient model of turbulent transport through direct Lagrangian simulation. AIAA J. **27**, 1290 (1989a)

Bernard, P.S., Ashmawey, M.F., Handler, R.A.: An analysis of particle trajectories in computer simulated turbulent channel flow. Phys. Fluids A. **1**, 1532 (1989b)

Bernard, P.S., Handler, R.A.: Reynolds stress and the physics of turbulent momentum transport. J. Fluid Mech. **220**, 99 (1990)

Bradshaw, P.: AGARDograph, No. 169. Nato Science and Technology Organisation, USA (1973)

Broer, H.W., Dumortier, F., van Strien, S.J., Takens, F.: Structures in Dynamics. Elsevier, Amsterdam (1991). ISBN 0-444-89257-5

Cantwell, B.J.: Exact solution of a restricted Euler equation for the velocity gradient tensor. Phys. Fluids A. **4**(4), 782 (1992)

Collet, P., Eckmann, J.-P.: Instabilities and Fronts in Extended Systems, Princeton Series in Physics. Princeton University Press, Princeton, USA (1990). ISBN 0-691-08568-4

Corrsin, S.: Transport model limitations in turbulence. Adv. Geophys. **18A**, 25 (1974)

Crow, S.C.: Viscoelastic properties of fine grained incompressible turbulence. J. Fluid Mech. **33**(1), 1 (1968)

Daly, B.J., Harlow, F.H.: Transport equations in turbulence. Phys. Fluids. **13**, 2634 (1970)

Egolf, P.W., Hutter, K.: From linear and local to nonlinear and nonlocal zero equation turbulence models. In: Proceeding of IMA Conference for Turbulence, Waves and Mixing, Kings College, pp. 71–74, Cambridge, UK (2016). 6–8 July

Egolf, P.W., Hutter, K.: Fractional turbulence models. In: Peintke, J., et al. (eds.) Progress in Turbulence VII, Springer Proceedings in Physics, vol. 165. Springer, Heidelberg (2017). ISBN 978-3-319-57933-7

Einstein, A.: Über die von der molekularkinetischen Theorie der Wärme geforderte Bewegung von in ruhenden Flüssigkeiten suspendierten Teilchen. Ann. Phys. **17**, 549 (1905). (in German)

Einstein, A.: Investigations on the Theory of the Brownian Movement. Menthuen Press, London (1926). (reprinted by Dover Press, New York)

Eringen, A.C.: Nonlocal Continuum Field Theories. Springer, New York (2002). ISBN 0-387-95275-6

Fiedler, B.H.: An integral closure model for vertical turbulent flux of a scalar in a mixed layer. J. Atm. Sci. **41**, 674 (1984)

Glöckle, W.G., Nonnenmacher, T.F.: Fractional integral operators and Fox functions in the theory of viscoelasticity. Macromolecules. **24**, 6426 (1991)

Guckenheimer, J., Holmes, P.: Nonlinear Oscillations, Dynamical Systems, and Bifurcations of Vector Fields, Applied Mathematics Sciences, vol. 42. Springer, New York (1983). ISBN 0-387-90819-6

Haken, H.: Synergetik: Eine Einführung. Springer, Berlin (1983). ISBN 3-540-12597-3 (in German)

Haken, H.: Advanced Synergetics: Instability, Hierarchies of Self-Organizing Systems and Devices. Springer, Berlin (1987). ISBN 3-540-12162-5

Hamba, F.: Nonlocal expression for scalar flux in turbulent shear flow. Phys. Fluids. **16**, 1493 (2004)

Hamba, F.: Nonlocal analysis of the Reynolds stress in turbulent shear flow. Phys. Fluids. **17**, 115102 (2005)

Hamba, F.: Exact transport equation for eddy diffusivity in turbulent shear flow. Theor. Comput. Fluid Dyn. **27**, 651 (2013)

Handler, R.A., Bernard, P.S., Rovelstad, A.L., Swearingen, J.D.: On the role of accelerating fluid particles in the generation of Reynolds stress. Phys. Fluids A. **4**(6), 1317 (1992)

Hinze, J.O.: Turbulence, 2nd edn. McGraw-Hill, New York (1975). ISBN 0-07-029037-7

Hinze, J.O., Sonnenberg, R.E., Builtjes, P.J.H.: Memory effect in a turbulent boundary-layer flow due to a relatively strong axial variation of the mean-velocity gradient. Appl. Sci. Res. **29**, 1 (1974)

Huang, Y.-N.: On modeling the Reynolds stress in the context of continuum mechanics. Commun. Nonlinear Sci. Numer. Simul. **9**, 543 (2004)

Hutter, K., Jöhnk, K.: Continuum Methods of Physical Modeling. Springer, Berlin (2004). ISBN 13-978-364-205-831-8

Kampé de Fériet: Sur le spectre de la turbulence homogène. J. Acad. Sci. Paris. **208**, 772 (1939). (in French)

Kraichnan, R.H.: The structure of isotropic turbulence at very high Reynolds numbers. J. Fluid Mech. **5**(4), 495 (1959)

Kraichnan, R.H.: The Closure Problem of Turbulent Theory. Research Report of Office of Naval Research (1961). No. HSN-3, USA, pp. 1–48

Kraichnan, R.H.: Direct-interaction approximation for shear and thermally driven turbulence. Phys. Fluids. **7**(7), 1048 (1964)

Lawson, J.M., Dawson, J.R.: On velocity gradient dynamics and turbulent structure. J. Fluid Mech. **780**, 60 (2015)

Lettau, H.: On eddy diffusion in shear zones. Geophys. Res. Pap. **19**, 437 (1952)

Liepmann, H.W.: Mécanique de la turbulence. C.N.R.S, Paris (1961). (in French)

Lumley, J.L.: Toward a turbulent constitutive relation. J. Fluid Mech. **41**, 413 (1970)

Mickens, R.E.: An Introduction to Nonlinear Oscillations. Cambridge University Press, Cambridge, UK (1981). ISBN 0-521-22208-9

Mompean, G., Jongen, T., Deville, M.O., Gatski, T.B.: On algebraic extra-stress models for the simulation of viscoelastic flows. J. Non-Newtonian Fluid Mech. **79**, 261 (1998)

Mompean, G., Qiu, X., Schmitt, F.G., Thompson, R.: Turbulence modeling based on non-Newtonian constitutive law. In: 13th European Turbulence Conference, Journal of Physics: Conference Series, vol. 318, p. 042030. IOP Publishing, Bristol (2011)

Monin, A.S., Yaglom, A.M.: Statistical Hydromechanics. Nauka, Moscow (1965)

Monin, A.S., Yaglom, A.M.: Statistical Fluid Mechanics: Mechanics of Turbulence, vol. I and II. MIT Press, Massachusetts (1971). ISBN 0-262-1-3062-9

Nieckele, A.O., Thompson, R.-L., Mompean, G.J.: Anisotropic Reynolds stress tensor representation in shear flows using DNS and experimental data. J. Turbul. **17**(6), 602 (2016)

Novikov, E.A.: Doklady Akad. Nauk. SSSR. **177**(2), 299 (1967). English translation: Kinetic equations for a vortex field. 1968, Soviet physics-Doklady **12**(11), 1006

Ohkitani, K., Kishiba, S.: Nonlocal nature of vortex stretching in an inviscid fluid. Phys. Fluids. **7** (2), 411 (1995)

Ott, E., Grebogi, C., Yorke, J.A.: Controlling chaos. Phys. Rev. Lett. **64**, 1196 (1990)

Picu, R.C.: On the functional form of non-local elasticity kernels. Mech. Phys. Solids. **50**, 1923 (2002)

Pope, S.B.: A more general effective-viscosity hypothesis. J. Fluid Mech. **72**(2), 331 (1975)

Pope, S.B.: Turbulent Flows. Cambridge University Press, Cambridge, UK (2000). ISBN 978-0-521-59125-6

Qiu, X., Mompean, G., Schmitt, F.G., Thompson, R.L.: Modeling turbulent-bounded flow using non-Newtonian viscometric functions. J. Turbul. **12**(15), 1 (2011)

Rabotnov, Y.N.: Elements of Hereditary Solid Mechanics. MIR, Moscow (1977)

Richardson, L.F.: The supply of energy from and to atmospheric eddies. Proc. Roy. Soc., Ser. A. **97**, 354 (1920)

Rhines, P.B., Holland, W.R.: A theoretical discussion of eddy-driven mean flows. Dyn. Atmos. Oceans. **3**, 289 (1979)

Rivlin, R.S.: The relation between the flow of non-Newtonian fluids and turbulent Newtonian fluids. Q. Appl. Math. **15**, 212 (1957)

Roberts, P.H.: Analytical theory of turbulent diffusion. J. Fluid Mech. **11**, 257 (1961)

Romanof, N.: Non-local models in turbulent diffusion. Z. Meteorol. **39**(2), 89 (1989)

Ruelle, D.: Chaotic Evolution and Strange Attractors, Accademia Nazional dei Lincei. Cambridge University Press, Cambridge, UK (1992). ISBN 0-521-36830-8

Saffman, P.G.: Application of the Wiener-Hermite expansion to the diffusion of a passive scalar in a homogeneous turbulent flow. Phys. Fluids. **12**, 1786 (1969)

Schmitt, F.G.: Direct test of a nonlinear constitutive equation for simple turbulent shear flows using DNS data. Commun. Nonlinear Sci. Numer. Simul. **12**, 1251 (2007)

Schmitt, F.G., Merci, B., Dick, E., Hirsch, C.: Direct investigation of the K-transport equation for a complex turbulent flow. J. Turbul. **4**, 21 (2003)

Schroeder, M.: Fractals, Chaos, Power Laws: Minutes from an Infinite Paradise. W.H. Freeman & Comp, New York (1991). ISBN 0-7167-2136-8

Schuster, H.G.: Deterministic Chaos: An Introduction, 2nd edn. Verlagsgesellschaft mbH (VGH), Weinheim (1988). ISBN 3-527-26862-6

Scott, B., Veinoglou, S.G.B.C., Daffyn, J.E.: Limitations of the Newtonian time scale in relation to non-equilibrium rheological states and a theory of quasi-properties. Proc. R. Soc. Lond. A. **189**, 69 (1947)

Spencer, A.J.M., Rivlin, R.S.: The theory of matrix polynomials and its application to the mechanics of isotropic continua. Arch. Ration. Mech. Anal. **2**, 309 (1959)

Spencer, A.J.M., Rivlin, R.S.: Further results in the theory of matrix polynomials. Arch. Ration. Mech. Anal. **4**, 214 (1960)

Speziale, C.G.: Invariance of turbulent closure models. Phys. Fluids. **22**(6), 1033 (1979)

Speziale, C.G.: On nonlinear k-l and k-ε models of turbulence. J. Fluid Mech. **178**, 459 (1987)

Strogaz, S.H.: Nonlinear Dynamics and Chaos: with Applications to Physics, Biology, Chemistry and Engineering, Studies in Nonlinearity. Westview, Cambridge, MA (1994). ISBN 13-978-0-7382-0453-6

Takayama, H.: Cooperative dynamics in complex physical systems. In: Proceedings of the Second Yukawa International Symposium, Kyoto (1989). 24–27 August, Reprint Edition Springer, Berlin. ISBN 3-540-508-65-1

Taylor, G.I.: The transport of vorticity and heat through fluids in turbulent motion. Proc. R. Soc. Lond. **135A**, 685 (1932)

Tsinober, A.: An Informal Conceptual Introduction to Turbulence. Springer, Cham (2009). ISBN 978-90-481-3174-7

West, B.J.: Colloquium: fractional calculus view of complexity: a tutorial. Rev. Mod. Phys. **86**, 1169 (2014)

Wilczek, M., Meneveau, C.: Pressure Hessian and viscous contributions to velocity gradient statistics based on Gaussian random fields. J. Fluid Mech. **756**, 191 (2014)

Zeilinger, A.: Dance of the Photons: From Einstein to Quantum Teleportation. MacMillan, New York (2010). ISBN 1-429-963-794

Zhu, J., Crawford, J.W., Palfreyman, J.W.: A general formulation of the reversible stress tensor for a nonlocal fluid. Int. J. Eng. Sci. **70**, 124 (2013)

# Chapter 7
# The Difference-Quotient Turbulence Model (DQTM)

## 7.1 The Discovery and Prandtl's Models

In Sect. 9.1, the turbulent shear stress will be determined for wake flow behind a cylinder (in the downstream or $x_1$-direction and its perpendicular, $x_2$-direction), by employing a momentum conservation consideration of the momentum in the $x_1$-direction. The result of this demonstration is (with a slightly different notation for the mean velocity $\overline{u}_1$ and $\overline{u}_{1_{\min}} = 0$) (see Hinze 1975),

$$\frac{\tau_{21}}{\rho U_G^2} = \frac{1}{2} \frac{x_2}{x_1 - p} \frac{\overline{u}_1}{U_G}. \tag{256}$$

The velocity $U_G$ denotes the basic stream velocity, with a rectangular averaged profile, of a turbulent flow approaching the cylinder from far upstream; $p$ is a pole distance (see also in Sect. 9.1). Equation (256) is a result, which is based only on fundamental laws of physics, mainly mass conservation and Newton's second law. Therefore, it is a cornerstone of turbulence modeling. Only turbulence models containing this important result (e.g., as a special case) can withstand critical scrutiny. To the authors' best knowledge, up-to-present only the DQTM, to be introduced in this section, fulfills this requirement.

The general acceptance of Boussinesq's constitutive equation and Prandtl's mixing length model were so strong that Eq. (256) was adapted to these findings. By forcing Eq. (54) to apply, a combination of this equation with (256) leads to the compulsory eddy diffusivity) (see Hinze 1975[1])

The original version of this chapter was revised. A correction to this chapter can be found at https://doi.org/10.1007/978-3-030-26033-0_14

[1]This obdurate defense of Boussinesq's Ansatz was at the origin of the discovery of the DQTM.



P. W. Egolf, K. Hutter, *Nonlinear, Nonlocal and Fractional Turbulence*,
https://doi.org/10.1007/978-3-030-26033-0_7

$$\varepsilon_{\mathrm{m}}=\frac{1}{2}\frac{x_2}{x_1-p}U_G\frac{\overline{u}_1}{\frac{\partial\overline{u}_1}{\partial x_2}}. \tag{257}$$

By substituting this eddy diffusivity into Eq. (53a), the turbulent shear stress presents itself in an artificially blown-up gradient-law form

$$\tau_{21}=\frac{1}{2}\rho\frac{x_2}{x_1-p}U_G\frac{\overline{u}_1}{\frac{\partial\overline{u}_1}{\partial x_2}}\frac{\partial\overline{u}_1}{\partial x_2}. \tag{258}$$

At this stage, in April 1985, Egolf was somewhat sceptical to take this particular result for granted and to accept it as a basis for further investigation. Favoring simplicity of physical laws, led to a rejection of Eq. (258) in favor of its simpler version (Eq. 256). The idea was to replace the most common and conventional path of a linear and local law by a nonlinear and nonlocal parameterization (see Egolf and Hutter 2016a). Thus, the question arose how Eq. (256) could be generalized to serve for the solution of other (also more complex) turbulent flow problems. Details of these efforts shall not be reported here. Taking symmetries, Galilean invariance, etc. (see also Eringen 2002) into consideration, led to the following final version, which was first published in a small note (see Egolf 1991) and 3 years later more detailed in Egolf (1994)

$$\tau_{21}=-\rho\overline{u_2'u_1'}=\rho\chi_2\frac{\mathrm{d}b}{\mathrm{d}x_1}\left(\overline{u}_1-\overline{u}_{1_{\min}}\right)\frac{\overline{u}_{1_{\max}}-\overline{u}_1}{x_{2_{\max}}-x_2}, \tag{259}$$

in which $\chi_2\in\left\{x_2-x_{2_{\min}},x_{2_{\max}}-x_2,x_{2_{\max}}-x_{2_{\min}}\right\}$ is a variable or characteristic overall length typical of the flow, perpendicular to the main flow direction (and can in many cases be determined by the geometry of the flow problem with its symmetries). Moreover, in expression (258), the gradient $\partial\overline{u}_1/\partial x_2$ is replaced by the mean velocity difference quotient, $(\overline{u}_{1_{\max}}-\overline{u}_1)/(x_{2_{\max}}-x_2)$, where the remaining terms compose together a quantity with the dimension of a viscosity. The quantities $\overline{u}_{1_{\min}}$ and $\overline{u}_{1_{\max}}$ are the flow-parallel minimum and maximum average velocities and $x_{2_{\min}}$ and $x_{2_{\max}}$ are the corresponding orthogonal flow positions within the cross sections, where the velocity assumes its minimum and maximum values of the mean velocity $\overline{u}_1$, respectively. Finally, $b$ is the (half) length or (half) size of the turbulent region in the cross section.

Note that this rather unusual model contains a difference quotient that gave it its name, it does not contain the eddy diffusivity concept with its "gradient" law (see Sect. 7.3)!

Three years after the introduction of the DQTM analogue features of Prandtl's shear layer model with the DQTM were demonstrated (see Egolf 1994).[2] By multiplying and dividing the term in Eq. (71) by the width $b$, one obtains

[2]It will be demonstrated that Prandtl's mixing-length model and Prandtl's shear-layer model are special cases of the DQTM.

$$\tau_{21} = \rho bc \underbrace{\frac{\partial \overline{u}_1}{\partial x_2} b}_{\left(\approx \overline{u}_{1_{\max}} - \overline{u}_{1_{\min}}\right)} \times \underbrace{\frac{\overline{u}_{1_{\max}} - \overline{u}_{1_{\min}}}{b}}_{\substack{\text{(Difference quotient} \\ \text{term (constant))}}} . \tag{260}$$

In elementary flow problems, the positions, where $\overline{u}_{1_{\min}}$ and $\overline{u}_{1_{\max}}$ arise, are usually just separated by the half width $b$ of the entire turbulent domain. Therefore, in this model a very stiff (constant) difference quotient appears as second fraction term. Furthermore, by demanding nonlocality in the eddy diffusivity, in Eq. (260) the remaining "gradient times $b$" is replaced by the maximum velocity difference $\overline{u}_{1_{\max}} - \overline{u}_{1_{\min}}$. So, by this relatively small change, the mean velocity difference model can be altered to take the form

$$\tau_{21} = \rho bc(\overline{u}_{1_{\max}} - \overline{u}_{1_{\min}}) \frac{\overline{u}_{1_{\max}} - \overline{u}_{1_{\min}}}{x_{2_{\max}} - x_{2_{\min}}}, \tag{261}$$

where in the denominator of Eq. (261) $b$ is replaced by

$$b = x_{2_{\max}} - x_{2_{\min}} . \tag{262}$$

This alteration, leading actually only to a constant closure, is close to the final form of the DQTM. However, with it Eq. (261) cannot serve as a useful and serious turbulence model. By comparing Eq. (261) with Eq. (259), one may identify the following two corresponding relations:

$$\chi_2 = b, \qquad c = \frac{\mathrm{d}b}{\mathrm{d}x_1} . \tag{263a, b}$$

Furthermore, one sees that in Eq. (261) two indices, *min* and one of the two indices *max,* ought to be omitted to obtain the DQTM given by Eq. (259).

If, in the new model, the intention is to stick to something close to the eddy diffusivity concept, a difference quotient is introduced to replace the gradient. Thus, one has the following shear stress formula with a kind of "proof" in Egolf (1998)

$$\tau_{21} = \rho \varepsilon_{\mathrm{m}} \frac{\overline{u}_{1_{\max}} - \overline{u}_1}{x_{2_{\max}} - x_2}, \tag{264}$$

with the eddy diffusivity

$$\varepsilon_{\mathrm{m}} = \chi_2 \frac{\mathrm{d}b}{\mathrm{d}x_1} (\overline{u}_1 - \overline{u}_{1_{\min}}) . \tag{265}$$

Notice that Prandtl did not make the step from Eq. (260) to Eq. (261) by consequently replacing both gradients and leaving the variable character in the eddy diffusivity and the difference quotient. Here, our intention is simply to demonstrate that only three further steps, namely:

1. An exchange of the role of the eddy diffusivity with that of the gradient.
2. A substitution of $b$ by $\chi_2$.
3. An omission of three indices (min, max) in the coordinate $x_{2_{\max}}$ and the averaged velocity components $\overline{u}_{1_{\min}}$ and $\overline{u}_{1_{\max}}$

would have led Prandtl to develop the DQTM more than 40 years before the DQTM was proposed. It is intriguing to realize how Prandtl's research efforts on turbulent momentum transfer in his late efforts (Prandtl 1925, 1942) point exactly in the direction of the DQTM and even approach it closely.

Finally, it is demonstrated that the DQTM has no deficiencies that occur in the older models as follows:

1. The DQTM is not a one-point closure with all the insufficiencies given by locality.
2. At space locations where the average velocity profile has an extremum (minimum or maximum), the eddy diffusivity (Eq. 265)—different from Prandtl's vanishing expression (54)—reaches its maximum value [to see this, substitute, e.g., the velocity $\overline{u}_1 = \overline{u}_{1_{\max}}$ into Eq. (265)].
3. The maximum "mixing length" is given by Eq. (263a). It is not of differential size, but instead a large-scale quantity, which may reach half the size of the overall flow regime.
4. In the DQTM, at the boundaries of the turbulent mixing zone, one has $\overline{u}_1 = \overline{u}_{1_{\min}}$, and with Eq. (265) this causes a vanishing eddy diffusivity $\varepsilon_m = 0$.
5. There are no problems of divergence at inflection points of the average velocity profile.
6. The space and mean velocity differences in the nonlocal DQTM guarantee that Galilean invariance is satisfied. On the other hand, Euclidian objectivity of this model remains a topic of further investigation and discussion.
7. As far as elementary turbulent flow problems were solved with the DQTM, no mathematical difficulties were encountered (see, e.g., Chap. 9).

A very important final remark to the development of the DQTM must be made already at this point. In this model, only one or two of the three scales $x_2 - x_{2_{\min}}$, $x_{2_{\max}} - x_{2_{\min}}$, $x_{2_{\max}} - x_2$ occur which at most locations is of non-differential size. So, it looks as if at a given location only eddies of two different sizes are active. However, this is not correct, because, by scaling, the net effect of eddies from all classes (the entire wavelength spectrum) is inherently present in the description of only two large-scale eddies (for details see Sects. 7.2.2 and 7.2.4).

## 7.2 Momentum Transfer Approach

### 7.2.1 Molecular Transport

The deformation of a fluid particle in a laminar flow can be successfully treated just as that of an elastic body in a solid (see, e.g., Landau and Lifshitz 1974). In an isotropic solid, the shear modulus $G$ is defined by (see Fig. 7.1)

$$G = \frac{\tau_{21}}{\gamma_{21}}, \tag{266}$$

where $\tau_{21}$ denotes the shear stress and $\gamma_{21}$ the shear angle, which is directly related to the strain by the following relation:

$$\tau_{21} = \frac{\delta F_1}{\Delta A_2}, \qquad \gamma_{21} \cong \tan(\gamma_{21}) = \frac{\delta x_1}{\Delta x_2}. \tag{267a – c}$$

In these equations $\delta F_1$ is the force increment in the $x_1$-direction, $\Delta A_2$ the surface element with its normal vector $n_2$ in the $x_2$-direction, which defines its index. The increment $\Delta x_1$ is the width of the cuboid and $\Delta x_2$ its height, whereas $\delta x_1$ denotes the displacement under forcing or stress, respectively, in the $x_1$-direction. In the model the height difference $\delta x_2$ occurring during deformation is neglected (see in Fig. 7.2).

In quantities as those of Eqs. (266) and (267a–c), the first index gives the direction of the normal vector of the surface element and the second index the direction of the force leading to the shear stress. The initial element is described by length differences $\Delta x_1$, $\Delta x_2$. They are usually at least one order larger than the displacements that are denoted by $\delta' s$, e.g., $\delta x_1$ and $\delta x_2$. However, in their limits both are actually "fluid dynamic differentials," applied in the sense of differential calculus.

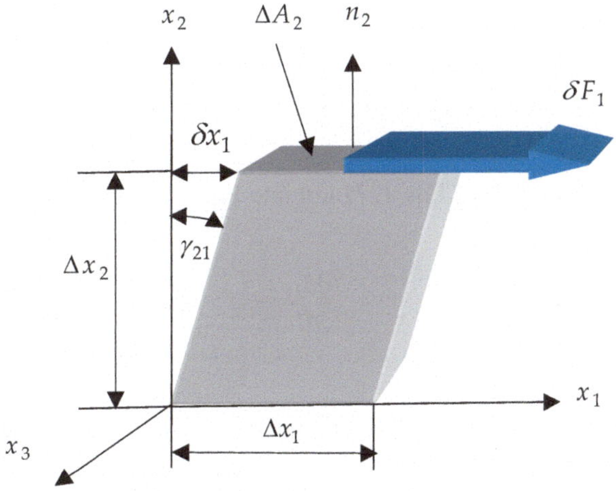


**Fig. 7.1** Static deformation of a cuboid elastic body of volume $\Delta V$. (Courtesy of Egolf and Hutter (2016b), © Springer Edition)

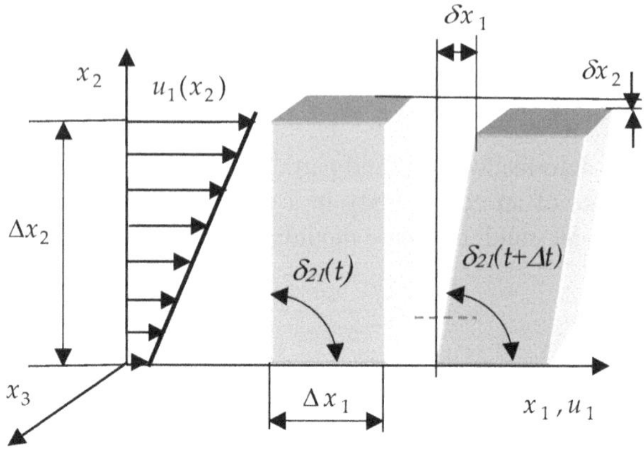


**Fig. 7.2** Dynamic deformation, described by the angle $\delta_{21}(t)$ of a cuboid fluid element between time $t$ and $t + \Delta t$. (Courtesy of Egolf and Hutter (2016b), © Springer Edition)

Combining Eqs. (266) and (267a,c), the shear modulus takes the form

$$G = \frac{\delta F_1 \Delta x_2}{\Delta A_2 \delta x_1}. \tag{268}$$

In the dynamic deformation of a laminar flowing fluid element, the rate of shear angle is important (see Fig. 7.2)

$$\frac{\mathrm{d}\gamma_{21}}{\mathrm{d}t} = -\frac{\mathrm{d}\delta_{21}}{\mathrm{d}t}. \tag{269}$$

From the same figure, it can be deduced that

$$\frac{\mathrm{d}\delta_{21}}{\mathrm{d}t} = \lim_{\Delta t \to 0} \frac{\delta_{21}(t+\Delta t) - \delta_{21}(t)}{\Delta t}, \tag{270}$$

with the angles

$$\delta_{21}(t) = \frac{\pi}{2}, \quad \delta_{21}(t+\Delta t) = \frac{\pi}{2} - \arctan\left(\frac{\delta x_1}{\Delta x_2}\right). \tag{271a, b}$$

Substitution of Eqs. (271a,b) into Eq. (270) leads to the following relation:

$$\frac{\mathrm{d}\delta_{21}}{dt} = -\lim_{\substack{\Delta t \to 0 \\ \Delta x_2 \to 0}} \frac{\arctan\left(\frac{\delta x_1}{\Delta x_2}\right)}{\Delta t} \tag{272}$$

in which "arctan" is the inverse function of "tan." These limits are of utmost importance when studying turbulent flows in detail (see in Sect. 7.2.2). Notice that with $\Delta t \rightarrow 0$, for a bounded velocity ($|u_1| < \infty$), it follows that the differential increment $\delta x_1$ simultaneously tends to zero. In this limit process, it is required that the third quantity, $\Delta x_2$, also approaches zero, but this is independent of the limits of the previous two quantities.
The displacement increment $\delta x_1$ can be described by the $x_1$ component of the flow velocity field, $u_1$,

$$\delta x_1 = [u_1(x_2 + \Delta x_2) - u_1(x_2)]\,\Delta t. \tag{273}$$

Now, combining Eqs. (272) and (273) yields

$$\begin{aligned}\frac{\mathrm{d}\delta_{21}}{\mathrm{d}t} &= - \lim_{\substack{\Delta t \rightarrow 0 \\ \Delta x_2 \rightarrow 0}} \left[\frac{1}{\Delta t}\arctan\left(\frac{[u_1(x_2 + \Delta x_2) - u_1(x_2)]\Delta t}{\Delta x_2}\right)\right] \\ &= - \lim_{\Delta x_2 \rightarrow 0} \frac{u_1(x_2 + \Delta x_2) - u_1(x_2)}{\Delta x_2}.\end{aligned} \tag{274a, b}$$

Because the argument (arg) of the "arctan"-function is small for small deformations, it can be replaced by "tan(arg)," so that "arctan(tan(arg))=arg," which explains Eq. (274b). Consequently, it follows that

$$\frac{\mathrm{d}\delta_{21}}{\mathrm{d}t} = -\frac{\mathrm{d}u_1}{\mathrm{d}x_2}. \tag{275}$$

The viscosity is defined as the ratio of the shear stress to the rate of shear strain

$$\mu = \frac{\tau_{21}}{\frac{\mathrm{d}\gamma_{21}}{\mathrm{d}t}} = -\frac{\tau_{21}}{\frac{\mathrm{d}\delta_{21}}{\mathrm{d}t}} = \frac{\tau_{21}}{\frac{\mathrm{d}u_1}{\mathrm{d}x_2}}, \quad \nu = \frac{\mu}{\rho}, \tag{276a – d}$$

where Eqs. (269) and (275) were applied. By a small rearrangement, the well-known form of Newton's shear law is obtained. This is the constitutive equation of laminar flow

$$\tau_{21} = \mu\frac{\mathrm{d}u_1}{\mathrm{d}x_2}. \tag{277}$$

The three-dimensional generalization is called Stokes' and Navier's shear stress formula; it is the second-order tensor relation

$$\tau_{ij} = \mu\left(\frac{\partial u_i}{\partial x_j} + \frac{\partial u_j}{\partial x_i}\right), \quad i,j \in \{1,2,3\}, i \neq j. \tag{278}$$

Our goal is to figure out the highest possible analogy between the laminar and turbulent shear stresses. However, the above-presented macroscopic model does not yield all available information. More details are hidden in the physical quantity "viscosity"; and this is revealed by a statistical microscopic description, the kinetic theory of gases (see, e.g., Becker 1978; Reif 1975; Truesdell and Muncaster 1980).

In a fluid, the normal stress components are identical to the pressure $p = p_{11} = p_{22} = p_{33}$ in the bulk. If the fluid has a nonconstant velocity profile and a nonzero viscosity, then also shear stresses will occur (Fig. 7.3). A formula for this shear stress can be derived as explained in the kinetic theory of gases by studying the behavior of molecules jiggling in the fluid. In a model assumption, the momentum transfer is governed by the molecules moving in all directions. For $N$ molecules per cube, the number of molecules transporting momentum across the plane $\delta\Gamma$ with a unit surface $\Delta A_2$ per unit of time $\Delta t$ is

$$\widetilde{n} = \frac{1}{6}\frac{\Delta^2 N}{\Delta t\, \Delta A_2}, \tag{279}$$

because the cube has six side faces. Thus, of the two expressions,

$$\frac{\Delta^2 N}{\Delta t \Delta A_2} \quad \text{and} \quad \frac{\partial^2 N}{\partial t \partial A_2}, \tag{280a, b}$$

the first difference quotient (Eq. 280a) is the finite analogue of the limiting differential quotient, which is shown as Eq. (280b).

In three-dimensional flows under isotropic fluctuations only one-third of all the molecules are flying in the vertical direction. If also the flight directions to the top

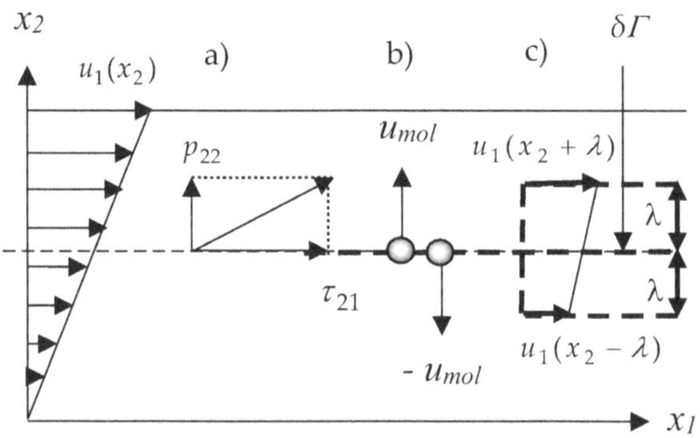


**Fig. 7.3** A non-zero directional derivative of the velocity profile leads to a momentum transfer and this is generated by a shear stress (**a**). The momentum transport by vertically moving molecules across the plane $\delta\Gamma$ is modeled as indicated by the geometrical situation shown in panels (**b**) and (**c**)

and to the bottom are distinguished then this factor is reduced by another factor 1/2, which explains the factor 1/6 occurring in Eq. (279). A more sophisticated derivation of this factor is found in Feynman et al. (1973). In their presentation the vertical velocity of molecules is the root mean square (rms) velocity of the molecules' velocity $u_{\mathrm{mol}}$ (compare with Eq. 281e). It is assumed that the flights extend over the distance $\Delta x_2$, identical to the mean free path length $\lambda$ (following a convention in the physics literature, we omit the sign for the average). Both quantities are assumed to be positive

$$\Delta x_2 = \lambda > 0, \qquad u_{\mathrm{mol}}^{(\mathrm{rms})} = \frac{\Delta x_2}{\Delta t} > 0, \qquad u_{\mathrm{mol}}^{(\mathrm{rms})} = \sqrt{\overline{u_{\mathrm{mol}}{}^2}}. \tag{281a – e}$$

Inserting Eq. (281c) into Eq. (279) leads to

$$\begin{aligned} \widetilde{n} &= \frac{1}{6}\frac{u_{\mathrm{mol}}^{(\mathrm{rms})}\Delta^2 N}{\Delta x_2 \Delta A_2} = \frac{1}{6}\frac{\Delta N}{\Delta V} u_{\mathrm{mol}}^{(\mathrm{rms})} = \frac{1}{6} n\, u_{\mathrm{mol}}^{(\mathrm{rms})}, \\ \Delta V &= \Delta A_2 \Delta x_2, \quad n = \frac{\Delta N}{\Delta V}. \end{aligned} \tag{282a – e}$$

These molecules, possessing their own small momenta, additionally transport fluid momentum from the low and high momentum regions to and across the plane $\delta\Gamma$. It is assumed that this occurs over a distance of the mean free path $\lambda$, defined as the mean distance travelled by a molecule between two consecutive encounters. The momentum transport by this normalized number of particles, $\widetilde{n}$, is by definition identical to the shear stress

$$\tau_{21} = \widetilde{n}\left[p_{\mathrm{mol}}(x_2 + \lambda) - p_{\mathrm{mol}}(x_2 - \lambda)\right], \quad p_{\mathrm{mol}}(x_2) = m_{\mathrm{mol}}\, u_1(x_2). \tag{283a, b}$$

Substitution of Eqs. (282c) and (283b) into Eq. (283a), leads to

$$\tau_{21} = \frac{1}{6} n\, m_{\mathrm{mol}}\, u_{\mathrm{mol}}^{(\mathrm{rms})}\left[u_1(x_2 + \lambda) - u_1(x_2 - \lambda)\right], \tag{284}$$

which equals for later use in the turbulence modeling case (see below)

$$\tau_{21} = \frac{1}{6} n\, m_{\mathrm{mol}}\, u_{\mathrm{mol}}^{(\mathrm{rms})}\left\{ \left[u_1(x_2 + \lambda) - u_1(x_2)\right] - \left[u_1(x_2 - \lambda) - u_1(x_2)\right] \right\}, \tag{285}$$

where in the transition from Eq. (284) to (285) $u_1(x_2)$ was added and subtracted. This is in agreement with Galilean invariance that demands that the shear stress does not depend on the applied inertial system. Performing a linear Taylor approximation

$$u_1(x_2 \pm \lambda) = u_1(x_2) \pm \frac{du_1}{dx_2}\lambda \tag{286}$$

and inserting these two expressions into Eq. (285) and comparing the resulting equation with Eq. (277), the shear stress formula with the dynamic viscosity $\mu$ is found as

$$\tau_{21} = \mu \frac{du_1}{dx_2}, \qquad \mu = \frac{1}{3} n\, m_{\text{mol}}\, \lambda\, u_{\text{mol}}^{(\text{rms})}, \tag{287a, b}$$

in which the product of the specific number of molecules $n$ with the mass of a molecule $m_{\text{mol}}$ is identical to the density $\rho$ of the fluid, viz.,

$$\rho = n m_{\text{mol}}. \tag{288}$$

Thus, the microscopic description of the shear stress of laminar flows takes the form

$$\tau_{21} = \sigma\, \rho\, \lambda\, u_{\text{mol}}^{(\text{rms})} \frac{du_1}{dx_2}, \qquad \sigma = \frac{1}{3}. \tag{289a, b}$$

For laminar flows, this model is very accurate and the gradient law has over and over been demonstrated to be adequate to describe the "driving force" for the molecular momentum transport processes.

Reif (1975) gives a simple, but very illustrative macroscopic analogy of the molecular momentum transport in a fluid. He looks at two compositions of train cars moving with constant, but different velocities on parallel rails in the same direction without a forcing by locomotives. On the cars of the trains, workmen are standing and they are throwing sandbags onto the neighboring train. By this, the higher velocity train is *de*celerated, and the lower velocity train of equal mass is correspondingly *ac*celerated. In the analogy, two adjacent fluid layers are compared with two parallel moving compositions of freight cars and the atoms or molecules, moving transversely to the main direction of the flow, play the role of the flying sandbags.

This molecular theory was reviewed in detail because it contains numerous elements that remain valid in the description of momentum exchange by turbulent eddies. This example will be introduced as our analogy in the next section with emphasis on the generalizations, which are important in understanding more complex memory-dependent and nonlocal transport phenomena as they occur in turbulent flows.

### 7.2.2 Transport by Eddies

The mean deformation of a turbulent flowing fluid with mean shear is analogous to that of a laminar flow derived in Sect. 7.2.1. However, still some important deviations shall be outlined. In this more complex case, Eq. (272)—with two modifications—remains valid

$$\frac{d\delta_{T_{21}}}{dt} = - \lim_{\substack{\Delta t \to 0 \\ \Delta x_2 \to \max}} \frac{\arctan\left(\dfrac{\delta \bar{x}_1}{\Delta x_2}\right)}{\Delta t} \tag{290}$$

Because the fluid motion fluctuates, the average deformation $\delta\bar{x}_1$ of a fluid element is considered. The velocities are also replaced by their averages (compare also with the Reynolds decomposition procedure, presented in Chap. 2). An important difference compared to Eq. (272) is that the third limit $\Delta x_2$ in the DQTM is chosen to tend instead to zero toward its maximum value. Indeed, there is no reason to demand this limit to go toward zero, if the theory shall not be of differential type. This is the subtle idea that opens the door for the derivation of a new nonlocal description of the turbulent shear stress! The averaged Eq. (273) remains valid and Eq. (274a), equipped with the index $T$ (turbulent), and Eqs. (274a,b), both upgraded with average signs, become

$$\begin{aligned} \frac{d\delta_{T_{21}}}{dt} &= - \lim_{\substack{\Delta t \to 0 \\ \Delta x_2 \to \max}} \left[\frac{1}{\Delta t}\arctan\left(\frac{[\bar{u}_1(x_2+\Delta x_2) - \bar{u}_1(x_2)]\,\Delta t}{\Delta x_2}\right)\right] \\ &\cong -\frac{\bar{u}_1(x_2+\Delta x_2) - \bar{u}_1(x_2)}{\Delta x_2} = -\frac{\Delta \bar{u}_1}{\Delta x_2} = -\frac{\bar{u}_{1_{\max}} - \bar{u}_1}{x_{2_{\max}} - x_2}. \end{aligned} \tag{291a – d}$$

In this equation, there is neither any reason why the spatial difference $\Delta x_2$ and the mean velocity difference $\delta\bar{u}_1$ may not be large-scale quantities! The limit $\Delta t \to 0$ causes $\Delta x_1$ also to be small for even very large $\delta\bar{u}_1$. It is only demanded that this velocity difference is limited. Because in the description of turbulent flows, a dominant effect of large-scale eddies occurs (see Sect. 7.2.3), it is here decided to let $\Delta x_2$ tend toward its maximum value instead toward its minimum (zero).

Plane turbulent Couette flows, to be derived in Sect. 9.3, yield the basic and most instructive shear situation. In this elementary flow type, the position of $x_{2_{\max}}$ is at the top boundary, where the maximum velocity $\bar{u}_{1_{\max}}$ occurs (see Fig. 9.25). The reason for the appropriateness of this procedure is the underlying self-similarity. Now, it follows that

$$\frac{d\delta_{T_{21}}}{dt} = -\frac{\overline{u}_1(x_{2_{max}}) - \overline{u}_1(x_2)}{x_{2_{max}} - x_2} = -\frac{\overline{u}_{1_{max}} - \overline{u}_1(x_2)}{x_{2_{max}} - x_2}, \qquad (292a-d)$$
$$\overline{u}_1(x_{2_{min}}) = \overline{u}_{1_{min}}, \quad \overline{u}_1(x_{2_{max}}) = \overline{u}_{1_{max}}.$$

We repeat that $x_{2_{max}}$ is the $x_2$ position where $\overline{u}_1$ is at its maximum (292d). Later on, definition (292c), for the minimum quantity, will be also required. In analogy to the laminar case, the turbulent dynamic viscosity is the ratio of turbulent shear stress to the rate of turbulent shear strain

$$\mu_T = \frac{\tau_{21}}{\frac{d\gamma_{T_{21}}}{dt}} = -\frac{\tau_{21}}{\frac{d\delta_{T_{21}}}{dt}} = \frac{\tau_{21}}{\frac{\overline{u}_{1_{max}} - \overline{u}_1(x_2)}{x_{2_{max}} - x_2}}, \quad \nu_T = \frac{\mu_T}{\rho}, \qquad (293a-d)$$

where $\nu_T$ denotes the effective turbulent kinematic viscosity. From Eq. (293c), it follows that

$$\tau_{21} = \mu_T \frac{\overline{u}_{1_{max}} - \overline{u}_1(x_2)}{x_{2_{max}} - x_2}. \qquad (294)$$

The term on the right-hand side is the replacement proposed in lieu of the gradient occurring in Boussinesq's approximation. It is (at most locations) a large-scale difference quotient.

We assume that the turbulent momentum transport is analogous to the momentum transport described by the kinetic theory of a monatomic gas. Then the question that remains is only to ask which modifications or generalizations are necessary to obtain the right closure condition. Again, we are interested in developing an underlying theoretical model that gives us here, instead of the dynamic viscosity $\mu$, the effective turbulent dynamic viscosity $\mu_T$ in terms of the density of the fluid, a length scale and a characteristic velocity. Naturally, the moving objects that transport turbulent (fluctuating) momentum are not microscopic molecules, but an infinite number of classes of eddies with different sizes instead.

In the following it is assumed, Egolf (2009), that, because of the fractal nature of turbulence and the application of self-similarity and scaling laws, the effect of an infinite number of classes of eddies of different sizes can mathematically be described just by a single class of largest eddies. For readers not acquainted with the concept of self-similarity, we provide here a simple and illustrative explanation of convincing analogy.

Thus, consider one very strong workman, who performs one work unit, $W_0$, per unit of time, and two smaller workers, who each perform one half of a working unit, $W_1 = W_0/2$, in the same time. Then two small workers together show the same performance as the single big counterpart. Now, it follows that the total work of the three workmen is equal to twice the work of the strongest workman, $W_{tot} = 2W_0$. In a description of the action of a multitude of workers with scaling properties (see Eq. 295), in the end, we can only refer to the data of the strongest, respectively, largest, workman (see also Fig. 7.4). However, this is only possible if the scaling law

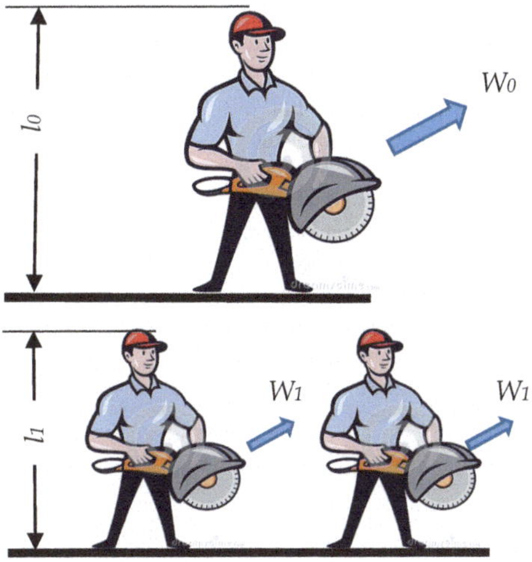


**Fig. 7.4** A workman performs equal work as two smaller ones. It is assumed that the performance of the workers is proportional to the third power of their height or, equivalently, proportional to their volume

for all classes with different members (of different sizes) is known. In our case, the work, $W_n$, of the different classes, $n$, is chosen to be related to the size of the workers, $l_n$, by the following power law:

$$W_n = W_0 \left( \frac{l_n}{l_0} \right)^{\alpha}. \tag{295}$$

The result $W_0 = 2W_1$, e.g., is given by $l_0 = 2.10$ m, $l_1 = 1.67$ m, and the scaling exponent $\alpha = 3$.

In a model analogy with turbulent flows, such a worker must be compared with a single eddy in a turbulent fluid. Different in this analogy is that eddies usually show a very large number of different classes or generations of eddies. For infinite Reynolds number flows, this number is even infinite.

Furthermore, eddies do not perform work, but instead obtain energy from and lose energy to larger size and smaller size eddies, which are phenomena that correspond to in- and outflowing energy fluxes, obeying energy conservation laws. It is possible to extend the analogy and to further introduce a work lifetime of workers (decay rate) and an occupation rate, which is the number of workers per unit volume (a number density), etc. It is clear that scaling laws can also be established for these quantities bringing thereby different classes of workers in relation to one another.

With the general idea of self-similarity being explained, we now address the description of interacting eddies. To this end, consider a population of eddies of many classes; of these we assume to know specific qualifications of the largest class (e.g., size (diameter), birth rate, death rate, lifetime, space occupation rate (density), linear (and eventually) angular momentum, turbulent kinetic energy, etc.) (see Fig. 7.5). Physical laws allow us to establish scaling laws, so that we gain knowledge of these quantities for the classes with eddies of smaller size. In this case, one can

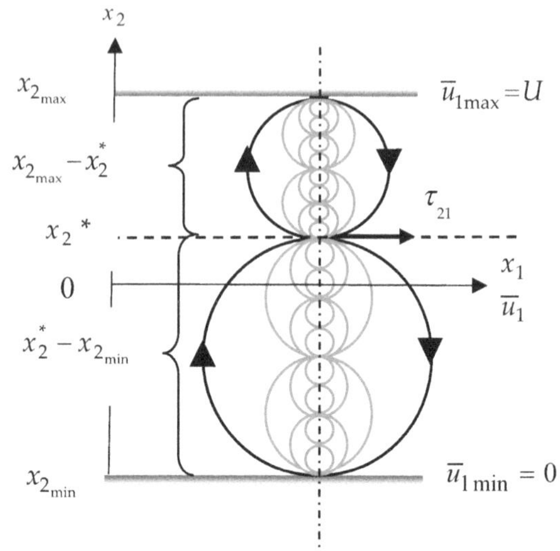


**Fig. 7.5** Turbulent momentum is transported by eddies of numerous sizes (also by grey eddies), but mathematically the transfer can be represented by an exchange of only two types of (black) eddies per vertical space location $x_2^*$. After Kolmogorov's simplifying idea, eddies of a succeeding class have exactly half a diameter of those of their predecessors. In this figure, only four classes are drawn

mathematically describe the actions of an infinite set of different eddies by the characteristic values of the class with eddies of largest size only!

Such fractal modeling is successful and yields a basis to physically interpret the DQTM. For a more detailed description see Sect. 7.2.4 or/and consult Egolf (2009). Finally, it follows that—based on Lévy flights, which describe clustering phenomena, fractal geometry, and intermittency—a one-dimensional theory for turbulent momentum transport exists for shear flows with fractal eddies. Fractional momentum transport is described by scaling laws, and this explains why the class of eddies of largest size suffices, in the framework of the DQTM, to yield a complete picture of turbulent momentum transport by all occurring eddies belonging to the different eddy classes.

Based on these briefly outlined ideas, it is now assumed that, just as presented in Fig. 7.5, the transport of momentum can be described by only two classes of different eddies. This is in agreement with the procedure outlined by Eq. (290) where the processes tending toward the maximum, instead to the minimum scale as in the laminar case, lead to the correct result. Then, in analogy to the main results of the kinetic theory of gases (see Eq. 284), it is concluded that

$$\tau_{21} = \frac{1}{6}\left(n_{\max}\,\overline{u}^*{}_{2_{\min}}\overline{p}_{1_{\max}} - n_{\min}\,\overline{u}^*{}_{2_{\max}}\,\overline{p}_{1_{\min}}\right), \tag{296}$$

where $n_{\min}$ and $n_{\max}$ are the eddy densities of the two classes of black eddies shown in Fig. 7.5. The excess momenta $\overline{p}_{1_{\min}}$ and $\overline{p}_{1_{\max}}$ are defined below. Notice that the main difference to the kinetic theory of monatomic gases is that one does not distinguish between the velocity of underlying entities (molecules) and the velocity field of the fluid, because the movement of eddies inherently leads to the velocity field! This equation requires some further explanation.

At first, one may expect the same indices with three "min" and three "max" each in the multiplicative terms. To demonstrate that the alternative conditions "max"–"min"–"max" and "min"–"max"–"min" are appropriate, let us see in Fig. 7.5, e.g., look at the lower boundary $x_2 = x_{2_{\min}}$. Here, the largest eddies created between the plates, which enter from the near fluid region of the upper plate with maximum momentum excess, are acting. However, by hindrance the vertical fluctuation velocity must diminish toward the lower wall. This important feature can only be described by $\overline{u}^*_{2_{\min}}$! The analogous argument for the situation at the upper plate guides us to select $\overline{u}^*_{2_{\max}}$ for this second location. The large fluctuation velocities [with our own not so strict notation in, e.g., Eq. (296)] are positive quantities, defined by the root-mean square velocities,

$$u_{\alpha\beta} = \overline{u}_{\alpha\beta} + u'_{\alpha\beta} \Rightarrow \overline{u}^*{}_{\alpha\beta} = \sqrt{\overline{\left(u'_{\alpha\beta}\right)^2}}, \quad \alpha \in \{1,2\}, \beta \in \{\min, \max\}. \quad (297\text{a, b})$$

The next steps will only be roughly explained, because they are identical to those given by Taylor and Prandtl. The first argument is the proportionality of the fluctuation quantities in the two space directions

$$\overline{u}^*_2 = \text{const} \times \overline{u}^*_1. \quad (298)$$

In homogeneous turbulence they are anyway identical. In shear flows, Eq. (298) is also a very good approximation to the occurring fluctuation phenomenon.

Now, we describe the largest fluctuation velocities by the velocity differences of the mean flow field

$$\overline{u}^*_{1_{\min}} = \overline{u}_1 - \overline{u}_{1_{\min}}, \; \overline{u}^*_{1_{\max}} = \overline{u}_{1_{\max}} - \overline{u}_1. \quad (299\text{a, b})$$

With Eqs. (298) and (299a,b), for the turbulent shear stress (Eq. 296), it follows that

$$\tau_{21} = \frac{\text{const}}{6}\left[n_{\max}\left(\overline{u}_1 - \overline{u}_{1_{\min}}\right)\overline{p}_{1_{\max}} - n_{\min}\left(\overline{u}_{1_{\max}} - \overline{u}_1\right)\overline{p}_{1_{\min}}\right]. \quad (300)$$

The excess momenta are also large size quantities

$$\overline{p}_{1_{\min}} = m_{\text{eddy, min}}\left(\overline{u}_{1_{\min}} - \overline{u}_1\right), \qquad \overline{p}_{1_{\max}} = m_{\text{eddy, max}}\left(\overline{u}_{1_{\max}} - \overline{u}_1\right), \quad (301\text{a, b})$$

where $m_{\text{eddy, min}}$ and $m_{\text{eddy, max}}$ denote each the mass of the corresponding eddy. It is assumed that

$$\rho = n_{\min} m_{\text{eddy, min}} = n_{\max} m_{\text{eddy, max}}. \quad (302\text{a, b})$$

Substituting Eqs. (301a,b) and (302a,b) into Eq. (300), the two terms become identical so that in the next formula the factor 1/6 is also replaced by 1/3, as it was replaced in Eq. (287b) by another argument,

$$\tau_{21} = \frac{\text{const}}{3} \rho (\overline{u}_1 - \overline{u}_{1_{\min}})(\overline{u}_{1_{\max}} - \overline{u}_1). \tag{303}$$

Next, Eq. (303) is transformed as in the section before. There are two equivalent developments depending upon which term in parentheses is chosen to be described by a difference quotient. We give preference to the development of the second one, as our target is to obtain Eq. (259). So, we write

$$\overline{u}_{1_{\max}} = \overline{u}_1 + \frac{\overline{u}_{1_{\max}} - \overline{u}_1}{x_{2_{\max}} - x_2} (x_{2_{\max}} - x_2), \tag{304}$$

which is a simple identity and insert it into the second parenthesis of Eq. (303); this yields

$$\tau_{21} = \sigma \, \rho \, (x_{2_{\max}} - x_2) \, (\overline{u}_1 - \overline{u}_{1_{\min}}) \frac{\overline{u}_{1_{\max}} - \overline{u}_1}{x_{2_{\max}} - x_2}, \qquad \sigma = \frac{\text{const}}{3}. \tag{305a, b}$$

The value of the coefficient $\sigma$ must be determined. It will be shown that this is possible by other Reynolds number-dependent constants of an underlying fractal theory, as outlined in Sect. 7.2.4. By comparing Eq. (305a) with Eq. (259), it is concluded that

$$\tau_{21} = \mu_T \frac{\overline{u}_{1_{\max}} - \overline{u}_1}{x_{2_{\max}} - x_2}, \qquad \mu_T = \sigma \rho \Lambda_T (\overline{u}_1 - \overline{u}_{1_{\min}}), \qquad \Lambda_T = (x_{2_{\max}} - x_2). \tag{306a – c}$$

The combination of these three quantities forms the DQTM. In complete analogy to the mean free path $\lambda$, also here, we find a characteristic exchange length, which in turbulence theory is denoted mixing length $\Lambda_T$. In plane Couette flow it is just the distance to the wall. Furthermore, the equally variable fluid velocity difference $\overline{u}_1 - \overline{u}_{1_{\min}}$ occurs in analogy to the constant velocity of the molecules. As already mentioned, Egolf (1991) developed the Difference-Quotient Turbulence Model (DQTM), which is identical to Eqs. (306a–c), by different reasoning. In Egolf and Weiss (1995), the second-order mean velocity fluctuation correlation

$$\overline{u_2' u_1'} = -\frac{\tau_{21}}{\rho} = -\sigma \chi_2 (\overline{u}_1 - \overline{u}_{1_{\min}}) \frac{\overline{u}_{1_{\max}} - \overline{u}_1}{x_{2_{\max}} - x_2}, \qquad \chi_2 = (x_{2_{\max}} - x_2), \tag{307a – c}$$

was applied in the Reynolds momentum equation to obtain the differential equations of the turbulent plane Couette flow (Figs. 7.5 and 7.6). Notice that in this development, the characteristic length $\chi_2$ is the distance of the point $x_2$ from the wall. In this case, a simplification, already exhibited by Eq. (303), is

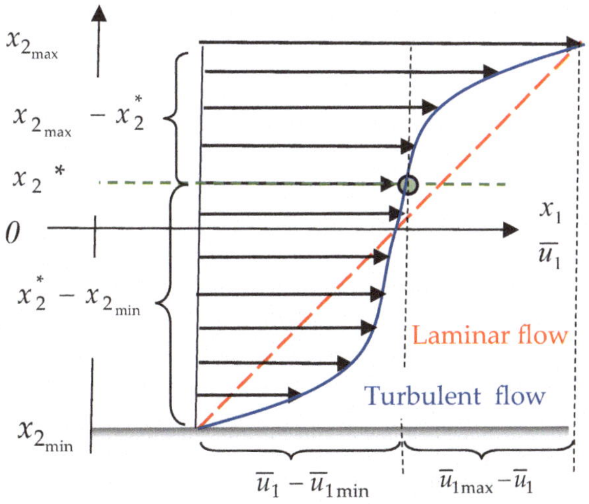


**Fig. 7.6** Average downstream velocity profiles of plane laminar and turbulent Couette flow. The laminar case shows a linear profile (- - - -) and the turbulent case shows a symmetric *S* shape profile (—). In Sect. 9.3, this profile is derived with the DQTM for different Reynolds numbers. The location of the coordinate system leads to antisymmetric averaged velocity profiles

$$\tau_{21} = \sigma\,\rho\,(\overline{u}_1 - \overline{u}_{1_{min}})\,(\overline{u}_{1_{max}} - \overline{u}_1). \tag{308}$$

This is the simplest representation of the shear stress of this fundamental flow type.

To mathematically symmetrize the flow solution, the origin of the coordinate system is relocated at the centerline between the two plates, where $\overline{u}_1 = U/2$. The shear stress, which is Galilean invariant, does not depend on a constant velocity shift, and therefore the altered version also applies. The advantage of this transformation is that with it the mean downstream velocity becomes antisymmetric

$$\begin{aligned} \overline{u}_1(-x_2) = -\overline{u}_1(x_2) \quad &\Rightarrow \overline{u}_{1_{min}} = \overline{u}_1(x_{2_{min}}) = \overline{u}_1(-x_{2_{max}}) \\ &= -\overline{u}_1(x_{2_{max}}) = -\overline{u}_{1_{max}}. \end{aligned} \tag{309a – e}$$

If one switches the index "min" to "max" and vice versa, then the sign of $\overline{u}_1$ changes. Together with Eq. (309a), we apply this to Eq. (308) and find

$$\begin{aligned} \frac{\tau_{21}(-x_2)}{\sigma\,\rho} &= [\overline{u}_1(-x_2) + \overline{u}_{1_{max}}] \cdot [-\overline{u}_{1_{min}} - \overline{u}_1(-x_2)] \\ &= [\overline{u}_{1_{max}} - \overline{u}_1(x_2)] \cdot [\overline{u}_1(x_2) - \overline{u}_{1_{min}}] = \frac{\tau_{21}(x_2)}{\sigma\,\rho}. \end{aligned} \tag{310a – c}$$

This demonstrates that in plane turbulent Couette flow the shear stress, different from the antisymmetric averaged downstream velocity, is symmetric.

### 7.2.3 Comparison of Laminar and Turbulent Flows

An attempt to perform a comparison between laminar and turbulent flows reveals an analogy. The most important analogy is the spatial differential quotient of the velocity field driving laminar flows compared to the difference quotient occurring in the theory of turbulent flows. The viscosity and the effective eddy or turbulent viscosity are built in the same way, namely each with a characteristic length and a characteristic velocity. The main difference occurs in the entities that transport the momentum. In laminar flows, these are molecules of a given diameter, of which we also assume that they have an infinite lifetime and, as a consequence, a birth and death rate equal to zero. On the other hand, eddies show a very large number of classes with entities related to the different sizes and also a size-dependent lifetime. Finally, the molecules obey normal flight conditions with a constant mean velocity leading to Brown diffusion, whereas eddies are assumed to move fluid lumps obeying the Lévy flight statistics as we shall just see. It follows that the laminar case is a linear and local problem, whereas the turbulent case is nonlinear and nonlocal. If time-dependent effects occur, it is assumed that in analogy to nonlocality also memory effects are playing a crucial role. The shear strain/stress relations can be described by an integral mathematical description containing a kernel, see Sect. 7.3. Interesting is that the relation for laminar flows leads to a kernel which is a Dirac delta distribution, whereas a Heaviside distribution determines the kernel of the turbulent flow description. All these analogies are summarized in Table 7.1.

### 7.2.4 Lévy Flight Turbulence Model and K41

#### 7.2.4.1 Introduction

Physicists are familiar with Brownian motion of small particles in a liquid. Originally pollen were observed which wiggle, because the atoms in the fluid are knocking them. This phenomenon may be observed under a microscope and was described by Einstein (1905, 1906). Lévy's (1886–1971) generalization of Brownian motion, especially $\alpha$-stable Lévy processes, also called Lévy walks (see Lévy 1925, 1937, 1948) later with the work of Mandelbrot (1977) showed to be of highest importance in numerous scientific domains. Alpha-stable Lévy processes are observed in wildlife, where, e.g., Albatrosses search for pray. Another application in nature is the human brain accessing an elusive memory, e.g., connecting a face to a name (West 2014).

It was remarked that a fluid lump undergoing a Lévy motion, describes the trajectory of a tracer particle in a turbulent flow. Processes based on Lévy statistics perform scale-free complex dynamics and non-Poissonian statistics. They are nonstationary, non-ergodic and, what is also very important, describe intermittent processes (West 2014).

**Table 7.1** Comparison of some features of laminar and turbulent flow are listed (see also Egolf and Hutter 2016b)

| Topic | Laminar flow | Turbulent flow |
|---|---|---|
| Name | Newton's shear law[a] | Difference-Quotient Turbulence Model |
| Extensions | Stokes viscosity law | Multidimensional DQTM |
| Driving quantity | Differential quotient | Difference quotient |
| Viscosity | Physical viscosity $\mu$ | Turbulent or eddy viscosity $\mu_T$ |
| Thermodynamics | Close-to-equilibrium | Nonequilibrium |
| Objects, size | Molecules (a single type) | Eddies (infinite number of sizes) |
| Number | $N$ molecules | $N$ eddies |
| Number density | Density of molecules $n$ | Distribution of eddy density $n$ |
| Birth rate | Zero | Lévy distribution |
| Death rate of object | Zero | Fractal Beta Model (Kolmogorov/Obukov) |
| Lifetime | Infinite | Finite |
| Character of objects | Particle (solid boundaries)[b] | Wave like (diffuse boundaries)[b] |
| Size | Diameter of molecule $d$ | Diameters of eddies $d_n$ |
| Movement | Brownian walk | Lévy flight |
| Characteristic length | Mean-free path length $\lambda$ | Turbulent mixing length $\Lambda_T$ |
| Characteristic velocity | Velocity of molecules $u_{mol}$ | Velocity of largest eddies $u_0$ |
| First condition: | $d \ll \lambda$ | $d \approx \Lambda_T$ |
| Second condition: | $\lambda \ll L$<br>$L$: size of container | $\Lambda_T \approx L$<br>$L$: overall size of flow domain |
| Spatial limit | Minimum | Maximum |
| Math. Complexity | Linear | Nonlinear, fractal, fractional |
| Time behavior | Instantaneous | Memory |
| Spatial behavior | Local | Nonlocal |
| Math. Relations | Identity rel. | Scaling laws |
| Correlations | No/zero | Yes/non-zero |
| Mathematical kernel | Fermi's Delta distribution $\delta$ | Heaviside distribution $\theta$[c] |
| Statistics | Brown's mov., Einstein diff. | Lévy statistics, anomalous diffusion |
| Characteristic length | Standard deviation $\sigma$ | None ($\sigma \to \infty$) |
| Critical phenomenon | No | Yes |

[a]The law is found in Sir Isaac Newton's (1643–1727) Philosophia Naturalis Principia Mathematica (1687)
[b]In the context of classical physics and not of quantum mechanics
[c]See Egolf and Weiss (1996a)

Today, the macroscopic transport processes, which are caused by the microscopic Lévy walks, are named anomalous diffusion (Bouchard and Georges 1990). Based on Lévy's work, Shlesinger et al. (1986) developed a statistical theory for motions with fractal features that applies to turbulent flows. Important for the present model is that its motions assume larger persistence times for larger jump sizes. It is clear that the statistics describing fluid lump motions in turbulence may be taken to also calculate turbulent momentum transfer by fluctuating eddies.

#### 7.2.4.2 Lévy Walks on a One-Dimensional Lattice

Consider a random walk with steps of different lengths $f$. The following power-law scaling is assumed to hold:

$$f(b) = b^n, \quad b > 0, \quad 0 \leq n \leq N. \tag{311}$$

Therefore, $N + 1$ different step sizes occur. Later we will speak of $N + 1$ different classes by relating each class of jumps with a certain step size to a class of eddies of a certain size. The smallest step has a length $b^0 = 1$ and the largest $b^N$. The step size increases with growing $n$, if the basic value is $b > 1$. Considering the one-dimensional case is sufficient, because the intention in further sections is to investigate basic turbulent flows with a pronounced directional derivative perpendicular to the main flow direction (free and "wall-turbulent" shear flows). Then, only momentum transport in this perpendicular direction is of importance, and a one-dimensional situation may be extracted from the three-dimensional problem. In Fig. 7.7, a four-class Lévy jump ($N = 3$, $N + 1 = 4$), or Lévy walk example is presented graphically.

The scaling law of Eq. (311) implies that

$$f(\lambda b) = (\lambda b)^n = \lambda^n b^n = \lambda^n f(b). \tag{312a – c}$$

The corresponding discrete jump probability distribution is defined as (Shlesinger et al. 1993)

$$p(s) = \frac{1}{2} \frac{a^{N+1} - a^N}{a^{N+1} - 1} \left[ \sum_{n=0}^{N} \frac{1}{a^n} \left( \delta_{s,b^n} + \delta_{s,-b^n} \right) \right]. \tag{313}$$

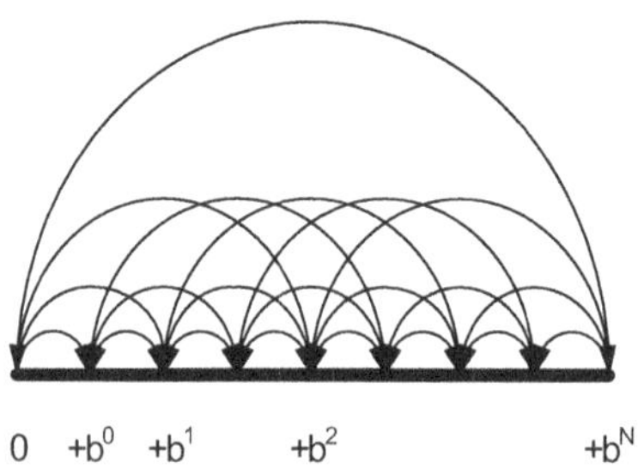


**Fig. 7.7** Schematic drawing of the possible Lévy walks on a finite lattice to the left and right. Here, a case with $b = 2$ is shown. For simplicity in the drawing, the largest value is only $N = 3$, corresponding to a series of $b^N = 2^3 = 8$ smallest jumps. (Courtesy of Egolf (2009), © Int. J. Refrigeration, reproduced with permission by Elsevier Science & Technology)

It contains in the $(\delta_{s,b^n})$-expressions (rather cryptically) the probabilities for the jumps of different sizes, $b^n$. The Kronecker delta symbol guarantees the correct link between a continuous and a discrete description. It is defined by

$$\delta_{\alpha,\beta} = \begin{cases} 1, & \alpha = \beta \\ 0, & \alpha \neq \beta. \end{cases} \tag{314}$$

In Eq. (313), apart from $b$, now also a second basic quantity, $a$, occurs. It determines the connection between the occurring probabilities for jumps of eddies of neighboring size. One can easily verify that the probability distribution is correctly normalized (see Egolf 2009 or Appendix A).

In the limit of an infinite number of step sizes, from Eq. (313) one obtains in agreement with Shlesinger et al. (1993), since $a > 1$,

$$p(s) = \frac{a-1}{2a}\left[\sum_{n=0}^{\infty} \frac{1}{a^n}\left(\delta_{s,b^n} + \delta_{s,-b^n}\right)\right]. \tag{315}$$

Equation (315) is identical to

$$p(s) = \frac{a-1}{2a}\left[\left(\delta_{s,1} + \delta_{s,-1}\right) + \frac{1}{a}\left(\delta_{s,b} + \delta_{s,-b}\right) + \frac{1}{a^2}\left(\delta_{s,b^2} + \delta_{s,-b^2}\right) + \ldots\right]. \tag{316}$$

In the limit as $a \to \infty$, there follows:

$$p(s) = \frac{1}{2}\left(\delta_{s,1} + \delta_{s,-1}\right). \tag{317}$$

This is the probability distribution describing the smallest jumps with a probability of 1/2 that the movement occurs to the left and 1/2 to the right, respectively. This movement is called Brownian motion (see Fig. 7.8). Therefore, the statistics of Brownian motion is a special case of the statistics of Lévy walks. A typical Lévy walk, namely one corresponding to the Kolmogorov/Obukhov turbulent flow can be seen in Fig. 7.9. In this second kind of movement clusters of different sizes occur. A typical cluster is shown inside the dotted circle. More such clusters, especially of smallest size steps are found in the figure. For a comparison of Lévy motion in turbulent flows see Weeks et al. (1994).

If one considers in Eq. (316) the symmetry of these jumps, it follows for $a \gg 1$,

$$2\frac{a-1}{2a} = \frac{a-1}{a} = 1 - \frac{1}{a} \approx 1. \tag{318a – c}$$

Equation (318c) approximately describes the probability of a jump of unit length to the left or right,

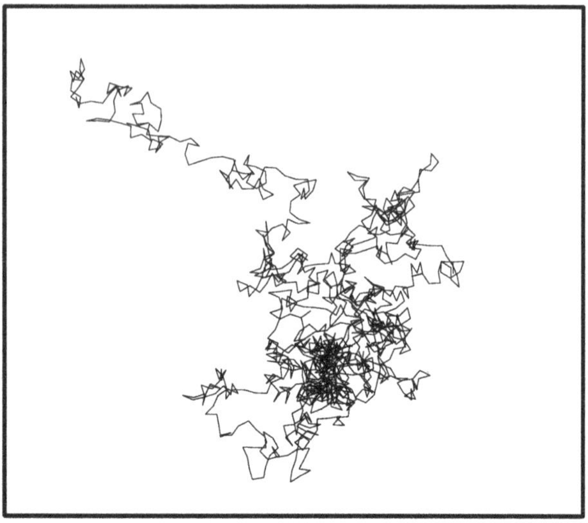

**Fig. 7.8** Two thousand two-dimensional Brownian jumps of equal size, describing, e.g., the path of an atom in gas, are shown [calculated with Eq. (317)]. (Courtesy of Egolf (2009), © Int. J. Refrigeration, reproduced with permission by Elsevier Science & Technology)

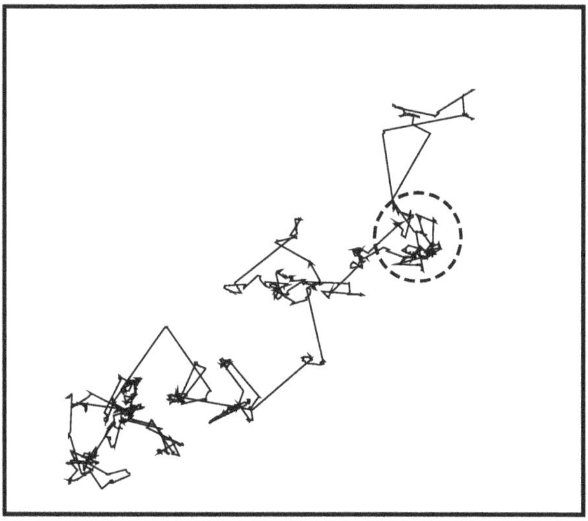

**Fig. 7.9** Two thousand two-dimensional Lévy jumps are shown (calculated with Eq. (313) and $a = 2$, $b = 2$), corresponding to a trajectory in a Kolmogorov/Obukhov turbulent flow. (Courtesy of Egolf (2009), © Int. J. Refrigeration, reproduced with permission by Elsevier Science & Technology)

$$s \cong |\pm 1| = 1. \tag{319a, b}$$

Analogously, again for $a \gg 1$, one obtains for the class performing the second largest jumps

$$\frac{a-1}{a}\frac{1}{a} = \left(1 - \frac{1}{a}\right)\frac{1}{a} = \frac{1}{a} - \frac{1}{a^2} \approx \frac{1}{a}. \tag{320a – c}$$

Equation (320c) approximately describes the probability of a jump of length

$$s \cong |\pm b| = b. \tag{321a, b}$$

These considerations can be continued. Then, one immediately realizes that for large $a$'s, large jumps occur rather seldom. Roughly, one may say that one observes on average $a$ jumps of length 1 before a jump of length $b$ occurs. Approximately, $a$ such clusters, divided by the length $b$, are created before a jump of strength $b^2$ occurs. This jump probability leads to clusters with self-similarity features (for self-similarity see, e.g., Mandelbrot 1977).

The variance of a Lévy walk process is (see Egolf 2009 or Appendix B)

$$\langle s^2 \rangle = \frac{a-1}{a}\left[\sum_{n=0}^{\infty}\left(\frac{b^2}{a}\right)^n\right] \to \infty, \quad b \geq \sqrt{a}. \tag{322a, b}$$

We summarize the characteristics of Lévy walks:

1. The probability distribution is not narrow, respectively, the second moment is infinite.
2. There are long-range correlations present.
3. There is no characteristic length of the process.
4. The signals are temporarily and spatially intermittent.

The structure function leads to the Weierstrass function, which is continuous, but nowhere differentiable (see Egolf 2009 or Appendix C). A continuous non-differentiable function with self-similarity features is called a fractal (for its definition see Mandelbrot 1977).

#### 7.2.4.3 Lévy Walks, Lévy Flights, Lévy Pairs, and Eddies in Turbulence

First, we deal with the terminology of Lévy-type movements. In the static picture, describing probability distributions, we used the terminology "Lévy walk." This remains also correct in a dynamical description if the times for single walks (ev. of different step sizes) remain unchanged. However, for the description of turbulence this would not be sufficient. For our purposes, we need to introduce special step size-dependent times for the different jumps. And following advice of Shlesinger et al. (1986), we may distinguish this second case from the first one, described above, by using the terminology "Lévy flight."

Next, if instead of single steps of a Lévy flight pairwise jumps are considered, then one is confronted with the situation shown in Fig. 7.10a. It is evident that two Lévy flights—one from the left to the right and another one of equal size in the counter direction—show the same effect of transporting a quantity as an eddy making half a turn (Fig. 7.10b). Mass conservation in fluid dynamics anyway favors a model with the occurrence of pairwise flights. This corresponds to a transfer of a mass element $\delta m$ to the left and an equivalent mass element $\delta m$ over the same line and distance to the right. Note that the transferred mass element in the counter direction is identical, but it is not necessary that other transported quantities, such as turbulent momentum or particle concentration, are also equal. A full rotation then corresponds to four linear Lévy jumps (see Fig. 7.10c).

An important assumption is that the probability distribution does not show any space dependence. This is not so evident for a shear flow. However, when this is assumed, and if we couple a jump to the right to one to the left, then one has to be aware that these two jumps—corresponding to a Lévy pair—are perfectly correlated. Therefore, the probability for the occurrence of a Lévy pair of class $n$ is

$$p_n^{(\text{Lévy pair})} = \frac{a^{N+1} - a^N}{a^{N+1} - 1} \frac{1}{a^n}, \quad 0 \le n \le N. \tag{323}$$

Because two jumps, which are perfectly correlated, occur, these probabilities are twice as large as the one for a single Lévy step (compare with Eq. 313).

A Kolmogorov–Oboukov eddy is defined by the Lévy parameters $a = 2$ and $b = 2$. The first condition, $a = 2$, leads to the statement that a single step occurs with half probability compared to a step of half-step size. In the eddy terminology, this means that eddies of half-size occur at double rate. The second condition, $b = 2$, leads to step sizes of always half size compared to neighboring classes of lower index number. In the eddy representation, this guarantees that an eddy of a higher class number has half the size, or diameter, compared to an eddy being a member of the lower number class. This case is realized in the random process of Fig. 7.12.

Problematic is that Lévy flight distributions and turbulence theory were developed independently and that they now, with their different specific notations, need to be merged together. Fact is that the index in turbulence research has been chosen opposite to that of Lévy flight modeling. This is outlined in more detail in Sect. 7.2.4.6. Therefore, we are obliged to change the indices in Eq. (323) by the following transformation rules: $n \rightarrow N - n$, where $0 \le n \le N$. For example, a Lévy jump of the largest size $b^N$ (highest index number) is related to an eddy of diameter $l_{N-n} = l_{N-N} = l_0$ (largest diameter, lowest index number). Furthermore, a smallest

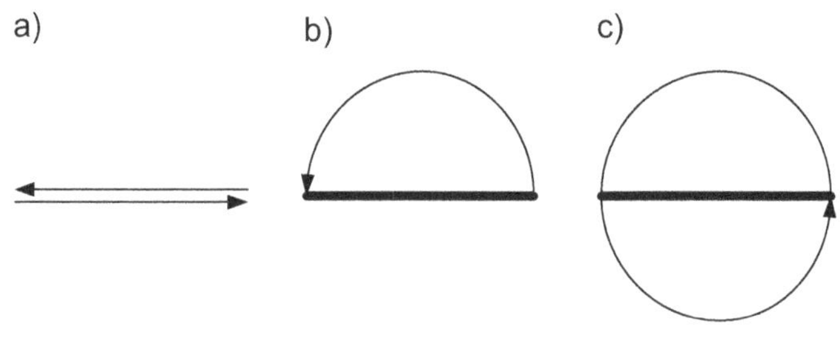


**Fig. 7.10** A Lévy flight to the left and one to the right is called a Lévy pair (**a**). Such a Lévy pair corresponds to an eddy, which turns half a period (**b**). A full turn of an eddy corresponds to four linear Lévy flights, two to the left and two to the right (**c**). (Courtesy of Egolf (2009), © Int. J. Refrigeration, reproduced with permission by Elsevier Science & Technology)

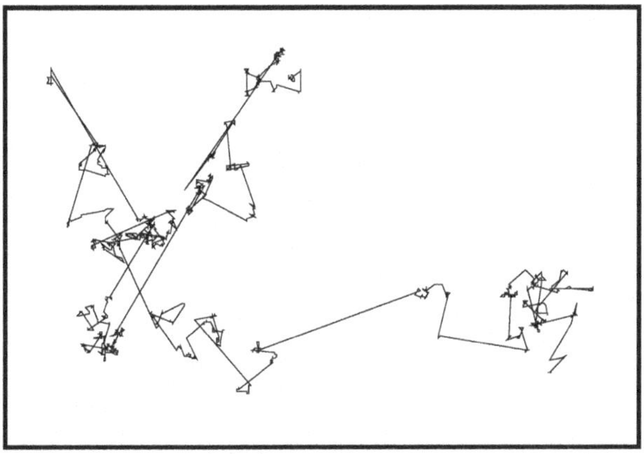

**Fig. 7.11** A Lévy flight distribution, shown for $a = b = 2$. Here 2000 Lévy jumps are plotted. Most of them are very small and hardly observable. (Courtesy of Egolf (2009), © Int. J. Refrigeration, reproduced with permission by Elsevier Science & Technology)

Lévy jump of size $1 = b^0$ (lowest index number) corresponds to an eddy of smallest size $l_{N-n} = l_{N-0} = l_N$ (highest index number). Mathematically this is outlined in Sect. 7.2.4.6. We now apply this transformation rule to Eq. (323). If we rigidly couple two Lévy pairs (Fig. 7.11) to create eddies (Fig. 7.12), the probability distribution, excluding the indexing, remains unchanged,

$$p_n^{(\mathrm{eddy})} = \frac{a^{N+1} - a^N}{a^{N+1} - 1} \frac{1}{a^{N-n}} = \frac{a-1}{a^{N+1}-1} a^n,$$
$$p_n^{(\mathrm{eddy})} = p_0^{(\mathrm{eddy})} a^n, \qquad p_0^{(\mathrm{eddy})} = \frac{a-1}{a^{N+1}-1}. \tag{324a – d}$$

In a simple model of turbulence, one assumes that a large eddy decays into two smaller ones. Each of these splits up again into two even smaller eddies. Therefore, this splitting process leads to a cascade of eddies of different sizes belonging to different classes. As the splitting is performed by a factor two, one immediately sees that this case corresponds to $b = 2$. The occurrence of twice as many eddies of a neighboring class with smaller eddies is also a result of the splitting process[3] and leads to the value $a = 2$. Therefore, the case $a = b = 2$, which is called the Kolmogorov/Obukhov limit, is a special case of the present theory. Note that an infinite Reynolds number flow does not show any intermittency (see Chap. 10). An example with $N = 2$, which corresponds to three classes of eddies, is graphically presented in Fig. 7.13. The quantity $m_n$ denotes the number of eddies of class $n$, whereas $M$ is the total number of eddies of all classes gathered together in one ensemble ($m_0 = 1$, $m_1 = 2$, $m_2 = 4$, $M = 7$).

[3] In reality eddies split and merge together. The net effect is assumed to be only a splitting effect.

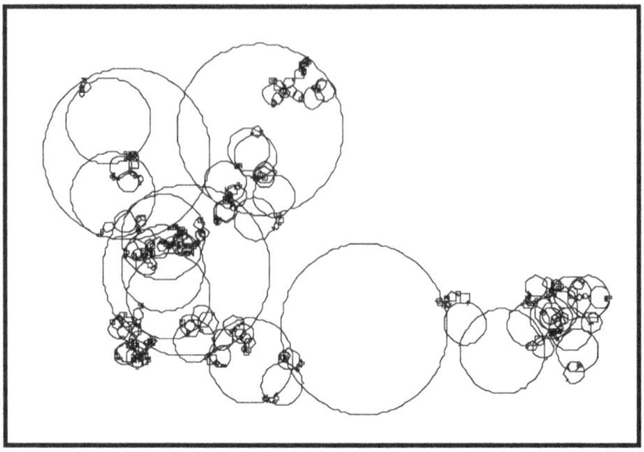

**Fig. 7.12** A line presenting a linear Lévy jump in Fig. 7.11 corresponds in this image to a diameter of a related eddy. Therefore, large Lévy jumps lead to large eddies (Courtesy of Egolf (2009), © Int. J. Refrigeration, reproduced with permission by Elsevier Science & Technology)

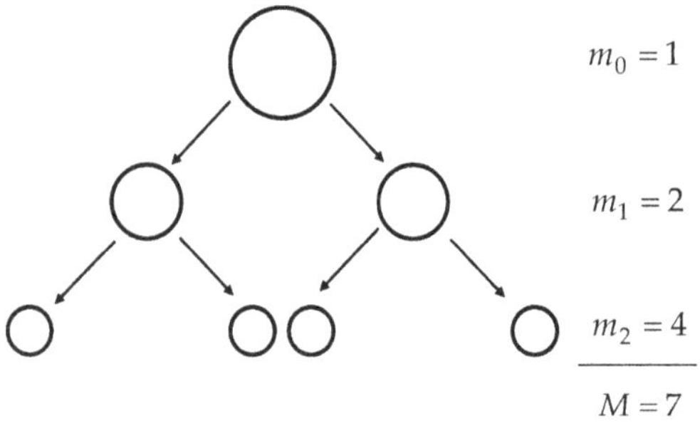


**Fig. 7.13** The decay of turbulent eddies following a Kolmogorov/Obukhov cascade (note that there are researchers who doubt the physical relevance of turbulent kinetic energy cascades in physical space (Sagaut and Cambon 2018)). In this case, the parameter $a = 2$ is leading to an integer splitting process, where one eddy in each cascade step splits into two new ones of the next class, which are just of half the size (Courtesy of Egolf (2009), © Int. J. Refrigeration, reproduced with permission by Elsevier Science & Technology)

#### 7.2.4.4 Eddy Class Statistics

The probability of observing two Lévy flights or a Lévy pair of class number $n$ is given by Eq. (323). As we have learned, the occurrence of a Lévy pair may be fully correlated with the occurrence of an eddy of class $n$. Therefore, the Lévy statistics directly leads to the turbulent eddy class statistics. Now, the probability to have a creation of an eddy of class $n$ is $p_n$, where we omit from now on the superscript "(eddy)" until to the end of our treatise.

The condition

$$\sum_{n=0}^{N} p_n = 1 \tag{325}$$

normalizes the probabilities $p_n$, which contain the information how many eddies of class $n$ are created in relation to eddies of another size, e.g., one of class $n + 1$. Next, the creation time intervals $\tau_n$ for two succeeding eddies of class $n$ is

$$\tau_n = \frac{\Delta t}{p_n}, \tag{326}$$

where $\Delta t$ is a characteristic time, which is freely adjustable to the problem under consideration.

Then, from Eqs. (325) and (326), we conclude that

$$\sum_{n=0}^{N} \frac{\Delta t}{\tau_n} = 1 \quad \Rightarrow \quad \Delta t = \frac{1}{\sum\limits_{n=0}^{N} \frac{1}{\tau_n}}. \tag{327a, b}$$

The eddy creation frequency follows from Eq. (326)

$$f_n = \frac{1}{\tau_n} = \frac{p_n}{\Delta t}. \tag{328a, b}$$

As we know that in Lévy flights larger jumps are less frequent, so larger eddies are also less numerously created, and, therefore, for larger eddies, their probability, $p_n$, must be smaller. This leads for larger eddies to a smaller production frequency.

If we now consider a time span $T$, the total number of eddies of all classes $n$ created in this interval of time is

$$M = \frac{T}{\Delta t}. \tag{329}$$

With Eq. (326), this implies

$$M = \frac{T}{p_n \tau_n}. \tag{330}$$

From Eq. (330), one immediately deduces

$$p_n = \frac{T}{M \tau_n}. \tag{331}$$

The quantity

$$m_n = \frac{T}{\tau_n} \tag{332}$$

is the number of eddies of class $n$ created in the time interval $T$. Thus, from Eqs. (331) and (332) it follows that

$$p_n = \frac{m_n}{M}. \tag{333}$$

The generation probability is correctly the number of created eddies of class $n$ divided by all created eddies of any size $M$. Note that this probability directly corresponds to the probability of step creation in the Lévy statistics. This is the consequence of the analogy of a Lévy flight of a certain step size with an eddy of corresponding diameter.

#### 7.2.4.5 The Lifetime of Eddies

If one now asks how many eddies are present in a quasi-steady turbulent flow field, then not only the creation frequency of eddies is an important quantity. Also, the lifetime of the eddies influences this number. Eddy lifetimes shall be approximately determined in this section.

The lifetime of an eddy of class $n$ is denoted by $t_n$ (see Fig. 7.14). This time may be related to the turnover time of an eddy. It is the time to transform turbulent kinetic energy from an eddy of class $n$ to such of class $n + 1$. The time where an eddy of class $n$ is present is

$$T_n^{(\mathrm{eddy})} = m_n\, t_n, \tag{334}$$

whereas its ratio with the basic time period $T$ is

$$\frac{T_n^{(\mathrm{eddy})}}{T} = \frac{m_n\, t_n}{T} = \frac{t_n}{\tau_n}, \tag{335a, b}$$

where Eq. (332) has been substituted. This equation shows that three cases exist:

**Case 1 ($t_n < \tau_n$):**

$$\frac{t_n}{\tau_n} < 1 \quad \Rightarrow \quad T_n^{(\mathrm{eddy})} < T. \tag{336a, b}$$

As may be seen in Fig. 7.14, an eddy of class $n$ is not present at all times.

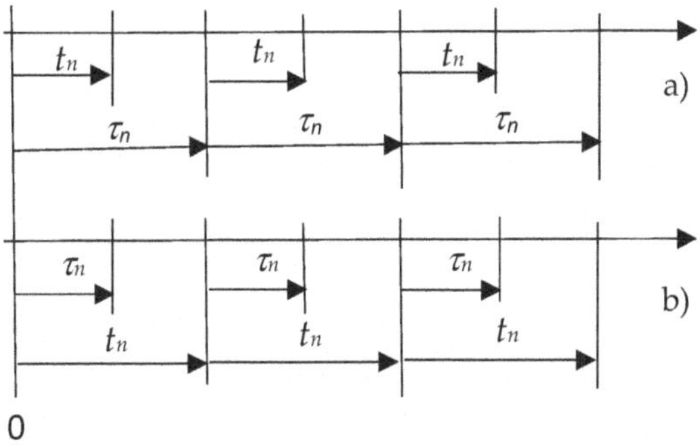


**Fig. 7.14** The creation time interval of eddies $\tau_n$ and their lifetime $t_n$ must be distinguished. In (**a**) the creation time is larger than the lifetime and in (**b**) it is reverse. The second case is that of turbulence

**Case 2 ($t_n = \tau_n$):**

$$\frac{t_n}{\tau_n} = 1 \quad \Rightarrow \quad T_n^{(\text{eddy})} = T. \tag{337a, b}$$

At all times, exactly one eddy is present.

**Case 3 ($t_n > \tau_n$):**
This is the case of importance in a turbulent flow. The eddy lifetime is larger than the time between two creation events of an eddy

$$\frac{t_n}{\tau_n} > 1 \quad \Rightarrow \quad T_n^{(\text{eddy})} > T. \tag{338a, b}$$

This situation permits to have a large number of eddies in a turbulent flow field at a certain time.

#### 7.2.4.6 The Eddy Diameters

The power $n$, characterizing the jump length $b^n$ ($b$ is a non-dimensionalized length) of a Lévy flight, is inversely defined when compared with the index characterizing the diameter of an eddy

$$b^0 \quad \leftrightarrow \quad l_N, \tag{339}$$

which denotes the characteristic length (respectively, the diameter) of the class with the smallest eddies. The size of eddies of this class is defined by Kolmogorov's dissipation length scale (see Sect. 7.2.4.11). Similarly for the largest eddies, it follows that

$$b^N \leftrightarrow l_0. \tag{340}$$

Therefore, the appropriate association of this inverse power and index behavior, respectively, is

$$b^n \leftrightarrow l_{N-n}. \tag{341}$$

One immediately sees that Eq. (341) fulfills Eqs. (339) and (340). Subsequently, the eddy diameters $l_n$ will be applied and only in cases of comparison with the Lévy flight statistics the quantity $b^n$ will appear again.

If one needs to transform in the reverse direction, the relation

$$l_n \leftrightarrow b^{N-n} \tag{342}$$

is useful. Again, Eqs. (339) and (340) are fulfilled. From Eq. (342) some valuable relations are derived

$$\frac{l_{n+1}}{l_n} = \frac{b^{N-n-1}}{b^{N-n}} = \frac{1}{b} \Rightarrow l_n = b\, l_{n+1}. \tag{343a – c}$$

For $b > 1$, eddies of a class of smaller index number are larger, which corresponds with the convention of turbulent eddy models.
By applying the rule (Eq. 343c) twice, one obtains

$$\frac{l_{n+1}}{l_{n-1}} = \frac{l_{n+1}}{l_n}\frac{l_n}{l_{n-1}} = \frac{1}{b^2}, \tag{344a, b}$$

from which one easily deduces

$$\frac{l_p}{l_q} = \frac{1}{b^{p-q}}, \qquad p \geq q. \tag{345}$$

With $p = n$ and $q = 0$ this implies

$$\frac{l_n}{l_0} = \frac{1}{b^n} \Rightarrow l_n = \frac{1}{b^n} l_0. \tag{346a, b}$$

It is again seen how simply and naturally the size of the Lévy flights enters turbulence modeling by just describing the diameters of the occurring eddies.

#### 7.2.4.7 A Fractal Eddy Cascade Model

In this section, a scaling model with intermittency, namely the fractal $\beta$-model (see Frisch 1995), is introduced. In the $\beta$-model, small eddies do not fully occupy the space. In each step a reduction by the factor $\beta$ $(0 < \beta < 1)$ occurs, which gave the model its name. The active space fraction is defined as

$$p_n^{(\text{active})} = \beta^n := \left(\frac{l_n}{l_0}\right)^{d-D}, \tag{347a, b}$$

where $d$ is the spatial dimension and $D$ the fractal dimension. The difference $d$-$D$ is called codimension. Usually, $d$ is equal to 3 and $D$ is smaller than or equal to 3. In this latter case, it follows that

$$p_0^{(\text{active})} = \beta^0 = 1. \tag{348a, b}$$

This first class is assumed to be fully present and completely space filling. Beyond that, an increasing intermittency to smaller scales is observed.

Next, $E_n$ is introduced, which denotes the turbulent kinetic energy for eddies of class $n$

$$E_n \propto p_n^{(\text{active})} \rho\, {u_n}^2 \propto \rho\, {l_n}^{d-D} {u_n}^2. \tag{349a, b}$$

Then, the characteristic velocity for eddies of class $n$ is

$$u_n \propto \left(\frac{E_n}{\rho}\right)^{1/2} l_n^{(D-d)/2}. \tag{350}$$

The characteristic time, in which the energy contained in eddies of class $n$ is transferred to those of class $n + 1$, is denoted as turnover time

$$t_n \propto \frac{l_n}{u_n} \propto \left(\frac{\rho}{E_n}\right)^{1/2} {l_n}^{(2+d-D)/2}. \tag{351a, b}$$

It is assumed that at a certain time energy is stopped to be fed into the cascade at the large wavelength end; then after the time given by Eqs. (351a,b), the turbulent kinetic energy will be transported out of the eddies of class $n$ into those of class $n + 1$. Therefore, it is tempting to assume that this time is identical to the eddy lifetime $t_n$.

According to the locality hypothesis of turbulence (see, e.g., Hunt et al. 1991), the energy transfer rate from eddies of class $n$ to those of class $n + 1$ is constant, so that

$$\varepsilon_n \propto \frac{E_n}{t_n} \propto \varepsilon = \text{const.} \tag{352a – c}$$

The regime, where this is the case, is called inertial range. Then it follows

$$E_n \propto \varepsilon\, t_n \propto \varepsilon \left(\frac{\rho}{E_n}\right)^{1/2} l_n^{(2+d-D)/2}, \tag{353a, b}$$

in which Eq. (351b) has been substituted. Now, it immediately follows that

$$E_n^{3/2} \propto \varepsilon\, \rho^{1/2} l_n^{(2+d-D)/2}, \tag{354}$$

or

$$E_n \propto \left(\rho\, \varepsilon^2\right)^{1/3} l_n^{(2+d-D)/3}. \tag{355}$$

For $n = 0$ this yields

$$E_0 \propto \left(\rho\, \varepsilon^2\right)^{1/3} l_0^{(2+d-D)/3}, \tag{356}$$

or

$$\left(\rho\, \varepsilon^2\right)^{1/3} \propto \frac{E_0}{l_0^{(2+d-D)/3}}, \tag{357}$$

respectively. The proportionality sign will be replaced by introducing a constant in front of the corresponding terms. However, it is clear that this constant would be the same in Eqs. (355) and (356). Therefore, if Eq. (357) with this constant is substituted into Eq. (355) one obtains

$$E_n = E_0 \left(\frac{l_n}{l_0}\right)^{(2+d-D)/3}. \tag{358}$$

Next, the right-hand side of Eq. (355) is substituted into Eq. (350); this yields

$$u_n \propto \left(\frac{\varepsilon}{\rho}\right)^{1/3} l_n^{(1+D-d)/3} = \left(\frac{\varepsilon}{\rho}\right)^{1/3} l_n^h, \tag{359a, b}$$

in which $h$ is the commonly used scaling exponent called Hurst exponent (see, e.g., Frisch 1995),

$$h = \frac{1}{3}(1 + D - d), \tag{360}$$

which is defined on the set of fractal dimension $D$ in the space of dimension $d$, on which the turbulent eddy cascade accumulates. With the same procedure as before this transforms $u_n$ to

$$u_n = u_0 \left( \frac{l_n}{l_0} \right)^{(1+D-d)/3}. \tag{361}$$

By substituting the expression for $E_n$, given by Eq. (355), into Eq. (351b), the following law for the time $t_n$ is obtained

$$t_n \propto \left( \frac{\rho}{\varepsilon} \right)^{1/3} l_n^{(2+d-D)/3}, \tag{362}$$

which can also be written as

$$t_n = t_0 \left( \frac{l_n}{l_0} \right)^{(2+d-D)/3}. \tag{363}$$

For the specific energy flux, one confirms from Eq. (352a) by substituting Eqs. (355) and (362) that this quantity does not depend on the ratio $l_n/l_0$.

In a continuous system, the spectral energy density is defined by

$$\widehat{e} = \frac{\mathrm{d}\widehat{E}}{\mathrm{d}k}, \tag{364}$$

where the hat denotes a Fourier transform and $k$ is the wavenumber. In analogy to Eq. (364), we may write

$$\widehat{e}_n = \frac{\Delta \widehat{E}_n}{\Delta k_n} \propto \frac{\widehat{E}_n}{k_n} \propto \widehat{E}_n \, l_n. \tag{365a – c}$$

In Eq. (365c), $l_n$ is the wavelength corresponding to the wave number $k_n = 1/l_n$. Substitution of Eq. (355) into Eq. (365c) leads to

$$\widehat{e}_n \propto \left( \rho \varepsilon^2 \right)^{1/3} l_n^{(5+d-D)/3} = \left( \rho \varepsilon^2 \right)^{1/3} k_n^{-(5+d-D)/3}. \tag{366a, b}$$

As a special case ($d = D = 3$), this law contains the power law of Kolmogorov–Obukov (see also Hunt et al. 1991) with the exponent $-5/3$

$$\widehat{e}_n \propto \left(\rho\varepsilon^2\right)^{1/3} l_n{}^{5/3} \quad \Rightarrow \quad \widehat{e}(k) \propto \rho\left(\varepsilon^2\right)^{1/3} k^{-5/3}. \tag{367a, b}$$

#### 7.2.4.8 The Occupation Number

It is impossible that $T^{(\mathrm{eddy})}$ is larger than $T$. Therefore, we introduce the occupation number of eddies of class $n$, which we denote by $o_n$. This number describes the number of eddies of class $n$, which are present at a certain time $0 \leq t \leq T$ in a turbulent fluid domain,

$$o_n = \frac{t_n}{\tau_n}. \tag{368}$$

It is clear that the total number of eddies of all classes $n$, present at a time $0 \leq t \leq T$, is

$$O = \sum_{n=0}^{N} o_n. \tag{369}$$

The occupation probability of eddies of class $n$ is

$$q_n = \frac{o_n}{O}. \tag{370}$$

This probability is correctly normalized, because owing to Eq. (369), it simply follows that

$$\sum_{n=0}^{N} q_n = \sum_{n=0}^{N} \frac{o_n}{O} = \frac{1}{O} \sum_{n=0}^{N} o_n = 1. \tag{371a – c}$$

With the help of Eqs. (368) and (370), one obtains

$$q_n = \frac{1}{O} \frac{t_n}{\tau_n}. \tag{372}$$

The objective is now to introduce the Lévy statistics concept into this eddy occupation probability. The place where the Lévy statistics can enter is by introducing the creation time $\tau_n$. For this purpose Eq. (326) is combined with Eq. (324b) to yield

$$\tau_n = \frac{a^{N+1} - 1}{a - 1} \Delta t \frac{1}{a^n}. \tag{373}$$

Here too, the creation time with zero index is of interest

$$\tau_0 = \frac{a^{N+1} - 1}{a - 1} \Delta t. \tag{374}$$

This value can be substituted into Eq. (373) to obtain

$$\tau_n = \tau_0 \frac{1}{a^n}. \tag{375}$$

If $a$ and the number of the largest eddies that are present at a certain time is known, the number of eddies of each class can also be determined. This result is substituted into Eq. (368) to obtain

$$o_n = \frac{t_n}{\tau_0} a^n. \tag{376}$$

Now it becomes clear why the fractal $\beta$-model is useful. With the aid of Eq. (363), the eddy lifetime $t_n$ is applied to yield

$$o_n = \frac{t_0}{\tau_0} \left( \frac{l_n}{l_0} \right)^{(2+d-D)/3} a^n, \tag{377}$$

or specialized for $n = 0$,

$$o_0 = \frac{t_0}{\tau_0}, \tag{378}$$

which in this case is in conformity with Eq. (368). By combining Eqs. (377) and (378), one obtains

$$o_n = o_0 \left( \frac{l_n}{l_0} \right)^{(2+d-D)/3} a^n. \tag{379}$$

To go completely back to the description by the Lévy statistics, Eq. (346a) is applied to obtain

$$o_n = o_0 \left( \frac{1}{b^n} \right)^{(2+d-D)/3} a^n. \tag{380}$$

This can be simplified to yield

$$o_n = o_0 \left( \frac{a}{b^{(2+d-D)/3}} \right)^n. \tag{381}$$

The special case of Kolmogorov/Obukhov, for infinite Reynolds number flows ($a = b = 2$, $d = D = 3$), leads to

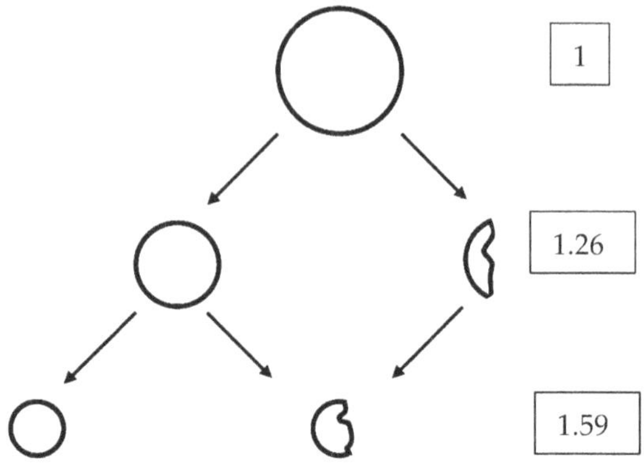


**Fig. 7.15** In a Kolmogorov/Obukhov process, one eddy decays into eddies of higher classes of space-filling nature. Notice that smaller eddies have smaller lifetimes and that this influences the occupation number. In this example, for the largest eddies (class zero), one assumes that $n = 0$, $o_0 = 1$, for the first class $n = 1$, $o_1 \cong 2^{1/3} = 1.26$ and for the second class $n = 2$, $o_2 \cong 2^{2/3} = 1.59$. The total number of eddies of all three classes adds up to $O \cong 3.85$ instead of 7 as in the situation of Fig. 7.13. Courtesy of Egolf (2009) © Int. J. Refrigeration, reproduced with permission by Elsevier Science & Technology

$$o_n = o_0 2^{n/3}. \tag{382}$$

For $o_0 = 1$ and $N = 2$ the solution is shown in Fig. 7.15. The presence of eddies of the first three classes is shown. The splitting process leads to a higher number of eddies of higher classes, which is mainly described by the creation dynamics of eddies as shown in Fig. 7.13. Because smaller eddies have a larger curvature, and a larger curvature leads to enhanced dissipation, the lifetimes of eddies of higher classes are smaller. Therefore, one experiences that the number of eddy occupation in a turbulent domain shows a smaller increase than that just taking the eddy creation into consideration.

#### 7.2.4.9 The Occupation Probability

The probability of the presence of eddies of class $n$ is given by Eq. (370) and Eq. (381)

$$q_n = \frac{o_0}{O}\left(\frac{a}{b^{(2+d-D)/3}}\right)^n. \tag{383}$$

With the abbreviation

$$\omega = \frac{a}{b^{(2+d-D)/3}}, \tag{384}$$

Equation (383) simplifies to

$$q_n = \frac{o_0}{O}\omega^n. \tag{385}$$

With Eqs. (369), (381), and (384), it follows that

$$O = o_0 \sum_{n=0}^{N} \omega^n = o_0 \frac{\omega^{N+1} - 1}{\omega - 1}, \tag{386a, b}$$

since $\omega < 1$. Substituting this equation into Eq. (385) yields

$$q_n = \frac{\omega - 1}{\omega^{N+1} - 1}\omega^n. \tag{387}$$

Next, one obtains the final probability distribution for the occupation of a turbulent domain with eddies being members of class number $n$

$$q_n = \frac{\left(\frac{a}{b^{(2+d-D)/3}}\right) - 1}{\left(\frac{a}{b^{(2+d-D)/3}}\right)^{N+1} - 1}\left(\frac{a}{b^{(2+d-D)/3}}\right)^n. \tag{388}$$

From Eq. (388), for $n = 0$, one deduces the relation

$$q_0 = \frac{\left(\frac{a}{b^{(2+d-D)/3}}\right) - 1}{\left(\frac{a}{b^{(2+d-D)/3}}\right)^{N+1} - 1}. \tag{389}$$

By a combination of Eqs. (388) and (389), one obtains

$$q_n = q_0\left(\frac{a}{b^{(2+d-D)/3}}\right)^n \tag{390}$$

and upon introducing the eddy length scale (see Eq. 346b), the result is

$$q_n = q_0\left(\frac{l_n}{l_0}\right)^{(2+d-D)/3} a^n. \tag{391}$$

#### 7.2.4.10 The Momenta of Eddies

In our modeling Lévy flights, or eddies turning 180°, respectively, are responsible for the transport of linear but not of angular momentum. Therefore, the linear momentum transfer of all eddies of class $n$ is

$$\widehat{P}_n = o_n p_n = \rho\ o_n u_n. \tag{392a, b}$$

Let "specific" mean per unit volume. Then the specific total momentum of all occurring eddies is described by

$$\widehat{P} = \sum_{n=0}^{N} \widehat{P}_n = \rho \sum_{n=0}^{N} o_n u_n. \tag{393a, b}$$

Then, the average specific momentum of a single eddy is

$$\widehat{p} = \frac{\widehat{P}}{O} = \frac{1}{O} \rho \sum_{n=0}^{N} o_n u_n. \tag{394a, b}$$

With Eq. (370), the occupation probability is introduced,

$$\widehat{p} = \rho \sum_{n=0}^{N} \frac{o_n}{O} u_n = \rho \sum_{n=0}^{N} q_n u_n. \tag{395a, b}$$

However, the average specific momentum is also given by the equation

$$\widehat{p} = \sum_{n=0}^{N} \widehat{p}_n. \tag{396}$$

Comparing Eq. (395b) with Eq. (396), the mean specific momentum of a single eddy class $n$ is given by the equation

$$\widehat{p}_n = \rho\ q_n u_n. \tag{397}$$

At last by combining Eqs. (361), (390), and (397), this quantity can be calculated to be

$$\widehat{p}_n = \rho q_0 \left( \frac{a}{b^{(2+d-D)/3}} \right)^n u_0 \left( \frac{1}{b^n} \right)^{(1+D-d)/3} = a^n b^{\left(-\frac{2}{3}-\frac{1}{3}d+\frac{1}{3}D-\frac{1}{3}-\frac{1}{3}D+\frac{1}{3}d\right)\,n} \rho\ q_0 u_0. \tag{398a, b}$$

The result of this consideration is of great simplicity, because the power of $b$ simplifies to $-n$, viz.

$$\widehat{p}_n = \left(\frac{a}{b}\right)^n \widehat{p}_0, \quad \forall n \in \{0, 1, \ldots, \mathrm{N}\}. \tag{399}$$

This main result shows how the momentum is distributed throughout the eddy classes and links it very directly to the two parameters of the Lévy flight statistics, $a$ and $b$ (compare with Eq. 313). When $a < b$, then the class of higher number contains less momentum than that with a lower number; however, when $a = b$ then all classes show identical momentum, and when $a > b$ then the momentum of the eddies increases toward the higher classes. This result corresponds well with the properties of Lévy statistics. When $a$ is large, then large eddies occur very seldom, and then the total momentum of the large eddy size classes is small. On the other hand, if $b$ is large, then the momentum of eddies of a class with small eddies is also small and, owing to this, contain smaller momentum than eddies of the larger class. It then follows for the momentum of all $N + 1$ classes

$$\widehat{P}_n = \sum_{n=0}^{N} \left(\frac{a}{b}\right)^n \widehat{p}_0 = \left[\sum_{n=0}^{N} \left(\frac{a}{b}\right)^n\right] \widehat{p}_0 = \kappa_{0,N} \widehat{p}_0,$$
$$\kappa_{0,N} := \frac{1 - \left(\frac{a}{b}\right)^{N+1}}{1 - \frac{a}{b}}, \quad \left|\frac{a}{b}\right| < 1. \tag{400a – e}$$

Applying the differentiation rule of Bernoulli-de l'Hôpital one obtains for the case of identical momenta exchanges

$$a = b \quad \Rightarrow \quad \kappa_{0,N} = N + 1. \tag{401a, b}$$

This special case (involving also the Kolmogorov/Obukhov special case: $a = b = 2$) was used to introduce and explain in Sect. 7.2.2 the scaling concept. Here, each class shows the same momentum exchange, even if the number of eddies of different classes are different. In the example in Fig. 7.4, it was the work that was put equal in the zeroth and first class. In the corresponding eddy concept, the total momentum transfer is just that of the zero-order class multiplied by the number of classes, $N + 1$. In the former example, with two classes of workers $n = 0$ and $n = N = 1$, it follows that twice the work of the strongest worker is performed in total.

#### 7.2.4.11 The Number of Eddy Classes

Kolmogorov (1941a, b) proposed *microscales* of turbulent flow fields, which are the ones where turbulent kinetic energy is dissipated into heat. The dissipation term in the NSEs [not shown in Eq. (16b)] contains a Laplacian (second-order derivative).

Therefore, this term is directly related to the curvature of the flow. The larger the curvature is, the larger will be the dissipation effect. In a model consideration, Kolmogorov stated that dissipation sets in at a sharp cut off given by his microscales, which he proposed by dimensional analysis using power combinations of $\varepsilon$ and $\nu$. They are the Kolmogorov micro length scale

$$\eta = \left(\frac{\nu^3}{\varepsilon}\right)^{1/4}, \tag{402}$$

Kolmogorov micro time scale

$$\tau_\eta = \left(\frac{\nu}{\varepsilon}\right)^{1/2}, \tag{403}$$

and Kolmogorov micro velocity scale

$$u_\eta = (\nu\varepsilon)^{1/4}. \tag{404}$$

The scale at which turbulent kinetic energy is dissipated depends on both, namely the dissipation rate $\varepsilon$ and the viscosity $\nu$.
From the above formulae, it follows that

$$u_\eta = \frac{\eta}{\tau_\eta}. \tag{405}$$

Furthermore, with Eqs. (402) and (404) it may be verified that the Kolmogorov or dissipation Reynolds number is identical to unity

$$\mathrm{Re}_\eta = \frac{\eta u_\eta}{\nu} = 1. \tag{406a, b}$$

With the assumption of the occurrence of an inertial range, $\varepsilon$ remains constant for all scales from the largest scale $L$ down to the microscale $\eta$ (see Eq. 352c). Therefore, the dissipation rate can be estimated by the large-scale flow features with length scale $L$ and velocity scale $U$. The kinetic energy of the flow is proportional to $U^2$. The time scale of the large-scale flow is $T = L/U$. It is tempting to assume that the kinetic energy supply rate is the inverse of this time scale. From this, it is concluded that

$$\varepsilon \cong \frac{U^2}{T} = \frac{U^2}{\frac{L}{U}} = \frac{U^3}{L}. \tag{407a – c}$$

With this estimate Eq. (402) becomes

$$\eta \cong \left(\frac{\nu^3 L}{U^3}\right)^{1/4}; \tag{408}$$

so the length-scale rate of largest to smallest occurring turbulent structures (eddies) is[4]

$$\frac{L}{\eta} \cong \left(\frac{UL}{\nu}\right)^{3/4} = \mathrm{Re}^{3/4}. \tag{409a, b}$$

It is seen that the spectral distance between the largest and smallest eddies, called the inertial range, increases with the Reynolds number Re[5].

By setting $L = l_0$, $\eta = l_N$, and $U = u_0$, this small model consideration can be directly applied to our fractal eddy model (see Samba 2016; Samba et al. 2019). Now, from Eqs. (409a,b) it follows that

$$\frac{l_0}{l_N} \cong \left(\frac{u_0 l_0}{\nu}\right)^{3/4} = \mathrm{Re}^{3/4}. \tag{410a, b}$$

Introducing the ratio (Eq. 346a) for $n = N$,

$$\frac{l_0}{l_N} = b^N, \tag{411}$$

and by substitution into Eqs. (410a,b), yields

$$b^N = \mathrm{Re}^{3/4}. \tag{412}$$

Therefore, the number of occurring classes is determined by (see Fig. 7.16)

$$N + 1 = \frac{3}{4}\frac{\log_{\mathrm{e}} \mathrm{Re}}{\log_{\mathrm{e}} b} + 1. \tag{413}$$

The number of classes of eddies naturally depends on the basic Lévy flight size $b$, but not on the inverse jump probability $a$.

---

[4]The special case of a flow at criticality fulfills $\eta = L$, which with Eq. (409a,b) leads to $L/L = \mathrm{Re}_c^{3/4} = 1 \rightarrow \mathrm{Re}_c = 1$. Therefore, it is concluded that also $L/\eta = (\mathrm{Re}/\mathrm{Re}_c)^{3/4}$. Then for practical use and to adapt also cases where $\mathrm{Re}_c > 1$, in following formulas Re may be replaced by the generalization $\mathrm{Re}/\mathrm{Re}_c$.

[5]The number of degrees of freedom is $(L/\eta)^3 = \mathrm{Re}^{9/4}$.

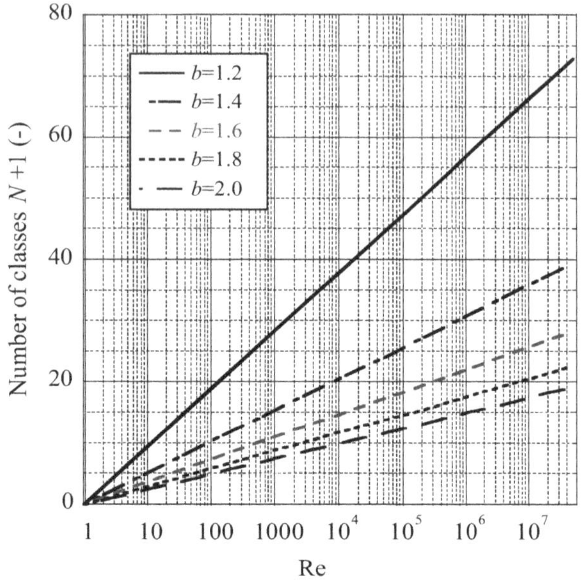


**Fig. 7.16** The number of eddy classes $N + 1$ as a function of the overall Reynolds number Re for different basic step sizes $b$ (Samba et al. 2019)

### 7.2.4.12 Lévy Flight Statistics, $\beta$-Fractal Model, and the DQTM

Up to the present time (2019), the application of a turbulence model actually is still an empirical procedure, because of a lacking proof of the validity of the closure technique in general. Nowadays, statistical methods, including fractal scaling properties, support the fact that such turbulence models could have a sound basis. Therefore, a simple zero-equation turbulence model, which is in agreement with statistical and nonlinear theories, maybe more powerful than an empirical higher order turbulence model (compare with Chap. 3). It is our goal here to provide support of this statement. Being still inspired by the ideas of Prandtl, the transfer of momentum by fluid elements is considered. However, now a control volume of a fluid lump, which causes the transport of momentum, is related to transport by fractal eddies as discussed in the previous subsections. It was shown that the entire transport of momentum can be described by considering only two classes of eddies, namely for the "driving force" the eddies of size $l_m$ and for the eddy diffusivity by the eddies of size $l_k = l_0 - l_m$ (Fig. 7.17).

Then, for a turbulent shear flow, as shown in Fig. 7.17, for the total linear excess momentum $\Delta M$, transported from the high-velocity region to the plane with coordinate $x_2^*$, it follows that

$$\overline{\Delta M} = \widehat{P}_{m,N} = \sum_{n=m}^{N} \widehat{P}_n, \qquad (414a, b)$$

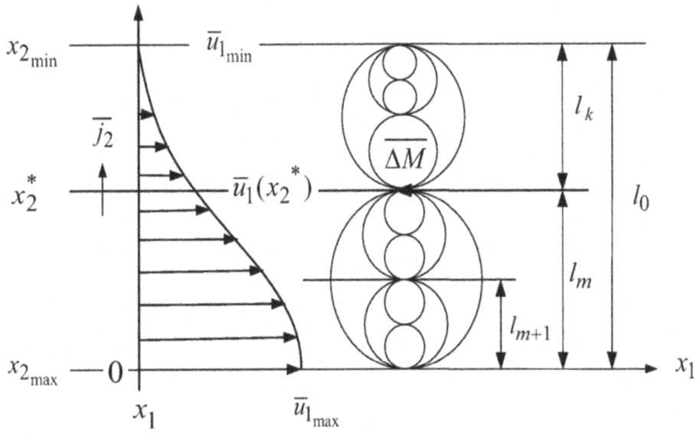


**Fig. 7.17** The mean velocity of a turbulent shear flow in the downstream direction with coordinates $x_1$ and $x_2$ and its comparison with the corresponding eddy distribution is shown. The largest eddies are of size $l_0 = |x_{2_{max}} - x_{2_{min}}|$ and the smallest of size $l_N$, from Egolf (2009), reproduced with permission by Elsevier Science & Technology

in which the eddies of the largest eddy class have the diameter $l_m$ and the ones of the smallest class, $l_N$. Now, the main result of Sect. 7.2.4.10 is applied, namely that the effect of a series of eddies of different classes may be described by the features of the eddies of only a single class. Applying Eqs. (400c,d) twice and simplifying the obtained relation leads to

$$\widehat{P}_{m,N} = O\sum_{n=m}^{N} \widehat{p}_n = O\kappa_{m,N}\widehat{p}_0 = \kappa_{m,N}\widehat{P}_0$$

$$\kappa_{m,N} = \kappa_{0,N} - \kappa_{0,m} = \frac{\left(\frac{a}{b}\right)^m - \left(\frac{a}{b}\right)^N}{\frac{b}{a} - 1}. \qquad (415\text{a} - \text{e})$$

It is clear that, if the largest size eddies would have a diameter of the Kolmogorov dissipation length ($m = N$), there would be no momentum transfer $\Delta M = 0$ (compare with Eqs. (415c,d)). Furthermore, because in turbulent flows usually $b > a$, for small $N$ it follows that there is a reduced momentum transfer. This results in a higher dissipation rate acting on more classes of small-size eddies. Then more of these are inexistent and consequently unable to participate in the momentum transfer.

Now, this model is applied to investigate the Reynolds number dependency of the Reynolds shear stress. By making the NSE dimensionless, it follows that between the dissipation term and the Reynolds shear stress there is a multiplicative factor, which is the inverse overall Reynolds number, $1/\text{Re}_0$. In Chap. 11, it will be demonstrated that this quantity is the stress parameter of a fluid dynamic system; it states that the higher the Reynolds number is, the more the system will be agitated. The question that arises is, whether there are more influences given by the Reynolds number than just this rudimentary dependence. The result of Eq. (413), with its logarithmic dependence on the Reynolds number, in fact suggests that this is so.

From Eqs. (414a) and (415c) it follows that

$$\overline{\Delta M} = \kappa_{m,N} \widehat{P}_0, \tag{416}$$

with the dimensionless constant (see Eq. 415c)

$$\kappa_{m,N} = \frac{\widehat{P}_{m,N}}{\widehat{P}_0}. \tag{417}$$

Now, the flux of momentum, which is mainly related to the eddy diffusivity, must be written in a convenient form. The turbulent momentum flux in the transverse direction is

$$j_2 = \frac{\chi_2}{x_2} u_2, \quad \Rightarrow \quad \bar{j}_2 = \frac{\chi_2}{x_2} \overline{u}_2, \tag{418a, b}$$

where the quantity $\chi_2$ depends on the symmetry of the problem

$$\begin{aligned} &1)\ \chi_2 = a \qquad \text{antisymmetric shear stress} \\ &2)\ \chi_2 = x_2 \qquad \text{symmetric shear stress,} \end{aligned} \tag{419a, b}$$

in which $a$ is a constant.
Next, the turbulent shear stress is defined by the following formula:

$$\tau_{21} = -\rho \overline{u_2{}' u_1{}'} = -\overline{j_2 \Delta M}. \tag{420a, b}$$

This is rewritten to yield

$$\tau_{21} = -c\, \bar{j}_2 \overline{\Delta M} = -\bar{j}_2 \overline{\Delta M}, \tag{421a, b}$$

where in Eq. (421a) a correlation coefficient $0 \leq c \leq 1$ was inserted. The correlation is strong; so, we set in Eq. (421b) $c = 1$ suggesting a perfect correlation.

A turbulent field spreads out by its transverse momentum transport and so gains width. This can be described by the formulas

$$\sigma = \frac{\mathrm{d}b}{\mathrm{d}x_1} = \frac{\overline{u}_2}{\overline{u}_1} \quad \Rightarrow \quad \overline{u}_2 = \sigma \overline{u}_1, \tag{422a – c}$$

with the width of the turbulent domain $b(x_1)$, which increases with the downstream distance $x_1$. In most simple turbulent shear flows, e.g., in a plane shear flow or an axi-symmetric jet (Sect. 9.2), the spreading of the turbulent domain is linear in the downstream direction $x_1$, implying that $\sigma$ is a constant. However, there are other cases, e.g., turbulent plane wake flows behind a cylinder (Sect. 9.1), where this is not the case, in which case $\sigma$ shows a spatial dependence on $x_1$.

As already stated in Sect. 7.2.2, the eddy diffusivity is related to the second main eddy shown in Fig. 7.17 (see upper eddy positioned between $x_{2_{\min}}$ and $x_2^*$). Then, with

Eq. (418b), with diameter $l_k$, and Eq. (422c), in the Lévy-flight/fractal-$\beta$-model, it follows that

$$\bar{j}_2 = \sigma\frac{\chi_2}{x_2}\overline{u}_1 = \sigma\frac{\chi_2}{x_2}\frac{1}{\rho}\widehat{P}_{k,N}. \tag{423a, b}$$

From Eqs. (414a), (421b), and (423) one may deduce

$$\tau_{21} = -\sigma\frac{\chi_2}{x_2}\frac{1}{\rho}\widehat{P}_{k,N}\widehat{P}_{m,N}, \tag{424}$$

and with Eq. (417) one concludes that

$$\tau_{21} = -\sigma\frac{\chi_2}{x_2}\frac{1}{\rho}\kappa_{k,N}\kappa_{m,N}\widehat{P}_0^2. \tag{425}$$

This equation can be rewritten as

$$\tau_{21} = -\sigma\frac{\chi_2}{x_2}\frac{1}{\rho}\lambda_{k,N}\lambda_{m,N}\kappa_{m,\infty}\kappa_{k,\infty}\widehat{P}_0^2, \tag{426}$$

in which

$$\lambda_{\alpha,\beta} = \frac{\kappa_{\alpha,\beta}}{\kappa_{\alpha,\infty}}, \qquad \alpha \in \{k,m\}, \beta = N. \tag{427a, b}$$

From Fig. 7.17, it follows that

$$\widehat{P}_0 = \rho(\overline{u}_{1_{\max}} - \overline{u}_{1_{\min}}), \tag{428}$$

and again from Fig. 7.17 and Eq. (428)

$$\kappa_{k,\infty} = \frac{\widehat{P}_{k,\infty}}{\widehat{P}_0} = \frac{\overline{u}_1 - \overline{u}_{1_{\min}}}{\overline{u}_{1_{\max}} - \overline{u}_{1_{\min}}}, \quad \kappa_{m,\infty} = \frac{\widehat{P}_{m,\infty}}{\widehat{P}_0} = \frac{\overline{u}_{1_{\max}} - \overline{u}_1}{\overline{u}_{1_{\max}} - \overline{u}_{1_{\min}}}. \tag{429a – d}$$

Substituting Eqs. (428) and (429b,d) into Eq. (426), introducing $x_{2_{\max}} = 0$ (see Fig. 7.17) and dropping the asterisk in $x_2^*$ leads to

$$\tau_{21} = \rho\sigma\chi_2\lambda_{k,N}\lambda_{m,N}(\overline{u}_1 - \overline{u}_{1_{\min}})\frac{\overline{u}_{1_{\max}} - \overline{u}_1}{x_{2_{\max}} - x_2}. \tag{430}$$

This is the DQTM in a generalized form. Compared to its previous version, it now contains two additional lambda functions, which in the remainder shall be discussed in detail. To do so, Eq. (427a) with (415e) is applied to derive

$$\lambda_{k,N} = \frac{\kappa_{k,N}}{\kappa_{k,\infty}} = \frac{\left(\frac{a}{b}\right)^k - \left(\frac{a}{b}\right)^N}{\left(\frac{a}{b}\right)^k - \lim\limits_{N\to\infty}\left(\frac{a}{b}\right)^N}. \tag{431a, b}$$

We now restrict ourselves to the case $a < b$ for which one concludes that

$$\lambda_{k,N} = 1 - \left(\frac{a}{b}\right)^{N-k}. \tag{432}$$

The first of two special cases is

$$\lim_{N\to\infty} \lambda_{\alpha,N} = 1, \quad \alpha \in \{k, m\}. \tag{433a, b}$$

In this case, corresponding to the high Reynolds number limit, from the generalized form (Eq. 430), with Eqs. (433a,b), the previous version (Eq. 307b) follows. This is the DQTM which—at the time of its derivation in 1985—was developed by different reasoning. First publications arrived soon after (see Egolf 1991, 1994). One notices that in the DQTM only two very large scales over important distances of the turbulent field are visible in its mathematical formulation.

The second special case is (compare with Eq. (432))

$$\lambda_{N,N} = 0, \tag{434}$$

which is the laminar case with vanishing Reynolds shear stress. The product of the two lambda factors is given by a twofold multiplicative application of Eq. (432)

$$\lambda_{k,N}\lambda_{m,N} = \left[1 - \left(\frac{a}{b}\right)^{N-k}\right] \cdot \left[1 - \left(\frac{a}{b}\right)^{N-m}\right]. \tag{435}$$

From Fig. 7.17, the following discrete relation can be read off:

$$l_k + l_m = l_0. \tag{436}$$

After a division by $l_0$, two ratios of length scales add up to unity,

$$\frac{l_k}{l_0} + \frac{l_m}{l_0} = 1. \tag{437}$$

With the help of Eq. (346a), it is concluded that

$$\frac{l_\alpha}{l_0} = b^{-\alpha}, \quad \alpha \in \{k, m\}. \tag{438a, b}$$

The continuous version of Eq. (436) is (see Fig. 7.17)

$$(x_2 - x_{2_{\min}}) + (x_{2_{\max}} - x_2) = x_{2_{\max}} - x_{2_{\min}}. \tag{439}$$

To obtain the analogous version of Eq. (437), this equation is divided by the half width of the flow domain. This yields

$$\frac{x_2 - x_{2_{\min}}}{x_{2_{\max}} - x_{2_{\min}}} + \frac{x_{2_{\max}} - x_2}{x_{2_{\max}} - x_{2_{\min}}} = 1. \tag{440}$$

By comparison of Eq. (437) with (440) and application of Eq. (438a,b), we deduce

$$b^{-k} = \frac{x_2 - x_{2_{\min}}}{x_{2_{\max}} - x_{2_{\min}}} \tag{441}$$

and

$$b^{-m} = \frac{x_{2_{\max}} - x_2}{x_{2_{\max}} - x_{2_{\min}}}. \tag{442}$$

Taking the logarithm of Eq. (441), for the index $k$, one then finds

$$k = -\frac{\log_e\left(\frac{x_2 - x_{2_{\min}}}{x_{2_{\max}} - x_{2_{\min}}}\right)}{\log_e(b)} \tag{443}$$

and similarly for the index $m$

$$m = -\frac{\log_e\left(\frac{x_{2_{\max}} - x_2}{x_{2_{\max}} - x_{2_{\min}}}\right)}{\log_e(b)}. \tag{444}$$

With Eqs. (413) and (443), we obtain the relation

$$N - k = \frac{\log_e\left(\mathrm{Re}^{3/4}\frac{x_2 - x_{2_{\min}}}{x_{2_{\max}} - x_{2_{\min}}}\right)}{\log_e(b)} \tag{445}$$

and similarly with Eqs. (413) and (444)

$$N - m = \frac{\log_e\left(\mathrm{Re}^{3/4}\dfrac{x_{2_{\max}} - x_2}{x_{2_{\max}} - x_{2_{\min}}}\right)}{\log_e(b)}. \tag{446}$$

Combining Eq. (430) with Eqs. (435), (445), and (446), the generalized DQTM becomes (see Samba et al. 2019)

$$\tau_{21} = \rho\sigma\chi_2\lambda(\mathrm{Re})\,(\overline{u}_1 - \overline{u}_{1_{\min}})\frac{\overline{u}_{1_{\max}} - \overline{u}_1}{x_{2_{\max}} - x_2},$$

$$\lambda(\mathrm{Re}) = \left[1 - \left(\frac{a}{b}\right)^{\log_e\left(\mathrm{Re}^{3/4}\frac{x_2 - x_{2_{\min}}}{x_{2_{\max}} - x_{2_{\min}}}\right)/\log_e(b)}\right] \tag{447a, b}$$

$$\times\left[1 - \left(\frac{a}{b}\right)^{\log_e\left(\mathrm{Re}^{3/4}\frac{x_{2_{\max}} - x_2}{x_{2_{\max}} - x_{2_{\min}}}\right)/\log_e(b)}\right].$$

The original DQTM is generalized here in a manner that the characteristics of the Lévy flight statistics is preserved. To this end, the Kolmogorov microscales were introduced. It is assumed that the Lévy flight characteristic values $a$ and $b$ can also be described as functions of Re, so that Eqs. (447a,b) may simply depend on this non-dimensional number (Yakhot 2018).

The two previous limiting cases shall again be verified in this final result:

1. The first limiting case $N \to \infty$ implies that also $\mathrm{Re} \to \infty$ (compare with Eq. 413). Since $a < b$ this limit applied to Eq. (447b) yields

$$\lim_{\mathrm{Re}\to\infty} \lambda(\mathrm{Re}) = 1. \tag{448}$$

In this special case, the inertial spectrum covers wave numbers from $k = 1/L$ up to infinity. Therefore, eddies belonging to the full wave number spectra are transferring momentum and there is no attenuation effect, which is described by the quantity $\lambda = 1$. Effects of the Lévy flight statistics and especially of the Kolmogorov microscales vanish. Therefore, the DQTM remains in its previous appearance, just as it was applied since 1991.

2. The second limiting case is when the turbulent (half) domain consists of a width of twice the Kolmogorov length scale. In this special case, it follows that $x_2 - x_{2_{\min}} = x_{2_{\max}} - x_2 = l_N$. Furthermore, it is evident that $x_{2_{\max}} - x_{2_{\min}} = l_0$. Then, from Eq. (447b) the result is

$$\lambda = \left[1 - \left(\frac{a}{b}\right)^{\log_e\left(\mathrm{Re}^{3/4} l_N / l_0\right)}\right]^2 = 0. \tag{449a, b}$$

The conclusion that in this case $\lambda = 0$ follows with the help of Eq. (410b) applied to the logarithmic exponent of Eq. (449a). In this special case even the largest

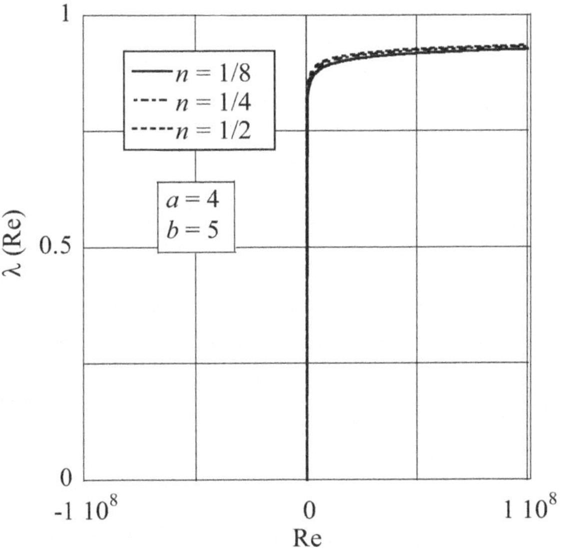


**Fig. 7.18** The correction factor $\lambda$ increases rapidly with the Reynolds number of the overall flow and in the main Reynolds number domain reaches immediately values of 90% and more. This is presented to match with the highest Reynolds number Princeton super pipe experiment (Zagarola et al. 1997). From Samba (2019)

eddies show a diameter that is smaller than the Kolmogorov dissipation length (that in this case may be very large) and thereby are destroyed by the transformation of turbulent kinetic energy into heat. Therefore, no eddies remain that can transfer momentum and a realistic closure must yield a shear stress that is zero. No zero-equation turbulence model is known that correctly describes this dissipation effect. The multiplicative factor $\lambda$ may be combined with any turbulence model to generalize it to fulfill this important requirement. In Fig. 7.18, the difference between the curves and the upper limiting value $\lambda \equiv 1$ describes the missing momentum transfer by the eddies that are destroyed by dissipation.

In this section, the importance of the correction term $\lambda$ to geophysical and technical turbulent flows shall be examined. A characteristic value of the fractal dimension $D$ (Hausdorff-Besicovitch dimension) for turbulent flows is

$$D = 2.\overline{6}. \tag{450}$$

This characteristic value is given also by Mandelbrot (1977) for Gauss-Kolmogorov turbulence. Following Hughes et al. (1981), a definition of the fractal dimension of a set may be explained as follows: *If a finite part $\Omega$ of a set can be divided into $\gamma$ identical parts, each of which is geometrically similar to $\Omega$, with a similarity ratio r, then its fractal dimension is*

$$D = \frac{\log_{e}\gamma}{\log_{e}\left(\frac{1}{r}\right)}. \tag{451}$$

*For our clustered random walk* [Lévy walk], *we note that each cluster can be divided into a set of clusters one order lower in the hierarchy and there are, on the average, about a of these lower order clusters, so that* $\gamma \cong a$. *Each lower order cluster, when expanded by a factor b, recovers the original cluster, giving* $r \cong 1/b$. *Then, it follows*:

$$D \cong \frac{\log_{e}a}{\log_{e}b} = \mu. \tag{452a, b}$$

We are working out a simplified one-dimensional model. Therefore, for this special case, we approximate the fractal dimension $D_1$ by

$$D_1 = \frac{D}{3} = 0.\overline{8}. \tag{453a, b}$$

A case that fulfills inequality (322b), $b \geq \sqrt{a}$, and Eq. (453b) with good accuracy is the one with the set of Lévy characteristic values: $a = 4$ and $b = 5$.

To apply our previous considerations on the $\lambda$ correction to a realistic situation of a turbulent flow, we now take the values $a = 4$ and $b = 5$ and substitute them into Eq. (447b). Furthermore, we introduce non-dimensional ratios of different distances $n = (x_2 - x_{2_{\min}})/(x_{2_{\max}} - x_{2_{\min}})$ and choose values $n = 1/8$, 1/4, and 1/2. In Fig. 7.18 for all these parameters $n$ the $\lambda$-curves are plotted.

It is realized that the correction factor increases very fast with the overall Reynolds number Re and is almost identical for different values of $n$. Therefore, it follows that in turbulent flows the correction factor $\lambda$ is only important at locations slightly above the critical Reynolds number. This is convenient, because with the generalized DQTM, containing the factor $\lambda$, hardly any analytical solutions are expected to be derivable. Therefore, in further work, that will be presented in Chap. 9, we concentrate on the earlier more rudimentary form of the DQTM.

Looking at the DQTM, now, one could conclude as in Tennekes and Lumley (1972): *In reality, turbulence consists of fluctuating motion in a broad spectrum of length scales. However, …, one may argue that large eddies contribute more to the momentum transfer than small eddies.*

By contrast, the new interpretation is that all scales contribute in an equally important manner to turbulent momentum transport. It is the scaling properties that allow to collect all important contributions also of the small- and medium-scale eddies in a description of only showing visibly two largest scales. This seems to be a very satisfying explanation why even simple turbulence models may lead to excellent results. It shall not be a serious criticism of Tennekes and Lumley's statement, and their natural interpretation, based on the knowledge at their time (≈1975). It was basically Richardson's and Mandelbrot's introduction of fractal and multi-fractal geometries, that makes it now possible to revise earlier statements as, e.g., that given above.

Starting with the Lévy-flight statistics and additionally applying the fractal $\beta$-model, which also implies making use of fractal theoretical tools, the DQTM was derived in a rather simple geometrical manner. Just as for this, long-range correlations and multi-scale elements are combined, as it is expected to occur in prospective new turbulence models. A newly derived $\lambda$-correction term to the DQTM allows to take the dissipation with its onset value, called Kolmogorov's dissipation length, into consideration. It was shown that this correction factor plays a crucial role in studies of turbulent flows only slightly above the critical Reynolds number. An advantage is that the multiplicative correction term can be split from the DQTM and also be applied in other turbulence models.

## 7.3 New Nonlocal Turbulence Models

### *7.3.1 Introduction*

In Chap. 6, from Kraichnan's convolution integral formulation of the Reynolds shear stress tensor, we have experienced that in turbulence the concept of nonlocality is built by collecting information not just at a single point at the actual time, but also in the neighborhood of this local point and not just at the present time, but also in the past.

To work on this topic and to make ideas clearer, we follow Herrmann (2011, 2014), Osterwalder (1980), Oldham and Spanier (1974), and Luo and Afraimovich (2010). Let us consider a position of a mass point in space $x(t)$. Now, to determine its velocity, it is necessary to have information of the position of this object at the two times $t - \delta t$ and $t + \delta t$. For spaces of higher dimensions, one can argue that some information inside a sphere of radius $r$ is required in order to determine this quantity. In the definition of the velocity as a derivative of spatial location $x$ as a function of time $t$, $r$ tends to zero. Therefore, velocity is an epsilon- or zero-nonlocal quantity, respectively, or simply a local quantity. The definition of the Weierstrass derivative is here modified to be twofold, namely

$$u(t) = \frac{\mathrm{d}x}{\mathrm{d}t} = \begin{cases} \lim\limits_{\tau \to 0} \dfrac{x(t) - x(t-\tau)}{\tau}, \text{ backward,} \\ \lim\limits_{\tau \to 0} \dfrac{x(t+\tau) - x(t)}{\tau}, \text{ forward,} \end{cases} \quad (454\mathrm{a} - \mathrm{c})$$

where $0 \leq \tau < \varepsilon$ and $\varepsilon$ is a small positive real number. Even if for a sufficiently steady mathematical function the two definitions lead to the same quantitative result, the first definition requires information backward in time or in the past and the second one forward in time or into the future. The latter definition is problematic, if the independent variable is time, because causality requires that a current state of a physical object is only determined by the present or/and past states, but cannot be influenced by its future states. Therefore, it violates causality (for a discussion of locality and causality see, e.g., Kant 1781; Einstein et al. 1935; von Weizsäcker 1939). The second definition of the derivative plays a crucial role in the description of anti-particle flights, particles known to fly backward in time.

It is also possible to define the derivative by a convolution integral of the form

$$u(t) = \frac{\mathrm{d}}{\mathrm{d}t}x(t) = 2\int\limits_{0}^{\infty} \delta(\tau)u(t-\tau)\mathrm{d}\tau, \tag{455a, b}$$

where we restrict ourselves only to the backward derivative. The symbol $\delta(t)$ denotes the Dirac distribution (generalized function) with the property

$$\int\limits_{-\infty}^{\infty} \delta(t)f(t)\mathrm{d}t = f(0). \tag{456}$$

For a definition of the Dirac distribution, consult Lighthill (1958), Gelfand and Schilov (1962) or Schwartz (1966). Numerous functions $\omega(\lambda,\tau)$ are valid candidates to create distributions with the Dirac delta distribution as limit, because they are everywhere differentiable and also obey the following attenuation behavior

$$O(|t|^{-n}) \to 0, \quad t \to \infty, \quad \forall n > 0. \tag{457}$$

These functions are also called test functions. Written with the dimensionless variable $|t|/\lambda$ some ideal candidates are, e.g., the exponential, the Airy function, and the sine function (see Table 7.2). Then, it follows for a smooth parameter $\lambda \geq 0$ independent of the particular test function and with the normalization (see Eq. 460)

$$\delta(t) = \lim_{\lambda \to 0} \omega(\lambda, t). \tag{458}$$

Without going into any mathematical depth, like $D$-convergence, dual space $D'$, regularity, etc. (Gelfand and Schilov 1962), in a straightforward manner, we substitute Eq. (458) into Eq. (455b) to obtain

$$\frac{\mathrm{d}}{\mathrm{d}t}x(t) = 2\lim_{\lambda \to 0} \int\limits_{0}^{\infty} \omega\,(\lambda, t)\; u(t-\tau)\;\mathrm{d}\tau. \tag{459}$$

At this point let us introduce higher nonlocality by omitting in the above formulae the limit operator $\lambda \to 0$. In this generalization, we no longer require Eq. (457) to be fulfilled. The factor 2 must be replaced by a normalization factor $\Omega$ defined such that

$$\frac{1}{\Omega}\int_{0}^{\infty} \omega(\lambda,\tau)\,\mathrm{d}\tau = 1. \tag{460}$$

With these specifications, a nonlocal derivative is obtained by the following operator assignment (see Herrmann 2011, 2014):

$$\frac{\mathrm{d}}{\mathrm{d}t}\bigg|x(t)_{\text{nonlocal}} = \frac{1}{\Omega}\int_{0}^{\infty} \omega\,(\lambda,\tau)\frac{\mathrm{d}}{\mathrm{d}t}\bigg|x(t-\tau)_{\text{local}}\,\mathrm{d}\tau. \tag{461}$$

A further generalization is to take over this recipe for arbitrary operators $\widehat{\mathrm{O}}$, which reads as

$$\widehat{\mathrm{O}}\bigg|(f(t))_{\text{nonlocal}} = \frac{1}{\Omega}\int_{0}^{\infty} \omega\,(\lambda,\tau)\,\widehat{\mathrm{O}}\bigg|\widehat{\mathrm{s}}_{-}(\tau)f(t)_{\text{local}}\,\mathrm{d}\tau, \tag{462}$$

where $\widehat{\mathrm{s}}_{-}(\tau)$ is the (negative) shift operator, obeying the following rule with the minus signs:

$$\widehat{\mathrm{s}}_{\pm}(f(t)) = f(t\pm\tau). \tag{463}$$

We will see below that non-commutativity is important in the ensuing analysis. An important fact is that not all operators commute with the shift operator. A lack of commutativity is described, e.g., by applying the commutator, that here leads to $\left\{\widehat{\mathrm{O}},\widehat{\mathrm{s}}\right\} = \widehat{\mathrm{O}}\widehat{\mathrm{s}} - \widehat{\mathrm{s}}\widehat{\mathrm{O}} \neq 0$, where in the classical mechanics this operation on $\widehat{\mathrm{O}}$ and $\widehat{\mathrm{s}}$ vanishes for commuting operators. Therefore, there are different possibilities of Eq. (462) with different orderings of the operators. In this equation, the most appropriate choice for nonlocal turbulence models has been chosen.

The assignment (Eq. 462) is a clear instruction how from a local a nonlocal operator can be created, namely by applying four rules (Herrmann 2011) :

R1) Starting with the local operator $\widehat{\mathrm{O}}_{\text{local}}$
R2) Choosing the appropriate shift operator $\widehat{\mathrm{s}}_{\pm}(\tau)$
R3) Choosing the appropriate weighting function $\omega\,(\lambda,\tau)$ with its norm $\Omega$
R4) Averaging over the appropriate domain.

These rules are an efficient tool to generalize local turbulence models to different nonlocal counterparts. If one counts all the local models that are available and takes only the fraction of possible weighting functions listed in Table 7.2 into consideration, and even when neglecting different forms of Eq. (462) with different orders of non-commuting operators, one is still confronted with an extremely large variety of new nonlocal turbulence models. They can be created by simply applying the techniques defined by the rules R1–R4, which actually are only routine assignments.

**Table 7.2** Weighting functions listed by their name, definitions of $\omega$, and norm $\Omega$ (from Herrmann 2011; Abramowitz and Stegun 1984)

| Name | Weight $\omega$ $(\lambda,\tau)$ | Norm $\Omega$ |
|---|---|---|
| Inverse parabolic | $\frac{1}{1+(\tau/\lambda)^2}$ | $\frac{\lambda\pi}{2}$ |
| Inverse cubic | $\frac{1}{1+(\tau/\lambda)^3}$ | $\frac{2\pi}{3\sqrt[3]{\lambda}}$ |
| Airy | $\mathrm{Ai}(\tau/\lambda)$ | $\frac{\lambda}{3}$ |
| Cosine | $\frac{1-\cos(\tau/\lambda)}{\tau^2}$ | $\frac{\pi}{2\lambda}$ |
| Inverse cosh | $\frac{1}{\cosh(\tau/\lambda)^2}$ | $\lambda$ |
| Exponential | $e^{-(\tau/\lambda)}$ | $\lambda$ |
| Gaussian | $e^{-(\tau/\lambda)^2}$ | $\lambda\,\Gamma(3/2)$ |
| Sine | $\frac{\sin(\tau/\lambda)}{\tau}$ | $\frac{\pi}{2}$ |
| Liouville | $\tau^{\lambda-1}$ | $\Gamma(\lambda)$ |
| Heaviside (const.) | $\theta(\lambda-\tau)$ | $\lambda$ |
| Linear | $(\lambda-\tau)\,\theta\,(\lambda-\tau)$ | $\frac{\lambda^2}{2}$ |
| Parabola | $(\lambda-\tau)^2\theta\,(\lambda-\tau)$ | $\frac{\lambda^3}{3}$ |
| Caputo | $\tau^{\lambda-1}\theta\,(\lambda-\tau)$ | $\Gamma(\lambda)$ |

$\Gamma$ is the Gamma function (Abramovitz and Stegun 1984)

This method shall now be applied to numerous problems, a purely mathematical one and some in the new area of nonlocal turbulence modeling.

### 7.3.2 *Liouville Fractional Derivative*

We choose a commuted Eq. (462), which is also presented in Herrmann (2011)

$$\widehat{\mathrm{O}}\Big|_{\text{nonlocal}}(f(t)) = \frac{1}{\Omega}\widehat{\mathrm{O}}\Big|_{\text{local}} \int\limits_{0}^{\infty} \omega\,(\lambda,\tau)\,\widehat{\mathrm{s}}_{-}(\tau)\,f(t)\,\mathrm{d}\tau \tag{464}$$

and consider the Liouville weight $\omega(\lambda,\tau) = \tau^{\lambda-1}(0 < \lambda < 1)$, norm $\Omega = \Gamma(\lambda)$ (Table 7.2), the first derivative and the negative shift operator $\widehat{\mathrm{s}}_{-}$. This leads to

$$\frac{\mathrm{d}}{\mathrm{d}t}\Big|_{\text{nonlocal}}(f(t)) = \frac{1}{\Gamma(\lambda)}\frac{\mathrm{d}}{\mathrm{d}t}\Big|_{\text{local}} \int\limits_{0}^{\infty} \tau^{\lambda-1} f(t-\tau)\mathrm{d}\tau. \tag{465}$$

By substituting $\lambda = 1 - \alpha$ and $\xi = t - \tau$, it follows that

$$\left.\frac{\mathrm{d}}{\mathrm{d}t}\right|_{\text{nonlocal}}(f(t)) = \frac{1}{\Gamma(1-\alpha)}\left.\frac{\mathrm{d}}{\mathrm{d}t}\right|_{\text{local}}\int_{-\infty}^{t}(t-\xi)^{-\alpha}f(\xi)\mathrm{d}\xi = {}_{\mathrm{L}}\mathrm{D}_{+}^{\alpha}f(t). \tag{466a, b}$$

${}_{\mathrm{L}}\mathrm{D}_{+}^{\alpha}f(t)$ is the left-handed Liouville fractional derivative (see Herrmann 2011) . A fractional derivative is a derivative of non-integer order.

### 7.3.3 *Overview of the Derivation of Important Nonlocal Turbulence Models*

It is evident that this method can be applied to derive the Hinze-Sonnenberg-Bultjes convolution integral model (see Hinze et al. 1974) with a memory kernel $M$ as a function of time, which is the equivalent of the weighting function $\omega$. Therefore, in this general derivation from Table 7.2, not yet a concrete $\omega$ needs to be selected. In time-dependent problems of turbulence research, one has to be very careful with all the occurring time scales, e.g., that of averaging the velocity field and now, additionally, also by taking into consideration the time scale of the fading or memory process, with its characteristic time.

The most general nonlocal models with a time and three-dimensional space dependence lead directly to Kraichnan's convolution integral model (compare with Eq. 222). Similarly, here too, this general solution is not further developed to a special application, so that neither in this example a weighting function $\omega = \nu_{ij}$ is chosen at this moment.

The analogue examples, exhibiting space instead of a time dependence, are, e.g., nonlocal generalizations of the main three Prandtl zero-equation turbulence models, namely the mixing length model (Prandtl 1925), the mean-gradient model (Prandtl 1942), and the (modified) free shear-layer model (Egolf and Hutter 2016c, 2017). For the purposes in this text, the first and third one will be further investigated. This will show high systematics (see Table 7.3), which reveals that a local model is actually missing. In Sect. 7.3.6 such a model will be supplemented.

Prandtl's mixing-length model contains two gradients. In Table 7.3, we denote them by the identification (G-G), where G denotes "Gradient." This zero-equation turbulence model is a fully local model. On the other hand, Prandtl's shear-layer model contains in its eddy diffusivity a velocity difference, which, in combination with a mixing-length in the denominator, can be interpreted to represent a difference quotient. This is shown in a detailed derivation in the remainder of this section. Just as the Liouville fractional derivative also the difference quotient is a nonlocal operator. In this sense, Prandtl's shear-layer model may be interpreted to be built by a local and a nonlocal operator. Such kind of models are called semi-local models. In the denotation of our systematics Prandtl's shear-layer model consists now of a nonlocal difference (-quotient) (D) and a local gradient (G). Therefore, its abbreviation in the first column and second line is (D-G). Being very systematic, now, the question is whether there exists a (G-D) model, which shows a gradient in its eddy diffusivity, but none in the driving "gradient," here described by a nonlocal difference quotient acting over a possibly large distance. It is stated that in the last century this model was not derived nor published. Explicitly, it has the appearance

$$\tau_{21} = \rho\sigma\chi_2^2 \left|\frac{\partial \overline{u}_1}{\partial x_2}\right| \frac{\overline{u}_{1_{\max}} - \overline{u}_1}{x_{2_{\max}} - x_2}. \tag{467}$$

Notice that in this equation, the following changes from Prandtl's to our notation were performed: $c = \sigma$ and $l_m = \chi_2$. In Table 7.3, this model is called *Mean velocity difference model*. In some special cases of $\chi_2$, Prandtl's shear-layer model and this supplementary new model may become identical. From the discussed local model (G-G) and the two semi-local models [(D-G) and (G-D)], henceforth, new nonlocal zero-equation turbulence models will be developed.

### 7.3.4 Liouville–Prandtl Mixing Length Model

Liouville-type nonlocal models can be created by replacing a usual first-order derivative in a local turbulence model by the Liouville fractional derivative. Notice

**Table 7.3** Five important transitions from local to nonlocal turbulence models

| Local and semi-local models | Weighting functions | Nonlocal models |
|---|---|---|
| Prandtl's mixing-length model (G-G) | Twice Liouville function (L-L) | Liouville–Prandtl mixing-length model (F-F) |
| Prandtl's (modified) shear-layer model (D-G) | Heaviside distribution already applied. Liouville function (H-L) | Heaviside–Liouville–Prandtl shear-layer model (D-F) |
| Mean velocity difference model (G-D) | Liouville function. Heaviside distribution already applied. (L-H) | Liouville–Heaviside model (F-D) |
| Prandtl's (modified) shear-layer model (D-G) | Heaviside distribution already applied. Heavisde distribution (H-H) | Difference-Quotient Turbulence Model (DQTM) (D-D) |
| Prandtl's mixing-length model (G-G) | Twice Heaviside distribution (H-H) | Difference-Quotient Turbulence Model (D-D) |

The weighting functions that are taken into consideration are presented in the center row. The nonlocal models of this table show systematic behavior and are discussed in the main text of the chapter

*G* gradient, *D* difference, respectively, difference quotient, *F* fractional derivative of Liouville type, *L* Liouville function, *H* Heaviside distribution

that in this case, the following substitutions are applied to derive the nonlocal models:

1. $t \to x_2$ (time–space replacement)
2. $f \to \overline{u}_1$ (average downstream velocity)
3. $\widehat{\mathrm{O}} \to \frac{\partial}{\partial x_2}$ (operator: partial derivative)
4. $\omega \to \tau^{\lambda-1}$ (Liouville weighting function)
5. $\tau \to \xi_2$ (time–space replacement)
6. $0 \to -\infty$ (in space no causality problem).

By applying these substitutions in Eqs. (466a,b), the time-dependent fractional derivative becomes space coordinate dependent, viz.,

$$\frac{\mathrm{d}}{\mathrm{d}x_2}\bigg|_{\text{nonlocal}}(\overline{u}_2(x_2)) = \frac{1}{\Gamma(1-\alpha)}\frac{\mathrm{d}}{\mathrm{d}x_2}\bigg|_{\text{local}} \int_{-\infty}^{\infty} (x_2-\xi_2)^{-\alpha} f(\xi_2)\,\mathrm{d}\xi_2 = {}_{\mathrm{L}}D_{\pm}^{\alpha}\overline{u}_2(x_2) \tag{468a, b}$$

Different from a time integration, here also an integration over negative values is allowed. Such an adaptation does not violate the causality principle.

Now, we rewrite Eq. (53a), which yields

$$\tau_{21} = \rho\,\varepsilon_{\mathrm{m},\alpha}\frac{\partial \overline{u}_1}{\partial x_2}, \quad \alpha \in \{\mathrm{ml}, \mathrm{mg}, \mathrm{sl}\}. \tag{469}$$

The parameter $\varepsilon_{\mathrm{m,ml}}$ of the mixing-*l*ength theory is given in Eq. (54), $\varepsilon_{\mathrm{m,mg}}$ of the mean-*g*radient model in Eq. (68) and $\varepsilon_{\mathrm{m,sl}}$ for the shear-layer model in Eq. (70). By replacing the two derivatives in Prandtl's mixing-length model (Eq. 53b) by the left- and right-handed Liouville fractional derivative each, a first valuable nonlocal turbulence model is obtained

$$\tau_{21} = \rho\sigma\chi_2^{\alpha+\beta}\,|{}_{\mathrm{L}}D_{\pm}^{\alpha}\overline{u}_1(x_2)|\,{}_{\mathrm{L}}\mathrm{D}_{\pm}^{\beta}\overline{u}_1(x_2). \tag{470}$$

This nonlocal model is listed in the first line of Table 7.3, where it is named the *Liouville-Prandtl mixing-length model.* This model was developed from a two-gradient model (G-G) by applying twice the Liouville weighting function (L-L) resulting in a nonlocal model consisting of two Riemann-Liouville fractional derivatives (F-F) (compare with abbreviations in the second line of Table 7.3).

### 7.3.5 The Heaviside–Liouville–Prandtl Shear Layer Model

Now, in Prandtl's shear-layer turbulence model of 1942, we replace the single local operator, respectively, first-order derivative, acting on the average velocity by the Liouville fractional derivative.

Furthermore, we apply a small, but very important correction to Prandtl's eddy diffusivity (see Eq. (70)). This results in a modified version of $\varepsilon_{\mathrm{m}}$ of the shear-layer model, Prandtl's latest development in the area of zero-equation turbulence modeling[6]

$$\varepsilon_{\mathrm{m}} = \rho\sigma\chi_2(\overline{u}_1 - \overline{u}_{1_{\min}}). \tag{471}$$

The quantity $\sigma$ is a constant and $\chi_2$ is a length scale of the flow in the $x_2$ direction (Sect. 7.2.2).
Combining Eq. (53a) with (471) yields

$$\tau_{21} = \rho\sigma\chi_2(\overline{u}_1 - \overline{u}_{1_{\min}})\frac{\partial \overline{u}_1}{\partial x_2}. \tag{472}$$

Now also here, the local first-order derivative will be replaced by the nonlocal Liouville fractional derivative. This leads to

$$\tau_{21} = \rho\sigma\chi_2^{\alpha}(\overline{u}_1 - \overline{u}_{1_{\min}})_{\mathrm{L}}D_{\pm}^{\alpha}\overline{u}_2(x_2). \tag{473}$$

As first authors Egolf and Hutter (2017) proposed this nonlocal Liouville fractional derivative model for the Reynolds stress and emphasized the equivalence between nonlocal and fractional derivative turbulence models. They called this emerging new model the *Fractional-derivative turbulence model*. This new fractional turbulence model is listed in Table 7.3 in the third line, and is called here the *Heaviside-Liouville-Prandtl shear-layer model*.

### 7.3.6 The Liouville-Heaviside Turbulence Model

Now we have become familiar with the substitution methods. If in the semi-local model (Eq. 467) in the eddy diffusivity the first-order derivative is replaced by the fractional Liouville derivative, a next fully nonlocal turbulence model follows, viz.,

[6]By canceling in Eq. (70), the index max in $\overline{u}_{1\,\max}$, the stiffness (constance) of $\varepsilon_{\mathrm{m}}$ is removed, leading to better agreement of mean velocity profiles with experiments at the boundaries of turbulent domains, where $\varepsilon_{\mathrm{m}} = 0$, because of $\overline{u}_1 = \overline{u}_{1\,\min} = 0$.

$$\tau_{21} = \rho\sigma\chi_2^{1+\alpha}\left|{}_{\mathrm{L}}D_{\pm}^{\alpha}\overline{u}_2(x_2)\right|\frac{\overline{u}_{1_{\max}} - \overline{u}_1}{x_{2_{\max}} - x_2}. \tag{474}$$

This model may be interpreted to have been developed by a mixed application of two weighting functions, namely the Liouville function and the Heaviside distribution (L-H). The resulting model contains a fractional derivative and a difference quotient (F-D) and is called *Liouville-Heaviside turbulence model* (see Table 7.3).

Turbulence models created by two different kinds of weighting functions, like Eqs. (473) and (474), also when compared with data, do not seem to be realistic (see later).

### 7.3.7 The Difference-Quotient Turbulence Model

The basic transformation Eq. (461) is now applied to our situation in turbulence modeling, namely

$$\frac{\partial}{\partial x_2}\bigg|_{\text{nonlocal}}(\overline{u}_1(x_2)) = \frac{1}{\Omega}\int\limits_{-\infty}^{\infty} N(\lambda, \xi_2)\,\frac{\partial}{\partial x_2}\bigg|_{\text{local}}\widehat{\mathrm{s}}_{-}(\xi_2)\,\overline{u}_1(x_2)\mathrm{d}\xi_2, \tag{475}$$

where the weighting "function" is generalized to also depend on the spatial coordinate $x_2$. Different than in time integration, here also the integral is extended from $-\infty$ to $+\infty$, because there are no causality problems in a space description.
In the new notation, Eq. (460), describing the norm, transforms to

$$\Omega = \frac{1}{\int\limits_{-\infty}^{\infty} N(\lambda, \xi_2)\,\mathrm{d}\xi_2}. \tag{476}$$

With the substitution $\xi_2 = x_2 - x_2{}'$, see Eqs. (472) and (475), one obtains the following integral representation:

$$\overline{u_2{}'u_1{}'} = -\sigma\chi_2(\overline{u}_1 - \overline{u}_{1_{\min}})\,\frac{\int\limits_{-\infty}^{\infty} N(\lambda, x_2 - x_2{}')\,\frac{\partial \overline{u}_1}{\partial x_2'}(x_2{}')\mathrm{d}x_2{}'}{\int\limits_{-\infty}^{\infty} N(\lambda, x_2 - x_2{}')\,\mathrm{d}x_2{}'}. \tag{477}$$

This closure was foreseen and published by Egolf and Weiss (1996a, b, 1998). Actually, it was more a guess in analogy to a rheological constitutive equation containing a convolution integral. That this new method of fractional calculus leads to the same result, makes it even more satisfying, because it now sets the result (Eq. 477) on a solid mathematical–physical basis.

In spite of developments by Kraichnan (1964) and Hinze et al. (1974) about a quarter century earlier, Egolf and Weiss (1998) wrote: *"Today, in turbulence research, we have some conceptual understanding of approach* [here they relate to the analogous time-dependent version of Eq. (477) of Hinze et al. (1974)], *even though a detailed mathematical derivation from basic equations is, to our knowledge, yet to be found.* From today's point-of-view, it becomes clear that it was necessary to first apply fractional calculus to arrive at this stage of a deeper understanding and to prove the general closure (Eq. 477).

The next question is which weighting function shall be chosen to proceed with this closure. We have given preference to two weighting functions, which are the Liouville "function" and the Heaviside distribution, although there may be others that could also succeed. We shall give a preference to the Heaviside distribution, because of four reasons (not all shall be taken as proofs, but as supports for this strategy):

1. Prandtl intuitively improved his mixing length model toward his shear-layer model. It is assumed that the aerodynamicist probably did not have knowledge of the nonlocality theory and fractional calculus. However, his approach can be derived by these concepts by taking the Heaviside distribution into consideration. The incompleteness of the shear-layer model (too many indices), which otherwise was developed in the right direction, is the reason that its results are not superior compared to those derived by applying the mixing-length model.[7]
2. There is a strong physical interpretation of turbulence that supports the application of Heaviside distributions. This will be described just below this listing.
3. Applying Heaviside distributions to Prandtl's turbulence models directly leads to the DQTM that has proven to be very successful in describing elementary turbulent shear flows in best agreement with experimental data.
4. The Heaviside distribution leads to the simplest nonlocal turbulence model, requiring the least computation power (see Chap. 12).

Let us be inspired by the physical interpretations of momentum transport by eddies, also based on Prandtl's ideas, described in Sects. 7.2.2 and 7.2.4. In flows with an inertial range, where in- and outflows of energy into a small spectral domain are in equilibrium, over an eddy turning time turbulent kinetic energy does not change, so that a transport by large eddies over large distances acts as being nearly dissipation free and shows thereby practically no decay of excess momentum. This important insight favors very much the application of the Heaviside distribution (see Table 7.2), which (like a Fermi distribution) is constant over a correlation length $\lambda$ and then immediately decays to zero. In the field of turbulence research, this is like a paradigm change. Influenced by the long past period of development of linear and

[7]The stiffness of his model given by the indices 'min' and 'max' probably hindered him to transfer the second gradient to a mean velocity difference, because then his Reynolds shear stress would have simply become a constant. On the other hand, by doing so, he probably already in the 1940s would have discovered the DQTM. Speculation is also whether the second world war was a hindrance for a further development, however, this seems probable.

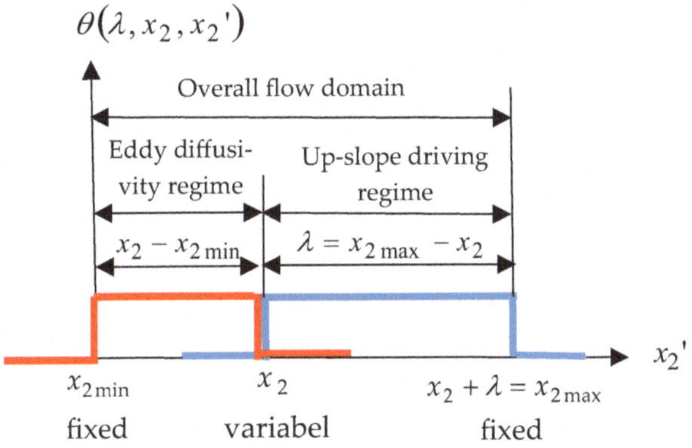


**Fig. 7.19** The two rectangular carrier domains correspond to the space occupied by the largest up- and down-gradient eddy (in Fig. 7.17 compare with the eddies of diameter $l_m$ and $l_k$)

local theories, almost all correlation functions, weighting functions, etc. were chosen to be of decaying type, e.g., inverse parabolic, exponential, and Gaussian (see Table 7.2). Such functions are appropriate for modeling turbulence in regimes dominated by dissipation, but rather questionable to describe inertial range turbulence, especially such of high Reynolds number flows.

Now, we describe the kernel $N(\lambda, x_2 - x_2')$ by a linear combination of Heaviside distributions (two-sided step "functions") (see Fig. 7.19)

$$N(\lambda, x_2 - x_2') = \theta(x_2' - x_2) \cdot \theta(x_2 - x_2' + \lambda). \tag{478}$$

Adapting this to our turbulence modeling (see again Fig. 7.19), with $x_{2_{\max}} = x_2 + \lambda$, it follows that

$$N(x_2 - x_2'; x_{2_{\max}}) = \theta(x_2' - x_2) \cdot \theta(x_{2_{\max}} - x_2'), \tag{479}$$

where by definition the Heaviside distribution is given by

$$\theta(x) = \begin{cases} 0, & x < 0, \\ 1 & x \geq 0. \end{cases} \tag{480a, b}$$

Substituting this into Eq. (477) yields

$$\overline{u_2' u_1'} = -\sigma \chi_2 (\bar{u}_1 - \bar{u}_{1_{\min}}) \frac{\int\limits_{-\infty}^{\infty} \theta(x_2' - x_2) \cdot \theta(x_{2_{\max}} - x_2') \dfrac{\partial \bar{u}_1}{\partial x_2'}(x_2') \mathrm{d}x_2'}{\int\limits_{-\infty}^{\infty} \theta(x_2' - x_2) \cdot \theta(x_{2_{\max}} - x_2') \, \mathrm{d}x_2'}. \tag{481}$$

Some information in the two Heaviside distributions can be transferred to the integration limits. Notice that below $x_2$ and above $x_{2_{\max}}$, there is no contribution ($\theta = 0$) and between these values just the integrand is present ($\theta = 1$) with no modification by the weighting and nonlocality distribution, respectively; thus

**Table 7.4** Comparison of Prandtl's turbulence models with the DQTM

| Turbulence model | Eddy diffusivity | Upgradient driving term |
|---|---|---|
| Prandtl's mixing-length model, 1925 | Mean velocity gradient | Mean velocity gradient |
| Prandtl's shear-layer model, 1942 | Mean velocity difference | Mean velocity gradient |
| Egolf's difference quotient turbulence model, 1991 | Mean velocity difference | Mean velocity difference quotient |

Upgrading the first gradient in the mixing-length model with a Heaviside weighting distribution to a nonlocal derivative leads to the modified shear-layer model. Transferring both gradients in the mixing-length model or the remaining one in the modified shear-layer model finally leads to the nonlocal DQTM. Therefore, this newer model is the natural nonlocal generalization of two of Prandtl's zero-equation turbulence models

$$\overline{u_2'u_1'} = -\sigma\chi_2\left(\overline{u}_1 - \overline{u}_{1_{\min}}\right)\frac{\int\limits_{x_2}^{x_{2_{\max}}}\frac{\partial\overline{u}_1}{\partial x_2'}(x_2')\,\mathrm{d}x_2'}{x_{2_{\max}} - x_2}. \tag{482}$$

With the following implicit definitions for the two fixed space values $x_{2_{\min}}$ and $x_{2_{\max}}$:

$$\overline{u}_1(x_{2_{\max}}):=\overline{u}_{1_{\max}}, \qquad \overline{u}_1(x_{2_{\min}}):=\overline{u}_{1_{\min}}, \tag{483a, b}$$

Equation (482) is transformed into the final result

$$\overline{u_2'u_1'} = -\sigma\chi_2\left[\overline{u}_1(x_1,x_2) - \overline{u}_{1_{\min}}(x_1)\right]\frac{\overline{u}_{1_{\max}}(x_1) - \overline{u}_1(x_1,x_2)}{x_{2_{\max}} - x_2},$$

which is the original *Difference-Quotient Turbulence Model (DQTM)* identical to Eq. (307b).

*We thus have proven that the DQTM is the natural nonlocal generalization of the modified local Prandtl shear-layer model by applying as weighting "function" a Heaviside distribution.* This is shown in Table 7.3 in the fifth line.

By applying a Heaviside distribution in the nonlocalization process applied to Prandtl's mixing-length model, the modified Prandtl shear-layer model follows, and by a second application of a Heaviside distribution as demonstrated above, the DQTM is the direct consequence (see Table 7.3, sixth line and Table 7.4).

### *7.3.8 Summary*

In summary, we conclude that old physical ideas of Prandtl on turbulent momentum transfer combined with newest mathematical methods of fractional calculus lead to convincing new derivations of nonlinear and nonlocal zero-equation turbulence models. In this section, four favorites were derived from old local turbulence models that Prandtl had developed. They are listed in Table 7.3. Two of them used two different weighting functions in the eddy diffusivity and the driving gradient. This seems unusual and offers a criterion to favor the two remaining models, which are the Fractional-Derivative Turbulence Model (FDTM) and the Difference-Quotient Turbulence Model (DQTM). Taking Prandtl's intuitive development of a new eddy diffusivity in his 1942 shear-layer model as criterion to decide between the two remaining models, the DQTM must be favored (see also the ten arguments listed below).

Finally, it was possible to derive the DQTM from Prandtl's shear-layer model by removing a small inconsistency and applying four mathematical rules. These are instructions on how to generalize a local operator to different nonlocal versions, depending on the chosen weighting "function." It shows that the Reynolds shear stress is described by a conservative picture of momentum transfer over large flow regions. This physical picture is perfectly described by the (constant) Heaviside weighting distribution. *Therefore, a new insight is that the local derivative, e.g., in Prandtl's shear-layer model, is generalized not to become a decaying nonlocal and fractional derivative, but with a common origin and derivation to be one of the most nonlocal derivatives at all, namely instead of a local differential quotient a (large scale) difference quotient.* Therefore, this fluid dynamic example has also revealed an important mathematical special case of fractional and nonlocal calculus, namely constituting the simplest and most nonlocal first-order derivative, which is just the difference quotient!

The difference-quotient is special in the sense that information at small and intermediate space locations does not occur at all, because in the mathematical description only the extremities are visible. Therefore, the effect of small and intermediate locations is described, e.g., only by a constant or Reynolds-dependent amplifying factor in front of a low number of large-scale properties. However, by our Lévy flight and fractal $\beta$-model calculus, we have learned that small and intermediate locations correspond to small- and medium-sized eddies, which are also important with their net effect. In this sense, the DQTM corresponds well with a fractal turbulence analysis presented in the first part of this section.

Moreover, it is satisfying to remark that differential equations with fractional derivatives describe phenomena with no characteristic length scale, e.g., the search for pray of albatrosses or the human brain when accessing an elusive memory by connecting a face to a name, etc. These processes are related to Lévy flights, clustering processes, anomalous diffusion, long-range correlations, intermittency, etc. which are all observed in turbulent flows (see, e.g., West 2014). It is remarkable that with Non-Gaussian statistics, e.g., Lévy statistics, it is possible to derive a Langevin equation containing fractional derivatives.

If at earlier times, the nonlocal impetus ballistic theory was replaced by the correct local Newton's law of motion, in turbulence it is just opposite, namely that the local Prandtl mixing-length and shear-layer model is replaced by the nonlocal *Difference-Quotient Turbulence Model* (DQTM), leading to much more accurate results (see Chap. 9).

Summarizing, the strongest arguments in favor of the DQTM are:

1. Correct Reynolds stress for turbulent flow behind a cylindrical wake in agreement with Newton's second principle.
2. The DQTM Reynolds stress is in agreement with Kraichnan's field-theoretical convolution integral approach.
3. The DQTM Reynolds stress is in agreement with nonlocal empirical expressions applied in rheology.
4. The DQTM is a nonlocal generalization of Prandtl's mixing-length and shear-layer model.
5. The DQTM is a simple fractional turbulence model with highest nonlocality, but minimum calculation time requirement.
6. The DQTM can be derived by Lévy statistics and the fractal $\beta$-model and is in a special case in full agreement with Kolmogorov's 1941 law.
7. The DQTM leads to simple analytical solutions for elementary turbulent flow configurations, see Chap. 9.
8. The infinite Reynolds number solution of the mean downstream velocity of plane turbulent Poiseuille flow in its core region is a circle profile, see Fig. 9.50.
9. The solution of "wall" turbulent flows is a defect power law with a quadratic irrationality exponent in the case of infinite Reynolds number flow.
10. The DQTM reveals in a natural manner a theory of cooperative phenomena of turbulence in analogy, e.g., to magnetic systems (see Chap. 11).

In the first part of this review, points (1) to (6) have been derived and demonstrated. In the following sections, the DQTM is chosen as our prime nonlocal turbulence model that will be applied to numerous elementary shear flow configurations. A selection of such examples is also available in Hutter and Wang (2016). In the following second part of this review, the demonstration of points (7) to (10) will be given.

## References

Abramowitz, M., Stegun, I.A.: Pocketbook of Mathematical Functions. Verlag Harri Deutsch, Thun (1984). ISBN 3-87144-818-4

Becker, R.: Theorie der Wärme. Springer, Berlin (1978). ISBN 3-540-08988-8 (in German)

Bouchard, J.-P., Georges, A.: Anomalous Diffusion in Disordered Media: Statistical Mechanisms, Models and Physical Applications. Elsevier, North Holland (1990)

Egolf, P.W.: A new model on turbulent shear flows. Helv. Phys. Acta. **64**, 944 (1991)

Egolf, P.W.: Difference-quotient turbulence model: a generalization of Prandtl's mixing-length theory. Phys. Rev. E. **49**(2), 1260 (1994)

Egolf, P.W.: An attempt to prove the difference-quotient turbulence model. Helv. Phys. Acta. **71**(1), 7 (1998)

Egolf, P.W.: Lévy statistics and beta model: a new solution of "wall" turbulence with a critical phenomenon. Int. J. Refrig. **32**, 1815 (2009)

Egolf, P.W., Hutter, K.: From linear and local to nonlinear and nonlocal zero equation turbulence models. In: Proceeding of IMA Conference for Turbulence, Waves and Mixing, Kings College, pp. 71–74, Cambridge, UK (2016a). 6–8 July

Egolf, P.W., Hutter, K.: Turbulent shear flow described by the algebraic difference-quotient turbulence model. In: Peintke, J., et al. (eds.) Progress in Turbulence VI, Springer Proceedings in Physics, vol. 165. Springer, Heidelberg (2016b). ISBN 978-3-319-29129-1

Egolf, P.W., Hutter, K.: A nonlocal zero-equation turbulence model and a deficit-power law of the wall with a dynamical critical phenomenon. In: Proceedings of the Joint DMV (German Association of Mathematicians) and GAMM (Association for Applied Mathematics and Mechanics) Annual Meeting and International Conference, Session 24–3, History of Mechanics (2016c). 7–11 March, Braunschweig

Egolf, P.W., Hutter, K.: Fractional turbulence models. In: Peintke, J., et al. (eds.) Progress in Turbulence VII, Springer Proceedings in Physics, vol. 165. Springer, Heidelberg (2017). ISBN 978-3-319-57933-7

Egolf, P.W., Weiss, D.A.: Model for plane turbulent Couette flow. Phys. Rev. Lett. **75**(16), 2956 (1995)

Egolf, P.W., Weiss, D.A.: A non-local method leading to the difference-quotient turbulence model. Helv. Phys. Acta. **69**(2), 43 (1996a)

Egolf, P.W., Weiss, D.A.: Analytical Reynold's stresses of axi-symmetric jets. Helv. Phys. Acta. **69** (2), 45 (1996b)

Egolf, P.W., Weiss, D.A.: Difference-quotient turbulence model: the axi-symmetric isothermal jet. Phys. Rev. E. **58**(1), 459 (1998)

Einstein, A.: Über die von der molekularkinetischen Theorie der Wärme geforderte Bewegung von in ruhenden Flüssigkeiten suspendierten Teilchen. Ann. Phys. **17**, 549 (1905). (in German)

Einstein, A.: Zur Theorie der Brownschen Bewegung. Ann. Phys. **19**, 371 (1906). (in German)

Einstein, A., Podolski, B., Rosen, N.: Can quantum-mechanical description of physical reality be considered complete? Phys. Rev. **47**, 777 (1935)

Eringen, A.C.: Nonlocal Continuum Field Theories. Springer, New York (2002). ISBN 0-387-95275-6

Feynman, R.P., Leighton, R.B., Sands, M.: In: Oldenburg, R. (ed.) Feynman – Mechanik, Strahlung, Wärme, vol. 1, 2nd edn, Germany (1973). ISBN 3-486-34161-8 (Bilingua in German and English)

Frisch, U.: Turbulence – The Legacy of A.N. Kolmogorov, 1st edn. Cambridge University Press, Cambridge, UK (1995). ISBN 0-85-403-441-2

Gelfand, I.M., Schilov, G.E.: Verallgemeinerte Funktionen. VEB Deutscher Verlag der Wissenschaften, Berlin (1962). (in German)

Herrmann, R.: Fractional Calculus. World Scientific, New Jersey (2011). ISBN 13-978-981-4340-24-3

Herrmann, R.: Fraktionale Infinitesimalrechnung, 2nd edn. Books on Demand, Norderstedt (2014). ISBN 978-3-7357-4109-7 (in German)

Hinze, J.O.: Turbulence, 2nd edn. McGraw-Hill, New York (1975). ISBN 0-07-029037-7

Hinze, J.O., Sonnenberg, R.E., Builtjes, P.J.H.: Memory effect in a turbulent boundary-layer flow due to a relatively strong axial variation of the mean-velocity gradient. Appl. Sci. Res. **29**, 1 (1974)

Hughes, B.D., Shlesinger, M.F., Montroll, E.W.: Random walks with self-similar clusters. Proc. Nat. Acad. Sci. **78**(6), 3287 (1981)

Hunt, J.C.R., Phillips, O.M., Williams, D.: Turbulence and Stochastic Processes: Kolmogorov's Ideas 50 Years on, The Royal Society, London. University Press, Cambridge, UK (1991). ISBN 0-854-03-441b-2

Hutter, K., Wang, Y.: Fluid and thermodynamics. In: Advanced Fluid Mechanics and Thermodynamic Fundamentals, vol. 2. Springer, Berlin (2016). ISBN 978-3-319-33635-0

Kant, I.: Anmerkungen zur dritten Antinomie: Die Antinomie der reinen Vernunft. Akademie Verlag, Berlin (1781). (in German)

Kolmogorov, A.N.: The local structure of turbulence in incompressible viscous fluid with very large Reynolds numbers. Dokl. Akad. Nauk. SSSR Seria fizichka. **30**, 301 (1941a). (in Russian). English translation in Proc. R. Soc. Lond. A **434,** 9

Kolmogorov, A.N.: Dissipation of energy in the locally isotropic turbulence. Dokl. Akad. Nauk SSSR. **32**, 16 (1941b). (in Russian). English translation in Proc. R. Soc. Lond. A **434,** 15

Kraichnan, R.H.: Direct-interaction approximation for shear and thermally driven turbulence. Phys. Fluids. **7**(7), 1048 (1964)

Landau, L.D., Lifshitz, E.M.: Lehrbuch der Theoretischen Physik, Hydrodynamik, vol. VI. Akademie–Verlag, Berlin (1974). (in German)

Lévy, P.: Calcules des probabilités. Gauthier-Villars, Paris (1925). (in French)

Lévy, P.: Théorie de l'addition des variables aléatoires. Gauthier-Villars, Paris (1937). (in French)

Lévy, P.: Processus stochastique et mouvement brownien. Gauthier-Villars, Paris (1948). (in French)

Lighthill, M.J.: Introduction to Fourier Analysis and Generalized Functions. Cambridge University Press, Cambridge, UK (1958). ISBN 9781139171427

Luo, A.C.J., Afraimovich, V.: Long-Range Interactions, Stochasticity and Fractional Dynamics. Springer, Heidelberg (2010). ISBN 978-3-642-12342-9

Mandelbrot, B.B.: The Fractal Geometry of Nature. W.H. Freeman and Company, New York (1977). ISBN 3-7643-2646-8

Oldham, K.B., Spanier, J.: The Fractional Calculus: Theory and Applications of Differentiation and Integration to Arbitrary Order. Dover, Mineola (1974). ISBN 13-978-0-486-45001-8

Osterwalder, K.: Methoden der Mathematischen Physik, Lecture Notes ETHZ. Department Mathematics and Physics (1980). (in German)

Prandtl, L.: Bericht über Untersuchungen zur ausgebildeten Turbulenz. ZAMM. **5**(2), 136 (1925). (in German)

Prandtl, L.: Bemerkungen zur Theorie der freien Turbulenz. ZAMM. **22**(5), 241 (1942). (in German)

Reif, F.: Grundlagen der physikalischen Statistik und der Physik der Wärme. Walter de Gruyter, Berlin (1975). ISBN 3-11-004103-0 (in German)

Sagaut P., Cambon, C.: Homogeneous Turbulence Dynamics. Springer Int. Publ. AG (2018). ISBN 978-3-319 73 162-9

Samba, F.: Turbulence as a dynamical phase change concept. Bachelor Diploma Work, Supervision by P.W. Egolf and K. Hutter, University of Applied Sciences of Western Switzerland, Yverdon-les-Bains (2016)

Samba, F., Egolf, P.W., Hutter, K.: Nonlocal turbulence modeling close to criticality involving Kolmogorov's dissipation microscales. In: Peintke, J., et al. (eds.) Progress in Turbulence VII, Proceedings in Physics. Springer, Heidelberg (2019)

Schwartz, L.: Théorie des distributions. Hermann, Paris (1966). (in French)

Shlesinger, M.F., Klafter, J., West, B.J.: Lévy walks with applications to turbulence and chaos. Physica. **140A**, 212 (1986)

Shlesinger, M.F., Zaslavsky, G.M., Klafter, J.: Strange kinetics. Nature. **363**, 31 (1993)

Tennekes, H., Lumley, J.L.: A First Course in Turbulence. MIT Press, Cambridge, MA (1972). ISBN 0-262-200-19-8

Truesdell, C.A., Muncaster, R.G.: Fundamentals of Maxwell's Kinetic Theory of a Simple Monoatomic Gas: Treated as a Branch of Rational Mechanics. Academic, Cambridge, MA (1980). ISBN 0-12-701350-4

von Weizäcker, C.F.: Der zweite Hauptsatz und der Unterschied zwischen Vergangenheit und Zukunft. Ann. Phys. **428**, 275 (1939). (in German)

Weeks, E.R., Solomon, T.H., Urbach, J.S., Swinney, H.L.: Lévy flights and related topics in physics. In: Shlesinger, M.F., Zaslavsky, G.M., Frisch, U. (eds.) Proceeding of International Workshop, p. 51, Nice (1994). 27–30 June

West, B.J.: Colloquium: fractional calculus view of complexity: a tutorial. Rev. Mod. Phys. **86**, 1169 (2014)

Yakhot, A.: Remark at iTi conference (interdisciplinary turbulence initiative) in Bertinoro, Italy (2018). 7–9 September

Zagarola, M.V., Perry, A.E., Smits, A.J.: Log laws or power laws: the scaling in the overlap region. Phys. Fluids. **9**(7), 2094 (1997)

# Chapter 8
# Self-Similar RANS

Quasi-steady and quasi-two-dimensional flows are grouped into two main classes, namely free turbulent flows, which are not directly influenced by boundaries, and "wall-turbulent" flows, which may exhibit boundary layer effects (see Hinze 1975). The first group can be further split into wake flows and jet flows, plumes, and mixing layers (see Fig. 8.1).

These flows show an initial average downstream velocity that is oriented in the $x_1$ direction. It is assumed that the spreading of such a flow is not too large, so that the downstream directional derivative of the averaged downstream velocity changes rather slowly (see Fig. 5.2). On the other hand, it is observed that a strong shearing leads to a large directional derivative of the average downstream velocity in the $x_2$-direction perpendicular to the main $x_1$-direction. In a flow between two plates (plane turbulent Couette or Poiseuille flows) no spreading of the turbulent domain occurs. Therefore, there is only a gradient in one direction, namely that perpendicular to the fluid-bounding walls. Thus, all these quasi-steady and quasi-two-dimensional flows are characterized by the following properties:

1. The average downstream velocity is much larger than the average velocity in the perpendicular direction.
2. The directional derivative is much larger in the transverse direction than in the downstream direction.

These conditions mainly define the boundary-layer approximations, known for their substantial simplification of the set of differential equations. In Sect. 5.2.6, condition (1) was already formulated and used as Eq. (77) and condition (2) led to Eq. (78). The strong gradient of the mean downstream velocity perpendicular to the main direction and the related mean shear, lead to a production of turbulent fluctuations, which together with the occurrence of dissipation produces a quasi-steady-state turbulent flow. This type of turbulent motion shall be examined in detail in Chaps. 9 and 10. Conditions (1) and (2) can be supplemented by a third one, which also has the positive effect of allowing a substantial simplification of the mathematical and physical problems that will be presented:

---

The original version of this chapter was revised. A correction to this chapter can be found at https://doi.org/10.1007/978-3-030-26033-0_14



P. W. Egolf, K. Hutter, *Nonlinear, Nonlocal and Fractional Turbulence*,
https://doi.org/10.1007/978-3-030-26033-0_8

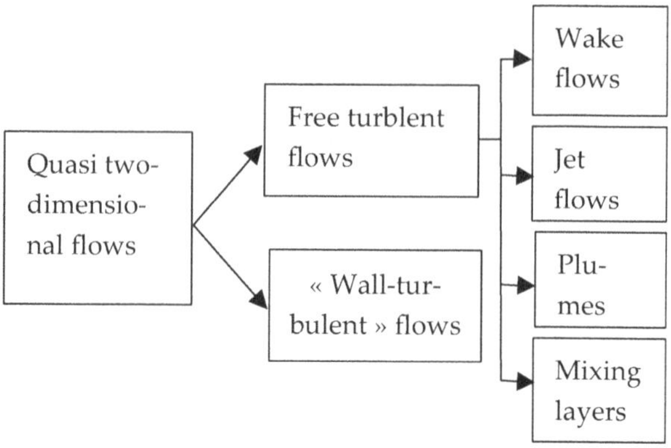


**Fig. 8.1** The main groups of quasi-two-dimensional turbulent flows are presented. All these flows can show different geometrical configurations, e.g., plane, spherical, and cylindrical ones. This leads to a multitude of elementary turbulent flow configurations. A variety of such flows are discussed and solved in Chap. 9

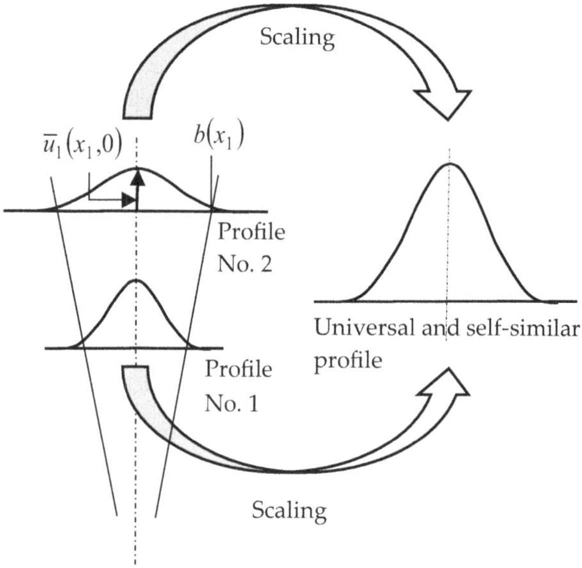


**Fig. 8.2** The average profiles, e.g., the mean velocity profile of a round jet beyond a certain distance downstream, are self-similar. If corresponding profiles (see on the left) are scaled in the $x_1$-direction by the width of the jet $b(x_1)$ and the average velocity $\bar{u}_1(x_1, x_2)$ by its maximum value $\bar{u}_1(x_1, 0)$ on the centerline, they all collapse on a single universal self-similar profile (velocity profile shown on the right)

3. The average pressure variation across the flow is mainly determined by the intensity of the turbulent velocity. In the downstream direction, it varies effectively as in the undisturbed flow. This pressure distribution is usually uniform. So, in some cases, it can be assumed that the average pressure distribution is uniform in the entire turbulent flow domain. More fundamentally expressed: the pressure in the turbulent flow is impressed by the outer flow conditions.

Another important feature of the ensuing developments is the effect of self-similarity, often also called self-preservation. This property indicates that the turbulent structure is largely maintained in its downstream flow direction. It manifests itself by scaling properties of the average velocity and velocity correlation profiles (see, e.g., Fig. 8.2).

Therefore, a fourth flow property reads:

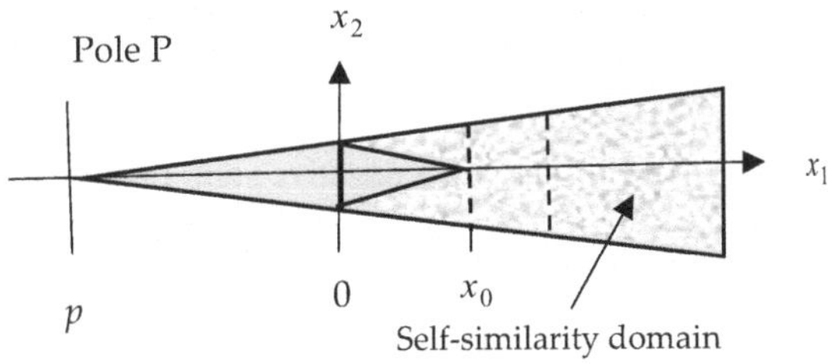


**Fig. 8.3** Self-similarity (here of a turbulent axisymmetric jet) originates in the virtual pole P, where $p$ denotes the coordinate of the pole, which is to the left of the origin of the coordinate system, located at the position of the nozzle. Different than in other references, we work with the negative pole coordinate $p < 0$. The coordinate $x_0$ denotes the end of the theoretical core region, where the turbulent conical domain also loses its core

4. Self-similarity offers the possibility to reduce a two-dimensional problem to a one-dimensional one. Thus, a set of partial differential equations is transformed to a single ordinary differential equation. This is an important step if the goal is to find the right analytical solutions of an otherwise rather complex elementary turbulent flow problem.

Two flow patterns are called self-similar, if each pattern can be normalized by a single length $L_0$ and velocity scale $U_0$ so that the flow patterns expressed in the reduced quantities become identical.

We will study cases for which these length and velocity scales are space dependent with a virtual origin of self-similarity (see Fig. 8.3). If self-similarity shall be preserved, the velocity scale must be inversely proportional to the length scale $x_1$, because only then the Reynolds number is conserved in the downstream direction (compare with Eq. (143))

$$\mathrm{Re}\,(x_1) = \frac{u_1(x_1) \cdot x_1}{\nu} = \text{const.} \tag{484}$$

It can be also shown that the pressure must be constant or will vary with $x_1^{-2}$ (see Hinze 1975).

In this section, the theoretical tools for self-similarity are developed. The spanwise coordinate is scaled and becomes dimensionless by taking

$$\eta(x_2) = \frac{x_2}{L_0}. \tag{485}$$

The two dimensionless average velocities of a quasi-two-dimensional flow are

$$f_k = \frac{\overline{u}_k}{U_0}, \qquad k \in \{1, 2\}, \tag{486a, b}$$

and the dimensionless Reynolds shear stress and the normal turbulence intensities are

$$f_{kl} = -\frac{\overline{u'_k u'_l}}{U_0^2}, \qquad k, l \in \{1, 2\}, \tag{487a – c}$$

and by convention has been chosen with a negative sign. Finally, the dimensionless average pressure is defined as

$$\Pi = \frac{\overline{p}}{\rho U_0^2}, \tag{488}$$

where the denominator can be interpreted as being twice the dynamic pressure of a uniform flow with constant velocity $U_0$.

However, experimental experience shows that condition (484) is very restrictive and that the advantages of self-similarity are still preserved, if condition (484) is not fulfilled. Then, the flow is called incompletely self-similar and its length and velocity scales will depend on the downstream coordinate $x_1$.

By modifying Eq. (485), we now take the dimensionless space variable

$$\eta(x_1, x_2) = \frac{x_2}{b(x_1)}, \tag{489}$$

where the integral length scale $L_0$ has been replaced by a (half-) width $b(x_1)$ of the entire flow domain. We have learned that this width can be constant or will depend on the downstream direction $x_1$ in the following way:

$$b(x_1) = \beta \left( \frac{x_1 - p}{l_0} \right)^{p_0} l_0. \tag{490}$$

The quantity $l_0$ is a characteristic length of the turbulent flow under consideration, e.g., in a wake flow behind a cylinder, it is proportional to the diameter $d$ of the cylinder (see Sect. 9.1).

The average downstream velocity $\overline{u}_1$ and transverse velocity $\overline{u}_2$ scale with the following power law relations:

$$\overline{u}_k(x_1, \eta) = u_0 \left( \frac{x_1 - p}{l_0} \right)^{p_k} f_k(\eta), \qquad k \in \{1, 2\}, \tag{491a, b}$$

where $u_0$ denotes a characteristic velocity of the turbulent flow, e.g., in a wake flow behind a cylinder the undisturbed and constant upstream velocity $U$ in front of the cylinder. In this formula, the dimensionless self-similarity function $f_k(\eta)$ are functions with one variable less than for the original average velocities. Similarly, the second-order velocity correlations are described by

$$\overline{u_k' u_l'}(x_1,\eta) = -u_0^2 \left(\frac{x_1 - p}{l_0}\right)^{p_{kl}} f_{kl}, \quad k,l \in \{1,2\}. \tag{492a – c}$$

Important in this context is the derivative of the width of the turbulent domain given by Eq. (490),

$$\frac{\mathrm{d}b}{\mathrm{d}x_1} = \beta\, p_0 \left(\frac{x_1 - p}{l_0}\right)^{p_0 - 1}, \tag{493}$$

and

$$\frac{1}{b}\frac{\mathrm{d}b}{\mathrm{d}x_1} = p_0 \frac{1}{x_1 - p}. \tag{494}$$

Next, we calculate the derivative of the right-hand term of Eq. (489) with respect to $x_1$. This yields

$$\frac{\partial \eta(x_1,x_2)}{\partial x_1} = -\frac{1}{b}\frac{\mathrm{d}b}{\mathrm{d}x_1}\frac{x_2}{b} = -\frac{1}{b}\frac{\mathrm{d}b}{\mathrm{d}x_1}\eta, \tag{495a, b}$$

or with Eq. (494)

$$\frac{\partial \eta(x_1,x_2)}{\partial x_1} = -p_0 \frac{1}{x_1 - p}\eta. \tag{496}$$

On the other hand, the derivative of the right-hand term of Eq. (489), with respect to $x_2$, is

$$\frac{\partial \eta(x_1,x_2)}{\partial x_2} = \frac{1}{b} = \frac{1}{\beta\, l_0}\left(\frac{l_0}{x_1 - p}\right)^{p_0}. \tag{497a, b}$$

Because in the following sections the continuity equation, the momentum equation, and the nonlocal DQTM shall be transformed to self-similar forms, the following transformation rules will be applied, which follow from Eqs. (496) and (497b):

$$\frac{\partial}{\partial x_1} \quad \rightarrow \quad -p_0 \frac{1}{x_1 - p}\eta \frac{\mathrm{d}}{\mathrm{d}\eta} \tag{498}$$

and

$$\frac{\partial}{\partial x_2} \quad \rightarrow \quad \frac{1}{\beta\, l_0}\left(\frac{l_0}{x_1 - p}\right)^{p_0} \frac{\mathrm{d}}{\mathrm{d}\eta}. \tag{499}$$

We will experience that, in order to guarantee self-similarity, the powers $p_0$, $p_1$, $p_2$, and $p_{12}$, etc. must have well-defined values. However, this will become more transparent below, where this method will be applied to different turbulent flow problems. With these preliminary results, we are now ready to tackle elementary turbulent flow configuration problems.

## Reference

Hinze, J.O.: Turbulence, 2nd edn. McGraw-Hill, New York (1975). ISBN 0-07-029037-7

# Chapter 9
# Elementary Turbulent Shear Flow Solutions

## 9.1 Plane Wake Flows

The following elementary flow configuration and its solutions were originally published by Egolf (1994). A large part of the basic theory originates from Hinze (1975). A reproduction with improvements is found in Hutter and Wang (2016).

Consider a constant parallel flow with velocity $U_G$ in the downstream direction. Now, a cylinder of diameter $d$ is mounted into the fluid that is passed by the fluid. In laminar flows, the streamlines are smoothly surrounding the cylinder in a symmetric manner. In transitional flows, periodic von Kármán eddies are created, which are swept leeward of the cylinder (see, e.g., Van Dyke 1982 and Fig. 9.1). In a turbulent situation, a turbulent wake occurs behind the cylinder (Fig. 9.2). On average, a symmetric downstream velocity profile develops as shown in Fig. 9.3. This is described by the expansion

$$\overline{u}_1 = U_G - \overline{u}_1^*, \quad \overline{u}_1^* \geq 0, \quad \overline{u}_2^* = \overline{u}_2, \tag{500a – c}$$

in which $\overline{u}_1^*$ denotes the reduction part of the mean flow velocity behind the cylinder. To guarantee fully turbulent velocity profiles $\overline{u}_1^*$, experience tells us that the global Reynolds number Re must fulfil the following criterion (see Hinze 1975, who based this limit on observations by Townsend 1947):

$$\mathrm{Re} = \frac{U_G d}{\nu} > 800. \tag{501}$$

Moreover, at locations far downstream, i.e., for

$$\xi_1 =: \frac{x_1 - p}{d} = c \gg 1, \tag{502a – c}$$

The original version of this chapter was revised. A correction to this chapter can be found at https://doi.org/10.1007/978-3-030-26033-0_14



P. W. Egolf, K. Hutter, *Nonlinear, Nonlocal and Fractional Turbulence*,
https://doi.org/10.1007/978-3-030-26033-0_9

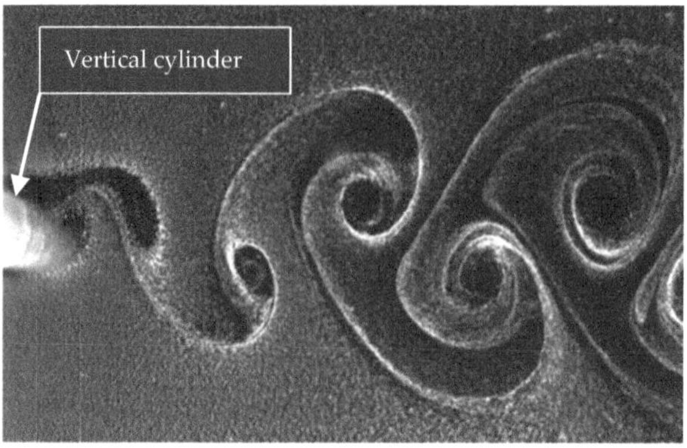


**Fig. 9.1** A von Kármán vortex street shows in the downstream direction alternatively spatiotemporal periodic eddies to the left (in the photo to the top) and the right (in the photo to the bottom). Reproduced with changes from Resonance (2019)

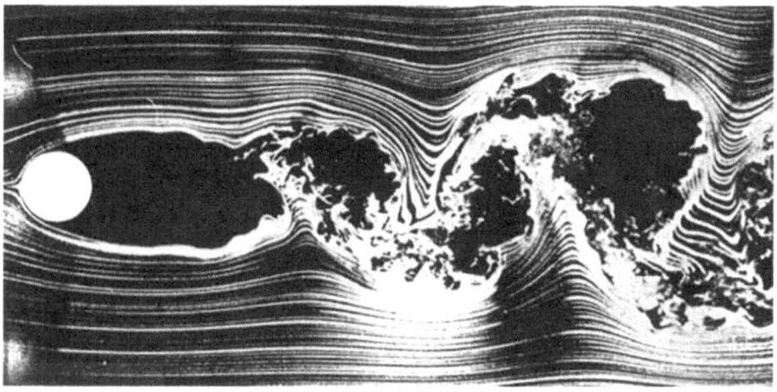

**Fig. 9.2** A highly excited von Kármán vortex street decays into a turbulent wake still showing the underlying spatiotemperal structure of the von Kármán street, however, now with high irregularities. Reproduced with changes from van Dike (1982)

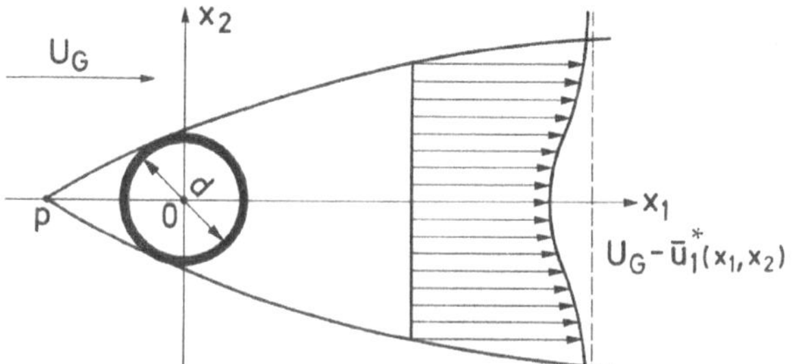


**Fig. 9.3** Turbulent plane wake flow behind a cylinder. The width of the turbulent regime grows downstream with a certain power of the downstream distance with its origin at $p$ (see main text). The average perturbation velocity is symmetrically bell shaped, decreasing to the two sides. Therefore, the minimum average velocity occurs on the axis of the flow. Courtesy of Egolf (1994) © Phys. Rev. E., reproduced with permission by APS

it can be assumed that the velocity profiles are Reynolds number self-similar. Hinze (1975) lists the value $c = 90$ for this specific turbulent flow configuration. This regime has its origin at $p$ slightly upstream of the center of the cylinder. The positioning of the $x_1$ and $x_2$ Cartesian coordinate axes is sketched in Fig. 9.3.

The conservation laws of mass and momentum are applied to a quasi-steady flow situation and to a density preserving fluid, in the absence of a gravity field, and also by neglecting the Newtonian viscosity (high Reynolds number flow). This yields for the continuity and Euler equations (see Egolf 1994)

$$\begin{aligned} \frac{\partial}{\partial x_1}\left(\frac{\overline{u}_1^*}{U_G}\right) - \frac{\partial}{\partial x_1}\left(\frac{\overline{u}_2^*}{U_G}\right) &= 0, \\ \frac{\partial}{\partial x_1}\left(\frac{\overline{u}_1^*}{U_G}\right) - \frac{\partial}{\partial x_2}\left(\frac{\overline{u_2'u_1'}}{U_G^2}\right) &= 0 \end{aligned} \qquad \text{(503a, b)}$$

Now, the self-similarity relations (489) to (499) are applied to Eq. (503a). The basic length $l_0$ is taken as proportional to the diameter of the cylinder weighted with the constant $k$: $l_0 = kd$. The basic velocity $u_0$ is set equal to the center line velocity $\overline{u}_1(x_1, 0)$. This calculus, which is shown in detail in Egolf (1994) and in the textbook on advanced fluid dynamics (see Hutter and Wang 2016), by multiplying their result with $kd$, leads to

$$\left(\frac{x_1 - p}{kd}\right)^{p_1 - 1}\left[p_1 f_1(\eta) - p_0\eta\frac{\mathrm{d}f_1(\eta)}{\mathrm{d}\eta}\right] - \left(\frac{x_1 - p}{kd}\right)^{p_2 - p_0}\frac{1}{\beta}\frac{\mathrm{d}f_2(\eta)}{\mathrm{d}\eta} = 0. \qquad (504)$$

Self-similarity, by its nature, demands that there occurs no dependence on the downstream coordinate $x_1$. This restriction is fulfilled if the exponents of the parentheses containing $x_1$ in Eq. (504) are the same: $p_1 - 1 = p_2 - p_0$. Then it follows that

$$p_0 + p_1 - p_2 - 1 = 0, \qquad (505)$$

so that Eq. (504) simplifies to

$$p_1 f_1(\eta) - p_0\eta\frac{\mathrm{d}f_1(\eta)}{\mathrm{d}\eta} - \frac{1}{\beta}\frac{\mathrm{d}f_2(\eta)}{\mathrm{d}\eta} = 0. \qquad (506)$$

Applying the same self-similarity transformations to the momentum Eq. (503b) generates the equation

$$\left(\frac{x_1 - p}{kd}\right)^{p_1 - 1}\left[p_1 f_1(\eta) - p_0\eta\frac{\mathrm{d}f_1(\eta)}{\mathrm{d}\eta}\right] + \left(\frac{x_1 - p}{kd}\right)^{p_{21} - p_0}\frac{1}{\beta}\frac{\mathrm{d}f_{21}(\eta)}{\mathrm{d}\eta} = 0. \qquad (507)$$

Here, self-similarity demands that $p_1 - 1 = p_{21} - p_0$, or

$$p_0 + p_1 - p_{21} - 1 = 0. \tag{508}$$

This is applied to simplify Eq. (507), with the result

$$p_1 f_1(\eta) - p_0 \eta \frac{\mathrm{d}f_1(\eta)}{\mathrm{d}\eta} + \frac{1}{\beta}\frac{\mathrm{d}f_{21}(\eta)}{\mathrm{d}\eta} = 0. \tag{509}$$

The above derivations have led to three self-similarity functions $f_1$, $f_2$ and $f_{21}$, defined in Eqs. (486a,b) and (487c). With Eqs. (506), (509), and a self-similar version of the Reynolds shear stress (turbulence model), which will be derived in an instance, three equations will be available to determine the three self-similar functions. This seems a well-defined mathematical problem; however, two problems remain in this context that will now be discussed. Less complete is the available system of two equations for the four power exponents $p_0$, $p_1$, $p_2$ and $p_{21}$, consisting only of Eqs. (505) and (508). Therefore, two further equations are required to determine these exponents of the power laws. The following two postulates provide help:

(1) It is requested that the dimensionless turbulent shear stress is also self-similar (compare with Eq. (492)). This leads to

$$\frac{\mathrm{d}}{\mathrm{d}x_1}\left(\frac{\overline{u_2'u_1'}}{\overline{u}_1^2(x_1,0)}\right) = 0. \tag{510}$$

Inserting Eqs. (491a) and (492c) into Eq. (510) yields

$$\frac{\mathrm{d}}{\mathrm{d}x_1}\left[\frac{-U_G^2\left(\frac{x_1-p}{kd}\right)^{p_{21}}}{U_G^2\left(\frac{x_1-p}{kd}\right)^{2p_1}}\right] = 0. \tag{511}$$

The requirement that this expression is independent of $x_1$ implies

$$2p_1 - p_{21} = 0. \tag{512}$$

(2) Owing to Eq. (503b), it is concluded that

$$\frac{\mathrm{d}}{\mathrm{d}x_1}\int\limits_{-\infty}^{\infty}\left(\frac{\overline{u}_1^*}{U_G}\right)\mathrm{d}x_2 = -\left.\frac{\tau_{21}}{\rho U_G^2}\right|_{-\infty}^{\infty} = 0 \tag{513}$$

and that the integral

$$\int_{-\infty}^{\infty} \left(\frac{\overline{u}_1^*}{U_G}\right) \mathrm{d}x_2 = \text{const} \tag{514}$$

along a path perpendicular to the downstream direction is a constant, since the shear stress vanishes at $x_2 = \pm\infty$. Self-similarity occurs only if the average disturbance velocity is an order of magnitude smaller than the free-stream velocity, which is a rather weak restrictive condition that reads

$$\frac{\overline{u}_1^*}{U_G} \ll 1. \tag{515}$$

Next, mass and momentum fluxes are studied through a fictitious rectangle, thought to be symmetrically immersed into the fluid domain relative to the cylinder (see Fig. 9.4).

In a quasi-steady flow, the horizontal flux difference between the in- and outlet is

$$\int_{-\infty}^{\infty} \rho \left[U_G^2 - \left(U_G - \overline{u}_1^*\right)^2\right] \mathrm{d}x_2 \approx 2\rho\, U_G \int_{-\infty}^{\infty} u_1^* \mathrm{d}x_2, \tag{516a, b}$$

where approximation (515) was applied. Notice that the contributions along the flow parallel paths vanish at $x_2 = \pm\infty$, because the unit normal to the two horizontal boarders of the rectangle is perpendicular to the flow path.

The mass flux entering through the boundary from the left-hand side is greater than the outflowing mass flux leaving the domain on the right. This mass flux is pushed through the lower and upper boundary out of the domain (see grey arrows). The related momentum loss by this process is

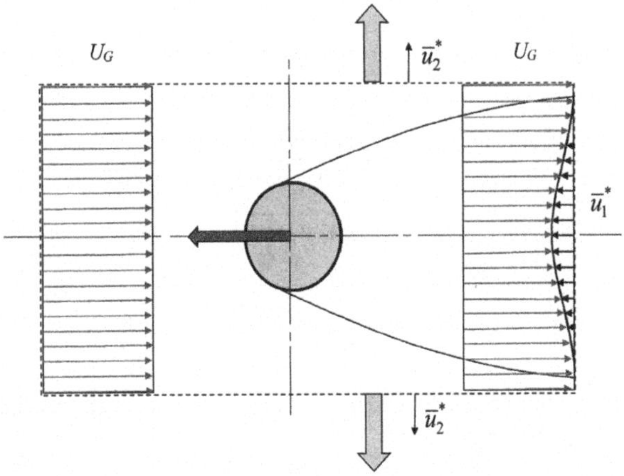


**Fig. 9.4** Mass and momentum fluxes through the cutout rectangle, which is vertically stretched to $x_2 = \pm\infty$. At these positions no horizontal momentum fluxes can occur, since the mean velocity is tangential to the averaged flow path. Courtesy of Hutter and Wang (2016) © Springer edition

$$-\rho\, U_G \int_{-\infty}^{\infty} u_1^* \mathrm{d}x_2. \tag{517}$$

According to Newton's second principle of (fluid) mechanics, the total momentum flux, being the sum of Eqs. (516b) and (517), equals the averaged and normalized specific drag force $\overline{F}$ acting on the cylinder, viz.,

$$\frac{\overline{F}}{d\rho\, U_G^2} = \frac{1}{d} \int_{-\infty}^{\infty} \frac{u_1^*}{U_G}\, \mathrm{d}x_2. \tag{518}$$

In a steady-state flow, this average force is constant. Introducing self-similar coordinates and functions ((490) and (491a,b)) the constant relation emerges

$$\frac{\overline{F}}{d\rho\, U_G^2} = \beta k \left(\frac{x_1 - p}{kd}\right)^{p_0 + p_1} \underbrace{\int_{-\infty}^{\infty} f_1(\eta)\, \mathrm{d}\eta}_{\text{constant}} \tag{519}$$

The request that this expression is independent of $x_1$ leads to the fourth relation

$$p_0 + p_1 = 0. \tag{520}$$

It is now possible to solve the system of four Eqs. (505), (508), (512), and (520) for the four exponents. The results are

$$p_0 = 1/2, \quad p_1 = -1/2, \quad p_2 = -1, \quad p_{21} = -1. \tag{521a – d}$$

As it was already announced, the width of the turbulent domain increases with a square root dependence. With the values (521a–d) one obtains

$$\begin{aligned}
\overline{u}_1^* &= U_G \left(\frac{kd}{x_1 - p}\right)^{1/2} f_1(\eta), \\
\overline{u}_2^* &= U_G \left(\frac{kd}{x_1 - p}\right) f_2(\eta), \\
\overline{u_2' u_1'} &= -U_G^2 \left(\frac{kd}{x_1 - p}\right) f_{21}(\eta), \\
\eta &= \frac{x_2}{b(x_1)}, \quad b(x_1) = \beta \left(\frac{x_1 - p}{kd}\right)^{1/2} kd,
\end{aligned} \tag{522a – e}$$

with the exponents (521a–d). The two main differential equations, Eqs. (506) and (509), then reduce to

$$f_1(\eta) + \eta \frac{\mathrm{d}f_1(\eta)}{\mathrm{d}\eta} + \frac{2}{\beta}\frac{\mathrm{d}f_2(\eta)}{\mathrm{d}\eta} = 0, \qquad (523\mathrm{a, b})$$
$$f_1(\eta) + \eta \frac{\mathrm{d}f_1(\eta)}{\mathrm{d}\eta} - \frac{2}{\beta}\frac{\mathrm{d}f_{21}(\eta)}{\mathrm{d}\eta} = 0.$$

The singularities at the position of the pole $p$ of the functions (522a–e) are irrelevant, since the similarity solutions are not physically representative there. The two ordinary differential Eqs. (523a,b) are, however, insufficient to determine the three functions $f_1, f_2$ and $f_{21}$ and, thus, require additionally a closure scheme. In a zero-equation turbulence model this is simply a relation of the form $f_{21}(f_1)$. Surprisingly, though, this condition cannot come from a postulate on the turbulent shear stress because $f_{21}$ can be directly expressed as a function of $f_1$ (see also Hutter and Wang 2016). Indeed, if one writes (523b) as

$$\frac{\mathrm{d}f_{21}}{\mathrm{d}\eta} = \frac{\beta}{2}\left[f_1 + \eta\frac{\mathrm{d}f_1}{\mathrm{d}\eta}\right] = \frac{\beta}{2}\frac{\mathrm{d}}{\mathrm{d}\eta}(\eta f_1), \qquad (524\mathrm{a, b})$$

integration leads to

$$f_{21} = \frac{\beta}{2}\eta f_1, \qquad (525)$$

which by itself corresponds to a zero-equation closure scheme $f_{21}(f_1)$. Notice also, that the antisymmetric property of the Reynolds shear stress made it possible to set the constant of integration in Eq. (525) equal to zero.

With the result (525), it is stated that, by an additional introduction of a turbulence closure, two possibilities exist:

(1) The closure contradicts relation (525).
(2) The closure coincides with relation (525).

Because the closure (525) was derived only on the basis of mass and momentum conservation (Newton's second principle), it is a result based on a sound physical basis. Therefore, a contradicting closure must, in any case, be rejected and conditions (1) and (2) yield a rigorous test for any kind of turbulence models (of low- or high order and these with and without transport equations).

Before this test shall be applied to the DQTM, we remark that the same structure of Eq. (523b) compared to (523a), with the antisymmetric property of $f_2$ and with the aid of Eq. (525), leads to

$$f_2 = -f_{21} = -\frac{\beta}{2}\eta f_1. \qquad (526\mathrm{a, b})$$

Transforming this closure back to the original non-self-similar coordinates shows Eq. (525) in the form

$$\tau_{21} = -\rho\overline{u_2' u_1'} = \rho\, U_G^2 \left(\frac{kd}{x_1 - p}\right) \frac{\beta}{2} \eta f_1. \tag{527a, b}$$

By substituting (522a,d, and e), it follows that

$$\frac{\tau_{21}}{\rho\, U_G^2} = \frac{\beta}{2} \left(\frac{kd}{x_1 - p}\right) \frac{x_2}{\beta\, kd} \left(\frac{kd}{x_1 - p}\right)^{1/2} \left(\frac{x_1 - p}{kd}\right)^{1/2} \frac{u_1^*}{U_G}, \tag{528}$$

which simplifies to

$$\frac{\tau_{21}}{\rho\, U_G^2} = \frac{1}{2} \frac{x_2 - x_{2\,\min}}{x_1 - p} \frac{\overline{u}_1^* - \overline{u}_{1\,\min}^*}{U_G}, \tag{529}$$

where the minimum values in space $x_{2\min}$ and average velocity $\overline{u}_{1\,\min}$ (which are zero) were introduced to make Galilean invariance visible.

This important result leads us to the following conceptually significant:

**Theorem 2** *There exists an elementary turbulent flow configuration (wake flow behind a cylinder) where basic physical laws prove that its Reynolds shear stress is a long-range nonlocal functional of the space coordinates and the average downstream velocity.*

We take this theorem as the main argument for our extensive explanations and support of nonlocal Reynolds shear stress modelling in previous sections.

The following evident relations are substituted into the DQTM (307b,c):

$$\begin{aligned} \chi_2 &= x_2 - x_{2\,\min}, \quad x_{2\,\min} = 0, \quad x_{2\,\max} = b, \\ \overline{u}_1^* &= U_G - \overline{u}_1^*, \quad \overline{u}_{1\,\min}^* = 0, \quad \overline{u}_{1\,\max}^* = U_G, \end{aligned} \tag{530a – f}$$

to obtain, with $\sigma = \mathrm{d}b/\mathrm{d}x_1$ and Eq. (500a),

$$\begin{aligned} \tau_{21} &= \rho \frac{\mathrm{d}b}{\mathrm{d}x_1} (x_2 - x_{2\,\min}) \left(U_G - \overline{u}_1^*\right) \\ &\times \frac{\left(U_G - \overline{u}_{1\,\min}^*\right) - \left[\left(U_G - \overline{u}_{1\,\min}^*\right) - \left(\overline{u}_1^* - \overline{u}_{1\,\min}^*\right)\right]}{b - x_2}. \end{aligned} \tag{531}$$

With Eq. (515) this can, alternatively, be written as

$$\frac{\tau_{21}}{\rho\, U_G^2} = \frac{1}{b} \frac{\mathrm{d}b}{\mathrm{d}x_1} \frac{x_2 - x_{2\,\min}}{1 - \frac{x_2}{b}} \frac{\overline{u}_1^* - \overline{u}_{1\,\min}^*}{U_G}, \tag{532}$$

with an error of order $O\left(\overline{u}_1^*/U_G\right)$.

Next, from Eq. (522e) it is concluded that

$$\frac{1}{b}\frac{\mathrm{d}b}{\mathrm{d}x_1} = \frac{1}{2}\frac{1}{x_1 - p}, \tag{533}$$

so that (532) transforms to

$$\frac{\tau_{21}}{\rho\, U_G^2} = \frac{1}{2}\frac{1}{1 - \frac{x_2}{b}}\frac{x_2 - x_{2\,\mathrm{min}}}{x_1 - p}\frac{\overline{u}_1^* - \overline{u}_{1\,\mathrm{min}}^*}{U_G}. \tag{534}$$

In this formula, there is some freedom in the choice of $b$. It can be defined as the width $b$ corresponding to a decay of the average velocity profile down to some ratio: $r = \overline{u}_1^*(b)/\overline{u}_1^*(0) = f_1(1)/f_1(0)$. Physicists usually work with $r = 1/\mathrm{e}$, whereas in the engineering literature decadal ratios, e.g., $r = 0.1$ and $r = 0.01$ are preferred. There is no objection to demand this value to be very small or even to tend to zero. This would correspond to a large width, which in the limit tends to infinity. Therefore, in the limit as $b \to \infty$ and $x_2/b \to 0$ the DQTM result (534) simplifies to the final physically exact form, already presented as Eq. (529). Based on an extensive literature review on turbulence modeling and the above rather elementary mathematical analysis, it can be stated that:

*To our best knowledge, at present the DQTM is the only turbulence model that for plane turbulent wake flows produces the correct nonlocal Reynolds shear stress, which can be derived by basic physical laws (without any empiricism).*

Summarizing the above findings, for the three functions $f_1$, $f_2$ and $f_{21}$, the following three simple equations have been derived:

$$\begin{aligned} &\text{Mass conservation}: f_2 = -\frac{\beta}{2}\eta f_1, \\ &\text{Momentum conservation}: f_{21} = \frac{\beta}{2}\eta f_1, \\ &\text{DQTM}: f_{21} = \frac{\beta}{2}\eta f_1. \end{aligned} \tag{535a – c}$$

At first sight, the above result seems to be disturbing, namely that in this approximate theory the momentum conservation equation and the DQTM closure relation yields the same expression for $f_{21}$. This implies that for the three functions $f_1$, $f_2$, and $f_{21}$ only two independent equations remain. Therefore, in this approximate theory these three functions cannot be determined. *However, positive is that this derivation leads to a testing example for all turbulence models. Only turbulence models which withstand this test of producing result (529), are judged to be physically accurate and acceptable. From the explanations in this section, it becomes also evident that any turbulence model of Boussinesq type, because of its lack of nonlocality, cannot pass this test.*

## 9.2 Axi-Symmetric Jets

### *9.2.1 Jets in Quiescent Surroundings*

An axisymmetric (round) quasi-steady turbulent jet of an incompressible viscous fluid is flowing out of a circular orifice into an infinite extended domain (see Fig. 9.5). In a first case, in this domain the fluid is assumed to be quiescent. The turbulent domain starts at the border of the orifice and increases linearly until the turbulent areas merge together on the axis of the jet. This position is denoted by $x_0$ and the distance to the orifice is called (laminar) core length. This is a rather theoretical consideration. It is clear that this laminar domain at its right end is very thin there and must become unstable. Therefore, at some position downstream a breakthrough occurs. This makes a real core length $x_1'$ to occur, i.e., smaller than its theoretical counterpart $x_0$. The real core region shows a parabolic shape to the right (see Fig. 9.6 (black region)). However, this small detail is of no importance in our considerations. In the core region, the velocity is identical to that in the orifice and, therefore, remains constant throughout this region. In the region between the real core length $x_1'$ and $x_1''$, the average velocity starts to decrease. This domain is called transition region. The theoretical core length ends in this region: $x_1' < x_0 < x''$. At $x_1$ values equal to the right of the real core end position, $x \geq x''$, the averaged center line velocity decreases hyperbolically, a result that will be derived below together with the linear spreading of the turbulent domain of the jet. At a position equal to or somewhat larger than $x_1''$ the jet is assumed to be self-similar. To observe this, the Reynolds number must fulfil the condition (Egolf and Weiss 1998)

$$\mathrm{Re} = \frac{u_0\, d_0}{\nu} \geq 25{,}000. \tag{536}$$

To reduce the mathematical load of this review, we shall not present the continuity and momentum equations in the $x_1$ and $x_2$ directions in cylindrical coordinates in their full forms. Readers interested in the full mathematical analyses are asked to

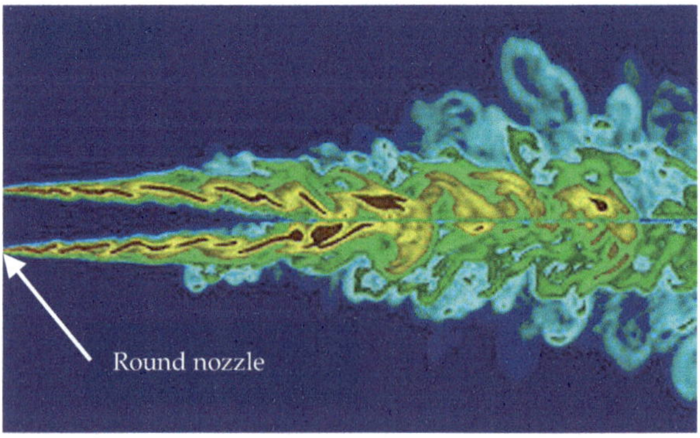


**Fig. 9.5** A vertical cut through an axi-symmetric jet flowing out of a circular orifice (shown on the left-hand side of the picture) into a quiescent environment. The triangular laminar core zone, shown in dark blue color, can be clearly identified. Reproduced with changes from Yangate (2019)

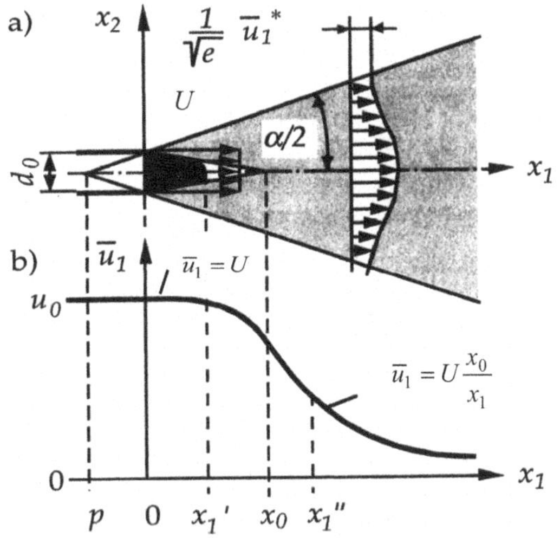


**Fig. 9.6** In panel (**a**) a turbulent round jet emerging from a nozzle into an infinitely extended fluid domain is shown. In panel (**b**), the average velocity on the center line is presented, which far downstream decays hyperbolically with the horizontal distance from the nozzle. Courtesy of Egolf and Weiss (1998) © Phys. Rev. E., reproduced with permission by APS

consult Hinze (1975), Egolf (1994), Egolf and Weiss (1998) or for the newest version Hutter and Wang (2016).

To arrive at a reduced form of the NSEs, the following simplifications are made, which are not presented here in detail, because they can be consulted in Hinze (1975):

(1) It is assumed that the jet is swirl free, which implies that all averaged quantities do not depend on the azimuthal coordinate $x_3$,

$$\frac{\partial \bar{\chi}}{\partial x_3} = 0, \quad \bar{\chi} \in \left\{\bar{u}_1, \bar{u}_2, \overline{u_1' u_2'}, \ldots\right\}. \tag{537}$$

Notice that this is not the case for fluctuation quantities, because the turbulent flows by definition are three dimensional.

(2) Because of the large Reynolds numbers (inequality (536)) viscous terms are negligible.
(3) The boundary layer approximations (77) and (78) allows to neglect terms in the NSEs.
(4) Orthotropy of the fluctuations is assumed to hold

$$\overline{u_2'^2} = \overline{u_3'^2}. \tag{538}$$

(5) Condition (4) relates the pressure $p$ to the constant pressure $p_0$ far from the jet and provides the possibility to omit the pressure term throughout the flow domain.

With these simplifications, the basic system of equations (mass and momentum conservation) are (see Hinze 1975)

$$\begin{aligned} &\frac{\partial \overline{u}_1}{\partial x_1} + \frac{1}{x_2}\frac{\partial}{\partial x_2}(x_2\overline{u}_2) = 0, \\ &\overline{u}_1\frac{\partial \overline{u}_1}{\partial x_1} + \overline{u}_2\frac{\partial \overline{u}_1}{\partial x_2} + \frac{1}{x_2}\frac{\partial}{\partial x_2}\left(x_2\overline{u_2'u_1'}\right) = 0. \end{aligned} \tag{539a, b}$$

According to the general concept of Chap. 8, the self-similar functions are analogous to those in plane wake flows and are of the form

$$\begin{aligned} \overline{u}_1 &= U_G + U\left(\frac{x_1-p}{x_0-p}\right)^{p_1} f_1(\eta), \\ \overline{u}_2 &= U\left(\frac{x_1-p}{x_0-p}\right)^{p_2} f_2(\eta), \\ \overline{u_2'u_1'} &= -U^2\left(\frac{x_1-p}{x_0-p}\right)^{p_{21}} f_{21}(\eta), \\ \eta &= \frac{x_2}{b(x_1)}, \quad b(x_1) = \beta\left(\frac{x_1-p}{x_0-p}\right)^{p_0}(x_0-p). \end{aligned} \tag{540a – e}$$

These expressions are substituted into Eqs. (539a,b), where a decomposition of the downstream velocity $\overline{u}_1$ into a horizontal flow of constant velocity $U_G$ and its perturbation $\overline{u}_1^*$ is applied

$$\overline{u}_1 = U_G + \overline{u}_1^*, \quad \overline{u}_2 = \overline{u}_2^*, \quad \overline{u_2'u_1'} = \overline{u_2^{*\prime}u_1^{*\prime}}. \tag{541a – c}$$

Employing this self-similarity method, these equations also imply the case of a round jet emerging into a parallel horizontal co-flow of basic velocity $U_G$, being the example treated below (see also Fig. 9.21).

Now, again the standard procedure of self-similarity transformations is applied with the characteristic velocity $U$ and length $l_0 = x_0 - p$. The result for the continuity Eq. (539a) is

$$\left(\frac{x_1-p}{x_0-p}\right)^{p_1-1}\left(p_1 f_1 - p_0\eta\frac{df_1}{d\eta}\right) + \frac{1}{\beta}\left(\frac{x_1-p}{x_0-p}\right)^{p_2-p_0}\frac{1}{\eta}\frac{d(\eta f_2)}{d\eta} = 0. \tag{542}$$

To guarantee self-similarity, this suggests

$$p_0 + p_1 - p_2 - 1 = 0. \tag{543}$$

This condition reduces Eq. (542) to the form

$$p_1 f_1 - p_0 \eta \frac{\mathrm{d}f_1}{\mathrm{d}\eta} + \frac{1}{\beta\eta}\frac{\mathrm{d}(\eta f_2)}{\mathrm{d}\eta} = 0. \tag{544}$$

The analogue procedure applied to the momentum Eq. (539b) leads to

$$\begin{aligned} &\left[\frac{U_G}{U} + \left(\frac{x_1 - p}{x_0 - p}\right)^{p_1} f_1\right]\left(\frac{x_1 - p}{x_0 - p}\right)^{p_1 - 1} \times \\ &\left(p_1 f - p_0 \eta \frac{\mathrm{d}f_1}{\mathrm{d}\eta}\right) + \frac{1}{\beta}\left(\frac{x_1 - p}{x_0 - p}\right)^{-p_0 + p_1 + p_2} \times \\ &f_2 \frac{\mathrm{d}f_1}{\mathrm{d}\eta} - \frac{1}{\beta}\left(\frac{x_1 - p}{x_0 - p}\right)^{-p_0 + p_{21}} \frac{1}{\eta}\frac{\mathrm{d}(\eta f_{21})}{\mathrm{d}\eta} = 0. \end{aligned} \tag{545}$$

It is clear that a jet in a quiescent surrounding corresponds to the limit of the special case

$$\frac{U}{U_G} \gg 1. \tag{546}$$

The largest perturbation velocity is

$$\max_{x_1}\left\{\overline{u}_1^*\right\} = U. \tag{547}$$

If we demand (546), it does not follow that

$$\frac{\overline{u}_1^*}{U_G} \gg 1; \tag{548}$$

but this is not needed. It becomes clear that $U_G/U$ in Eq. (545) can be neglected; consequently

$$\begin{aligned} &f_1\left(\frac{x_1 - p}{x_0 - p}\right)^{2p_1 - 1}\left(p_1 f_1 - p_0 \eta \frac{\mathrm{d}f_1}{\mathrm{d}\eta}\right) + \\ &\frac{1}{\beta}\left(\frac{x_1 - p}{x_0 - p}\right)^{-p_0 + p_1 + p_2} f_2 \frac{\mathrm{d}f_1}{\mathrm{d}\eta} - \\ &\frac{1}{\beta}\left(\frac{x_1 - p}{x_0 - p}\right)^{-p_0 + p_{21}} \frac{1}{\eta}\frac{\mathrm{d}(\eta f_{21})}{\mathrm{d}\eta} = 0. \end{aligned} \tag{549}$$

To guarantee self-similarity, it must be demanded that

$$p_0 + p_1 - p_2 - 1 = 0 \tag{550}$$

and

$$p_0 + 2p_1 - p_{21} - 1 = 0. \tag{551}$$

The fourth equation follows from the requirement that the scaled Reynolds shear stress is equally self-similar. This leads to the fraction

$$\frac{\overline{u_2^{*\prime} u_1^{*\prime}}}{\overline{u}_1^{*2}} = \frac{\left(\dfrac{x_1 - p}{x_0 - p}\right)^{p_{21}} f_{21}}{\left(\dfrac{x_1 - p}{x_0 - p}\right)^{2p_1} f_1^2}, \tag{552}$$

which becomes $x_1$-independent if

$$2p_1 - p_{21} = 0. \tag{553}$$

Solving Eqs. (543), (550), (551), and (553) leads to

$$p_0 = 1, \quad p_1 = -1, \quad p_2 = -1, \quad p_{21} = -2. \tag{554a – d}$$

These relations give information on the decrease of the functions by the multiplicative power laws. We experience that the average downstream and transverse velocities are decreasing hyperbolically. On the other hand, the width of the jet is increasing linearly. The Reynolds shear stress decreases inversely to $x_1$ by the power two.
With the relations (554a–d) the expressions (540a–e) transform to

$$\begin{aligned}
\overline{u}_1^* &= U\left(\frac{x_0 - p}{x_1 - p}\right) f_1(\eta), \\
\overline{u}_2^* &= U\left(\frac{x_0 - p}{x_1 - p}\right) f_2(\eta), \\
\overline{u_2^{*\prime} u_1^{*\prime}}^* &= -U^2\left(\frac{x_0 - p}{x_1 - p}\right)^2 f_{21}(\eta), \\
\eta &= \frac{x_2}{b(x_1)}, \qquad b(x_1) = \beta\,(x_1 - p).
\end{aligned} \tag{555a – e}$$

With Eqs. (554a–d) the momentum equation (549) is simplified to the ODE

$$f_1^2 + \eta f_1 \frac{\mathrm{d}f_1}{\mathrm{d}\eta} - \frac{1}{\beta}\left[f_2 \frac{\mathrm{d}f_1}{\mathrm{d}\eta} - \frac{1}{\eta}\frac{\mathrm{d}(\eta f_{21})}{\mathrm{d}\eta}\right] = 0 \tag{556}$$

and the closure scheme DQTM (307b) yields

$$f_{21} = -\beta \frac{1}{\eta} f_1 (1 - f_1). \tag{557}$$

This equation remains to be proved. Again, we recognize that the closure scheme is a relation connecting the second-order correlation $f_{21}$ directly with the first-order function $f_1$. It is satisfying that in this case, three independent equations could be derived for the three unknown functions $f_1$, $f_2$, and $f_{21}$, namely (544), (556), and (557); this lets us hope to find well-defined solutions.

Next, the DQTM (see Eq. (307b)) is applied for the axi-symmetric jet in a quiescent surrounding by substituting the evident relations (compare with Fig. 9.6)

$$x_{2\,\max} = 0, \quad \overline{u}^*_{1\,\min} = 0, \quad \overline{u}^*_{1\,\max} = \overline{u}^*_1(x_1, 0) \tag{558a – c}$$

and

$$\chi_2 = b, \quad \sigma = \frac{\mathrm{d}b}{\mathrm{d}x_1} = \beta \quad \text{with} \quad \beta = \tan\left(\frac{\alpha}{2}\right). \tag{559a – d}$$

With these relations, we obtain

$$\overline{u^{*\prime}_2 u^{*\prime}_1} = -\beta b \frac{1}{x_2}\, \overline{u}^*_1(x_1, x_2)\, \left[\overline{u}^*_1(x_1, 0) - \overline{u}^*_1(x_1, x_2)\right]. \tag{560}$$

It is straightforward to see that this relation, written in self-similar variables, leads to Eq. (557), which has now been proven.

The normalized average velocity in the radial direction is derived by integration by parts of a rearranged Eq. (544). This equation reads

$$f_2 = \beta \left[ \eta f_1 - \frac{1}{\eta} \int_0^\eta \xi f_1(\xi)\, \mathrm{d}\xi \right]. \tag{561}$$

Now Eqs. (557) and (561) are substituted into Eq. (556), which yields the following integro-differential equation for the function $f_1$:

$$\int_0^\eta f_1(\xi) \xi \, \mathrm{d}\xi = 1 - 2f_1 - \eta \frac{f_1^2}{f_1{}'}, \tag{562}$$

where the abbreviation $f_1{}' = \mathrm{d}f_1/\mathrm{d}x_1$ was introduced. By multiplying this equation with $f_1{}'$ and differentiating the result yields the highly nonlinear ordinary differential equation

$$\eta f_1^2 f_1'' - 2\,(f_1')^3 - 3\eta f_1 (f_1')^2 - f_1^2 f_1' = 0. \tag{563}$$

There are no standard procedures to solve such complicated ODEs. A successful trial solution is the Ansatz

$$f_1(\eta) = \exp\,[g(\eta)]. \tag{564}$$

Then, it follows that

$$f_1' = e^g g' \quad \text{and} \quad f_1'' = e^g g'^2 + e^g g''. \tag{565a, b}$$

Substituting these equations into Eq. (563) generates the equation

$$\eta\, e^{2g}\left(e^g g'^2 + e^g g''\right) - 2\,(e^g g')^3 - 3\eta\, e^g (e^g g')^2 - e^{2g} e^g g' = 0. \tag{566}$$

Very convenient is that all the powers of the exponential function are of order three, so that alternatively (566) takes the form

$$\left(\eta\, g'^2 + \eta\, g'' - 2\, g'^3 - 3\eta\, g'^2 - g'\right) e^{3g} = 0. \tag{567}$$

Because the exponential function is larger than zero for all $g$ the expression in parenthesis must vanish. Its terms are regrouped to yield the second-order ODE for $g$,

$$\eta\, g'' - 2\, g'^3 - 2\eta\, g'^2 - g' = 0. \tag{568}$$

With the substitution

$$h(\eta) = g'(\eta), \tag{569}$$

Eq. (568) is transformed to the nonlinear first-order ordinary differential equation

$$\eta\, h' - 2\, h^3 - 2\eta\, h^2 - h = 0. \tag{570}$$

By substitution, one easily corroborates that

$$h(\eta) = -\eta \tag{571}$$

solves Eq. (570). From Eq. (569), one concludes that

$$g(\eta) = \int h(\xi)\, d\xi = -\frac{1}{2}\eta^2 + C_1, \tag{572}$$

with the constant of integration $C_1$. With Eq. (564), the self-similar average velocity is found as

$$f_1(\eta) = \exp\left(-\frac{1}{2}\eta^2 + C_1\right) = C_2 \exp\left(-\frac{1}{2}\eta^2\right), \tag{573a, b}$$

with $C_2 = \exp(C_1)$. From Eq. (555a) it follows that

$$\overline{u}_1(x_0, 0) = U f_1(0) = U \quad \Rightarrow \quad f_1(0) = 1. \tag{574a – c}$$

This condition determines the second constant to be $C_2 = 1$, and finalizes Eq. (573b) to the form

$$f_1(\eta) = \exp\left(-\frac{1}{2}\eta^2\right). \tag{575}$$

In summary: *For the axisymmetric jet emerging into a quiescent ambient fluid, for the averaged downstream velocity, the DQTM produces a highly complex nonlinear differential equation with the Gaussian function as solution. This result was experimentally discovered hundred years ago and confirmed in a multitude of different experiments over many decades (e.g., Townsend* 1949*), so that it can be taken as a further very strong argument for a confirmation of the proposed nonlocal Reynolds shear stress modelling by the DQTM.*

The good agreement between our theoretical results and data of hot wire anemometry experiments by Wygnanski and Fiedler (1969) is displayed in Fig. 9.7.

Because $f_1$ is determined, now from (575), via Eq. (561), also $f_2$ is available and expressible as

$$f_2(\eta) = \beta\left\{\eta \exp\left(-\frac{1}{2}\eta^2\right) - \frac{1}{\eta}\left[1 - \exp\left(-\frac{1}{2}\eta^2\right)\right]\right\}. \tag{576}$$

Comparison of Eq. (576) with Eq. (575) yields

$$f_2(\eta) = \beta\left[\eta f_1 - \frac{1}{\eta}(1 - f_1)\right]. \tag{577}$$

In Fig. 9.8, the average transverse velocity distribution in the radial direction is shown and also compared with measurements of Wygnanski and Fiedler (1969) for the same (relative) downstream locations $x_1/d_0$. Because this velocity is an order of magnitude smaller than the average axial velocity, it is no surprise that the agreement between theory and measurements is of slightly minor quality than that shown in Fig. 9.7. It is assumed that the deviations from the exact solution are rather due to

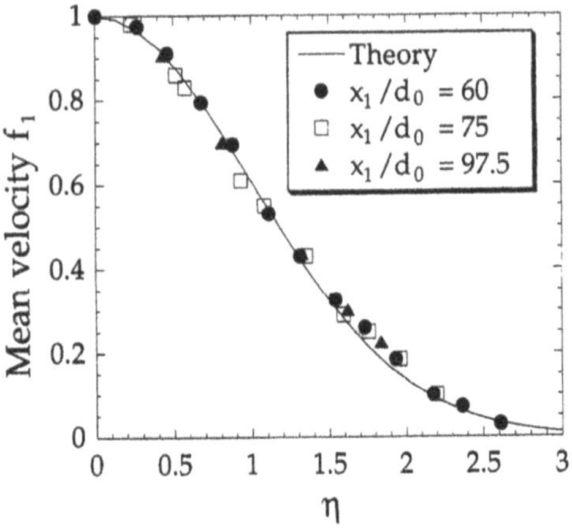


**Fig. 9.7** Theoretical solid line curve and experimental data symbols by Wygnanski and Fiedler (1969) of the average axial velocity as function of the radial direction for different downstream locations. The measured points lie very close to the Gaussian distribution function (575). Courtesy of Egolf and Weiss (1998) © Phys. Rev. E., reproduced with permission by APS

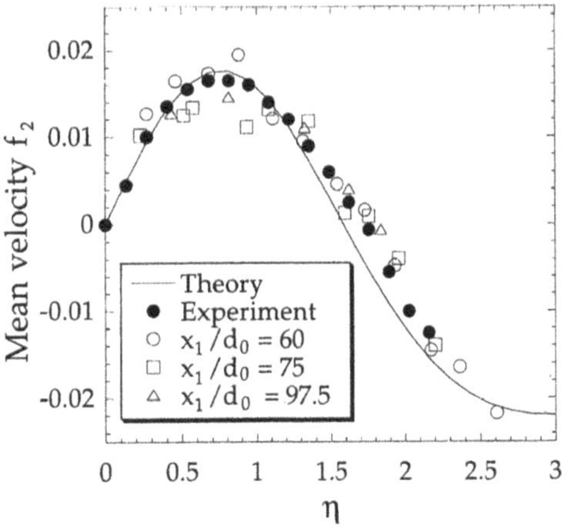


**Fig. 9.8** The radial average velocity of a round jet represented by the present theory (solid curve) in comparison with experimental data from Wygnanski and Fiedler (1969) for three downstream locations. A symmetry requirement demands that the radial flow on the axis is zero. Negative values stand for entrainment of ambient fluid into the jet. Courtesy of Egolf and Weiss (1998) © Phys. Rev. E., reproduced with permission by APS

higher uncertainties in the experimental information than a lack of quality of the present theory.

Interesting is that for a large distance from the axis ($\eta \geq 1.6$) the average radial velocity is negative. This means that beyond $\eta = 1.6$ the fluid flows toward the jet. The jet attracts surrounding fluid; this phenomenon is called entrainment. It will be treated in more detail below.

The closure relation (557) allows us to calculate and plot the Reynolds shear stress as a function of $\eta$ (see Fig. 9.9)

$$f_{21}(\eta) = -\beta \frac{1}{\eta} \left[ \exp\left( -\frac{1}{2}\eta^2 \right) - \exp\left( -\eta^2 \right) \right]. \tag{578}$$

Egolf and Weiss (1998) originally showed these graphical presentations, including that of the Reynolds shear stress (Fig. 9.9) together with the different experimental data of Wygnanski and Fiedler (1969). Those data represented by a symbol "*D*" indicate directly measured values, whereas those denoted by "*I*" were determined by Egolf and Weiss (1998). They took experimental data of $f_1$ and by applying Eq. (578) extracted values of the Reynolds shear stress $f_{21}$.

The mass flux in sections perpendicular to $x_1$,

$$m(x_1) = 2\,\pi\rho \int_0^\infty \bar{u}_1(x_1, x_2)\, x_2 \mathrm{d}x_2, \tag{579}$$

is increasing with the downstream coordinate $x_1$; that at the outlet is

$$m(0) = m_0 = \rho\pi \frac{d_0^2}{4} u_0. \tag{580a, b}$$

Now, the entrainment rate $e$ follows from

$$e(x_1) = \frac{m(x_1)}{m_0} - 1. \tag{581}$$

In the following treatment, we replace $x_1 - p$ by $x_1$ and $x_0 - p$ by $x_0$, because the large distances downstream, where the self-similarity domain begins, guarantee a satisfactory fulfillment of these approximations.

By inserting Eq. (540a), with $U_G = 0$, and (575) into (579), integrating the emerging expression and dividing it by (580b), leads to

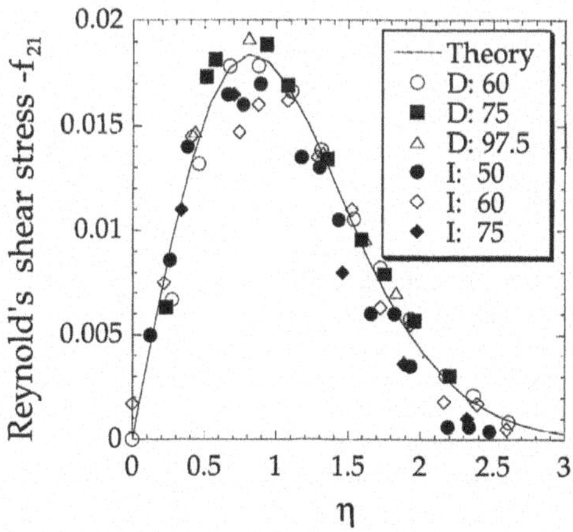


**Fig. 9.9** Reynolds shear stress with two types of experimental results *D* and *I* (see main text). Also, in this case, the agreement between theory and experiments is excellent. The Reynolds shear stress shows a maximum at $\eta = 0.8$ and decreases for $\eta \to \infty$ to zero. Courtesy of Egolf and Weiss (1998) © Phys. Rev. E., reproduced with permission by APS

$$\frac{m(x_1)}{m_0} = k\frac{x_1}{d_0}, \quad k = 8\frac{\beta^2}{m}, \qquad (582a, b)$$

with the mixing number

$$m = \frac{d_0}{x_0}. \qquad (583)$$

The mixing number is the dimensionless ratio of the nozzle diameter $d_0$ to the fictitious core distance $x_0$. For the round jet, it possesses values in the interval [0.16, 0.19] (see Egolf 1992).

From Eqs. (582a) and (581), it is concluded that the mass flux $m_1$ and the entrainment rate $e$ are linear functions of the $x_1$ coordinate (see Fig. 9.10).

Please notice that a dimensional analysis suffices to predict the linear behavior of the mass flux. On the other hand, the integral in Eq. (579) is influenced by the solution of the average downstream velocity. Therefore, the constant in (582a) depends on the obtained solution, which in our case is the outcome of the DQTM. As a result, we may state that the inclination of the linear curve in Fig. 9.10 confirms the correctness of the Gaussian distribution function for $f_1$, leaving open also the correctness of some other functions, which possess the same numerical value of the integral in (579).

*The mass flow of an axi-symmetric jet merging into a quiescent surrounding confirms the solution for the average downstream velocity, which is obtained with the DQTM. However, in the evaluation of the mass flow, there is a small uncertainty. Let us consider a selection of other functions for the scaled average downstream velocity, showing several analogous features as the Gaussian distribution, e.g., the position of the maximum, its integration value. Then, even if it is highly improbable, also such a function could be the correct one.*

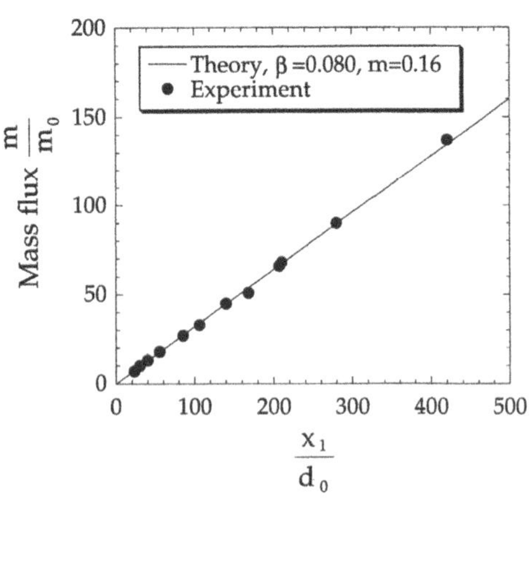


**Fig. 9.10** Linear dependence of the mass flow in the axial direction plotted against the distance downstream of the origin of the jet; and its theoretical prediction by the DQTM. These results are compared with experimental data of Ricou and Spalding (1961), which very convincingly corroborate the theoretical prediction. Courtesy of Egolf and Weiss (1998) © Phys. Rev. E., reproduced with permission by APS

Ricou and Spalding (1961), after a reviewing process, reported values for the constant $k$ to range from 0.22 to 0.404 according to different authors listed by them. Their own determined numerical value is $k = 0.32$. Later, it will be shown that

$$m = 2\beta \Leftrightarrow k = 4\beta. \tag{584a, b}$$

Equation (584b) follows by substituting (584a) into (582b). Ricou and Spalding (1961) do not mention the spreading angle or spreading parameter. However, from $k = 0.32$ it follows that $\beta = k/4 = 0.08$. This is in reasonable agreement with values published by different authors, e.g., by Wygnanski and Fiedler (1969), $\beta = 0.074$, or Panchapakesan and Lumley (1993), $\beta = 0.082$. Because different jets from different orifices show slightly different mixing numbers, there is no doubt that relations (584a,b) are correct.

An interesting question is whether a turbulence model, which predicts correctly the Reynolds shear stress, which is a second-order cross correlation, can also make useful statements concerning the second-order autocorrelations, which in turbulence are the normal stresses. In this section, it will be demonstrated that for the DQTM this is the case.

Therefore, the aim is to also produce a good closure for the normal Reynolds stress components $\overline{u_i^{*\prime 2}}$, $i \in \{1, 2, 3\}$. Egolf and Weiss (1998), to derive an expression for $\overline{u_1^{*\prime 2}}$, proposed to simply replace in the DQTM (Eq. (307b)) the index "2" by "1". The result, written in its briefest form, is

$$\overline{u_i^{*\prime 2}} = -\sigma \chi_1 \left( \overline{u}_1^* - \overline{u}_{1\,\min}^* \right) \frac{\overline{u}_{1\,\max}^* - \overline{u}_1^*}{x_{1\,\max} - x_1}. \tag{585}$$

In analogy to $x_{2\max}$, the quantity $x_{1\max}$ is defined as

$$x_{1\,\max} = \left\{ x_1 \middle| \, \overline{u}_1^* = \max_{x_1} \left\{ \, \overline{u}_1^* \right\} \right\}. \tag{586}$$

From Fig. 9.6 the obvious relations

$$\chi_1 = x_0, \quad x_{1\,\max} = x_0, \quad \overline{u}_{1\,\min}^* = 0, \quad \overline{u}_{1\,\max}^* = U \tag{587a – d}$$

are extracted. The only characteristic length in the $x_1$-direction is the core length $x_0$. Thus, Eq. (585) with $\sigma = \beta$ (see Eq. 559c) transforms to

$$\overline{u_1^{*\prime 2}} = -\beta \, x_0 \, \overline{u}_1^* \frac{U - \overline{u}_1^*}{x_0 - x_1}. \tag{588}$$

Introducing the self-similar function for the average streamwise velocity (540a), with $U_G = 0$ and Eq. (554b), implies

$$\overline{u_1^{*\prime 2}} = -\beta\, x_0\, U\frac{x_0}{x_1}f_1\frac{U - U\dfrac{x_0}{x_1}f_1}{x_0\left(1 - \dfrac{x_1}{x_0}\right)}. \tag{589}$$

With the normalization

$$\overline{u}_1^*(0) = \overline{u}_1^*(x_1, 0) = U\frac{x_0}{x_1}f_1(0) = U\frac{x_0}{x_1}, \tag{590a – c}$$

the normal stress component in the $x_1$–direction is obtained as

$$f_{11} = \frac{\overline{u_1^{*\prime 2}}}{\overline{u}_1^*(0)^2} = \beta\frac{\dfrac{x_1}{x_0} - f_1}{\dfrac{x_1}{x_0} - 1}f_1, \tag{591a, b}$$

or expressed in terms of the turbulent intensity,

$$g_{11} = \sqrt{f_{11}} = \sqrt{\beta}\frac{\sqrt{\dfrac{x_1}{x_0} - f_1}}{\sqrt{\dfrac{x_1}{x_0} - 1}}\sqrt{f_1}, \tag{592a, b}$$

respectively. For the axis of the jet, where $f_1(0) = 1$, this generates the result

$$g_{11}(0) = \sqrt{\beta}. \tag{593}$$

Finally, a result of great simplicity has been revealed that directly links the turbulent intensity on the axis of the jet to its spreading angle. Remarkable is that only the turbulent intensity at this single location determines the spreading of the jet. In our opinion, this is also a result of self-similarity that allows a simpler description of a seemingly more complex situation. Now, this leads us to formulate a first of four propositions:

**Proposition 1** *The relative turbulent intensity in the downstream direction and in the self-similarity domain of a round jet is identical to the root of the spreading parameter,* $g_{11} = \sqrt{\beta}$.

It is difficult to quantitatively evaluate this result experimentally, because in works were the turbulent intensities were determined, usually the related spreading angles were not evaluated. However, Fig. 9.11 qualitatively shows that relation (593) is correct. The idea presented in this figure was successfully applied by Egolf (1992) in a quantitative modeling of the core domain of a round jet. In this article also a semiempirical modeling of the mean velocity in the transition region $x_1' \le x \le x_1''$ is presented that shows good agreement with experiments. However, in

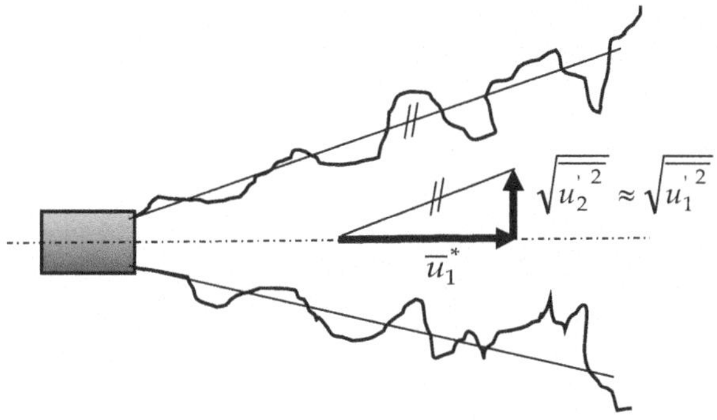


**Fig. 9.11** Spreading of a round jet into its environment. The higher the turbulent intensities, the more a fluid lump will jiggle also to the side, which enlarges the spreading angle. Furthermore, the smaller the average downstream velocity is on the jet axis, the larger will also be the spreading angle in this case

this treatise the objective is to avoid empirical relations as much as possible, which are so much present in the literature of turbulence. This is the reason that this transition region model is here not further outlined.

The theoretical curves for the downstream turbulent intensities show a maximum at approximately $x_2/x_1 = 0.08$, which decreases with the relative longitudinal coordinate $x_1/x_0$. For large values, the intensity distribution converges to a limiting curve, which now shows its maximum on the axis of the jet (see Fig. 9.12).

A jet investigated by Corsin and Uberoi (1949) was substantially narrower than that experimentally investigated by Wygnanski and Fiedler (1969) (see Fig. 9.13). Remarkable is that in the broader jet the relative turbulent intensity, $g_{11}$, is larger than in the narrower jet, an observation that qualitatively confirms the first proposition. Convincing for the present theory is that these experiments show also off-axis peaks to occur.

More reliable experimental data of the turbulent intensity in the axial direction, $g_{11}$, as a function of the nondimensional space coordinate $\eta$, is shown in Fig. 9.14. Figure 9.15 shows the turbulent intensity, $g_{22}$, in the radial and Fig. 9.16, $g_{33}$, in the azimuthal directions. These experimental data sets were again extracted from Wygnanski and Fiedler (1969). It can be seen that all these quantities basically show the same behavior. Only $g_{11}$ shows larger values at the position where the theory predicts maxima for smaller downstream locations. Different than in the theory, in the experiments the lift of the experimental values in the domain around $\eta = 1$ does not seem to depend on the relative downstream position $x_1/d_0$. For the value $x_1/d_0 = 97.5$, the theory shows no longer larger values, whereas the experimental data are still elevated. The deviation of the measurements from the theoretical results varies to a great extent on the fact that experimental work is taken into consideration. For example, Townsend (1976) reports that in many experiments self-similarity was not attained completely. According to him, a lack of self-similarity leads to visible deviations. It is believed that these deviations, occurring at medium values of $x_2/x_1$ only, are caused by the underlying production of turbulent kinetic energy and that fluctuation energy has not been perfectly distributed over the whole width of the jet by transportation of the mean motion and turbulent convection. For example, better agreement was achieved in experiments performed with a helium jet

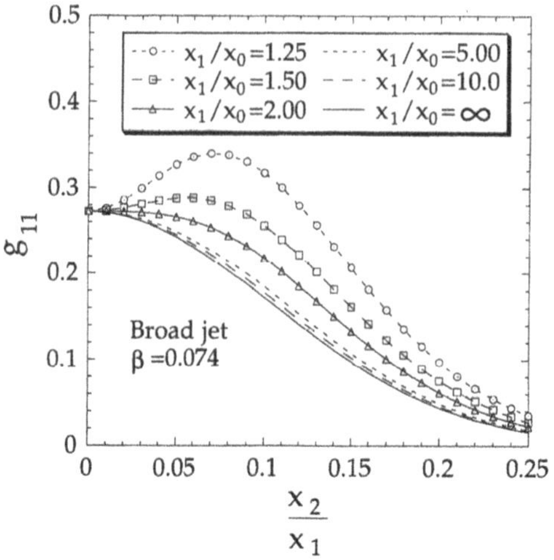


**Fig. 9.12** Downstream turbulent intensities $g_{11}$ for different normalized longitudinal directions $x_1/x_0$ as functions of the relative distance from the axis of the jet $x_2/x_1$. The theoretical results predict an off-axis peak, which depends on the downstream distance and which vanishes for large distances. Courtesy of Egolf and Weiss (1998) © Phys. Rev. E., reproduced with permission by APS

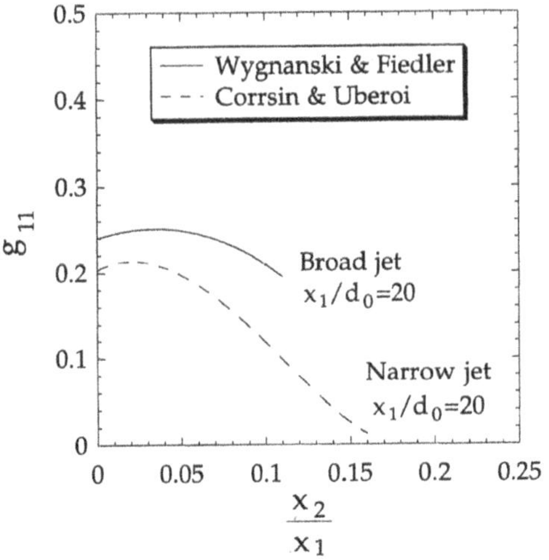


**Fig. 9.13** A broader jet shows a larger turbulent intensity, $g_{11}$, compared to a narrower one. The observed maxima beside the jet axis, which our theory predicts, is also experimentally confirmed. Courtesy of Egolf and Weiss (1998) © Phys. Rev. E., reproduced with permission by APS

(Panchapakesan and Lumley 1993). Wygnanski and Fiedler (1969) interpreted the deviations as follows:

*The fact that the longitudinal intensity contains more of its energy at the lower part of the spectrum is indicative of the manner in which the energy is being transferred, namely from the mean motion to* $\sqrt{\overline{u_1^{*\prime 2}}}$, *and only then to* $\sqrt{\overline{u_2^{*\prime 2}}}$ *and* $\sqrt{\overline{u_3^{*\prime 2}}}$.

Because the agreement is good at $x_2 = 0$, from measured data the following results are obtained, which provide a confirmation of "Proposition 1" by experiments. From the average velocity profile (see Fig. 9.7), the value of $\beta$ was determined by curve fitting to be $\beta = 0.074$. On the other hand, from the turbulent

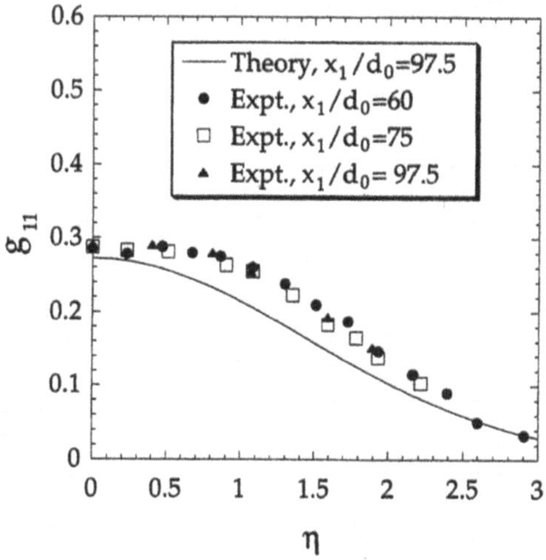


**Fig. 9.14** The turbulent intensity, $g_{11}$, shows a slight lift of the experimental points compared to the theoretical curve. Courtesy of Egolf and Weiss (1998) © Phys. Rev. E., reproduced with permission by APS

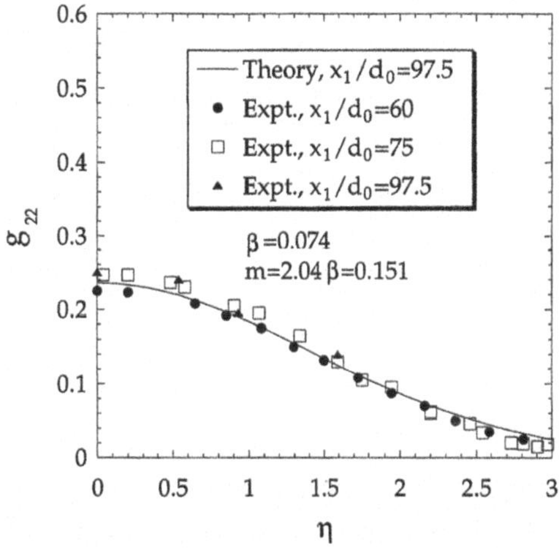


**Fig. 9.15** The theoretical radial turbulent intensity is in very good agreement with experiments. Courtesy of Egolf and Weiss (1998) © Phys. Rev. E., reproduced with permission by APS

intensity, $g_{11}$, (see Fig. 9.14) the value on the axis ($\eta = 0$) leads to $g_{11} = \sqrt{\beta} = 0.28$. From this, it follows that $\beta = 0.078$. Following "Proposition 1," from one set of experimental data, $\beta$ was derived in two different ways. The deviation of the two results is marginal and the results can be taken as a confirmation of "Proposition 1" and a rewarding corroboration of the value of $\beta$ by two independent approaches.

Comparing Figs. 9.15 and 9.16 with Fig. 9.14, it is recognized that the turbulent intensity on the axis of the jet in the downstream direction is larger than in the two other directions. On the other hand, the two values of the radial and azimuthal components are identical. Based on these observations, we write

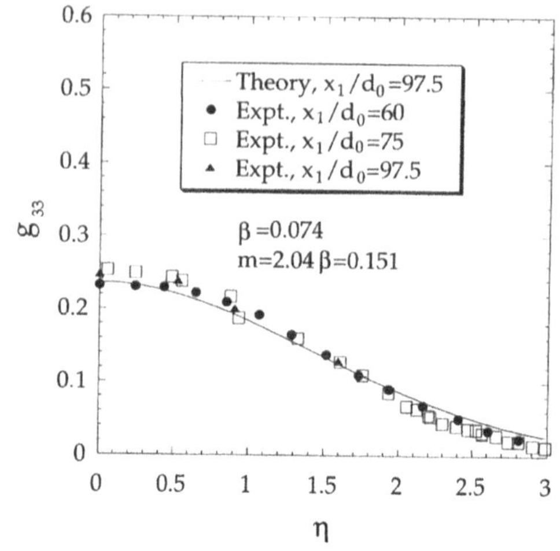


**Fig. 9.16** The theoretical azimuthal turbulent intensity is also in good agreement with experiments. Courtesy of Egolf and Weiss (1998) © Phys. Rev. E., reproduced with permission by APS

$$\overline{u_2^{*'2}} = \overline{u_3^{*'2}} = \frac{\gamma}{\beta}\overline{u_1^{*'2}}, \quad \beta \geq \gamma. \tag{594a – c}$$

In the case of isotropic turbulence, it follows that

$$\overline{u_1^{*'2}} = \overline{u_2^{*'2}} = \overline{u_3^{*'2}} \quad \Leftrightarrow \quad \beta = \gamma. \tag{595a – c}$$

The momentum conservation equation is (see Panchapakesan and Lumley 1993)

$$M(x_1) = 2\pi\,\rho \int\limits_0^\infty \left( \overline{u_1^{*}}^2 + \overline{u_1^{*'2}} - \frac{\overline{u_2^{*'2}} + \overline{u_3^{*'2}}}{2} \right) x_2 \,\mathrm{d}x_2. \tag{596}$$

These authors have found experimentally that this quantity is conserved within a ±5% relative error interval related to $M(0)$. Therefore, we may assume that

$$M(x_1) = M(0), \quad 30 \leq x_1/d_0 \leq 150, \tag{597}$$

where $M(0)$ is the momentum flux out of the orifice, which is defined by

$$M(0) = \frac{\pi\, d_0^2}{4}\rho\, U^2. \tag{598}$$

For the self-similar domain, by substituting Eq. (540a) and (541a), with $U_G = 0$, into Eq. (596), and after a division by $\rho\, \overline{u}_1^*(0)^2$, it follows that

$$\frac{M(x_1)}{\rho\,\overline{u}_1^*(0)^2} = 2\pi\, b^2 \int\limits_0^\infty \left(f_1^2 + f_{11} - \frac{f_{22} + f_{33}}{2}\right) \eta\, \mathrm{d}\eta, \tag{599}$$

where in correspondence with (591a) the following self-similar functions were introduced

$$f_{22} = \frac{\overline{u_2'^2}}{\overline{u}_1^*(0)^2}, \quad f_{33} = \frac{\overline{u_3'^2}}{\overline{u}_1^*(0)^2}. \tag{600a, b}$$

Now, Eq. (591b) is inserted into (599) and (594a,b) is applied. This yields

$$\frac{M(x_1)}{\rho\,\overline{u}_1^*(0)^2} = 2\pi\, b^2 \int\limits_0^\infty \left(f_1^2 + (\beta - \gamma)\frac{\frac{x_1}{x_0} - f_1}{\frac{x_1}{x_0} - 1} f_1\right) \eta\, \mathrm{d}\eta. \tag{601}$$

From Eq. (598) and (590c), it is concluded that

$$\frac{M(0)}{\rho\,\overline{u}_1^*(0)^2} = \frac{\pi\, d_0^2}{4}\frac{\mathrm{U}^2}{\overline{u}_1^*(0)^2} = \frac{\pi\, d_0^2}{4}\left(\frac{x_1}{x_0}\right)^2. \tag{602a, b}$$

Equation (601) contains Eq. (597) as a special case. With Eq. (602b) and $b = \beta\, x_1$ it leads to the formula

$$\int\limits_0^\infty \left(f_1^2 + (\beta - \gamma)\frac{\frac{x_1}{x_0} - f_1}{\frac{x_1}{x_0} - 1} f_1\right) \eta\, \mathrm{d}\eta = \frac{d_0^2}{8b^2}\left(\frac{x_1}{x_0}\right)^2 = \frac{1}{8\beta^2}\left(\frac{d_0}{x_0}\right)^2. \tag{603a, b}$$

This is transformed to

$$\int\limits_0^\infty \left[\left(\frac{x_1}{x_0} - 1\right) f_1^2 + (\beta - \gamma)\left(\frac{x_1}{x_0} - f_1\right) f_1\right] \eta\, \mathrm{d}\eta = \frac{1}{8\beta^2}\left(\frac{d_0}{x_0}\right)^2\left(\frac{x_1}{x_0} - 1\right), \tag{604}$$

and

$$\int\limits_0^\infty \left\{(\gamma - \beta - 1) f_1^2 + \frac{x_1}{x_0}\left[f_1^2 + (\beta - \gamma) f_1\right]\right\} \eta\, \mathrm{d}\eta = \frac{1}{8\beta^2}\left(\frac{d_0}{x_0}\right)^2\left(\frac{x_1}{x_0} - 1\right), \tag{605}$$

respectively. Then this is altered to

$$(\gamma - \beta - 1)\int_0^\infty f_1^2\,\eta\,\mathrm{d}\eta + \frac{x_1}{x_0}\int_0^\infty \left[f_1^2 + (\beta - \gamma)f_1\right]\eta\,\mathrm{d}\eta = \frac{1}{8\beta^2}\left(\frac{d_0}{x_0}\right)^2\left(\frac{x_1}{x_0} - 1\right), \tag{606}$$

After the introduction of the Gaussian distribution function (575) into Eq. (606), with $\xi = \eta^2$, the following equation is obtained:

$$(\gamma - \beta - 1)\underbrace{\int_0^\infty \mathrm{e}^{-\xi}\,\mathrm{d}\xi}_{=1} + \frac{x_1}{x_0}\left[\underbrace{\int_0^\infty \mathrm{e}^{-\xi}\,\mathrm{d}\xi}_{=1} + (\beta - \gamma)\underbrace{\int_0^\infty \mathrm{e}^{-\xi/2}\mathrm{d}\xi}_{=2}\right] = \frac{1}{4\beta^2}\left(\frac{d_0}{x_0}\right)^2\left(\frac{x_1}{x_0} - 1\right), \tag{607}$$

or after explicit evaluation

$$(\gamma - \beta - 1) + \frac{x_1}{x_0}[1 + 2\,(\beta - \gamma)] = \frac{1}{4\beta^2}\left(\frac{d_0}{x_0}\right)^2\left(\frac{x_1}{x_0} - 1\right). \tag{608}$$

This is rearranged to

$$(\gamma - \beta)\left(1 - 2\frac{x_1}{x_0}\right) = \frac{1}{4\beta^2}\left(\frac{d_0}{x_0}\right)^2\left(\frac{x_1}{x_0} - 1\right) + \left(1 - \frac{x_1}{x_0}\right) \tag{609}$$

or

$$(\gamma - \beta)\left(1 - 2\frac{x_1}{x_0}\right) = \left[1 - \frac{1}{4\beta^2}\left(\frac{d_0}{x_0}\right)^2\right]\left(1 - \frac{x_1}{x_0}\right), \tag{610}$$

which yields

$$\gamma = \beta - \frac{\left[1 - \frac{1}{4\beta^2}\left(\frac{d_0}{x_0}\right)^2\right]\left(1 - \frac{x_1}{x_0}\right)}{2\frac{x_1}{x_0} - 1}. \tag{611}$$

By substituting the mixing number $m$ (see Eq. (583)), this equation becomes

$$\gamma = \beta - \frac{\left(1 - \frac{m^2}{4\beta^2}\right)\left(1 - \frac{x_1}{x_0}\right)}{2\frac{x_1}{x_0} - 1}. \tag{612}$$

Isotropy leads to $\beta = \gamma \Leftrightarrow m = 2\beta$ (compare with Eq. (584a)). This equation was introduced in Fig. 9.10 to calculate the entrainment rate $e$, where the important relation $m = 2\beta$ was used without the proof given above. An incomplete self-similar state is characterized by $m > 2\beta$. In Figs. 9.12 and 9.13, $m$ is 2% larger than $2\beta$. Comparing the results of Fig. 9.14 with Figs. 9.15 and 9.16, we note that the experiments do not indicate isotropy on the axis of the jet, which is in contrast with observations made by other investigators, e.g., Corrsin (1943) and Townsend (1976). These authors found isotropic behavior in some domain around the jet axis. Nonisotropy is to be expected since production of turbulence in the middle of the jet is weak and the presence of turbulence is mainly due to transport of mean and turbulent motion. Because of a lack for self-similarity, the additional energy still contained in the axial turbulent intensity at intermediate values of $\eta$ (see Fig. 9.14) is missing in the intensities in the radial and azimuthal directions (Figs. 9.15 and 9.16). This is why these contributions, which have reasonable profiles when compared with model calculations, show smaller amplitudes in comparison to those shown in Fig. 9.14.

Finally, from the last section we draw this main conclusion:

*The DQTM predicts also correctly second-order autocorrelations, which in turbulence modelling are the turbulent normal stresses in the three coordinate directions. The agreement between theory and experiments is convincing.*

Isotropy (see Eq. (595a,b)) simplifies the momentum Eq. (596) considerably, which reduces to

$$M(x_1) = 2\pi\,\rho \int_0^\infty \overline{u}_1^{*2} x_2\,\mathrm{d}x_2. \tag{613}$$

This relation is frequently used by fluid dynamicists and engineers (see, e.g., Abramovich 1963); however, often without mentioning that it is an approximation by assuming isotropy of turbulent intensities.

The equation for the total turbulent energy of an axi-symmetric flow can be straightforwardly derived from the momentum equation (see, e.g., Hinze 1975; Tennekes and Lumley 1972; Hutter and Jöhnk 2004). The arising terms are summarized as follows:

$$pr + co + df + pd + di = 0, \tag{614}$$

in which

*pr*: Production of turbulent kinetic energy
*co*: Convection of turbulent kinetic energy
*df*: Diffusion of turbulent kinetic energy

*pd*: Pressure diffusion
*di*: Dissipation of turbulent kinetic energy.

All these terms can be found in Hinze (1975), Egolf and Weiss (1998), and Hutter and Wang (2016). They consist of quantities that have been derived with the DQTM. Therefore, they can be further evaluated and compared with quantities obtained by measurements. To demonstrate these applications, the two first turbulent energy terms will be the topics of more detailed investigations.

Concerning the production of turbulent kinetic energy, a second proposition, that will be proven is:

**Proposition 2** *The normalized turbulent production pr in the self-similarity domain on the axis of the jet is identical to minus one half of the spreading parameter:* $\pi(0) = -\beta/2$.

The normalized form of the production, *pr*, is

$$\pi = \frac{pr\ x_1}{\bar{u}_1^*(0)^3}, \tag{615}$$

and can be straightforwardly derived from

$$pr = \overline{u_2' u_1'}\frac{\mathrm{d}\bar{u}_1}{\mathrm{d}x_2} + \overline{u_1'^2}\frac{\mathrm{d}\bar{u}_1}{\mathrm{d}x_1} + \overline{u_2'^2}\frac{\mathrm{d}\bar{u}_2}{\mathrm{d}x_2}. \tag{616}$$

The self-similar functions $f_1, f_2, f_{11}$, and $f_{22}$, according to our previous analyses, are now all available. The result of this derivation is (see Egolf 1998; Hutter and Wang 2016)

$$\pi = x_1\left\{-f_{21}\frac{\mathrm{d}f_1}{\mathrm{d}\eta}\frac{\partial\eta}{\partial x_2} + f_{11}\frac{1}{\bar{u}_1^*(0)}\left[\frac{\mathrm{d}\bar{u}_1^*(0)}{\mathrm{d}x_1}f_1 + \bar{u}_1^*(0)\frac{\mathrm{d}f_1}{\mathrm{d}\eta}\frac{\partial\eta}{\partial x_1}\right] + f_{22}\frac{\mathrm{d}f_2}{\mathrm{d}\eta}\frac{\partial\eta}{\partial x_2}\right\}, \tag{617}$$

which, with the expressions

$$\frac{\partial\eta}{\partial x_1} = -\eta\frac{1}{x_1}, \quad \frac{\partial\eta}{\partial x_2} = \frac{1}{\beta\, x_1}, \quad \frac{\mathrm{d}\bar{u}_1^*(0)}{\mathrm{d}x_1} = -\frac{1}{x_1}\bar{u}_1^*(0), \tag{618a -- c}$$

takes the form

$$\pi(\eta) = -\frac{1}{\beta}f_{21}\frac{\mathrm{d}f_1}{\mathrm{d}\eta} - f_{11}\left(f_1 + \eta\frac{\mathrm{d}f_1}{\mathrm{d}\eta}\right) + \frac{1}{\beta}f_{22}\frac{\mathrm{d}f_2}{\mathrm{d}\eta}. \tag{619}$$

Equation (591b) for $x_1 \gg x_0$ simplifies to

$$f_{11} = \beta f_1. \tag{620}$$

Expression (576) is now inserted for $f_2$ and Eq. (578) yields the expression for $f_{21}$. Furthermore, Eq. (591b), with the isotropy condition (595a,b), yields the expression for $f_{11}$ and $f_{22}$. After cancelling several terms, the new expression takes the form

$$\pi(\eta) = \left(\frac{1}{\eta} f_1 \frac{\mathrm{d}f_1}{\mathrm{d}\eta} + \beta \frac{1}{\eta^2} f_1\right)(1 - f_1) + \beta \frac{1}{\eta} f_1 \frac{\mathrm{d}f_1}{\mathrm{d}\eta}. \tag{621}$$

From Eq. (575), it is concluded that the derivative of $f_1$ is given as

$$\frac{\mathrm{d}f_1}{\mathrm{d}\eta} = -\eta f_1. \tag{622}$$

With this identification, Eq. (621) transforms to the form

$$\pi(\eta) = f_1^3 - f_1^2\left(1 + \beta + \beta\frac{1}{\eta^2}\right) + \beta\frac{1}{\eta^2} f_1. \tag{623}$$

This equation is graphically represented in Fig. 9.17. The theoretical curve is compared with experimental data from Wygnanski and Fiedler (1969). Notice that the open circles are not directly measured experimental results; they only represent a mean curve given by measurements. For more detailed information, we refer to the original publication. Substituting the Gauss distribution function, represented by Eq. (575) into Eq. (623) and performing a Taylor series expansion of the emerging expression about $\eta = 0$ produces

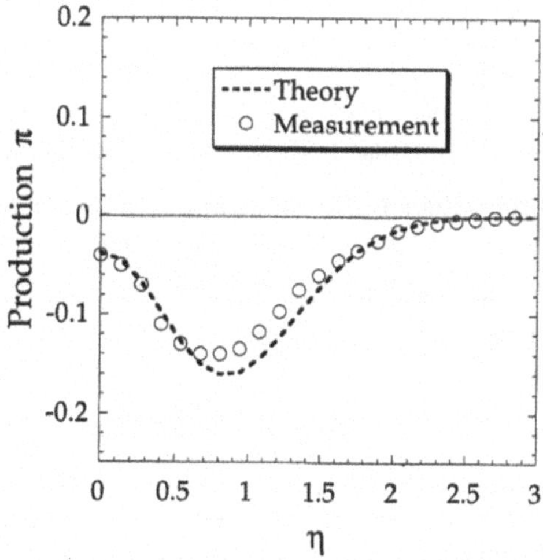


**Fig. 9.17** Dimensionless turbulent kinetic energy production $\pi(\eta)$ plotted against $\eta$. The dashed line is from the DQTM modeling, open circles were extracted from measurements of Wygnanski and Fiedler (1969). Courtesy of Egolf and Weiss (1998) © Phys. Rev. E., reproduced with permission by APS

$$\pi \approx \eta^2\left(\beta - \frac{1}{2}\right) - \frac{\beta}{2}, \quad \eta \ll 1. \tag{624}$$

It follows for $\eta = 0$

$$\pi(0) = -\frac{\beta}{2}, \tag{625}$$

a result postulated as "Proposition 2." The numerical value is $\beta = 0.08$, a value identical to its actual value deduced from experiments.

The next example leads to:

**Proposition 3** *In a round jet far downstream and on its axis the dimensionless turbulent isotropic kinetic energy density, takes the constant value* $\kappa(0) = 3/2\ \beta$.

The dimensionless turbulent kinetic energy density is given by (Hutter and Wang 2016)

$$\kappa = \frac{k}{\overline{u}_1^*(0)^2} = \frac{1}{2}\frac{\overline{u_1'^2} + \overline{u_2'^2} + \overline{u_3'^2}}{\overline{u}_1^*(0)^2} = \frac{1}{2}(f_{11} + f_{22} + f_{33}). \tag{626a, b}$$

By inserting Eq. (591b) and (594a,b), with $f_{22} = f_{33}$, the following expression is derived (Egolf and Weiss 1998):

$$\kappa = \left(\frac{1}{2} + \frac{\gamma}{\beta}\right)\beta\,\frac{\frac{x_1}{x_0} - f_1}{\frac{x_1}{x_0} - 1} f_1, \quad \kappa = \frac{3}{2}\beta f_1, \tag{627a, b}$$

where Eq. (627b) holds for the special case of isotropy ($\beta = \gamma$ ), and large distances downstream ($x_1 \gg x_0$). On the jet axis, where $f_1(0) = 1$, Eq. (627b) yields the result of Proposition 3.

A further quantity is the turbulent kinetic energy convection term *co*, listed in Eq. (614). It also leads to a simple relation, which is stated as:

**Proposition 4** *The turbulent convection in the self-similarity domain on the axis of the jet is identical to minus three times the spreading parameter,* $\chi(0) = -3\beta$.

The convection equation of axi-symmetric turbulent flows takes the form (see Hinze 1975; Egolf and Weiss 1998; Hutter and Wang 2016)

$$co = \frac{\partial}{\partial x_1}\left(\overline{u}_1\overline{u}_1^*(0)^2\kappa\right) + \frac{1}{x_2}\frac{\partial}{\partial x_{21}}\left(x_2\overline{u}_2\overline{u}_1^*(0)^2\kappa\right). \tag{628}$$

This is transformed to the self-similar coordinate $\eta$ and the dimensionless functions $f_1$ and $f_2$

$$co = \overline{u}_1^*(0)^3 \left[ \frac{df_1}{d\eta}\frac{\partial \eta}{\partial x_1}\kappa + 3f_1 \frac{1}{\overline{u}_1^*(0)}\frac{d\overline{u}_1^*(0)}{dx_1}\kappa + f_1 \frac{d\kappa}{d\eta}\frac{\partial \eta}{\partial x_1} + \frac{1}{x_2}f_2\kappa + \frac{df_2}{d\eta}\frac{\partial \eta}{\partial x_2}\kappa + f_2 \frac{d\kappa}{d\eta}\frac{\partial \eta}{\partial x_2} \right], \tag{629}$$

as shown, e.g., in Egolf and Weiss (1998), Hutter and Jöhnk (2004) and Hutter and Wang (2016).

The dimensionless turbulent convection term takes the form

$$\chi = \frac{co\ x_1}{\overline{u}_1^*(0)^3}. \tag{630}$$

Thus, Eq. (629) can alternatively be written in self-similar coordinates as

$$\chi(\eta) = -\eta \frac{df_1}{d\eta}\kappa - 3f_1\kappa - \eta f_1 \frac{d\kappa}{d\eta} + \frac{1}{\beta}\frac{1}{\eta}f_2\kappa + \frac{1}{\beta}\frac{df_2}{d\eta}\kappa + \frac{1}{\beta}f_2\frac{d\kappa}{d\eta}. \tag{631}$$

If Eq. (577) with (627b) are substituted into this expression, a number of terms cancel out; the new form for $\chi$, thus, is

$$\chi(\eta) = \frac{3}{2}\beta\left(2\frac{1}{\eta}f_1\frac{df_1}{d\eta} - f_1^2 - \frac{1}{\eta}\frac{df_1}{d\eta}\right). \tag{632}$$

Substituting the representation (622) further simplifies expression (632) to

$$\chi(\eta) = \frac{3}{2}\beta f_1(1 - 3f_1). \tag{633}$$

This result is in excellent agreement with measurements derived by Wygnanski and Fiedler (1969) (Fig. 9.18).
Because $f_1(0) = 1$, the special algebraic expression on the axis is

$$\chi(0) = -3\beta. \tag{634}$$

This result terminates the proof of "Proposition 4" on the basis of the DQTM. Notably, the result was obtained without the use of a single empirical constant. If $\beta = 0.074$, then $\chi(0) = -0.222$, a value that is in very good agreement with the experimentally determined value observable in Fig. 9.18. Finally, the following important conclusion may be drawn:

*The DQTM of a round jet emerging into a quiescent ambient fluid shows that only three parameters must be known to determine the entire jet behavior. These parameters are the diameter of the orifice $d_0$, the outlet velocity U, and the spreading parameter $\beta$ or the spreading angle $\alpha$.*

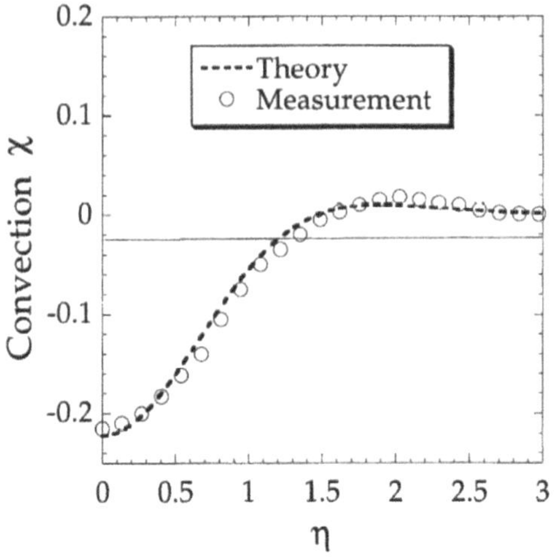


**Fig. 9.18** The dimensionless turbulent convection term $\chi$ is negative toward the axis of the jet and positive for large $\eta$. Interesting is that several dimensionless quantities, e.g., $\chi(0)$ on the axis of the jet, show a universal ($x_1$-coordinate independent) value, for different downstream locations. Courtesy of Egolf and Weiss (1998) © Phys. Rev. E., reproduced with permission by APS

The spreading parameter is defined by the technical realization of the orifice (see, e.g., Egolf 1992).

The terms *df*, *pd*, and *di* in Eq. (614) are not further explored; they are found, e.g., in Egolf and Weiss (1998).

In Egolf (1992), a rigorous result was derived that can be taken as a further test of a turbulence model. In the following, we demonstrate that the DQTM also excellently passes this test.

Starting point is Eq. (539b). This equation is evaluated only on the jet axis. Because $\overline{u}_2 = 0$ there (compare with Fig. 9.8), this yields

$$\overline{u}_1^*(0)\frac{\partial \overline{u}_1^*(0)}{\partial x_1} + \frac{\partial}{\partial x_2}\overline{u_2' u_1'}\bigg|_{x_2=0} + \left(\frac{1}{x_2}\overline{u_2' u_1'}\right)\bigg|_{x_2=0} = 0. \tag{635}$$

It is clear that (compare with Fig. 9.9)

$$\overline{u_2' u_1'}\big|_{x_2=0} = 0. \tag{636}$$

Furthermore, from basic differential calculus and the fact that the function $\overline{u_2' u_1'}$ vanishes at $x_2 = 0$ identically, we conclude, on using the rule of Bernoulli–de l'Hôpital, that

$$\lim_{x_2 \to 0} \frac{\overline{u_2' u_1'}}{x_2} = \frac{\mathrm{d}}{\mathrm{d}x_2}\overline{u_2' u_1'}\bigg|_{x_2=0}. \tag{637}$$

Thus, the last two terms on the left-hand side of (635) are identical at $x_2 = 0$, implying

$$\overline{u}_1^*(0)\frac{\partial \overline{u}_1^*(0)}{\partial x_1} + 2\frac{\partial}{\partial x_2}\overline{u_2' u_1'}\bigg|_{x_2=0} = 0. \tag{638}$$

From Eq. (540a), by employing the approximation $|p| \ll x_0$ and using $f_1(0) = 1$ for the location on the jet axis, we obtain

$$\overline{u}_1^*(0) = \overline{u}_1^*(x_1, 0) = U\left(\frac{x_1}{x_0}\right)^{p_1}, \tag{639a, b}$$

where we know from momentum considerations that $p_1 = -1$. By substituting Eq. (639b) into Eq. (638), and dividing the emerging relation by $\overline{u}_1^*(0)^2$, yields

$$\frac{\mathrm{d}}{\mathrm{d}\left(\frac{x_2}{x_1}\right)}\frac{\overline{u_2' u_1'}}{\overline{u}_1^*(0)^2}\Bigg|_{x_2=0} = -\frac{p_1}{2} = \frac{1}{2}. \tag{640a, b}$$

This is generally a valid result that is correct also for other shear flows, e.g., a plane jet, an axi-symmetric, or plane jet along a wall. These cases have other $p_1$-values. The numerical value of the derivative of Eq. (640b) at $x_2 = 0$ , for a round jet, is experimentally confirmed with high accuracy, as shown in Fig. 9.19.

As for turbulent flows with a wake behind a cylinder, we have now derived from first principles a specific result for the Reynolds shear stress. A useful turbulence model must reproduce this result, which is based on fundamental physical laws. The DQTM achieves this objective (see below).

Let us now go back to Eq. (578), which is the self-similar solution of the Reynolds shear stress,

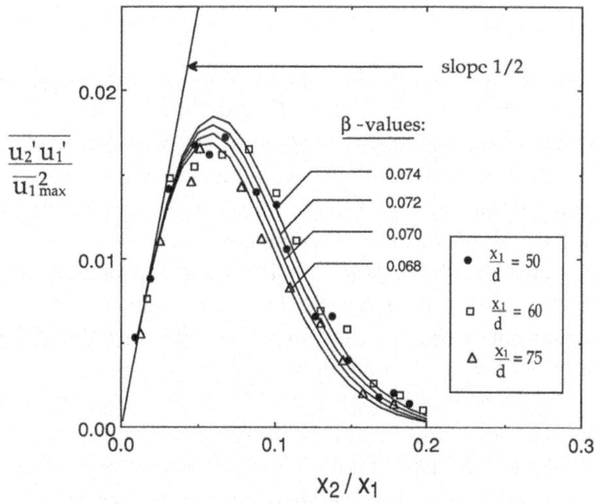


**Fig. 9.19** The Reynolds shear stresses measured by Wygnanski and Fiedler (1969) for different streamwise distances and spreading parameter $\beta$ on the jet axis; all show at $x_2 = 0$ a unique derivative of value 1/2. Reproduced from Egolf (1992)

$$\frac{\overline{u'_2 u'_1}}{\overline{u}_1^{*2}} = \beta\, e^{-\frac{1}{2}\eta^2}\, \frac{1 - e^{-\frac{1}{2}\eta^2}}{\eta}. \tag{641}$$

Now, a Taylor series expansion of this expression up to order $\eta^2$ is taken to determine the value near the axis. This implies

$$\frac{\overline{u'_2 u'_1}}{\overline{u}_1^{*2}} = \frac{\beta}{2}\eta + O(\eta^3). \tag{642}$$

With $\eta = x_2/b = x_2/(\beta\, x_1)$, this becomes

$$\frac{\overline{u'_1 u'_2}}{\overline{u}_1^{*2}} = \frac{1}{2}\frac{x_2}{x_1}, \tag{643}$$

which by differentiation directly establishes the proof of Eq. (640b). As confirmed by measurements, there is no influence on this quantity by the spreading angle of the jet (see Fig. 9.19). We, thus, conclude:

*The DQTM is in agreement with an expression based on first physical principles, namely the nondimensional derivative with respect to $x_2/x_1$ of the nondimensional Reynolds shear stress in the self-similarity domain on the axis of a round jet, which is identical to 1/2.*

The analysis with the DQTM of this section also allows an interesting interpretation of the physics of fluids occurring along the axis of a round jet.

The average velocity is interpreted as a mean velocity of a small fictitious fluid lump of unchanging shape (e.g., a sphere) moving downstream on the jet axis with the average velocity

$$\overline{u}_1^*(0) = \frac{d\overline{x}_1}{d\overline{t}}. \tag{644}$$

In the subsequent formulas in this section, the index "1" in the mean space coordinate $\overline{x}_1$ and the index "1" and the asterisk "$*$" as well as the argument "(0)" in the average centerline velocity $\overline{u}_1^*(0)$ will for simplicity henceforth be omitted. The fluid lump has a characteristic diameter that is large as compared to that of the fluid particles (atoms or molecules), but is small compared to the overall characteristic length of the entire flow region.

Naturally, the averages in Eq. (644) are actually ensemble averages of many single fluid particle flights. The mean values then correspond to the smoother movement of the fictitious fluid lump. Because this system is ergodic, we keep the overbars for the denotation of the average values.

If a single fluid particle is in focus, flying from the nozzle on the jet axis, its actual time is $t$, the position $x$, and the velocity $u$. The difficulty here is that it would not remain on the axis. However, if an ensemble average of many fluid particles is considered, the corresponding mean makes a very regular movement and because of symmetry reasons will neither leave the jet axis. Therefore, $\bar{t}$ denotes the average time when the center of gravity of a fluid lump arrives at the average position $\bar{x}$, where it shows the average velocity $\bar{u}$.

Based on these explanations, we assume that a fluid lump moves regularly with its average downstream velocity. Then it follows that

$$\bar{u}(\bar{t}) = \frac{\mathrm{d}\bar{x}}{\mathrm{d}\bar{t}}. \tag{645}$$

By substituting Eq. (640a) into (638) and employing (644) and (645), the nonlinear differential equation

$$\frac{\mathrm{d}^2\bar{x}}{\mathrm{d}\bar{t}^2} - \frac{p_1}{\bar{x}}\left(\frac{\mathrm{d}\bar{x}}{\mathrm{d}\bar{t}}\right)^2 = 0 \tag{646}$$

is derived. Notice that in Eq. (640a), $x_1$ has been replaced by $\bar{x}$ and $\bar{u}_1^*(0)$ by $\bar{u}$ and $\mathrm{d}\bar{x}/\mathrm{d}\bar{t}$, respectively. *In this nonlinear differential equation, the first term is the inertia term of the fluid lump. The second term is a fictitious friction term, given by the turbulent momentum exchange by fluid particles leaving the jet axis region toward areas of smaller momentum and others arriving from an environment of smaller momentum content toward the jet axis.*

Therefore, a fluid lump of unchanging geometrical shape is moving downstream. Notice that in this process a continuous exchange of fluid particles with the environment takes place. However, by mass conservation, the net particle flux in and out of the considered fluid lump is zero. From a thermodynamic point of view, the fluid lump is an open system. Note that the friction term decreases inversely with the mean downstream distance $\bar{x}$ and is proportional to the square of the mean velocity of the fluid lump. Because the mean velocity of a round jet decreases inversely to the distance from the nozzle, the fictitious force decreases inversely to the third power of the downstream distance. According to d'Alembert's principle, the balance of forces of a dynamical system, by taking the inertia force into account, reduces the dynamical to a quasi-static problem. Then, the equilibrium of all the occurring forces determines the dynamical state of the moving object. In our case it is the nonlinear ordinary differential Eq. (646), which as demonstrated below, possesses an analytical solution.

The same differential equation is valid for other types of jets, e.g., the axi-symmetric jet along a wall, the plane free jet and the plane wall jet (see Table 9.1). Among these, the only distinction is that momentum conservation leads to different values of $p_1$ and as a consequence to other fictitious friction forces. The constants $p_1$ of two free jets and two plane jets are listed in the second row of Table 9.1 and the technical references are found in the third column. These alternative jet types play an important role, e.g., in air-conditioning (see Egolf 1992).

Qualitatively, the physical explanation of the fictitious force by a momentum exchange by fluid particles works well. Geometrically, the free jet can exchange

**Table 9.1** Constants $p_1$ of two free jets and two wall jets. The negative value of $p_1$ leads to a negative fictitious force and makes the acceleration term in Eq. (646) negative

| Type of jet | Constant $n$ | References |
|---|---|---|
| Axi-symmetric free jet | $p_1 = -1$ | Landau and Lifshitz (1974) |
| "Axi-symmetric" wall jet | $p_1 = -2/3$ | Glauert (1956) |
| Plane free jet | $p_1 = -1/2$ | Katz (1974) |
| Plane wall jet | $p_1 = -3/8$ | Regenscheit (1971) |

A hypothetical fluid lump, moving in the downstream direction, is continuously and persistently decelerated. This occurs because of the strong entrainment effect of smaller momentum fluid toward the jet axis, where the fluid contains higher momentum

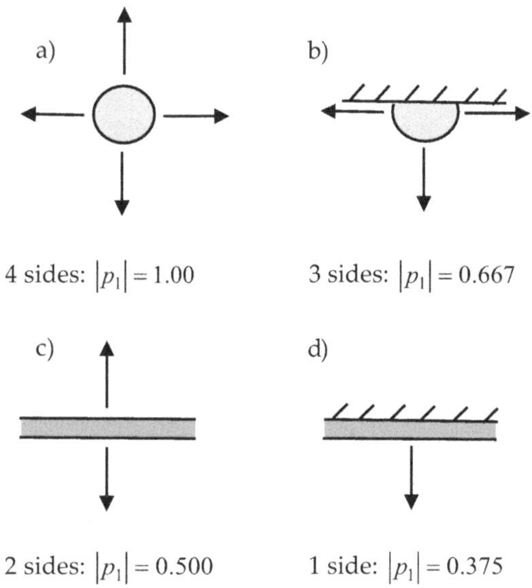


**Fig. 9.20** In the four panels a systematic behavior in the $|p_1|$ values is demonstrated, which is described in detail in the main text and can be easily explained in a pure geometrical–physical context

most fluid particles with its environment because there is an entrainment in an all-around manner toward the jet axis. Approximately, particle exchange occurs at all four sides (see Fig. 9.20a). This allows the highest possible momentum exchange, which results in the highest absolute value $|p_1| = 1.00$. The second case is the axi-symmetric jet along a wall (see Fig. 9.20b). Approximately, one may assume that substituent particle and momentum exchange occurs here in three directions, therefore, leading to a slightly lower absolute value $|p_1| = 2/3 = 0.667$. Then, the third case is the plane free jet (see Fig. 9.20c), which shows exchange in only two directions and has an absolute value of half the size of the free round jet: $|p_1| = 1/2 = 0.500$. The plane wall jet (see Fig. 9.20d) exhibits the poorest exchange, which is possible only in a single direction and, therefore, has the minimum absolute value $|p_1| = 3/8 = 0.375$. Qualitative or approximate quantitative results are obtained by this simple geometrical-physical model consideration.

The core region close to the nozzle is assumed to be free of turbulence (or at least it shows a very poor degree of turbulence). So, for this first region the turbulent friction term in Eq. (646) can be neglected. This corresponds to a value $p_1 = 0$. Therefore, for this region the basic differential equation and the two initial conditions are

$$\frac{d^2\overline{x}}{d\overline{t}^2} = 0, \quad 0 \le \overline{x} \le \overline{x}_0, \quad \overline{x}(0) = 0, \quad \overline{u}(0) = U. \qquad (647a - c)$$

To derive the solutions, the initial value is the constant outlet velocity $U$ in the nozzle, which applies at the location $\overline{x} = 0$. The (mean) velocity remains constant throughout the entire (theoretical) core region of length $\overline{x}_0$. By integration(s) and by also considering the initial conditions, the solution is obtained as

$$\frac{d\overline{x}}{d\overline{t}} = \overline{u} = U, \qquad \overline{x} = U\overline{t}, \qquad 0 \le \overline{x} \le \overline{x}_0. \qquad (648a, b)$$

In the self-similarity domain with turbulent mixing the full Eq. (646) (with nonzero $p_1$) must be solved. A simpler approach is to go back to Eq. (638), to combine it with (640a) and to divide this new equation by $\overline{u}_1^*(0)^2 = \overline{u}^2$. This yields

$$\frac{1}{\overline{u}(\overline{x})}\frac{d\overline{u}(\overline{x})}{d(\overline{x})} - p_1\frac{1}{\overline{x}} = 0, \quad \overline{u}(\overline{x}_0) = u_0 = U, \qquad (649a - c)$$

where the initial condition at the end of the core region has been added (see Eq. (649b)). An alternative form of this ordinary differential equation (ODE) is

$$d(\log_e \overline{u}) = p_1 d(\log_e \overline{x}) = d(\log_e \overline{x}^{p_1}). \qquad (650a, b)$$

Then the solution of Eq. (649a,b) is

$$\overline{u}(\overline{x}) = u_0\, \overline{x}_0^{-p_1}\overline{x}^{p_1} = u_0\left(\frac{\overline{x}}{\overline{x}_0}\right)^{p_1}, \quad \overline{x}_0 \le \overline{x} \le \infty. \qquad (651a, b)$$

For example, the average velocity of an axi-symmetric jet ($p_1 = -1$), decreases hyperbolically as our theory has already revealed.

Having knowledge of the average velocity, the average time of flight of a fluid lump is calculated by integrating the inverse average velocity for a variable average time interval. The average velocity is constant up to $\overline{x}_0$ and decays inversely to the downstream distance beyond this point. Finally, the result given below is valid for positions beyond the core distance $\overline{x}_0$; explicitly,

$$\bar{t}(\bar{x}) = \int_0^{\bar{x}} \frac{1}{\bar{u}(\xi)}\,\mathrm{d}\xi = \frac{1}{u_0}\int_0^{x_0} \mathrm{d}\xi + \frac{\bar{x}_0^{p_1}}{u_0}\int_{x_0}^{\bar{x}} \xi^{-p_1}\,\mathrm{d}\xi. \qquad (652\mathrm{a,b})$$

The solution is

$$\bar{t}(\bar{x}) = \frac{\bar{x}_0}{u_0} + \frac{\bar{x}_0^{p_1}}{u_0}\,\frac{\bar{x}^{1-p_1} - \bar{x}_0^{1-p_1}}{1-p_1}, \qquad \bar{x}_0 \le \bar{x} < \infty. \qquad (653)$$

Introducing

$$\bar{t}_0 = \frac{\bar{x}_0}{u_0} \qquad (654)$$

simplifies Eq. (653) to

$$\bar{t}(\bar{x}) = \frac{\left(\dfrac{\bar{x}}{\bar{x}_0}\right)^{1-p_1} - p_1}{1-p_1}\,\bar{t}_0, \qquad \bar{t}_0 \le \bar{t} < \infty, \qquad (655)$$

or when solved for the mean position,

$$\bar{x}(\bar{t}) = \left[(1-p_1)\frac{\bar{t}}{\bar{t}_0} + p_1\right]^{\frac{1}{1-p_1}} \bar{x}_0, \qquad \bar{x}_0 \le \bar{x} < \infty, \qquad (656)$$

Taking the derivative with respect to $\bar{t}$ leads to the average velocity as a function of the average time

$$\bar{u}(\bar{t}) = \frac{\mathrm{d}\bar{x}}{\mathrm{d}\bar{t}} = \left[(1-p_1)\frac{\bar{t}}{\bar{t}_0} + p_1\right]^{\frac{p_1}{1-p_1}} u_0, \quad \bar{t}_0 \le \bar{t} < \infty, \qquad (657\mathrm{a,b})$$

in which Eq. (654) was applied. The second derivative is the average acceleration, viz.,

$$\bar{a}(\bar{t}) = \frac{\mathrm{d}^2\bar{x}}{\mathrm{d}\bar{t}^2} = \frac{\mathrm{d}\bar{u}}{\mathrm{d}\bar{t}} = p_1\left[(1-p_1)\frac{\bar{t}}{t_0} + p_1\right]^{\frac{2p_1-1}{1-p_1}} \bar{a}_0, \qquad \bar{t}_0 \le \bar{t} < \infty, \qquad (658\mathrm{a-c})$$

with the abbreviation

**Table 9.2** List of the dynamical laws of a fictitious fluid lump moving downstream

| Symbol | $0 \leq \overline{x} \leq x_0$ | $x_0 \leq \overline{x} \leq \infty$ |
|---|---|---|
| $\overline{t}(\overline{x})$ | $\overline{t} = \frac{\overline{x}}{u_0}$ | $\overline{t} = \frac{\left(\frac{\overline{x}}{\overline{x}_0}\right)^{1-p_1} - p_1}{1-p_1}\overline{t}_0$ |
| $\overline{x}(\overline{t})$ | $\overline{x} = u_0\overline{t}$ | $\overline{x} = \left[(1-p_1)\frac{\overline{t}}{\overline{t}_0} + p_1\right]^{\frac{1}{1-p_1}} x_0$ |
| $\overline{u}(\overline{t})$ | $\overline{u} = u_0$ | $\overline{u} = \left[(1-p_1)\frac{\overline{t}}{\overline{t}_0} + p_1\right]^{\frac{p_1}{1-p_1}} u_0$ |
| $\overline{u}(\overline{x})$ | $\overline{u} = u_0$ | $\overline{u} = u_0 \left(\frac{\overline{x}}{\overline{x}_0}\right)^{p_1}$ |
| $\overline{a}(\overline{t})$ | 0 | $\overline{a} = p_1 \left[(1-p_1)\frac{\overline{t}}{\overline{t}_0} + p_1\right]^{\frac{2p_1-1}{1-p_1}} \overline{a}_0$ |
| $\overline{a}(\overline{x})$ | 0 | $\overline{a} = p_1 \left(\frac{\overline{x}}{\overline{x}_0}\right)^{2p_1-1} \overline{a}_0$ |

$$\overline{a}_0 = \frac{u_0}{\overline{t}_0} = \frac{\overline{x}_0}{\overline{t}_0^2}. \tag{659a, b}$$

Substituting Eq. (655) into (658c) leads to the average acceleration, which depends on the average downstream coordinate

$$\overline{a}(\overline{x}) = p_1 \left(\frac{\overline{x}}{\overline{x}_0}\right)^{2p_1-1} \overline{a}_0, \qquad \overline{x}_0 \leq \overline{x} < \infty. \tag{660}$$

A systematic overview of all these power laws for the time, location, velocity, and acceleration of the dynamics of a fluid lump is presented in Table 9.2.

Egolf (1992) produced a finer model that, instead of only having the two theoretical domains $0 \leq \overline{x} < \overline{x}_0$ and $\overline{x}_0 \leq \overline{x} < \infty$, takes account of the three distinct regions $0 \leq \overline{x} < \overline{x}'$, $\overline{x}' \leq \overline{x} < \overline{x}''$, and $\overline{x}'' \leq \overline{x} < \infty$ (see Fig. 9.6 where $\overline{x} = x_1$). The additional intermediate regime is called transition region. For this region, Egolf derived an empirical average velocity distribution that connects correctly both neighboring velocity profiles and does not show a kink of the profile at the partial domain limits. In this context, a very valuable expression of the theoretical core length $x_0 = f(x_1', x_1'')$ was derived. Turbulence models mostly contain empirical relations. On the other hand, this work is intended to be almost free of empiricism and, therefore, these results are not presented here. However, an interested reader may consult Egolf (1992).

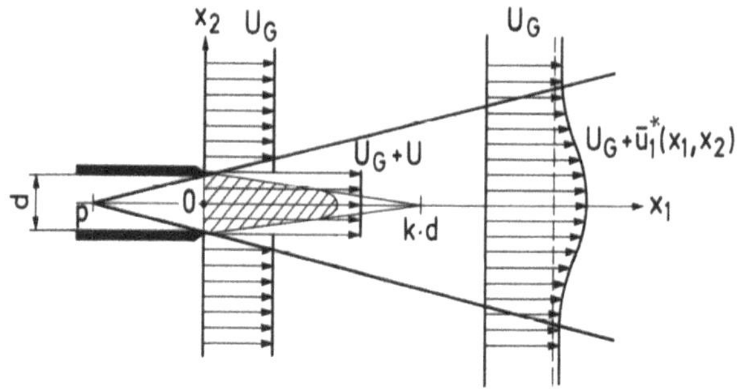


**Fig. 9.21** Turbulent circular jet entering an ambient region with a constant velocity $U_G$ parallel to the jet and in the same direction (co-flow). The center velocity on the axis at the outlet is $U_G + U$. The perturbation velocity, decreasing in the downstream direction, is $\overline{u}_1^*(x_1, x_2)$. The diameter of the nozzle is $d$ and the length of the theoretical core region is $x_0 = kd$. Courtesy of Egolf and Weiss (1998) © Phys. Rev. E., reproduced with permission

## 9.2.2 *Jets in Parallel Co-flows*

The jet model developed in the last subsection was restricted to the situation of a jet flowing into a quiescent ambient fluid. In an approximation cylindrical jets fulfilling condition (546) were treated. In this section, the reverse restrictive condition

$$\frac{U}{U_G} \ll 1, \quad \Leftrightarrow \quad \frac{\overline{u}_1^*}{U_G} \ll 1, \forall x_1 \tag{661a, b}$$

is applied. Here, $U + U_G$ denotes the outlet velocity in the orifice (see Fig. 9.21). Notice that self-similar solutions only occur in the two special cases (546) and (661a). In this new case, the relative jet velocity $\overline{u}_1^*(x_1, x_2)$ and even its maximum value, $U$, is small compared to the uniform and constant velocity of the ambient fluid with velocity $U_G$.

For axi-symmetric steady flows, the balances of mass and momentum of the mean turbulent motion are given by Eqs. (539a,b) and for the situation of Fig. 9.21 transform to

$$\begin{aligned} &\frac{\partial \overline{u}_1^*}{\partial x_1} + \frac{1}{x_2}\frac{\partial}{\partial x_2}\left(x_2 \overline{u}_2^*\right) = 0, \\ &\left(U_G + \overline{u}_1^*\right)\frac{\partial \overline{u}_1^*}{\partial x_1} + \overline{u}_2^*\frac{\partial \overline{u}_1^*}{\partial x_2} + \frac{1}{x_2}\frac{\partial}{\partial x_2}\left(x_2 \overline{u_2^{*\prime} u_1^{*\prime}}\right) = 0, \end{aligned} \tag{662a, b}$$

where $\overline{u}_1 = U_G + \overline{u}_1^*$, $\overline{u}_2 = \overline{u}_2^*$, and $U_G$ is the constant parallel speed of the surrounding fluid, while $\overline{u}_1^*$ is the mean perturbation velocity component. With $\overline{u_2' u_1'}^* := \overline{u_2^{*\prime} u_1^{*\prime}}$ the momentum equation can be rewritten as

$$\left(1+\frac{\overline{u}_1^*}{U_G}\right)\frac{\partial}{\partial x_1}\frac{\overline{u}_1^*}{U_G}+\frac{\overline{u}_2^*}{U_G}\frac{\partial}{\partial x_2}\frac{\overline{u}_1^*}{U_G}+\frac{1}{x_2}\left(x_2\frac{\partial}{\partial x_2}\frac{\overline{u_2'u_1'}^*}{U_G^2}\right)=0. \tag{663}$$

If we now use the product decompositions (491a,b) and (492b) together with (489) and (490) for the self-similar coordinate transformations, then the continuity equation leads to Eqs. (542) and (544), the same relations as in the last case. However, the axial momentum equation is different and requires a more detailed analysis. Equation (663) transforms to (see, e.g., Egolf 1994)

$$\begin{aligned}&\Bigg[\underbrace{1+\frac{U}{U_G}\left(\frac{x_1-p}{kd}\right)^{p_1}f_1}_{\text{neglected}}\Bigg]\left(\frac{x_1-p}{kd}\right)^{p_1-1}\Bigg(p_1f_1\\ &-p_0\eta\frac{\mathrm{d}f_1}{\mathrm{d}\eta}\Bigg)+\underbrace{\frac{1}{\beta}\frac{U}{U_G}\left(\frac{x_1-p}{kd}\right)^{-p_0+p_1+p_2}f_2\frac{\mathrm{d}f_1}{\mathrm{d}\eta}}_{\text{neglected}}\\ &-\frac{1}{\beta}\frac{U}{U_G}\left(\frac{x_1-p}{kd}\right)^{-p_0+p_{21}}\frac{1}{\eta}\frac{\mathrm{d}(\eta f_{21})}{\mathrm{d}\eta}=0.\end{aligned} \tag{664}$$

This equation is slightly more difficult to explore than that of the last case, and much care must be taken. A sophisticated analysis was introduced by Hutter and Wang (2016). If we just drop the terms with the small multiplicative factor $U/U_G$, we get a trivial solution and a nontrivial one. However, the second solution is unphysical (for more details on these special solutions we refer to Hutter and Wang (2016)). The only way to obtain a physical solution in Eq. (664) is to neglect the two underbraced terms, which are denoted by "neglected." Then, the $x_1$-independence requirement demands a balancing of the two terms

$$\left(\frac{x_1-p}{kd}\right)^{p_1-1} \quad \text{and} \quad \left(\frac{x_1-p}{kd}\right)^{-p_0+p_{21}}. \tag{665a, b}$$

This leads to the first equation for the exponents

$$p_0+p_1-p_{21}-1=0. \tag{666}$$

The second equation follows from the continuity equation and, as we already remarked, is unaltered from the case of a round jet flowing into a quiescent environment, viz.,

$$p_0+p_1-p_2-1=0. \tag{667}$$

Requesting again that the Reynolds shear stress, scaled with the principal mean velocity, is independent of $x_1$, leads to

$$2p_1 - p_{21} = 0. \tag{668}$$

Finally, one may equally request that the force on the nozzle (instead of the cylinder) induced by the flow is $x_1$-independent. The lengthy mass and momentum balance calculation yields (Egolf 1994)

$$2p_0 + p_1 = 0. \tag{669}$$

The solution of the last four equations is

$$p_0 = \frac{1}{3}, \quad p_1 = -\frac{2}{3}, \quad p_2 = -\frac{4}{3}, \quad p_{21} = -\frac{4}{3}. \tag{670a – d}$$

Next, it must be proven that with these solutions the neglected terms are smaller than the ones taken into consideration. By setting in Eq. (664) or Eqs. (665a,b), the sum of the exponents of the $(x_1 - p)/kd$ terms equal to zero, and applying Eqs. (670a–d) it follows for the remaining (not neglected) terms

$$p_1 - 1 = -p_0 + p_{21} = -\frac{5}{3}. \tag{671a, b}$$

For the neglected terms, this analysis yields

$$2p_1 - 1 = -p_0 + p_1 + p_2 = -\frac{7}{3}. \tag{672a, b}$$

With the help of Eqs. (665a,b), (671a,b), and (672a,b), the following inequality is obtained:

$$\left(\frac{x_1 - p}{kd}\right)^{-\frac{7}{3}} \ll \left(\frac{x_1 - p}{kd}\right)^{-\frac{5}{3}} \Rightarrow \left(\frac{kd}{x_1 - p}\right)^{\frac{2}{3}} \ll 1, \tag{673a, b}$$

which defines the domain of validity of the approximate solutions of this section. Because in Eq. (673a), the term on the left-hand side decreases faster in the downstream direction than that on the right-hand side, an asymptotic similarity is obtained. However, in each case it must be evaluated, whether this condition leads to a second restriction of the application domain that in Egolf (1994) was just the domain of self-similarity.

Based on this finer analysis, now the basic equations, namely the full continuity equation and a reduced momentum equation, can be presented. The continuity equation is

$$2f_1 + \eta\frac{\mathrm{d}f_1}{\mathrm{d}\eta} - \frac{3}{\beta}\frac{1}{\eta}\frac{\mathrm{d}(\eta f_2)}{\mathrm{d}\eta} = 0, \tag{674}$$

where the correct exponents (670a–d) have already been substituted. This ordinary differential equation, with its two dimensionless velocity functions, takes the form

$$\frac{\mathrm{d}}{\mathrm{d}\eta}(\eta f_2) = \frac{\beta}{3}\frac{\mathrm{d}}{\mathrm{d}\eta}(\eta^2 f_1). \tag{675}$$

By integration it follows that

$$f_2 = \frac{\beta}{3}\eta f_1. \tag{676}$$

A constant of integration is set to zero, as $\eta f_1$ and $f_2$ vanish simultaneously at $\eta = 0$. Furthermore, the momentum equation takes the form

$$2f_1 + \eta\frac{\mathrm{d}f_1}{\mathrm{d}\eta} + \frac{3}{\beta}\frac{U}{U_G}\frac{1}{\eta}\frac{\mathrm{d}(\eta f_{21})}{\mathrm{d}\eta} = 0. \tag{677}$$

By comparing Eq. (674) with (676), one concludes that

$$f_{21} = -\frac{U_G}{U} f_2 \tag{678}$$

and owing to (676)

$$f_{21} = -\frac{\beta}{3}\frac{U_G}{U}\eta f_1. \tag{679}$$

The differential Eqs. (674) and (677) contain three variables $f_1, f_2$, and $f_{21}$. To close the system of equations, the DQTM (307b) is added. Choosing the following quantities is self-explaining:

$$\begin{aligned} &\sigma = \frac{\mathrm{d}b}{\mathrm{d}x_1} = \frac{\beta}{3}b, \quad x_{2\,\mathrm{min}} = 0, \\ &\bar{u}_1 = U_G + \bar{u}_1^*(x_1, x_2), \quad \bar{u}_{1\,\mathrm{max}} = U_G + \bar{u}_1^*(x_1, 0). \end{aligned} \tag{680a – e}$$

Substituting these into Eq. (307b) yields

$$\begin{aligned} \overline{u_2' u_1'}^* = &-\frac{\beta}{3}b\left[U_G + \bar{u}_1^*(x_1, x_2)\right]\times \\ &\frac{\left[U_G + \bar{u}_1^*(x_1, 0)\right] - \left[U_G + \bar{u}_1^*(x_1, x_2)\right]}{x_2}. \end{aligned} \tag{681}$$

Applying Eq. (661b), the second term in the first bracket is omitted, and in the fraction term $U_G$ cancels out. A division of Eq. (681) by $U^2$ then yields the following relation:

$$\frac{\overline{u_2' u_1'}^*}{U^2} = -\frac{\beta}{3}\frac{U_G}{U}\left[\frac{\overline{u}_1^*(x_1,0)}{U} - \frac{\overline{u}_1^*(x_1,x_2)}{U}\right]\frac{b}{x_2}. \tag{682}$$

Now, taking

$$\overline{u}_1^*(x_1,0) = U \quad \text{and} \quad \overline{u}_1^*(x_1,x_2) = U f_1, \tag{683a, b}$$

it follows that

$$f_{21} = -\frac{\beta}{3}\frac{U_G}{U}\frac{1-f_1}{\eta}. \tag{684}$$

Substituting Eq. (684) into (677) yields

$$2f_1 + \left(\eta + \frac{1}{\eta}\right)\frac{\mathrm{d}f_1}{\mathrm{d}\eta} = 0, \tag{685}$$

which is a first-order ordinary differential equation for $f_1$ that possesses the solution

$$f_1 = \frac{1}{1+\eta^2}. \tag{686}$$

This is an elementary function that enjoys the symmetry property $f_1(-\eta) = f_1(\eta)$. This function is shown in Fig. 9.22.

With the help of Eqs. (676) and (686), the transverse average velocity function is derived to be

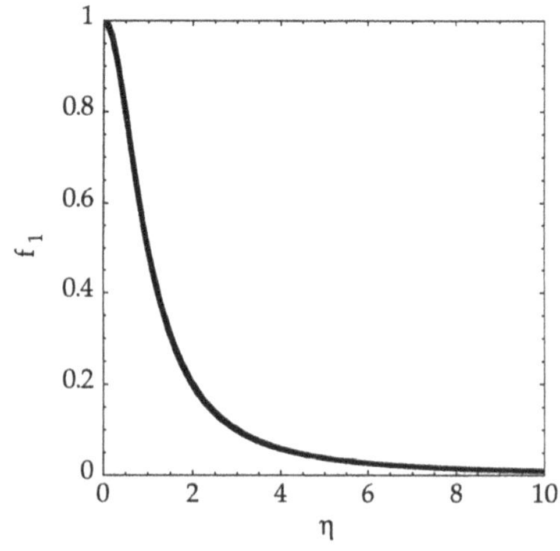


**Fig. 9.22** Averaged and scaled downstream velocity profile. It is symmetric in $\eta$ and does not exhibit entrainment effects. Courtesy of Egolf and Weiss (1998) © Phys. Rev. E., reproduced with permission by APS

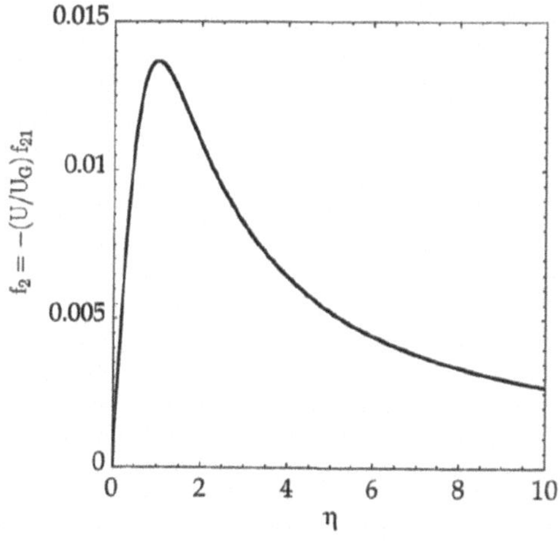


**Fig. 9.23** Averaged and scaled transverse velocity profile and the Reynolds shear stress; both are antisymmetric with changes $\eta \leftrightarrow -\eta$. Courtesy of Egolf and Weiss (1998) © Phys. Rev. E., reproduced with permission by APS

$$f_2 = \frac{\beta}{3} \frac{\eta}{1+\eta^2}. \tag{687}$$

It possesses nowhere negative values. Consequently, there is no fluid flux toward the axis of the jet and, therefore, also no entrainment.

The scaled Reynolds shear stress can be calculated by substituting Eq. (686) into (679) or (684) or by substituting (687) into (678). These are different calculation procedures that lead to the same result, namely

$$f_{21} = -\frac{\beta}{3} \frac{\mathrm{U_G}}{\mathrm{U}} \frac{\eta}{1+\eta^2}; \tag{688}$$

its graph, which shows a pronounced peak at the vertical distance $\eta = 1$ is sketched in Fig. 9.23.

## 9.3 Plane Couette Flows

Plane Couette flows are flows of a fluid between two parallel plates that are rectilinearly sheared. These flows were studied by applying the DQTM by Egolf and Weiss (1995a). The flow can be realized in many different ways, e.g., by keeping the lower horizontal plate at rest and moving the upper plate with a constant horizontal velocity $U$ (in our case to the right) (see Fig. 9.24) or by movement of the plates in opposite directions with velocities $\pm U/2$ relative to a laboratory system as shown in Fig. 9.25. Whereas in the first case the mean mass flux is positive, it is zero in the second case.

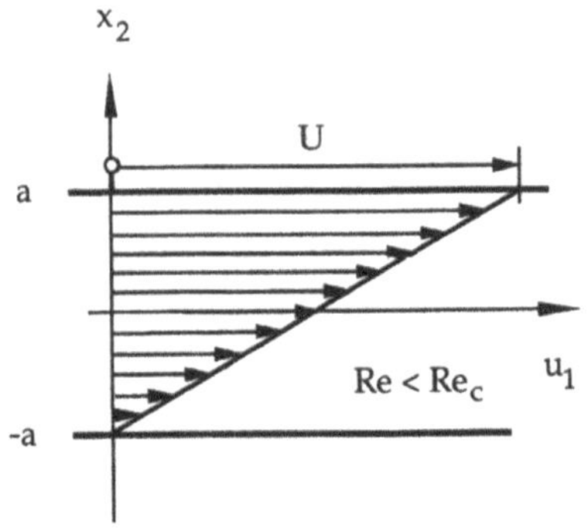


**Fig. 9.24** Plane laminar Couette flow between two rectilinearly sheared plates. The linear velocity profile of a laminar flow is shown in a system where the lower plate rests and only the upper plate is being moved. Courtesy of Egolf and Weiss (1995a) © Phys. Rev. Lett., reproduced with permission by APS

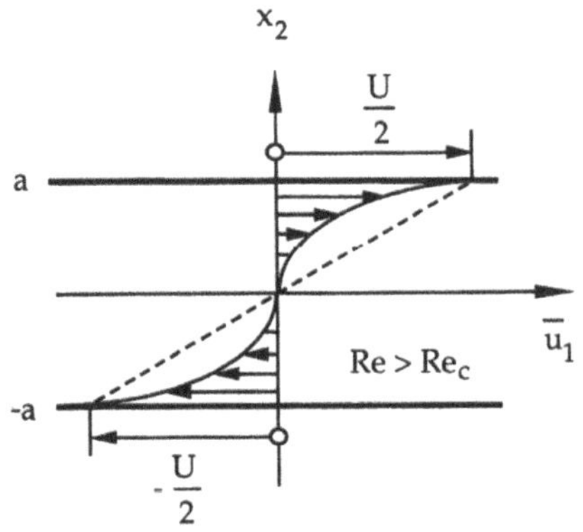


**Fig. 9.25** Turbulent variant of Fig. 9.24; there are two differences to that figure. At first, the flow is turbulent leading to an *S*-shaped average velocity profile. Secondly, the excitation of the system occurs by shearing the two plates with velocities $\pm U/2$ in opposite directions. Courtesy of Egolf and Weiss (1995a) © Phys. Rev. Lett., reproduced with permission by APS

Laminar plane Couette flows yield the basis to explain Newton's friction law. This fundamental law and constitutive equation of laminarly flowing Newtonian fluids are based on knowledge of a linear velocity profile occurring between adjacent parallel layers as observed in flows dominated by viscous forces. In turbulent states, the mean flow profiles are of *S*-shape with an increasing curvature with growing Reynolds number Re above the critical value, $\mathrm{Re}_c$. Measurements are reported by Reichardt (1956, 1959), Robertson and Johnson (1970), Aydin and Leutheusser (1991), El Telbany and Reynolds (1982) and Bech et al. (1995).

On the other hand, Couette flows are probably the simplest fundamental problems of wall turbulence just as the free shear layers are of free turbulence, and, therefore, their study is extremely valuable for a theoretical understanding of turbulent flows at medium or high Reynolds numbers. Notice that the Sect. 7.2.2, where a new "constitutive" equation of turbulence, the DQTM, was derived, contains much information that is overlapping with results of this section.

Since von Kármán pointed out the importance of plane Couette flows, they have intrigued a number of researchers such as Sommerfeld (1908), Heisenberg (1924), Kampé de Fériet (1939), Burgers and Mitchner (1953), Pearson (1959), Deissler (1970), etc. These and many more scientists were motivated to work on Couette flows by the desire to understand homogeneous turbulence. As we showed in Sect. 9.2.1, the production of turbulent kinetic energy, for finite Reynolds number flows, is nonzero and directly related to the product of the gradient of the mean velocity with the Reynolds shear stress. In the remainder of this section it is demonstrated that in plane turbulent Couette flows, the production of turbulent kinetic energy is to a good approximation constant over a large domain located symmetrically to the centerline between the two plane parallel plates. The higher the Reynolds number is, the more uniform will be the corresponding curve. This yields a homogeneous turbulence field, which, on the other hand, is non-isotropic because of its mean shear. Therefore, the results are quite different from those of plane turbulent Poiseuille flows (see Sect. 9.4), where the production of turbulent kinetic energy even vanishes on the centerline between the nonmoving plates (see Egolf and Weiss 1995b, 2000). Nowadays, by direct numerical simulations (DNS), inherent structures of turbulence in plane Couette flows at modest and high Reynolds numbers can be studied more easily; they comprise, e.g., large-scale structures and wall streaks (see Reynolds and Wieghardt 1995; Hamilton et al. 1995).

In the work of Douce (2016), the continuity and Navier Stokes equations were prepared in their RANS versions for Taylor-Couette flows, which are the cylindrical forms of plane Couette flows.

A Taylor-Couette flow apparatus consists of two concentric cylinders with a gap between them, into which the fluid is filled. Turning the two cylinders with different angular velocities leads to a shearing process in the fluid. This kind of fluid dynamic experiment became, similar as Rayleigh–Bénard convection, famous in the context of studying scenarios and routes to chaos and turbulence (see Sect. 1.2 and, e.g., Schuster 1988). In this case, care must be taken in defining the Reynolds number (or Rayleigh number) to avoid an influence of solid body rotation on this significant number. An adequate definition and many other relevant results of turbulent Taylor–Couette flows can be found, e.g., in Busse (1970), Weisshaar et al. (1991), Brauckmann and Eckhardt (2013).

Gollub and Swinney (1975) investigated the onset of turbulence in Taylor–Couette flows. Their observation was that these flows, in a spatial bifurcation, pile into azimuthal sheets (which are also called "fluid donuts"). A further increase of the angular rotation velocity makes these sheets to oscillate in the vertical direction and then additionally to twist. Accelerating the system even more, at highest rotation rates, the fluid system becomes fully turbulent. Gollup and Swinney's observation revealed the "Ruelle-Takens route to turbulence" (see Ruelle and Takens 1971a, b), where a stable fluid system by self-organization creates three internal frequencies before undergoing a transition to turbulence. High-quality visualization experiments by Fenstermacher et al. (1979) show four different flow regimes of Taylor–Couette flows (see Fig. 9.26).

An experimental apparatus for plane Couette flows is difficult to build. No vibrations of the moving plates should influence the observed fluid structures. A successful realization can be seen in Fig. 9.27, where a vertically positioned tape is moved around two rotating cylinders. The observed flow profile is sketched above the moving belt.

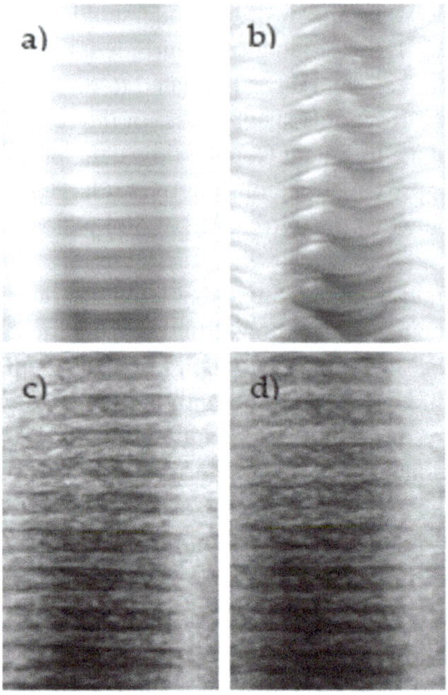


**Fig. 9.26** Four flow regimes of Taylor–Couette flows. In laminar flows no structure is seen. After a first bifurcation the rolls (donuts) are observed (**a**). At higher rotation these rolls start to oscillate (**b**) and additionally twist (**c**). Increasing the rotation further, the flow becomes turbulent (**d**), whereas at medium Reynolds numbers the roll-type structure still rests visible. Courtesy Fenstermacher et al. (1979) © J. Fluid Mech., reproduced with permission by Cambridge University Press

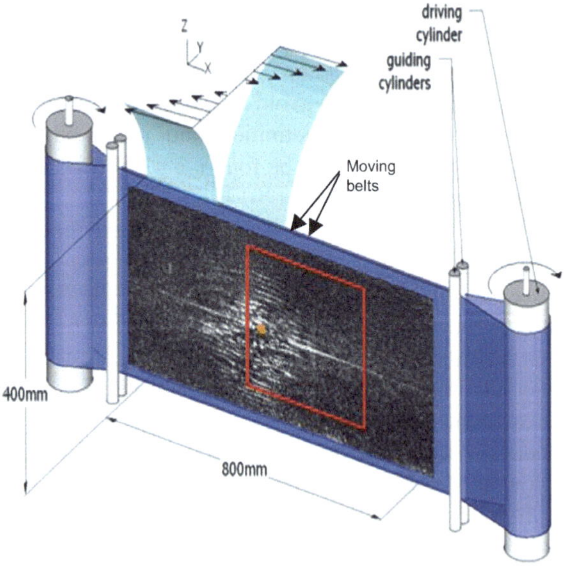


**Fig. 9.27** Plane Couette flow takes place ideally between two infinitely extended plates moving at the same velocity in opposite directions. Experimentally this is realized by a rotating belt, which constitutes the two moving boundaries. This difficult experimental setup is taken from Monchaux (2019a), reproduced with changes

In such experiments, the laminar-to-turbulent transition can be studied and also average profiles of laminar, transitional, and turbulent flows, as they are predicted by our theory, can be measured with good accuracy.

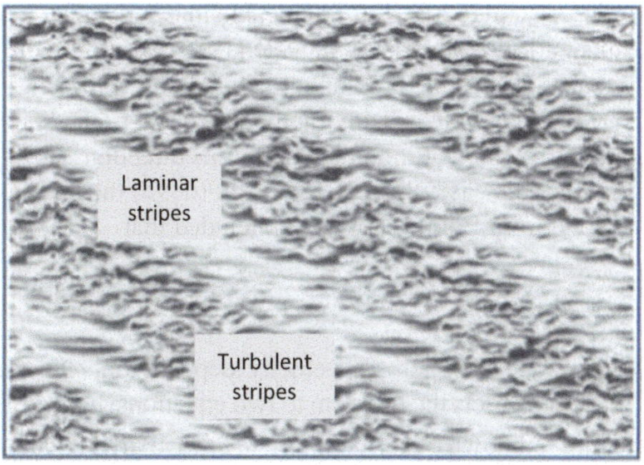


**Fig. 9.28** In plane Couette flows the coexistence of laminar and turbulent regions, in this figure occurring as diagonal stripes, can be observed. How important is the observation in the context of interpreting turbulence as a dynamical phase change concept (see Chap. 11) is at present not fully clear. From Monchaux (2019b), reproduced with changes

In this fluid flow system also a very interesting observation can be made. By a careful decrease of the Reynolds number from very high to medium and then close to the critical value, a coexistence of laminar and turbulent regions in stripes can be observed (see Fig. 9.28). The image in this figure is the result of numerical simulations. The laminar and turbulent stripes can be seen as inclined regions ranging from the left top to the right bottom of the image. The dynamical creation process, from an experimental arrangement, may also be clearly observed in a movie on the Internet: https://www.youtube.com/watch?v=PGA5RiRS4.

After a small tour d'horizon into the experimental area of Taylor–Couette and plane Couette flows, we now return to the theoretical development of the plane case. In the limit of infinite radii, the system of equations for Taylor–Couette flows reduces to those of the plane Couette case. This is rigorously demonstrated in Douce (2016) with the final result

$$
\begin{aligned}
&\frac{\partial \overline{u}_1}{\partial x_1} = 0, \\
&\frac{\partial}{\partial x_2}\overline{u'_2 u'_1} + \frac{1}{\rho}\frac{\partial \overline{p}}{\partial x_1} - \nu \frac{\partial^2 \overline{u}_1}{\partial x_2^2} = 0, \\
&\frac{\partial}{\partial x_2}\overline{u'^2_2} + \frac{1}{\rho}\frac{\partial \overline{p}}{\partial x_2} = 0, \\
&\frac{\partial}{\partial x_2}\overline{u'_3 u'_2} = 0.
\end{aligned}
\qquad (689\text{a} - \text{d})
$$

These equations represent in consecutive order (i) the mass balance with the assumption of incompressibility incorporated, (ii) the steady momentum equation parallel to the direction of the mean motion, (iii) the momentum equation perpendicular to the mean motion, and (iv) the nonvariability statement of the mean turbulent shear stress in the $x_2$-direction.

The system (689a–d) is identical to that derived by Pai (1953) and reprinted in the mathematical textbook of Stanišić (1988). The only difference is that Douce in his system neglected the pressure gradient in the $x_1$ direction. This is correct for Taylor–Couette flows with an azimuthally directed periodic boundary condition, leading to a constant pressure around the gap in this direction. However, in plane Couette flows this is not the case; this is why in Eq. (689b) the pressure term was kept.

Now, from Eq. (689a), it is concluded that

$$\overline{u}_1 = \overline{u}_1(x_2). \tag{690}$$

The average downstream velocity depends only on the transverse coordinate $x_2$. Then Pai restricts the problem to two equations, namely Eqs. (689b) and (689c), with the four unknown quantities $\overline{p}, \overline{u}_1,\ \overline{u_2'u_1'}$ and $\overline{u_2'^2}$. Again, we obtain a mathematically under-determined system of equations. However, following Stanišić's ideas, it is possible to derive certain relations between the unknowns. For this purpose, the self-similarity variables and functions are again helpful. In this case, the spatial coordinates in the two main directions are

$$\xi = \frac{x_1}{a}, \quad \eta = \frac{x_2}{a}. \tag{691a, b}$$

The characteristic velocity is chosen to be the friction velocity $u^*$ (see Eq. (136)). Then the following Reynolds number is defined:

$$\mathrm{Re}^* = \frac{u^* a}{\nu}. \tag{692}$$

To describe plane Couette flow, the usual Reynolds number is, however,

$$\mathrm{Re}^+ = \frac{u^* \delta}{\nu}, \tag{693}$$

where $\delta$ denotes the boundary layer thickness close to the wall (see Sect. 9.5). This Reynolds number is defined with the so called inner characteristic quantities and is usually applied together with inner variables. An alternative definition is

$$\mathrm{Re} = \frac{\overline{u}_{1\,\max}\, a}{\nu}, \tag{694}$$

where $a$ is the half distance between the plates. In some references, the full distance $2a$ is used instead. The average velocity is the maximum occurring downstream velocity $\overline{u}_{1\,\max}$. This Reynolds number is defined with the outer characteristic quantities and is usually applied with outer variables. Now, it is recognized that the definition (692) contains a merge of inner and outer quantities; this may be criticized, however, it does no harm. Nevertheless, we will stick to the definition of $\mathrm{Re}^*$, but later will use the two alternatives $\mathrm{Re}^+$ and Re. It is evident that

$$\mathrm{Re}^* = \frac{a}{\delta}\,\mathrm{Re}^+, \qquad \mathrm{Re}^* = \frac{u^*}{\overline{u}_{1\,\max}}\,\mathrm{Re}. \qquad (695\mathrm{a,b})$$

Now, in a self-similar representation, we have

$$f_1(\eta,\,\mathrm{Re}^*) = \frac{\overline{u}_1}{u^*} \qquad (696)$$

and

$$f_{21}(\eta,\,\mathrm{Re}^*) = -\frac{\overline{u_2'u_1'}}{u^{*2}}. \qquad (697)$$

Furthermore, we define

$$f_{22}(\eta,\,\mathrm{Re}^*) = \frac{\overline{u_2'^2}}{u^{*2}}, \qquad (698)$$

and the normalized pressure

$$\Pi(\xi,\eta,\,\mathrm{Re}^*) = \frac{\overline{p}-\overline{p}_0}{\rho\,u^{*2}}, \qquad (699)$$

where $\overline{p}_0$ is a reference pressure, e.g., the average pressure at the point (0,−a).

With these self-similar variables and functions, Eq. (689b) transforms to

$$\frac{\partial\Pi(\xi,\eta,\,\mathrm{Re}^*)}{\partial\xi} - \frac{1}{\mathrm{Re}^*}\frac{\mathrm{d}^2 f_1(\eta,\,\mathrm{Re}^*)}{\mathrm{d}\eta^2} - \frac{\mathrm{d}f_{21}(\eta,\,\mathrm{Re}^*)}{\mathrm{d}\eta} = 0. \qquad (700)$$

The second basic equation of importance, Eq. (689c), on the other hand, takes the form

$$\frac{\partial\Pi(\xi,\eta,\,\mathrm{Re}^*)}{\partial\eta} + \frac{\mathrm{d}f_{22}(\eta,\,\mathrm{Re}^*)}{\mathrm{d}\eta} = 0. \qquad (701)$$

It follows from Eq. (701) that $\Pi + f_{22} = F(\xi,\mathrm{Re}^*)$. Moreover, Eq. (700) states that $\partial\Pi/\partial\xi$ cannot be a function of $\xi$. Because $f_{22}$ is also independent of $\xi$, it follows that

$$\frac{\partial\Pi}{\partial\xi} = G(\eta,\,\mathrm{Re}^*) = \frac{\partial F}{\partial\xi}(\xi,\,\mathrm{Re}^*) \quad \Rightarrow \quad F = A(\mathrm{Re}^*)\xi + B(\mathrm{Re}^*) \qquad (702\mathrm{a}-\mathrm{c})$$

Then, since $\Pi + f_{22} = F$, the conclusion is that

$$\Pi(\xi, \eta, \mathrm{Re}^*) = -f_{22}(\eta, \mathrm{Re}^*) + A(\mathrm{Re}^*)\xi + C(\mathrm{Re}^*). \tag{703}$$

Next, this normalized pressure is substituted into (700); this yields

$$\frac{1}{\mathrm{Re}^*}\frac{\mathrm{d}^2 f_1(\eta, \mathrm{Re}^*)}{\mathrm{d}\eta^2} + \frac{\mathrm{d}f_{21}(\eta, \mathrm{Re}^*)}{\mathrm{d}\eta} = A(\mathrm{Re}^*), \tag{704}$$

where the pressure derivative has disappeared.
Now, Eq. (704) is integrated to yield

$$\frac{1}{\mathrm{Re}^*}\frac{\mathrm{d}f_1(\eta, \mathrm{Re}^*)}{\mathrm{d}\eta} + f_{21}(\eta, \mathrm{Re}^*) - A(\mathrm{Re}^*)\,\eta - B(\mathrm{Re}^*) = 0. \tag{705}$$

On the above path of derivation, the variables $\Pi$ and $f_{22}$ disappeared from Eqs. (704) and (705). This is very convenient, because, if an additional closure scheme in form of a functional $f_{21} = f(f_1)$ is at one's disposal, this single basic ordinary differential equation remains valid and may be solved.

Before we introduce a closure scheme, Eq. (705) is supplemented by its boundary conditions. At first, the no-slip boundary conditions are introduced at the lower and upper boundaries

$$f_1(-1, \mathrm{Re}^*) = 0, \quad f_1(+1, \mathrm{Re}^*) = \frac{\overline{u}_{1\,\max}}{u^*}. \tag{706a, b}$$

Moreover, by hindrance of fluctuations perpendicular to the wall, the turbulent motion vanishes at the boundary, leading to a Reynolds shear stress, which is zero there,

$$f_{21}(-1, \mathrm{Re}^*) = 0, \quad f_{21}(+1, \mathrm{Re}^*) = 0. \tag{707a, b}$$

It is possible to formulate boundary conditions for $f_{22}$ and $\Pi$. We omit them, because they are not relevant to our ensuing arguments.
By evaluating (705) at the boundaries, $\eta = \pm 1$, one obtains

$$\begin{aligned} \frac{1}{\mathrm{Re}^*}\frac{\mathrm{d}f_1}{\mathrm{d}\eta}\bigg|_{\eta=1} - A - B = 0, \\ \frac{1}{\mathrm{Re}^*}\frac{\mathrm{d}f_1}{\mathrm{d}\eta}\bigg|_{\eta=-1} + A - B = 0. \end{aligned} \tag{708a, b}$$

From Eq. (136), we conclude that

$$\frac{\tau_0}{\rho} = u^{*^2}. \tag{709}$$

Furthermore, close to the wall no turbulent shear stress occurs and Newton's friction law is valid

$$\tau_0 = \mu \frac{\mathrm{d}\overline{u}_1}{\mathrm{d}x_2}\bigg|_{x_2=\pm a}. \tag{710}$$

From Eqs. (709) and (710), one can deduce that

$$u^{*^2} = \frac{\mu}{\rho}\frac{\mathrm{d}\overline{u}_1}{\mathrm{d}x_2}\bigg|_{x_2=\pm a} = \nu \frac{\mathrm{d}\overline{u}_1}{\mathrm{d}x_2}\bigg|_{x_2=\pm a}, \tag{711a, b}$$

or

$$\frac{u^* a}{\nu} = \frac{\mathrm{d}\left(\dfrac{\overline{u}_1}{u^*}\right)}{\mathrm{d}\left(\dfrac{x_2}{a}\right)}\Bigg|_{x_2=\pm a} \quad \Leftrightarrow \quad \mathrm{Re}^* = \frac{\mathrm{d}f_1}{\mathrm{d}\eta}\bigg|_{\eta=\pm 1}. \tag{712a, b}$$

With the help of Eq. (712b), (708a,b) simplify to

$$\begin{aligned} 1 - A - B &= 0, \\ 1 + A - B &= 0, \end{aligned} \tag{713a, b}$$

and implies

$$A = 0, \quad B = 1. \tag{714a, b}$$

Thus, Eq. (705) reduces to

$$\frac{1}{\mathrm{Re}^*}\frac{\mathrm{d}f_1(\eta, \mathrm{Re}^*)}{\mathrm{d}\eta} + f_{21}(\eta, \mathrm{Re}^*) - 1 = 0. \tag{715}$$

Next, we introduce the DQTM (307b). The characteristic length $\chi_2$ is the distance to the wall,

$$\chi_2 = a - x_2. \tag{716}$$

The quantity $x_{2\max}$ denotes the transverse space coordinate at the position, where $\overline{u}_1$ takes its maximum as a function of the variable $x_2$; this is at the inner boundary of the upper plate. This is equal to the half distance between the two parallel plates,

$$x_{2\,\max} = a. \tag{717}$$

With these characteristic quantities, the "mixing length" cancels the denominator of the DQTM and a symmetric expression results

$$\overline{u_2' u_1'} = -\sigma\,(\overline{u}_1 - \overline{u}_{1\,\min})\,(\overline{u}_{1\,\max} - \overline{u}_1). \tag{718}$$

The minimum and maximum average velocities occur at the lower and upper plates (see Fig. 9.25), viz.,

$$\overline{u}_{1\,\min} = \overline{u}_1(x_1, -a), \quad \overline{u}_{1\,\max} = \overline{u}_1(x_1, +a). \tag{719a, b}$$

Introducing this into (718) leads to

$$\overline{u_2' u_1'} = -\sigma\,[\overline{u}_1(x_1, x_2) - \overline{u}_1(x_1, -a)] \cdot [\overline{u}_1(x_1, +a) - \overline{u}_1(x_1, x_2)]. \tag{720}$$

Furthermore, in agreement with Eqs. (706a,b), we conclude that

$$\overline{u}_1(x_1, -a) = 0, \quad \overline{u}_1(x_1, +a) = \overline{u}_{1\,\max}. \tag{721a, b}$$

Applying these two equations to (720) and dividing the resulting equation by $u^{*2}$ yields

$$\frac{\overline{u_2' u_1'}}{u^{*2}} = -\sigma\,\frac{\overline{u}_1(x_1, x_2)}{u^*}\left[\frac{\overline{u}_{1\,\max}}{u^*} - \frac{\overline{u}_1(x_1, x_2)}{u^*}\right]. \tag{722}$$

With the help of Eqs. (696) and (697), this is rewritten to reveal the normalized Reynolds shear stress expressed as

$$f_{21} = \sigma f_1(\eta)\,[f_1(1) - f_1(\eta)\,]. \tag{723}$$

This is now substituted into the RANS Eq. (715), resulting in

$$\frac{1}{\mathrm{Re}^*}\frac{\mathrm{d}f_1}{\mathrm{d}\eta} + \sigma f_1(\eta)\;[f_1(1) - f_1(\eta)\,] - 1 = 0. \tag{724}$$

Unlike Stanišić (1988), we now transform this equation to outer variables by introducing the new functions $g_1(\eta)$ and $g_{21}(\eta)$ as follows:

$$g_1(\eta) = \frac{f_1(\eta)}{f_1(1)}, \quad g_{21}(\eta) = f_{21}(\eta). \tag{725a, b}$$

Equation (724) then takes the form

$$\frac{f_1(1)}{\mathrm{Re}^*}\frac{\mathrm{d}g_1}{\mathrm{d}\eta}+\sigma f_1^2(1)\,g_1(\eta)[1-g_1(\eta)]-1=0. \tag{726}$$

With the parameters

$$\alpha=\frac{f_1(1)}{\mathrm{Re}^*},\quad \beta=\sigma f_1^2(1), \tag{727a, b}$$

the principal boundary value problem for plane Couette flows now reads

$$\begin{aligned}&\alpha\frac{\mathrm{d}g_1}{\mathrm{d}\eta}+\beta\,g_1(\eta)-\beta\,g_1^2(\eta)-1=0,\\ &g_1(-1)=0,\qquad g_1(+1)=1,\end{aligned} \tag{728a – c}$$

where the boundary conditions followed from Eqs. (706a,b) and the definition (725a).

To solve this two-point boundary-value problem, we make the Ansatz

$$g_1(\eta)=a\tan(b\eta)+c. \tag{729}$$

This implies

$$\frac{\mathrm{d}g_1}{\mathrm{d}\eta}=a\frac{\mathrm{d}}{\mathrm{d}\eta}[\tan(b\eta)]=ab\left[1+\tan^2(b\eta)\right]. \tag{730a, b}$$

Substitution of (729) and (730b) into (728a) yields the transcendental equation

$$\begin{aligned}\alpha ab+\alpha ab\tan^2(b\eta)+\beta a\tan(b\eta)+\beta c-\beta a^2\tan^2(b\eta)-2\beta ac\tan(b\eta)\\ -\beta c^2-1=0.\end{aligned} \tag{731}$$

Grouping the terms, according to the three powers of the tangent function, leads to

$$\begin{aligned}&\left(\alpha ab+\beta c-\beta c^2-1\right)+(\beta a-2\beta ac)\\ &\times\tan(b\eta)+\left(\alpha ab-\beta a^2\right)\tan^2(b\eta)=0.\end{aligned} \tag{732}$$

This equation must be fulfilled for all values of $\eta$. Therefore, each parenthesis must be identical to zero leading to the three coefficient equations

$$\begin{aligned}&\alpha ab+\beta c-\beta c^2-1=0,\\ &1-2c=0,\\ &\alpha b-\beta a=0.\end{aligned} \tag{733a – c}$$

From Eq. (733b) we deduce

$$c = \frac{1}{2}, \tag{734}$$

and from Eq. (733c), one concludes that

$$b = \frac{\beta}{\alpha} a, \tag{735}$$

so that (733a) implies

$$a^2 + \frac{1}{4} - \frac{1}{\beta} = 0 \tag{736}$$

or

$$a = \frac{1}{2}\sqrt{\frac{4-\beta}{\beta}}. \tag{737}$$

Therefore, with this result Eq. (735) leads to

$$b = \frac{\beta}{2\alpha}\sqrt{\frac{4-\beta}{\beta}}. \tag{738}$$

Now, the constants given by Eqs. (737), (738), and (734), are substituted into (729); this leads to

$$g_1(\eta) = \frac{1}{2}\left[1 + \sqrt{\frac{4-\beta}{\beta}} \tan\left(\frac{\beta}{2\alpha}\sqrt{\frac{4-\beta}{\beta}}\eta\right)\right]. \tag{739}$$

With the help of the boundary conditions, this equation can be further simplified. Let us start with the boundary condition at the lower plate, viz.,

$$g_1(-1) = \frac{1}{2}\left[1 + \sqrt{\frac{4-\beta}{\beta}} \tan\left(-\frac{\beta}{2\alpha}\sqrt{\frac{4-\beta}{\beta}}\right)\right] = 0. \tag{740}$$

The antisymmetry property of the tangent function, $\tan(-x) = -\tan(x)$, now yields

$$g_1(-1) = \frac{1}{2}\left[1 - \sqrt{\frac{4-\beta}{\beta}} \tan\left(\frac{\beta}{2\alpha}\sqrt{\frac{4-\beta}{\beta}}\right)\right] = 0. \tag{741}$$

Combining this equation with (728b) requests that

$$\sqrt{\frac{4-\beta}{\beta}} \tan\left(\frac{\beta}{2\alpha}\sqrt{\frac{4-\beta}{\beta}}\right) = 1. \tag{742}$$

This is identical to the relation

$$\tan\left(\frac{\beta}{2\alpha}\sqrt{\frac{4-\beta}{\beta}}\right) = \sqrt{\frac{\beta}{4-\beta}}, \tag{743}$$

or

$$\frac{\beta}{2\alpha} = \sqrt{\frac{\beta}{4-\beta}}\arctan\left(\sqrt{\frac{\beta}{4-\beta}}\right). \tag{744}$$

With this formula, we can express in solution (739) the term involving $\alpha$ such that it only contains $\beta$. The result reads

$$g_1(\eta) = \frac{1}{2}\left\{1 + \sqrt{\frac{4-\beta}{\beta}}\tan\left[\arctan\left(\sqrt{\frac{\beta}{4-\beta}}\right)\eta\right]\right\}. \tag{745}$$

There remains the question, whether this equation also fulfills the second boundary condition at the upper plate. Calculating $g_1(1)$ with the aid of Eq. (745) yields the correct boundary condition (see Eq. (728c)), because the tangents function acts directly on the arctangent function, so that the boundary condition (728c) is the consequence.

We, thus, have shown that both boundary conditions are fulfilled. Therefore, Eq. (745) yields the solution describing the average velocity profile of plane Couette flows (see Fig. 9.29). Figure 9.30 shows a comparison between the theoretical and some measured data.

Notice that Eq. (744) is a unique relation between $\alpha$ and $\beta$. Equation (727a) indicates that $\alpha$ is inversely proportional to the Reynolds number. Therefore, $\alpha$ is

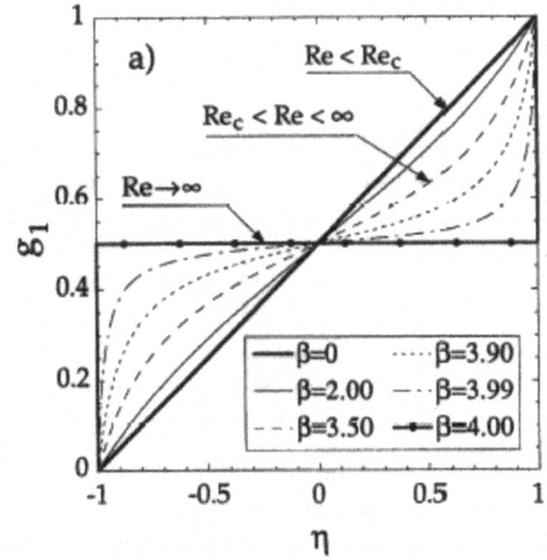


**Fig. 9.29** Average velocity profiles of laminar $(\alpha,\beta) = (2,0)$ to fully turbulent infinite Reynolds number $(\alpha,\beta) = (0,4)$ plane Couette flows. The two limiting cases are presented by thicker solid lines, which are linear in the laminar and constant in the highest Reynolds number regime. Courtesy of Egolf and Weiss (1995a) © Phys. Rev. Lett., reproduced with permission by APS

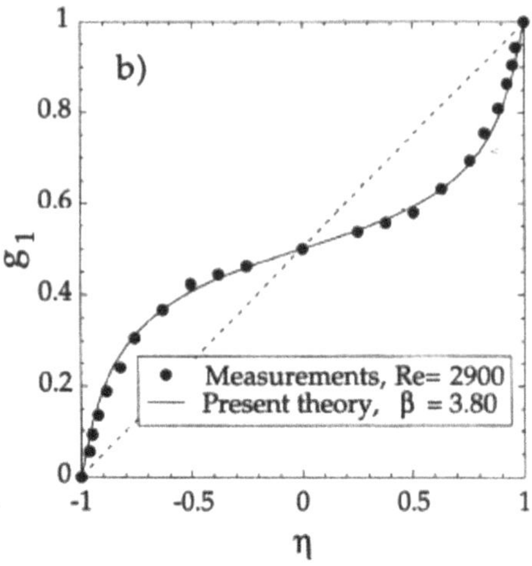


**Fig. 9.30** Comparison of experimental results, represented by full dots, for a turbulent flow with a Reynolds number Re = 2900 with theoretical calculations shown by the solid curve. The experimental data are from Reichardt (1959). Courtesy of Egolf and Weiss (1995a) © Phys. Rev. Lett., reproduced with permission by APS

also a stress or control parameter of the fluid dynamic system! Could it be that the Reynolds number should better have been defined inversely? The answer to this question will be given in Chap. 11. This theory reveals a cooperative phenomenon, and $\beta$ is the related order parameter of the system (see Chap. 11). We have seen that $\beta = 4$ is the maximum occurring value. Because an order parameter is always normalized to be at its maximum equal to unity, this number is scaled by a factor "1/4," viz.,

$$\chi = \frac{\beta}{4}. \tag{746}$$

With this definition result (745) is rewritten as

$$g_1(\eta) = \frac{1}{2}\left\{1 + \sqrt{\frac{1-\chi}{\chi}}\tan\left[\arctan\left(\sqrt{\frac{\chi}{1-\chi}}\right)\eta\right]\right\}. \tag{747}$$

The point-wise (p.w.) limits for $| \eta \leq 1 |$ are

$$\begin{aligned} &\beta, \chi \rightarrow 0 \;\Rightarrow\; \alpha \rightarrow 2, \\ &g_1(\eta) \rightarrow \frac{1}{2}(1+\eta) \;\text{ p.w. } |\eta| \leq 1, \\ &\beta \rightarrow 4,\; \chi \rightarrow 1 \;\Rightarrow\; \alpha \rightarrow 0, \\ &g_1(\eta) \rightarrow \frac{1}{2}[\theta(1-\eta) + \theta(1+\eta)] \;\text{ p.w. } |\eta| \leq 1, \end{aligned} \tag{748a, b}$$

where $\theta$ denotes the Heaviside distribution.

Now, a discussion of the laminar case ($\chi \rightarrow 0$) is necessary. For this we apply a Taylor series expansion of the arctangent function in the vicinity of $\chi$=0

$$\arctan\left(\sqrt{\frac{\chi}{1-\chi}}\right)=\sqrt{\frac{\chi}{1-\chi}}+\mathrm{O}\left(\sqrt{\frac{\chi}{1-\chi}}\right)^3. \tag{749}$$

Now, in the limit $\chi \to 0$, Eq. (747) reduces to

$$g_1(\eta)=\frac{1}{2}\left[1+\sqrt{\frac{1-\chi}{\chi}}\tan\left(\sqrt{\frac{\chi}{1-\chi}}\eta\right)\right]. \tag{750}$$

Substituting in this expression the McLaurin series,

$$\tan\left(\sqrt{\frac{\chi}{1-\chi}}\eta\right)=\sqrt{\frac{\chi}{1-\chi}}\eta+\mathrm{O}\left(\sqrt{\frac{\chi}{1-\chi}}\eta\right)^3, \tag{751}$$

reduces Eq. (750) to

$$g_1(\eta)=\frac{1}{2}(1+\eta). \tag{752}$$

This corresponds to the linear velocity profile of Newtonian shear flows (Fig. 9.29 and Table 9.3).

Next, let us combine Eqs. (723), (725b), and (727b), to obtain

$$g_{21}=\beta\, g_1(\eta)\,[1-g_1(\eta)\,]=4\chi\, g_1(\eta)\,[1-g_1(\eta)], \tag{753a, b}$$

in which, according to Eq. (746), $\beta=4\chi$ was applied.

The formula for the Reynolds shear stress is calculated by first introducing the abbreviation

$$\gamma=\sqrt{\frac{4-\beta}{\beta}}\tan\left[\arctan\left(\sqrt{\frac{\beta}{4-\beta}}\right)\eta\right]. \tag{754}$$

Then Eq. (745) reduces to

$$g_1(\eta)=\frac{1}{2}(1+\gamma). \tag{755}$$

In terms of $\gamma$ Eq. (753a) takes the form

$$g_{21}=\beta\,\frac{1}{2}(1+\gamma)\left[1-\frac{1}{2}(1+\gamma)\right]=\frac{\beta}{2}(1+\gamma)\frac{1}{2}(1-\gamma)=\frac{\beta}{4}(1-\gamma^2). \tag{756a – c}$$

Substitution of the quantity $\gamma$ from (754) into (756c) leads to

**Table 9.3** Here, the results of laminar and turbulent plane Couette flows, with their "constitutive equations," namely Newton's shear stress law and the DQTM, are summarized

| | Laminar flow | Turbulent flow |
|---|---|---|
| Char. length | $\lambda$ | $\Lambda_{T,2} = x_{2\max} - x_2$ |
| Char. velocity | $u_{\mathrm{mol}}$ | $\overline{u}_1 - \overline{u}_{1\min}$ |
| Shear stress | $\tau_{12} = \mu \frac{du_1}{dx_2}$ | $\tau_{T,12} = \mu_T \frac{\overline{u}_{1\max} - \overline{u}_1}{x_{2\max} - x_2}$ |
| Effective dynamic viscosity | $\mu = \sigma\rho\lambda u_{\mathrm{mol}}$ | $\mu_T = \sigma\rho\Lambda_{T,2}(\overline{u}_1 - \overline{u}_{1\min})$ |
| Dimensionless variable | $\eta = x_2/x_{2\max}$ | $\eta = x_2/x_{2\max}$ |
| Dimensionless velocity | $g_1 = u_1/U$ | $g_1 = u_1/U$ |
| Differential equation | $g_1'' = 0$ | $2(\mathrm{Re}_c/\mathrm{Re})g_1' + 4\chi$ $g_1(1-g_1) - 1 = 0$ |
| Stress parameter | $1/\mathrm{Re}_c \leq 1/\mathrm{Re} \leq \infty$ | $0 \leq 1/\mathrm{Re} \leq 1/\mathrm{Re}_c$ |
| Order parameter | $\chi(1/\mathrm{Re} > 1/\mathrm{Re}_c) = 0$ | $\mathrm{Re}_c/\mathrm{Re} = \sqrt{\chi(1-\chi)}/\arctan$ $\left[\sqrt{\chi/(1-\chi)}\right], 0 \leq 1/\mathrm{Re} \leq 1/\mathrm{Re}_c$ |
| Boundary conditions | $g_1(-1) = 0,\ g_1(+1) = 1$ | $g_1(-1) = 0,\ g_1(+1) = 1$ |
| Mean streamwise velocity | $g_1 = 1/2\ (1 + \eta)$ | $g_1 = 1/2\left\langle 1 + \sqrt{(1-\chi)/\chi}\right.$ $\left.\tan\left\{\arctan\left[\sqrt{\chi/(1-\chi)}\right]\eta\right\}\right\rangle$ |
| Dimensionless shear stress | $g_{21} = g_1' = 1/2$ | $g_{21} = 4\chi g_1\ (1 - g_1)$ |

Courtesy of Egolf and Hutter (2016b) © Springer Int. Publ. Switzerland, reproduced with changes

$$g_{21} = \frac{\beta}{4}\left\{1 - \frac{4-\beta}{\beta}\tan^2\left[\arctan\left(\sqrt{\frac{\beta}{4-\beta}}\right)\eta\right]\right\}. \tag{757}$$

In Fig. 9.31, the theoretical Reynolds shear stress (757) is plotted for different values of $\beta$, and in Fig. 9.32 a comparison with results from numerical simulation are shown.

For completeness the Reynolds shear stress $g_{21}$ is given as a function of the order parameter $\chi$

$$g_{21} = \chi\left\{1 - \frac{1-\chi}{\chi}\tan^2\left[\arctan\left(\sqrt{\frac{\chi}{1-\chi}}\right)\eta\right]\right\}. \tag{758}$$

From Eq. (744), it follows that

$$\alpha = \frac{1}{2}\frac{\sqrt{\frac{4-\beta}{\beta}}}{\arctan\left(\sqrt{\frac{\beta}{4-\beta}}\right)}\beta. \tag{759}$$

This relation is plotted as Fig. 9.33. When this figure was drawn for the first time, it became clear what is explicitly stated as:

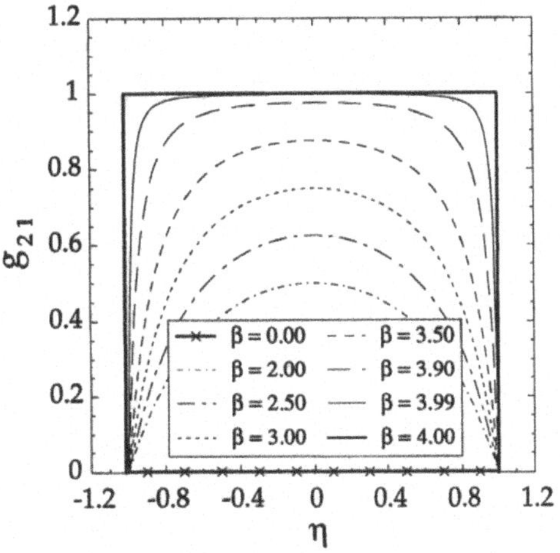


**Fig. 9.31** Reynolds shear stresses calculated with Eq. (753a) for different order parameters $\beta$. The limiting cases of laminar and infinite Reynolds number flows are shown by thick lines ($\beta = 0$ and $\beta = 4$). Courtesy of Egolf and Weiss (1995a) © Phys. Rev. Lett., reproduced with permission by APS

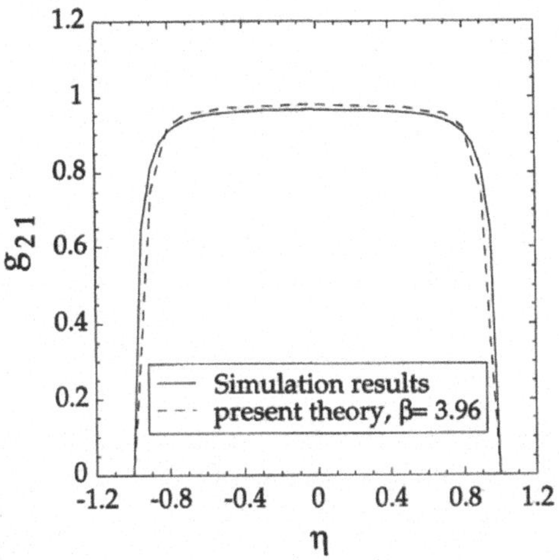


**Fig. 9.32** Reynolds shear stress comparison of analytical calculations of this section with simulation results from Lee and Kim (1991). The agreement is excellent. It is recognized that the profiles for larger Reynolds numbers converge for the infinite value to a rectangular profile. Courtesy of Egolf and Weiss (1995a) © Phys. Rev. Lett., reproduced with permission by APS

**Theorem 3** *In the DQTM a physical cooperative and critical phenomenon occurs in a natural manner, as it is the case in many other domains of physics, e.g., solid–liquid transitions of mixtures, magnetism, superconductivity, superfluidity, and liquid crystals.*

This discovery is a strong argument for support of the nonlocal and fractional DQTM.

*To the best of our knowledge, these results are the first theoretical proof that turbulent shear flows are described by a cooperative phenomenon showing criticality* (see Chap. 11).

Rearranging Eq. (727a) yields

$$\mathrm{Re}^* = \frac{f_1(1)}{\alpha}. \tag{760}$$

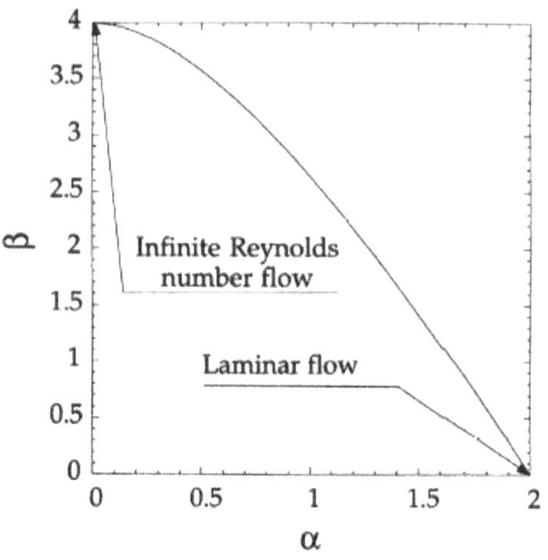


**Fig. 9.33** Functional relation between $\alpha$ and $\beta$, which was calculated by Eq. (759). The critical value, related to the critical Reynolds number $\mathrm{Re}_c$, is $\alpha_c = 2$. The highest order occurs for $\alpha = 0$, where $\beta$ takes its maximum "4." Courtesy of Egolf and Weiss (1995a) © Phys. Rev. Lett., reproduced with permission by APS

Equation (727b) leads to the relation

$$f_1(1) = \sqrt{\frac{\beta}{\sigma}}. \tag{761}$$

Combining the last two equations yields the result

$$\mathrm{Re}^* = \frac{1}{\alpha}\sqrt{\frac{\beta}{\sigma}}. \tag{762}$$

Inserting the value for $\alpha$, as given by Eq. (759), delivers $\mathrm{Re}^*$ as a function of $\beta$, and with $\beta = 4\chi$ as a function of $\chi$,

$$\mathrm{Re}^* = \frac{1}{\sqrt{\sigma}} \frac{\arctan\left(\sqrt{\frac{\chi}{1-\chi}}\right)}{\sqrt{1-\chi}}. \tag{763}$$

For the stress parameter $\alpha$, from Eq. (727a) it is evident that the Reynolds number $\mathrm{Re}^*$ acts inversely to $\alpha$. Thus, writing (760) for $(\alpha, \mathrm{Re}^*)$ and $(\alpha_c, \mathrm{Re}^*{}_c)$ and forming their ratio yields

$$\frac{\alpha}{\alpha_c} = \frac{\mathrm{Re}^*{}_c}{\mathrm{Re}^*} = \frac{\mathrm{Re}_c}{\mathrm{Re}}, \tag{764a, b}$$

where Eq. (695b) was applied to derive (764b). Therefore, from Eq. (759), with $\alpha_c = 2$, one has

$$\frac{\mathrm{Re}_c}{\mathrm{Re}} = \frac{\alpha}{2} = \frac{\beta}{4}\frac{\sqrt{\frac{4-\beta}{\beta}}}{\arctan\left(\sqrt{\frac{\beta}{4-\beta}}\right)} = \frac{\sqrt{\chi(1-\chi)}}{\arctan\left(\sqrt{\frac{\chi}{1-\chi}}\right)}. \quad (765a-c)$$

This function was the basis to calculate the results of Fig. 9.34.

The mean shear $S$ is defined by the following formula:

$$S = 2\frac{\mathrm{d}g_1}{\mathrm{d}\eta}\bigg|_{\eta=0}. \quad (766)$$

The derivative of the average velocity profile is obtainable from (745),

$$\begin{aligned}\frac{\mathrm{d}g_1}{\mathrm{d}\eta}\bigg|_{\eta=0} &= \frac{1}{2}\sqrt{\frac{4-\beta}{\beta}}\frac{\mathrm{d}}{\mathrm{d}\eta}\left\{\tan\left[\arctan\left(\sqrt{\frac{\beta}{4-\beta}}\right)\eta\right]\right\}\bigg|_{\eta=0} \\ &= \frac{1}{2}\sqrt{\frac{4-\beta}{\beta}}\arctan\left(\sqrt{\frac{\beta}{4-\beta}}\right)\left\{1+\tan^2\left[\arctan\left(\sqrt{\frac{\beta}{4-\beta}}\right)\eta\right]\right\}\bigg|_{\eta=0} \\ &= \frac{1}{2}\sqrt{\frac{4-\beta}{\beta}}\arctan\left(\sqrt{\frac{\beta}{4-\beta}}\right).\end{aligned} \quad (767a-c)$$

Combining Eqs. (766) and (767c), leads to the final result for the mean shear

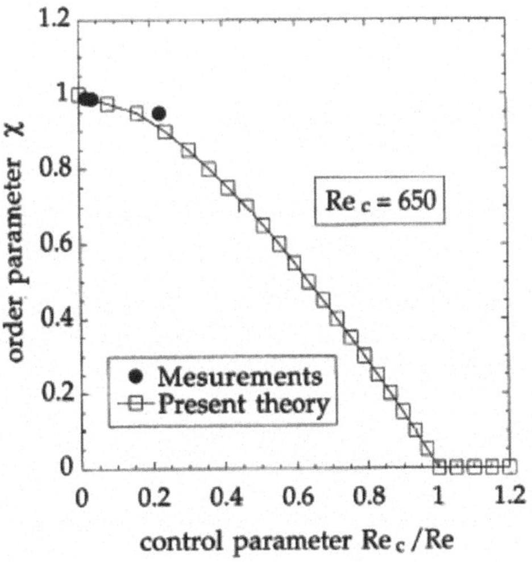


**Fig. 9.34** Order parameter $\chi$ as compared with experimental results from Reichardt (1959). There is a lack of good experimental results at excitations just slightly above the criticality $\mathrm{Re}_c$. Courtesy of Egolf and Weiss (1995a) © Phys. Rev. Lett., reproduced with permission by APS

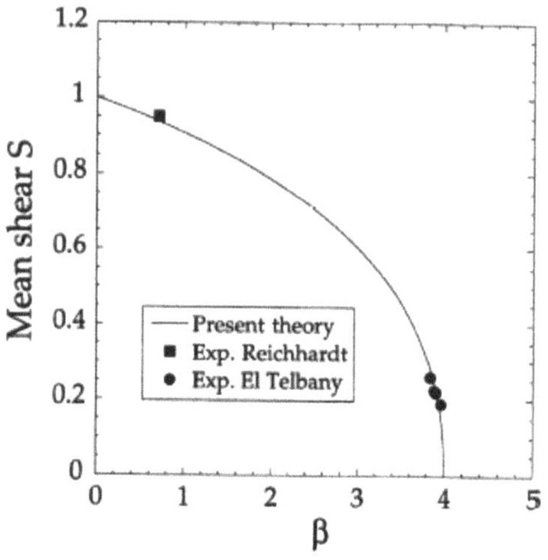


**Fig. 9.35** Calculated mean shear $S$ as compared with results from Reichardt (1959) and El Telbany and Reynolds (1982). Courtesy of Egolf and Weiss (1995a) © Phys. Rev. Lett., reproduced with permission by APS

$$\mathrm{S} = \sqrt{\frac{4-\beta}{\beta}}\arctan\left(\sqrt{\frac{\beta}{4-\beta}}\right). \tag{768}$$

The mean shear (768) is presented in Fig. 9.35 and compared with a small data set of measurements from two different laboratories.

Referring to Hinze (1975), the specific normalized production of turbulent kinetic energy is given by

$$P = g_{21}\frac{\mathrm{d}g_1}{\mathrm{d}\eta}, \tag{769}$$

which can easily be calculated. To simplify these calculations, we introduce the abbreviations

$$\delta = \arctan\sqrt{\frac{\beta}{4-\beta}}, \qquad \varepsilon = \sqrt{\frac{4-\beta}{\beta}}. \tag{770a, b}$$

Consider now Eq. (769). Multiply the derivative $\mathrm{d}g_1/\mathrm{d}\eta$, given by Eq. (767b) for $\eta \neq 0$, with the Reynolds shear stress, presented as Eq. (757); the result of this operation yields

$$\begin{aligned} P = \frac{\beta\delta\varepsilon}{8}\left[1-\varepsilon^2\tan^2(\delta\,\eta)\right]\cdot\left[1+\tan^2(\delta\eta)\right] = \\ \frac{\beta\delta\varepsilon}{8}\left[1+\left(1-\varepsilon^2\right)\tan^2(\delta\eta)-\varepsilon^2\tan^4(\delta\eta)\right]. \end{aligned} \tag{771a, b}$$

Substitution of (770b) transforms this to

$$P = \frac{\beta\delta}{8}\sqrt{\frac{4-\beta}{\beta}}\left[1+2\left(1-\frac{2}{\beta}\right)\tan^2(\delta\eta)-\frac{4-\beta}{\beta}\tan^4(\delta\eta)\right]. \tag{772}$$

By also substituting, the other remaining constant $\delta$ leads to

$$\begin{aligned}P(\eta) &= \frac{\beta}{8}\sqrt{\frac{4-\beta}{\beta}}\arctan\sqrt{\frac{\beta}{4-\beta}}\cdot \\ &\left[1+2\left(1-\frac{2}{\beta}\right)\tan^2\left(\arctan\sqrt{\frac{\beta}{4-\beta}}\eta\right)-\right. \\ &\left.\frac{4-\beta}{\beta}\tan^4\left(\arctan\sqrt{\frac{\beta}{4-\beta}}\eta\right)\right].\end{aligned} \tag{773}$$

This equation was the basis to calculate the specific kinetic energy production shown in Fig. 9.36.

The overall production of the turbulent kinetic energy between the two plates,

$$P = \int_{-1}^{+1} P(\eta)\,\mathrm{d}\eta, \tag{774}$$

would be a perfect order parameter of plane Couette flows. A stress parameter quantifies the external excitation of the system. Under its influence, a system response is determined only by its internal mechanisms, i.e., in a manner that cannot be influenced by an external experimenter. We found that $\beta$, respectively, its normalized counterpart $\chi$, is the order parameter of plane Couette flows. However, it must be confessed that this conclusion was only drawn by the observation that the

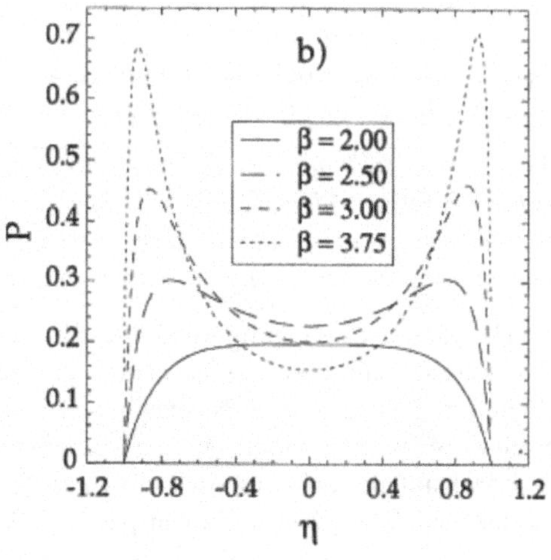


**Fig. 9.36** Specific production rate of the turbulent kinetic energy $P$. For large Reynolds numbers Re ($\beta$ values), this quantity is approximately constant in broad domains symmetrical to the centerline, leading to near-homogenous turbulent fields. Notice that when $\beta \to 4$, the specific production $P$ converges toward two delta distributions at the positions of the bounding plates. Courtesy of Egolf and Weiss (1995a) © Phys. Rev. Lett., reproduced with permission by APS

$\chi(\alpha)$ curve has a form analogous to the order parameter curves of other physical systems, exhibiting a cooperative phenomenon (e.g., compare with the magnetization curve of a ferromagnetic system). If a proportionality between the production rate $P$ and $\beta$ or $\chi$, respectively, could be proven, then this new mathematical discovery would receive its necessary solid physical basis (see below).

To calculate the specific production rate $P$, Eq. (773) could be substituted into Eq. (774) and the integration would reveal the result. However, there is a more elegant solution to this problem, namely, see (769)

$$P = \int_{-1}^{+1} g_{21} \frac{\mathrm{d}g_1}{\mathrm{d}\eta} \mathrm{d}\eta = \int_{0}^{1} g_{21} \, \mathrm{d}g_1, \tag{775a, b}$$

because $g_1(-1) = 0$ and $g_1(+1) = 1$ (compare with the two boundary conditions (728b,c)).

Substitution of the normalized Reynolds shear stress (753a) into (775b) leads to

$$P = \beta \int_{0}^{1} g_1(1 - g_1) \, \mathrm{d}g_1 = \beta \int_{0}^{1} \left(g_1 - g_1^2\right) \mathrm{d}g_1. \tag{776a, b}$$

Evaluation of the integrals yields

$$P = \beta \left( \frac{1}{2} g_1^2 \Big|_0^1 - \frac{1}{3} g_1^3 \Big|_0^1 \right), \tag{777}$$

with the final result

$$P = \frac{\beta}{6} \quad \Leftrightarrow \quad P = \frac{2}{3}\chi. \tag{778a, b}$$

This is taken as proof that turbulence is a cooperative or critical phenomenon, respectively, as already claimed above. The result of this is:

**Theorem 4** *With decreasing inverse Reynolds number (stress parameter),* $\mathrm{Re}_c/Re$*, below its critical value,* $\mathrm{Re}_c/Re = 1$*, the order parameter,* $\chi$*, which is three half of the production rate of turbulent kinetic energy,* $\chi = 3/2\ P$*, continuously increases to its maximum value "1," which it reaches at inverse Reynolds number zero.*

The parameter $\chi$ is proportional to the production of turbulent kinetic energy in the entire domain between the plates per unit length in the downstream direction. Therefore, this parameter exhibits a very meaningful property, with all the desired features of an order parameter characterizing turbulence. This is not very surprising as turbulent fields are composed of two phases, namely laminar streaks separated by fractal boundaries from turbulent patches. What we now can predict with general knowledge of phase transitions is that, at the critical Reynolds number, only a single laminar fluid domain exists. However, with increasing Reynolds number, turbulent patches are created and start to grow and become more numerous. Furthermore, the laminar streaks become smaller and less frequently observed, until in the infinite Reynolds number limit they completely disappear, and by this, in a statistical sense, obviously a higher symmetry is obtained again (Frisch 1995).

Turbulent plane Couette flows show striking analogous behavior as described by the theory of critical phenomena. Most experiments on Couette flows, which were reported in the past, are related to Reynolds numbers between 3000 and 35,000. Because the real stress parameter of this flow system is actually not the Reynolds number, but its inverse value, in this domain, the order parameter varies only in the extremely narrow range between 3.8 and 4.0! Therefore, our results give evidence that, at least from a disorder/order point of view, the stress parameter window of looking into turbulence should be differently stretched, and many more experiments should be performed at Reynolds numbers only slightly above criticality, which is a domain where experiments are more difficult to be performed. For plane turbulent Couette flows, this is the domain $650 \leq \text{Re} \leq 3000$. Only then, the presented theory could be more thoroughly tested.

It was demonstrated that Newton's shear stress law for low Reynolds numbers is a special case of the DQTM. Newton's shear stress law is the one-dimensional constitutive equation for laminar flowing Newtonian fluids. Therefore, it becomes evident that the DQTM is the analogous one-dimensional "constitutive equation" of turbulent shear flows. To stress the analogies between the laminar and turbulent case a little more, in Table 9.3 the results of this section are summarized. It is remarkable that the presented new nonlocal theory of turbulence still works with strong analogies to laminar theories. This is not so astonishing as momentum transport is described in laminar flows by Einstein–Brown diffusion and in turbulence by anomalous Lévy diffusion. As a result, local physics has to be replaced by nonlocal and fractional theories (see Sect. 7.3).

## 9.4 Plane Poiseuille Flows

Laminar plane Poiseuille flows are the flows between two plane parallel solid plates. Differently from plane Couette flows, where the excitation is given by a moving boundary, here the flow is driven by a pressure head, whereas the plates rest immobile. At small excitations, the flows are laminar and called Hagen–Poiseuille flows. This denotation is more frequently used for the cylindrical case, which is the flow through a long pipe, which has a high number of technical applications. However, also flows between plates occur in many technical realizations, e.g., in fin heat exchangers, in broad and not so high air-conditioning channels.

By increasing the driving of a flow, a laminar–turbulent transition is observed (see Fig. 9.37) and at even higher excitation rates, turbulent and then fully turbulent flow states develop, which in plane Poiseuille flows contain laminar streaks and turbulent spots. Therefore, in such flows also a cooperative phenomenon may be expected to occur.

For our modeling purposes, we define the problem geometrically by Fig. 9.38, showing the two plates apart by the distance 2a with a laminar flow represented by an average streamwise velocity profile of parabolic type; Fig. 9.39 is representing the turbulent counterpart with a flatter profile, which is the result of an increased momentum exchange by additional turbulent momentum transfer, given by the turbulent fluctuations. In turbulent quasi-steady flows, the profiles flatten more and more as the Reynolds number is increased.

The present theory of plane Poiseuille flows is due to Egolf and Weiss (2000). The theories of plane Couette and Poiseuille flows from Eqs. (689a–d) to Eq. (705) are identical. We work in our modeling of plane Poiseuille flows with Eqs. (703) and (705), which here are repeated with only small changes

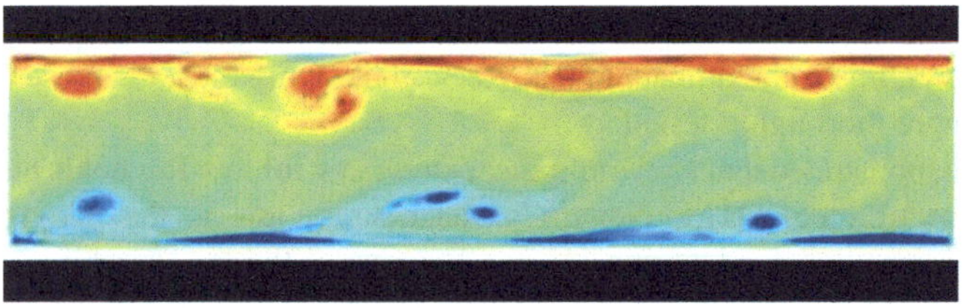

**Fig. 9.37** Plane turbulent Poiseuille flow from the left to the right-hand side, showing laminar–turbulent transition. The picture was taken from a video sequence of a flow with periodic boundary conditions at a Reynolds number of 10,000. The flow was calculated with a LES dynamic model using a stable Lattice Boltzmann Method. From Youtube (2019)

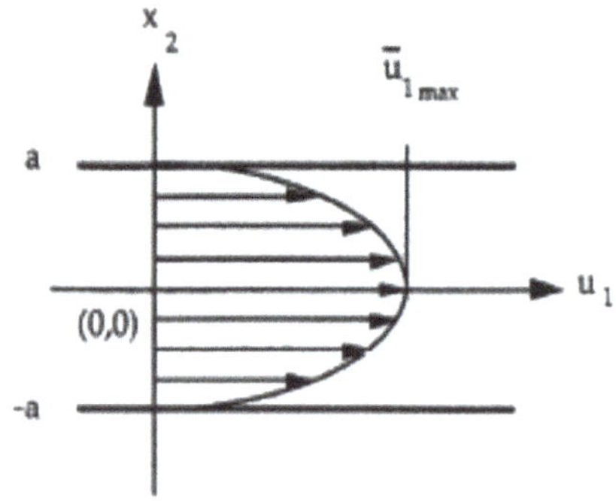


**Fig. 9.38** Plane laminar Poiseuille flow between two horizontally positioned planes of infinite extent. The coordinates are the real geometrical and physical ones. The maximum average velocity occurs on the centerline and is denoted by $\overline{u}_{1\,\mathrm{max}}$. Courtesy of Egolf and Weiss (2000) © Phys. Rev. E., reproduced with permission by APS

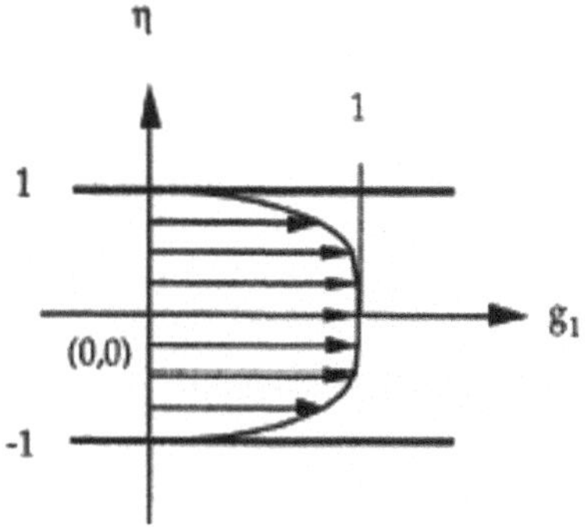


**Fig. 9.39** Plane turbulent Poiseuille flow between two plates with its bulbous average velocity profile. The coordinates are normalized, respectively, self-similar ones. Courtesy of Egolf and Weiss (2000) © Phys. Rev. E., reproduced with permission by APS

$$\Pi(\eta,\,\mathrm{Re}^*) + f_{22}(\eta,\,\mathrm{Re}^*) - A(\xi,\,\mathrm{Re}^*)\xi - C(\mathrm{Re}^*) = 0,$$
$$\frac{1}{\mathrm{Re}^*}\frac{\mathrm{d}f_1(\eta,\,\mathrm{Re}^*)}{\mathrm{d}\eta} + f_{21}(\eta,\,\mathrm{Re}^*) - A(\mathrm{Re}^*)\,\eta - B(\mathrm{Re}^*) = 0. \qquad (779\mathrm{a,b})$$

The boundary conditions are not all the same as in plane Couette flows; now they are

$$f_\alpha(-1,\,\mathrm{Re}^*) = 0, \qquad f_\alpha(+1,\,\mathrm{Re}^*) = 0, \qquad \alpha \in \{1, 21, 22\}. \qquad (780\mathrm{a-f})$$

Evaluating Eq. (779b) at $\eta = \pm 1$ and considering the boundary conditions (780a,b) for $\alpha = 21$, leads to

$$\frac{1}{\mathrm{Re}^*}\frac{\mathrm{d}f_1(+1,\,\mathrm{Re}^*)}{\mathrm{d}\eta} = A + B, \qquad \frac{1}{\mathrm{Re}^*}\frac{\mathrm{d}f_1(-1,\,\mathrm{Re}^*)}{\mathrm{d}\eta} = -A + B. \qquad (781\mathrm{a,b})$$

The total shear stress $\tau_{\mathrm{tot}}$ is given by the viscous shear stress $\tau_0$ plus the turbulent Reynolds shear stress, $\tau_{21}$,

$$\tau_{\mathrm{tot}} = \tau_0 + \tau_{21} = \mu\frac{\mathrm{d}\bar{u}_1}{x_2} - \rho\,\overline{u_2' u_1'} \qquad (782\mathrm{a,b})$$

By hindrance of the fluctuations near the wall, the turbulent part disappears, so that

$$\frac{\tau_{\mathrm{tot}}}{\rho} = \frac{\tau_0}{\rho} = \mu\frac{\mathrm{d}\bar{u}_1}{\mathrm{d}x_2}\bigg|_{x_2=\pm a}, \qquad (783\mathrm{a,b})$$

where $\mu$ is the fluid viscosity. With the help of Eq. (709) it follows that

$$\mu\frac{\mathrm{d}\bar{u}_1}{\mathrm{d}x_2}\bigg|_{x_2=\pm a} = \mp {u^*}^2, \qquad (784)$$

where the signs become evident by observing the velocity profiles in Figs. 9.38 or 9.39. Introducing as before, the self-similarity variable $\eta$ and the self-similarity function $f_1$ yields

$$\left.\frac{\mathrm{d}f_1}{\mathrm{d}\eta}\right|_{\eta=\pm 1} = \mp \mathrm{Re}^*. \tag{785}$$

These two relations transform Eqs. (781a,b) into

$$A + B = -1, \quad -A + B = 1, \tag{786a, b}$$

yielding the two solutions

$$A = -1, \quad B = 0. \tag{787a, b}$$

Substitution of Eqs. (787a,b) into (779a,b) results in

$$\begin{aligned} &\Pi(\eta, \mathrm{Re}^*) + f_{22}(\eta, \mathrm{Re}^*) + \xi - C(\mathrm{Re}^*) = 0, \\ &\frac{1}{\mathrm{Re}^*}\frac{\mathrm{d}f_1(\eta, \mathrm{Re}^*)}{\mathrm{d}\eta} + f_{21}(\eta, \mathrm{Re}^*) + \eta = 0. \end{aligned} \tag{788a, b}$$

Now, we fix the pressure, consistent with Eq. (699) as a kind of boundary condition, by requesting at $\xi = 0$ and $\eta = -1$ that

$$\Pi(0, -1, \mathrm{Re}^*) = 0. \tag{789}$$

This transforms Eq. (788a) at this boundary location to $f_{22}(-1, \mathrm{Re}^*) - C(\mathrm{Re}^*) = 0$. Equation (780c) states that $f_{22}(-1, \mathrm{Re}^*) = 0$ and from this it is concluded that $C = 0$. Therefore, Eqs. (788a,b) lead to

$$\begin{aligned} &\Pi(\eta, \mathrm{Re}^*) + f_{22}(\eta, \mathrm{Re}^*) + \xi = 0, \\ &\frac{1}{\mathrm{Re}^*}\frac{\mathrm{d}f_1(\eta, \mathrm{Re}^*)}{\mathrm{d}\eta} + f_{21}(\eta, \mathrm{Re}^*) + \eta = 0. \end{aligned} \tag{790a, b}$$

The second of these can be integrated subject to the boundary conditions $f_1(\pm 1, \mathrm{Re}^*) = 0$. This then yields

$$f_1(\eta, \mathrm{Re}^*) = -\mathrm{Re}^* \int_{-1}^{\eta} (f_{21}(\xi, \mathrm{Re}^*) + \xi)\, \mathrm{d}\xi. \tag{791}$$

This produces the main equation, in which a turbulence model of one's choice may be substituted. It has the form

$$f_1(\eta, \mathrm{Re}^*) = \mathrm{Re}^* \left[ c - \frac{1}{2}\eta^2 - \int_{-1}^{\eta} f_{21}(\xi, \mathrm{Re}^*)\, \mathrm{d}\xi \right]. \tag{792}$$

To fulfill the boundary condition in the laminar case [$f_{21}(\pm 1, \mathrm{Re}^*) = 0$], in Eq. (792) we set the integration constant $c$ equal to 1/2, so,

$$f_1(\eta, \mathrm{Re}^*) = \mathrm{Re}^* \left[ \frac{1}{2}(1-\eta^2) - \int_{-1}^{\eta} f_{21}(\xi, \mathrm{Re}^*)\, \mathrm{d}\xi \right]. \tag{793}$$

The velocity profile for laminar flows ($f_{21} \equiv 0$) is called the (parabolic) Hagen–Poiseuille flow profile given by

$$f_1(\eta, \mathrm{Re}^*) = \frac{\mathrm{Re}^*}{2}(1-\eta^2). \tag{794}$$

As next step there remains the implementation of the DQTM parameterization of the Reynolds shear stress into Eq. (790b). To this end, consider again Eq. (307b) and assign the following evident identifications:

$$\chi_2 = a, \qquad x_{2\,\max} = 0, \qquad \overline{u}_{1\,\min} = 0, \qquad \overline{u}_{1\,\max} = \overline{u}_1(0), \tag{795a – d}$$

which leads to

$$\tau_{21} = -\rho\overline{u_2' u_1'} = -\sigma\,\rho\,a\,\overline{u}_1 \frac{\overline{u}_1(0) - \overline{u}_1}{x_2}. \tag{796a, b}$$

This corresponds to

$$\frac{\overline{u_2' u_1'}}{u*^2} = \sigma\,\frac{\overline{u}_1}{u*}\,\frac{\dfrac{\overline{u}_1(0)}{u*} - \dfrac{\overline{u}_1}{u*}}{\dfrac{x_2}{a}}. \tag{797}$$

In terms of the self-similar coordinate $\eta$ and function $f_1$, the normalized Reynolds shear stress $f_{21}$ takes the form

$$f_{21}(\eta) = -\sigma\,\frac{1}{\eta} f_1(\eta)\,[f_1(0) - f_1(\eta)]. \tag{798}$$

With it, the differential equation (790b) becomes an ordinary differential equation (ODE) for $f_1$ alone

$$\frac{1}{\mathrm{Re}^*}\frac{\mathrm{d}f_1(\eta)}{\mathrm{d}\eta} - \sigma\,\frac{1}{\eta} f_1(\eta)\,[f_1(0) - f_1(\eta)] + \eta = 0, \tag{799}$$

or when incorporating $f_1(0)$,

$$\frac{f_1(0)}{\mathrm{Re}^*}\frac{\mathrm{d}}{\mathrm{d}\eta}\frac{f_1(\eta)}{f_1(0)} - \sigma f_1(0)^2\,\frac{1}{\eta}\frac{f_1(\eta)}{f_1(0)}\left[1 - \frac{f_1(\eta)}{f_1(0)}\right] + \eta = 0. \tag{800}$$

With the abbreviations

$$g_1(\eta) = \frac{f_1(\eta)}{f_1(0)}, \quad \alpha = \frac{f_1(0)}{\mathrm{Re}^*}, \quad \beta = \sigma f_1(0)^2, \tag{801a – c}$$

Equation (800) transforms to the ODE

$$\alpha\frac{\mathrm{d}g_1(\eta)}{\mathrm{d}\eta} - \beta\,\frac{1}{\eta}g_1(\eta)\,[1 - g_1(\eta)] + \eta = 0. \tag{802}$$

or

$$\alpha\,\eta\frac{\mathrm{d}g_1(\eta)}{\mathrm{d}\eta} - \beta\,g_1(\eta) + \beta\,g_1^2(\eta) + \eta^2 = 0. \tag{803}$$

This is the main differential equation of plane turbulent Poiseuille flows. A symmetry condition and the boundary conditions are

$$g_1(\eta) = g_1(-\eta), \quad g_1(0) = 1, \quad g_1(1) = 0. \tag{804a – c}$$

From Eqs. (785) and (801a), it is concluded that

$$\left.\frac{\mathrm{d}g_1(\eta)}{\mathrm{d}\eta}\right|_{\eta=\pm 1} = \mp\frac{\mathrm{Re}^*}{f_1(0)}. \tag{805}$$

With (801b), the quantity $\alpha$ receives its first physical interpretation

$$\frac{1}{\alpha} = \mp\left.\frac{\mathrm{d}g_1(\eta)}{\mathrm{d}\eta}\right|_{\eta=\pm 1}. \tag{806}$$

Owing to its definition in Eq. (801b), $\alpha$ is inversely proportional to the Reynolds number $\mathrm{Re}^*$, and in view of (806) it equals the boundary slope of the averaged and normalized velocity profile. Thus, the larger the Reynolds number is, the larger (by absolute value) will be the boundary slope of this profile. Moreover, the relation $\alpha \propto 1/\,\mathrm{Re}^*$ identifies $\alpha$ as stress parameter of this plane Poiseuille flow problem.

To solve the two-point boundary value problem (803) and (804a–c), the numerical shooting method has been used, i.e., for any chosen and fixed value of $\alpha$ forward integration is used, starting from $\eta = -1$ with a selected value of $\beta$ to $\eta = +1$ and varying $\beta$ until the boundary condition at $\eta = +1$ is satisfied. In detail, the procedure consists of the following steps:

(1) A value $\alpha$ is chosen ($0 \le \alpha \le ½$).
(2) A value $\beta$ is chosen ($0 \le \beta \le 4$).
(3) A numerical forward integration is started at the lower plate ($\eta = -1$) with $g_1(-1) = 0$, which reaches the upper plate ($\eta = +1$). Then the value $g_1(+1)$ is computed.
(4) If the calculation fulfills the boundary condition at the upper plate with a predefined accuracy, $g_1(+1) = 0$, the correct average velocity profile $g_1(\eta)$ is determined.
(5) If (4) is not fulfilled, then the $\beta$-value must be corrected and the process must be repeated, starting again with point 3, until point 4 is fulfilled with the demanded accuracy.

The procedure determines, apart from the function $g_1(\eta)$, $-1 \le \eta \le 1$, the pairs ($\alpha$, $\chi = \beta/4$), which solve the boundary value problem. The curve $\chi(\alpha)$ is shown in Fig. 9.40. For any physically possible value $\alpha$ ($0 \le \alpha \le \alpha_{\text{crit}}$), the corresponding $\chi$-values lying on this curve solve the turbulent plane Poiseuille flow problem. On the other hand, because $\chi$ is also the order parameter of the system, the same curve defines these states, which thermodynamically correspond to a stable coexistence of two phases of this flow system. In Chap. 11, it will be explained that in this turbulent cooperative and critical phenomenon these two phases correspond to laminar streaks and turbulent spots or patches. These findings are summarized by the following:

**Theorem 5** *In a turbulent field, the states with a stable coexistence of laminar streaks and turbulent patches correspond exactly to the dynamical conditions that fulfill the fluid dynamic no-slip boundary condition of the boundary value problem.*

The numerical results of Figs. 9.34 and 9.40 show different critical stress parameters for plane Couette flows ($\alpha_c = 2$) and plane Poiseuille flows ($\alpha_c = 1/2$), whereas the maximum non-normalized order parameters for infinite Reynolds number flows, $\alpha = 0$, are identical, namely $\chi = 1$ ($\beta = 4$). This upper limit will be proven analytically to be the correct one.

Because of the symmetry relation (804a), it is sufficient to study the system behavior for the positive part of the transverse normalized coordinate $0 \le \eta \le 1$. The stress parameter, $\alpha$, is positive and the spatial derivative of the average velocity profile at the boundary of this positive-valued interval is negative. Mathematically this reads as

$$\alpha \ge 0, \quad \frac{dg_1(\eta)}{d\eta} \le 0 \quad \Rightarrow \quad \alpha \frac{dg_1(\eta)}{d\eta} \le 0. \tag{807a – c}$$

Writing Eq. (790b) in terms of the self-similar coordinate $\eta$ and the two self-similar functions $g_1$ and $g_{21}$, yields

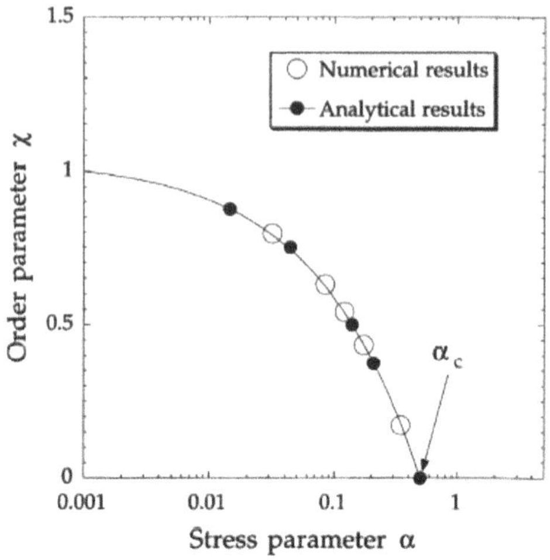


**Fig. 9.40** Analytically and numerically derived functional relation between the stress parameter $\alpha$, which is inversely related to the Reynolds number, $Re^*$, by Eqs. (805) and (806), and the quantity $\beta$, respectively, its normalized version, the order parameter $\chi$. The critical value is $\alpha_{crit} = 1/2$. Courtesy of by Egolf and Weiss (2000) © Phys. Rev. E., reproduced with permission by APS

$$\alpha \frac{dg_1(\eta)}{d\eta} + g_{21}(\eta) + \eta = 0. \tag{808}$$

This is rearranged as

$$\alpha \frac{dg_1(\eta)}{d\eta} = -g_{21}(\eta) - \eta \leq 0, \tag{809}$$

in which Eq. (807c) was applied. The self-similar version of the Reynolds shear stress (798) may also be written as

$$g_{21}(\eta) = -\beta \, \frac{1}{\eta} g_1(\eta) \, [1 - g_1(\eta)]. \tag{810}$$

This equation, inserted into (809), leads to

$$\beta \, g_1 \, (1 - g_1) - \eta^2 \leq 0, \tag{811}$$

or

$$\beta \, h(g_1) \leq \eta^2 \leq 1, \qquad 0 \leq \eta \leq 1, \tag{812a, b}$$

where

$$h(g_1) := g_1 (1 - g_1) = g_1 - g_1^2. \tag{813a, b}$$

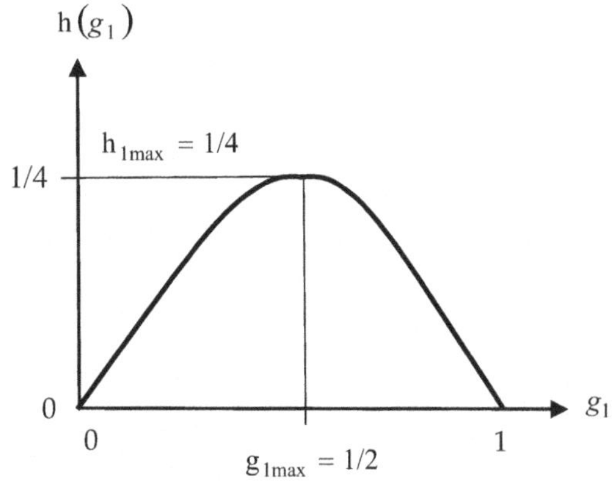


**Fig. 9.41** The symmetric function $h\ (g_1\ )$ has a maximum at $g_{1\max} = 1/2$ of positive value $h(g_{1\max}) = 1/4$

Expression (812a) is a positive parabola on $0 \leq \eta \leq 1$ (see Fig. 9.41) with its maximum at $g_{1\max}$, given by

$$g_{1\,\max} = \frac{1}{2}, \tag{814}$$

and

$$h_{1\,\max} = h(g_{1\,\max}) = h\left(\frac{1}{2}\right) = \frac{1}{4}. \tag{815a – c}$$

Therefore, we can now conclude that the function $h$ has an upper limit

$$h(g_1) = g_1(1 - g_1) \leq \frac{1}{4}. \tag{816a, b}$$

This result, inserted into Eq. (812a), yields

$$\frac{\beta}{4} \leq 1, \qquad \Leftrightarrow \qquad \beta \leq 4 \qquad \Leftrightarrow \qquad \chi \leq 1. \tag{817a – c}$$

The other corner value is the criticality $\alpha_c = 1/2$ (Fig. 9.40). This characteristic number will be derived analytically. Its value is revealed when the differential equation (802) for the laminar case ($\beta = 0$),

$$\alpha \frac{dg_1(\eta)}{d\eta} + \eta = 0, \qquad \alpha = \alpha_c, \tag{818a, b}$$

is solved. Rearranging this equation yields

$$\frac{dg_1(\eta)}{d\eta} = -\frac{1}{\alpha_c}\eta \tag{819}$$

and after integration

$$g_1(\eta) = c - \frac{1}{2\alpha_c}\eta^2. \tag{820}$$

With the boundary condition (804c) there follows

$$g_1(1) = c - \frac{1}{2\alpha_c} = 0 \quad \Rightarrow \quad c = \frac{1}{2\alpha_c} \tag{821a – c}$$

and consequently,

$$g_1(\eta) = \frac{1}{2\alpha_c}\left(1 - \eta^2\right). \tag{822}$$

The normalization condition (804b) yields

$$\alpha_c = \frac{1}{2}, \tag{823}$$

so that the velocity profile of plane laminar Poiseuille flow is

$$g_1(\eta) = \left(1 - \eta^2\right), \tag{824}$$

which is the parabolic Hagen–Poiseuille flow profile (see Fig. 9.42). This profile is valid for Reynolds numbers $0 \leq 1/\text{Re}_c \leq 1/\text{Re}$.

Similarly simple is the derivation of the solution for an infinite Reynolds number ($\alpha = 0, \beta = 4$). With these limiting values Eqs. (808) and (810) simplify to

$$g_1^2(\eta) - g_1(\eta) + \frac{1}{4}\eta^2 = 0, \tag{825}$$

with the two solutions

$$g_{1_\pm}(\eta) = \frac{1}{2}\left(1 \pm \sqrt{1 - \eta^2}\right), \tag{826a, b}$$

where only the solution with the positive sign is of physical relevance. It leads to the correct solution showing the maximum on the axis between the plates: $g_{1_+}(0) = 1$, whereas $g_{1_-}(0) = 0$ corresponds to an erroneous velocity profile (with a kink) and a minimum at $\eta = 0$.

Transforming Eq. (826a) to

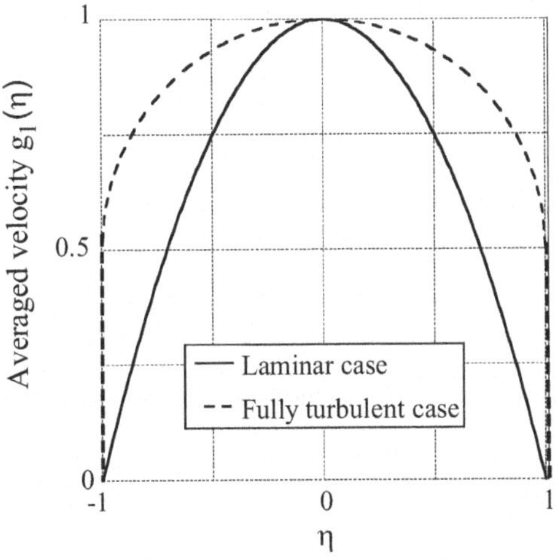


**Fig. 9.42** The two limiting-case solutions for the average velocity profiles, the laminar flow and the infinite Reynolds number solution of fully turbulent plane Poiseuille flows. They are both of simple geometric form, namely of parabolic and of circular type

$$2g_1 - 1 = \sqrt{1 - \eta^2}, \tag{827}$$

and employing the variable change

$$\varsigma = 2g_1 - 1, \tag{828}$$

we obtain the equation of a semicircle

$$\varsigma^2 + \eta^2 = 1. \tag{829}$$

*Thus, we have proven that the infinite Reynolds number plane Poiseuille flow shows an average circular velocity profile* (see Fig. 9.42). This profile is much flatter than the laminar parabolic profile. The explanation, which is an intensification of the momentum transport by additional transport of turbulent fluctuations has already been given before.

This discovery could easily be taxed as a *wonder of nature*. If one considers how irregular the spatiotemporal velocity field of a fully turbulent plane Poiseuille flow between the plates is and that a usual temporal averaging process gives birth to one of the most beautiful geometric figures, namely a circle, we may imagine that this result even would have fascinated some ancient Greek philosophers. In summary:

**Theorem 6** *Infinite Reynolds number plane Poiseuille flow reveals in its irregular and chaotic flow velocity profile by conventional time-averaging a circular average velocity profile.*

More effort is needed to solve the third case, which is the solution for moderate Reynolds number plane turbulent Poiseuille flows.

Consider the following function:

$$g_1(\eta) = H(\eta)\frac{h'(\eta)}{h(\eta)}. \tag{830}$$

Its derivative with respect to $\eta$ leads to

$$g_1'(\eta) = H'\frac{h'}{h} + H\frac{h''}{h} - H\frac{h'^2}{h^2}, \tag{831}$$

where for better visibility the functional dependence was dropped. With

$$H = \frac{\alpha}{\beta}\eta \tag{832}$$

Eq. (830) transforms to

$$g_1 = \frac{\alpha}{\beta}\eta\frac{h'}{h}, \tag{833}$$

whereas Eq. (831) becomes

$$g_1{}' = \frac{\alpha}{\beta}\frac{h'}{h} + \frac{\alpha}{\beta}\eta\frac{h''}{h} - \frac{\alpha}{\beta}\eta\frac{h'^2}{h^2}. \tag{834}$$

Now, substituting Eqs. (833) and (834) into (803) leads to the following expression:

$$\frac{\alpha^2}{\beta}\eta\frac{h'}{h} + \frac{\alpha^2}{\beta}\eta^2\frac{h''}{h} - \alpha\eta\frac{h'}{h} + \eta^2 = 0. \tag{835}$$

Multiplying this equation by $\beta h/\eta$ leads to

$$\alpha^2 h' + \alpha^2\eta h'' - \alpha\beta h' + \beta\eta h = 0, \tag{836}$$

which is better written as the following Bessel-type differential equation:

$$\alpha^2\eta h'' + \left(\alpha^2 - \alpha\beta\right) h' + \beta\eta h = 0. \tag{837}$$

Introducing the new variable

$$\psi = \lambda\eta \quad \text{with} \quad \lambda \neq 0 \tag{838a, b}$$

into (837), with the new abbreviation (different to $' := \mathrm{d}/\mathrm{d}\eta$ in earlier equations) $' := \mathrm{d}/\mathrm{d}\psi$, leads to

$$\alpha^2\psi\lambda^2 h'' + (\alpha^2 - \alpha\beta)\,\lambda^2 h' + \beta\psi h = 0 \tag{839}$$

or by dividing by $\alpha^2$,

$$\lambda^2\psi h'' + \left(1 - \frac{\beta}{\alpha}\right)\lambda^2 h' + \frac{\beta}{\alpha^2}\psi h = 0. \tag{840}$$

The freedom to choose the constant $\lambda$ in such a manner that Eq. (840) takes its simplest form leads us to the proposition

$$\lambda = \frac{\sqrt{\beta}}{\alpha}. \tag{841}$$

Then, Eq. (840) becomes

$$\psi h'' + \left(1 - \frac{\beta}{\alpha}\right)h' + \psi h = 0. \tag{842}$$

It is a second-order differential equation of Bessel type.

According to Abramowitz and Stegun (1984) the solution of this differential equation is

$$h(\psi) = \psi^{\kappa} C_{\kappa}\left(\tilde{\lambda}\psi\right), \tag{843}$$

where

$$\kappa = \frac{\beta}{2\,\alpha}, \quad \tilde{\lambda}^2 = 1, \quad C_{\kappa} \in \{J_{\kappa}, Y_{\kappa}\}. \tag{844a – d}$$

$J_{\kappa}$ denotes the Bessel function and $Y_{\kappa}$ the Weber function of order $\kappa$.
From Eq. (801c) it follows that

$$\sigma = \frac{\beta}{f_1(0)^2}. \tag{845}$$

Combining Eq. (727a), but replacing the maximum value of plane Couette flows, $f_1(1)$, by that of plane Poiseuille flows, $f_1(0)$, with Eq. (845), yields

$$\sigma = \frac{\beta}{\alpha^2\,\mathrm{Re}^{*2}}, \quad \alpha = \frac{f_1(0)}{\mathrm{Re}^*}. \tag{846a, b}$$

This result also corresponds to a combination of Eqs. (805) and (806). Furthermore, we deduce from Eq. (696), again for plane Poiseuille flows, that

$$f_1(0) = \frac{\overline{u}_{1\,\max}}{u^*}, \tag{847}$$

which, when introduced into Eq. (845), yields

$$\sigma = \beta \left( \frac{u^*}{\overline{u}_{1\,\max}} \right)^2. \tag{848}$$

In Eqs. (727a,b), replacing $f_1(1)$ by $f_1(0)$ and substituting these two equations into (844a), it follows that

$$\kappa = \frac{1}{2} \sigma f_1(0)\,\mathrm{Re}^*, \tag{849}$$

and after inserting Eq. (847),

$$\kappa = \frac{1}{2} \sigma \; \mathrm{Re}^* \frac{\overline{u}_{1\,\max}}{u^*}. \tag{850}$$

Changing the type of Reynolds number by applying Eq. (695b) yields

$$\kappa = \frac{1}{2} \sigma \; \mathrm{Re}. \tag{851}$$

Continuing with Eqs. (843) and (844), the following four independent trial assignments are now at our disposal, which are candidates to construct solutions of (842) of the form

$$\begin{aligned} h_1(\psi) &= \psi^\kappa J_\kappa(\psi), \qquad & h_2(\psi) &= \psi^\kappa J_\kappa(-\psi), \\ h_3(\psi) &= \psi^\kappa Y_\kappa(\psi), \qquad & h_4(\psi) &= \psi^\kappa Y_\kappa(-\psi). \end{aligned} \tag{852a – d}$$

The Bessel functions of the first kind, $h_2$, and the Weber function of the second kind, $h_4$, having negative arguments, are less convenient to deal with than those of positive argument. By applying an analytic continuation, these two types of functions are reduced to other Bessel functions with positive arguments. The respective relations are

$$h_2(\psi) = (-1)^n h_1(\psi), \quad h_4(\psi) = (-1)^n h_3(\psi) + 2\mathrm{i}(-1)^n h_1(\psi). \tag{853a, b}$$

These two equations apply if $\kappa$ has integer value, $\kappa = n$. In the case of rational indices $\kappa$, one has

$$h_2(\psi) = e^{i\pi\kappa} h_1(\psi),$$
$$h_4(\psi) = e^{-i\pi\kappa} h_3(\psi) + 2\mathrm{i}\,\sin(\pi\kappa)\cdot \qquad (854\mathrm{a,b})$$
$$\cot(\pi\kappa)\, h_1(\psi).$$

This generalization of Bessel- and Weber functions to such with real indices is a requirement originating from the fact that the index $\kappa$ is directly linked with the real-valued Reynolds number (compare with Eq. (851)).

Now, the general solution of the differential equation (842) can be written as a linear combination of $h_1(\psi)$ and $h_3(\psi)$

$$h(\psi) = p\psi^\kappa J_\kappa(\psi) + q\psi^\kappa Y_\kappa(\psi), \qquad (855)$$

in which $p$ and $q$ follow from boundary conditions. After Abramowitz and Stegun (1984) the derivative of Bessel- and Weber functions are given by a single formula, namely

$$\left(\frac{1}{\psi}\frac{\mathrm{d}}{\mathrm{d}\psi}\right)^k [\psi^\kappa \Im_\kappa(\psi)] = \psi^{\kappa-k}\Im_{\kappa-k}, \quad \Im \in \{J, Y\} \qquad (856\mathrm{a,b})$$

For a first-order derivative this is

$$\frac{\mathrm{d}}{\mathrm{d}\psi}[\psi^\kappa \Im_\kappa(\psi)] = \psi^\kappa \Im_{\kappa-1}. \qquad (857)$$

With the help of these auxiliary relations, the derivative of the function $h(\psi)$ (see Eq. (855)) is obtained as

$$\frac{\mathrm{d}h(\psi)}{\mathrm{d}\psi} = p\psi^\kappa J_{\kappa-1}(\psi) + q\psi^\kappa Y_{\kappa-1}(\psi). \qquad (858)$$

Now, we apply Eqs. (830) and (832),

$$g_1(\eta) = H(\eta)\frac{\frac{\mathrm{d}h}{\mathrm{d}\eta}}{h} = \frac{\alpha}{\beta}\eta\frac{\frac{\mathrm{d}h}{\mathrm{d}\psi}}{h}\frac{\mathrm{d}\psi}{\mathrm{d}\eta}. \qquad (859\mathrm{a,b})$$

With the aid of Eqs. (838) and (841) one derives

$$\eta = \frac{1}{\lambda}\psi = \frac{\alpha}{\sqrt{\beta}}\psi. \qquad (860\mathrm{a,b})$$

Furthermore, the derivative of (860b) is

$$\frac{\mathrm{d}\psi}{\mathrm{d}\eta} = \lambda = \frac{\sqrt{\beta}}{\alpha}. \tag{861a, b}$$

Substituting Eqs. (860b) and (861b) into (859b) delivers $g_1$ as a function of $\psi$,

$$g_1(\psi) = \frac{\alpha}{\beta}\psi \frac{\frac{\mathrm{d}h}{\mathrm{d}\psi}}{h}. \tag{862}$$

With the help of Eq. (844a), this can be written as

$$g_1(\eta) = \frac{\psi}{2\kappa} \frac{\frac{\mathrm{d}h}{\mathrm{d}\psi}}{h}. \tag{863}$$

By inserting the function $h$ (Eq. 855) and its derivative $h'$ (Eqs. 858), (863) takes the form

$$g_1(\psi) = \frac{\psi}{2\kappa} \frac{p\psi^{\kappa} J_{\kappa-1}(\psi) + q\psi^{\kappa} Y_{\kappa-1}(\psi)}{p\psi^{\kappa} J_{\kappa}(\psi) + q\psi^{\kappa} Y_{\kappa}(\psi)}, \tag{864}$$

in which $p$ and $q$ are constants. The normalization condition (boundary condition) is (compare with Eq. (804b))

$$\lim_{\eta\to 0} g_1(\eta) = \lim_{\psi\to 0} g_1(\psi) = 1. \tag{865a, b}$$

According to properties of the Bessel and Weber functions, from (864) with (865b), one concludes that (see also below)

$$q = 0. \tag{866}$$

Thus, Eq. (864) simplifies to

$$g_1(\psi) = \frac{\psi}{2\kappa} \frac{J_{\kappa-1}(\psi)}{J_{\kappa}(\psi)}. \tag{867}$$

Because the statement immediately below Eq. (865a,b) may not be so clear to some readers, we will now prove that Eq. (867) fulfills the boundary condition (865b). From Abramowitz and Stegun (1984), the following limiting form for $J_{\kappa}(\psi)$ holds for small arguments $\psi$

$$J_{\kappa}(\psi) \propto \frac{\left(\frac{1}{2}\psi\right)^{\kappa}}{\Gamma(\kappa+1)}, \quad \text{if} \quad \psi \to 0. \tag{868}$$

Here $\Gamma(x)$ denotes the Gamma function. Then, from Eqs. (867) and (868) it is concluded that the ratio of two expressions (868) of neighboring order is

$$g_1(\psi) = \frac{\psi}{2\kappa} \frac{\dfrac{\left(\frac{1}{2}\psi\right)^{\kappa-1}}{\Gamma(\kappa)}}{\dfrac{\left(\frac{1}{2}\psi\right)^{\kappa}}{\Gamma(\kappa+1)}} = \frac{1}{\kappa}\frac{\Gamma(\kappa+1)}{\Gamma(\kappa)}, \quad \psi \to 0. \tag{869a, b}$$

With the recurrence formula (see again Abramowitz and Stegun 1984)

$$\Gamma(\kappa+1) = \kappa\,\Gamma(\kappa), \tag{870}$$

it follows that the limit (865b) is valid, which corroborates the correctness of the reduced form (867).

According to Abramowitz and Stegun (1984), the following relation of continued fraction (see, e.g., Khinchin 1992) is valid:

$$\frac{J_\kappa(\psi)}{J_{\kappa-1}(\psi)} = \cfrac{1}{\cfrac{2\kappa}{\psi} - \cfrac{1}{\cfrac{2\kappa+1}{\psi} - \cfrac{1}{\cfrac{2\kappa+2}{\psi} - \dots}}}. \tag{871}$$

Evaluating the continued fraction yields

$$\frac{J_\kappa(\psi)}{J_{\kappa-1}(\psi)} = \frac{1}{\dfrac{2\kappa}{\psi} - \dfrac{J_{\kappa+1}}{J_\kappa}}. \tag{872}$$

Thus,

$$\frac{2\kappa}{\psi}\frac{J_\kappa(\psi)}{J_{\kappa-1}(\psi)} = \frac{1}{1 - \dfrac{\psi}{2\kappa}\dfrac{J_{\kappa+1}}{J_\kappa}}, \tag{873}$$

or

$$\left(1 - \frac{\psi}{2\kappa}\frac{J_{\kappa+1}}{J_\kappa}\right)\frac{2\kappa}{\psi}\frac{J_\kappa(\psi)}{J_{\kappa-1}(\psi)} = 1. \tag{874}$$

Furthermore, one may derive the identity

$$\frac{2\kappa}{\psi}\frac{J_\kappa(\psi)}{J_{\kappa-1}(\psi)} - \frac{J_{\kappa+1}(\psi)}{J_{\kappa-1}(\psi)} = 1, \tag{875}$$

and

$$\frac{2\kappa}{\psi} \frac{J_\kappa(\psi)}{J_{\kappa-1}(\psi)} = 1 + \frac{J_{\kappa+1}(\psi)}{J_{\kappa-1}(\psi)}, \tag{876}$$

respectively. The reciprocal relation reads

$$\frac{\psi}{2\kappa} \frac{J_{\kappa-1}(\psi)}{J_\kappa(\psi)} = \frac{1}{1 + \dfrac{J_{\kappa+1}(\psi)}{J_{\kappa-1}(\psi)}} = \frac{J_{\kappa-1}(\psi)}{J_{\kappa-1}(\psi) + J_{\kappa+1}(\psi)}. \tag{877a, b}$$

Comparison of Eq. (877a,b) with (867) yields an alternative version for the averaged velocity profile

$$g_1(\psi) = \frac{J_{\kappa-1}(\psi)}{J_{\kappa-1}(\psi) + J_{\kappa+1}(\psi)}. \tag{878}$$

The requirement of Eq. (804c) together with (860b) demands that

$$J_{\kappa-1}\left(\frac{\sqrt{\beta}}{\alpha}\right) = 0. \tag{879}$$

Therefore, the argument of the Bessel function (878) must correspond with the first zero $j_{\kappa-1,\,1}$ of this particular Bessel function

$$j_{\kappa-1,1} = \frac{\sqrt{\beta}}{\alpha}. \tag{880}$$

Combining this equation with

$$2\kappa = \frac{\beta}{\alpha} \tag{881}$$

(compare with Eq. (844a)), we find that

$$\alpha = \frac{2\kappa}{\left(j_{\kappa-1,1}\right)^2}, \tag{882}$$

and

$$\beta = \frac{4\kappa^2}{\left(j_{\kappa-1,1}\right)^2}. \tag{883}$$

In this way, for a given positive real value of $\kappa \in \mathbb{R}^+$, a related pair $(\alpha,\beta)$ can be calculated by means of tables and formulas presented in Abramowitz and Stegun (1984) that define the order parameter function $\beta(\alpha)$ or $\chi(\alpha)$, respectively (see Fig. 9.40).

In the same reference, also an alternative serial formulation of the Bessel function (containing the gamma function Γ) is presented, which is

$$J_\kappa(\psi) = \left(\frac{\psi}{2}\right)^\kappa \sum_{k=0}^{\infty} \frac{\left(-\frac{\psi^2}{4}\right)^k}{k!\Gamma(\kappa+k+1)}. \tag{884}$$

For negative arguments, $\psi$, instead of (884) an alternative expression must be applied. This is of the form

$$g_1(\psi) = \frac{1}{\kappa} \frac{\sum_{k=0}^{\infty} \frac{\left(-\frac{\psi^2}{4}\right)}{k!\Gamma(\kappa+k)}}{\sum_{k=0}^{\infty} \frac{\left(-\frac{\psi^2}{4}\right)^k}{k!\Gamma(\kappa+k+1)}}. \tag{885}$$

This equation satisfies the symmetry condition (804a).

From Eqs. (810) and (860a), it follows that

$$g_{21}(\psi) = -\beta \frac{\lambda}{\psi} g_1(\psi)\,[1 - g_1(\psi)]. \tag{886}$$

Moreover, by comparing (861a,b) and (880), one concludes that

$$\lambda = j_{\kappa-1,1}. \tag{887}$$

Thus, from (886), (887), and (878) we deduce the relation

$$g_{21}(\psi) = -\beta \frac{j_{\kappa-1,1}}{\psi} \left\{ \frac{J_{\kappa-1}(\psi)}{J_{\kappa-1}(\psi) + J_{\kappa+1}(\psi)} - \frac{J_{\kappa-1}(\psi)^2}{[J_{\kappa-1}(\psi) + J_{\kappa+1}(\psi)]^2} \right\}, \tag{888}$$

which, finally, results in

$$g_{21}(\psi) = -\beta\, j_{\kappa-1,1} \frac{1}{\psi} \frac{J_{\kappa-1}(\psi)\, J_{\kappa+1}(\psi)}{[J_{\kappa-1}(\psi) + J_{\kappa+1}(\psi)]^2}. \tag{889}$$

The functions (867) and (889) were also numerically calculated by the application of a simple first-order forward Euler integration scheme, which additionally was programmed as a shooting method. It was applied directly to the basic differential equation (803). The results for four different $\beta$ values are shown in Figs. 9.43, 9.44, 9.45, 9.46.

The higher the degree of turbulence ($\beta$ value) is, the flatter will be the averaged velocity profiles and the more will the Reynolds shear stress converge toward a linear distribution, which actually is given by two rotational symmetric triangles, where one triangle transforms into the other by a rotation of $\pi$ (see algebraic treatment below and Figs. 9.47b and 9.48).

The solutions of plane turbulent Couette and plane turbulent Poiseuille flows are correct in the core domain, which is symmetric to the centerline and shows a Reynolds number dependent width. However, in the boundary or overlap region (see

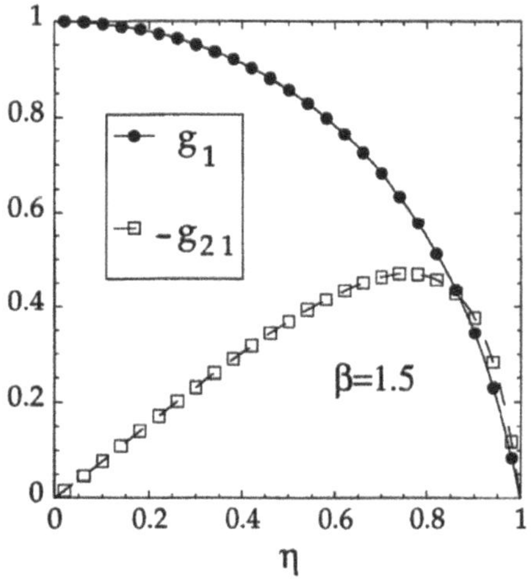


**Fig. 9.43** Average velocity, $g_1$, and Reynolds shear stress half profiles, $-g_{21}$, for $0 \leq \eta \leq 1$ for $\beta = 1.5$. Courtesy of Egolf and Weiss (2000) © Phys. Rev. E., reproduced with permission by APS

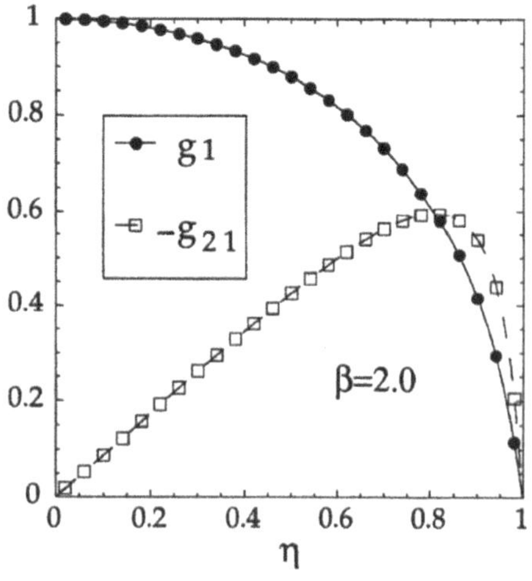


**Fig. 9.44** The same quantities as in Fig. 9.43, however, for $\beta = 2.0$. Courtesy of Egolf and Weiss (2000) © Phys. Rev. E., reproduced with permission by APS

Fig. 9.49), a refinement of the model, leading to a different limiting wall behavior, may be performed and is discussed below and in Sect. 9.5.

In Egolf and Weiss (2000) and Appendix D, it is demonstrated that the generalized rational-order Bessel functions, that describe the average velocity profiles $g_1$, for infinite Reynolds number, converges toward the circle profile (Eq. (829)) shown in Figs. 9.47a and 9.50, right.

The Reynolds shear stress for infinite Reynolds number is determined by inserting Eq. (827) into (810) and by setting $\beta = 4$. This leads to

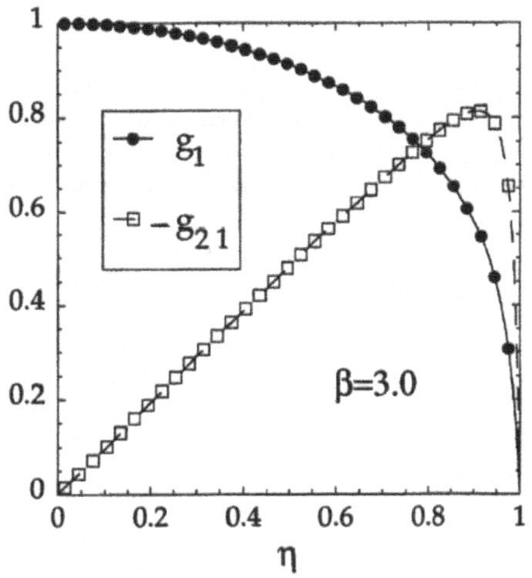


**Fig. 9.45** The average velocity and Reynolds shear stress for a medium excitation, $\beta = 3.0$. Courtesy of Egolf and Weiss (2000) © Phys. Rev. E., reproduced with permission by APS

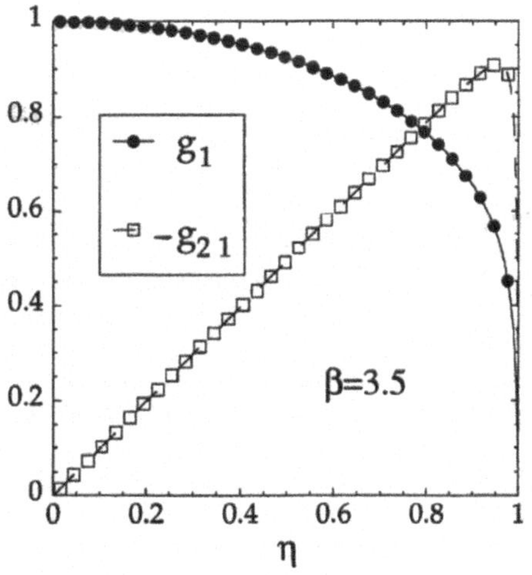


**Fig. 9.46** In the case of a higher excitation, $\beta = 3.5$, a flattening of the average velocity profile can be observed. Courtesy of Egolf and Weiss (2000) © Phys. Rev. E., reproduced with permission by APS

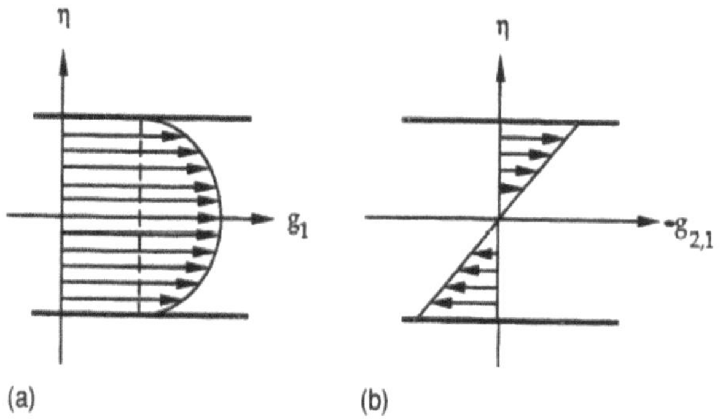


**Fig. 9.47** The average velocity profile of infinite Reynolds number flow consists of a rectangle and a semicircle (**a**). The second panel (**b**) shows the corresponding Reynolds shear stress, which is of linear or double-triangular type. Therefore, the two average profiles are of elementary geometric nature. Courtesy of Egolf and Weiss (2000) © Phys. Rev. E., reproduced with permission by APS

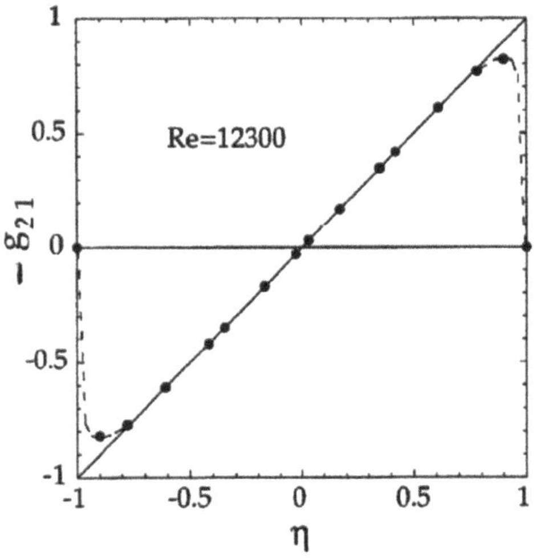


**Fig. 9.48** The Reynolds shear stress at a Reynolds number of 12,300 has converged very closely to the theoretical profile for the infinite Reynolds number limit. The experimental data, presented by Laufer (1954) and Pai (1953), with $\eta > 0$, have been reproduced for $\eta < 0$ by a rotation of 180° around the origin of the coordinate system to complete the triangular averaged Reynolds shear stress profile. There is no doubt that the DQTM predicts the right solution. Courtesy of Egolf and Weiss (2000) © Phys. Rev. E., reproduced with permission by APS

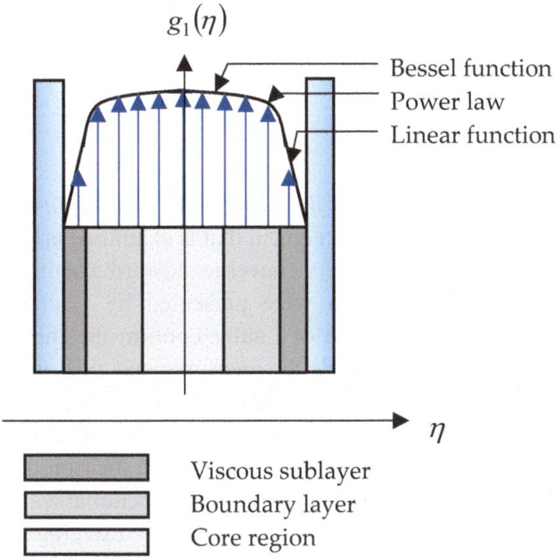


**Fig. 9.49** From the plate toward the centerline, the plane Poiseuille flow regions are as follows: the viscous sublayer, the boundary layer, and the core region (shaded domains). Between these, two small transition regions (1 and 2) occur. The related functions describing the average velocity profile are shown and described in this figure. The three functions add together to give a bulbous average velocity profile as occurring in highly turbulent Poiseuille flows. From Egolf and Weiss (2000), reproduced with changes

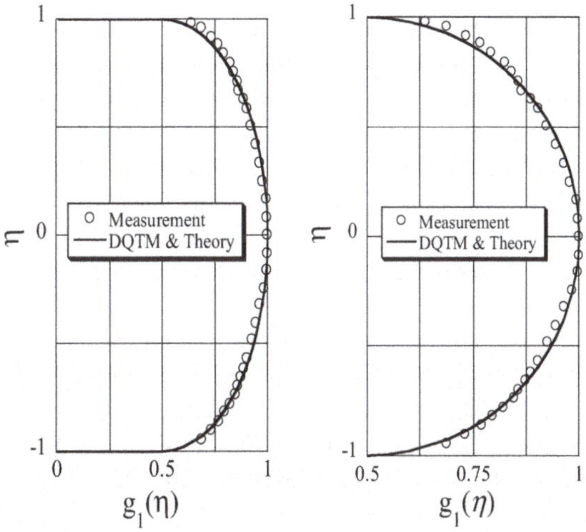


**Fig. 9.50** Two differently scaled presentations of the measured average velocity profile for fully developed turbulent flows with $u^*$=15.2 cm/s (from Reichardt 1951) compared with the circle solution of the present theory. The width and height of the channel was 98.0 cm and 24.6 cm, respectively. On the left is the conventional and on the right the new presentation. Courtesy of Egolf and Weiss (2000) © Phys. Rev. E., reproduced with permission by APS

$$
\begin{aligned}
g_{21}(\eta) &= -\beta\,\frac{1}{\eta}\frac{1}{2}\left(1+\sqrt{1-\eta^2}\right)\\
&\times\left[1-\frac{1}{2}\left(1+\sqrt{1-\eta^2}\right)\right] = -\frac{\beta}{4}\frac{1}{\eta}\left(1+\sqrt{1-\eta^2}\right)\\
&\times\left(1-\sqrt{1-\eta^2}\right) \overset{\beta=4}{=} -\eta, \quad |\,\eta\,| < 1.
\end{aligned}
\qquad (890\mathrm{a}-\mathrm{c})
$$

The final result describes a linear function as shown in Fig. 9.47b. This antisymmetric function partly represents two triangles. A more sophisticated treatment is based on the mathematical theory of distributions (see Gelfand and Schilow 1962), which has not been applied till date. In such a generalized description, the Reynolds shear stress would also fulfill the boundary condition (804c). By an application of (804a) and (804c), one concludes that also the following implication is valid:

$$g_1(-1) = 0. \tag{891}$$

In this section, the above theoretical predictions are also compared with experimental facts. A first statement that is examined is that the average velocity profile for high Reynolds numbers converges toward a semicircle or at least a part of it.

Experimental data were presented by Laufer (1954) and Reichardt (1951). Even though the data of Laufer confirm the theoretical results very well, for the average velocity profile the measurements of Reichardt are chosen for comparison.

As shown in Fig. 9.50, the experimental data confirm the theoretical model results convincingly. However, the very good results are a little misleading. In the domain beneath $\eta = -0.5$ and above 0.5 some measured quantities are somewhat smaller than the corresponding theoretical functional values. However, exactly here, for higher Reynolds numbers, the measured average velocity profile tends to higher values than those predicted by the theoretical circle profile. The reason for this will be discussed below.

The Princeton super pipe data (see Zagarola et al. 2018a, b) represent the highest Reynolds number flows, so far done. The measurements were performed for Reynolds numbers up to 35,259,000, which is the highest Re value reached in laboratory experiments. The average velocity profile between $\eta = -0.5$ and $\eta = 0.5$ also follows the circle profile with a maximum relative error of 1.2% (Egolf and Weiss 2000). The comparison of super-pipe data with the circle profile was also personally performed by M.V. Zagarola and estimated to be fairly good (see Zagarola 1998). Only in the turbulent boundary layer at larger values of $|\eta|$, the relative deviation reveals slightly higher values. From theory and experiments, it is known that closer to the wall the results of pipe and channel flows are practically identical. Comparison of experimental results of boundary layer and pipe flow is presented by Zagarola et al. (2018a, b). Therefore, in this domain equal behavior must be expected in plane channels and axisymmetric flows in pipes. So, also in the plane case at high Reynolds numbers, it is expected that the experimental values could exceed the theoretical ones shown in Fig. 9.50. However, the solutions in the core region, which is roughly defined by the interval $-0.5 < \eta < 0.5$, hardly alter when the excitation is further increased. Therefore, for this region the half-circle profile is securely confirmed.

The generalized mixing length in the DQTM of plane turbulent Poiseuille flows was assumed to be constant. This is only the case for the core region. Toward the wall, it decreases linearly and leads to deviating results. Libby (1996) writes: "*There is a temptation to assume that the mixing length is constant throughout the central portion of the flow and varies only within the wall layer.*" Furthermore, he states, "*Briefly, it is not possible to match a solution for the wall layer to that for an outer flow involving a constant mixing length. We thus conclude that at the edges of the outer flow the mixing length must vary as in channel flows, i.e. it must become*

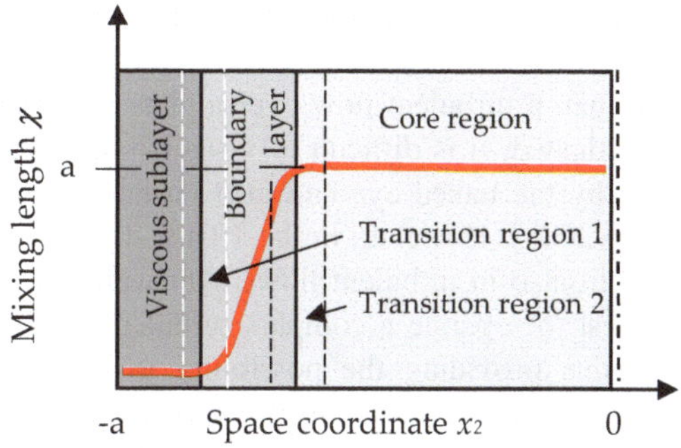


**Fig. 9.51** Continuous mixing length for the different flow regions. These regions can also be seen in Fig. 9.49. The bent parts of the red curve define the transition regions 1 (white dashed lines) and 2 (black dashed lines). They continuously connect the mixing length of two neighboring main regions

*proportional to the wall distance."* This corresponds fully with our findings. Furthermore, the linear mixing length must be matched to the vanishing value in the viscous sublayer and to the constant value in the core region without a kink (continuous first derivative). This requirement leads to a Reynolds shear stress dependence on the wall distance that is of higher order than the approximate models can produce (see Fig. 9.51 and Sect. 9.5).

Two reasons are responsible that the circle profile was not discovered earlier. Firstly, it is this restriction of application to a part of the core region only; secondly, in the literature on turbulence by convention the velocity coordinate is presented in a different manner (e.g., stretched by a factor two (see, e.g., Schlichting 1979) and Fig. 9.50, left).

A second crucial test is a comparison of the calculated Reynolds shear stress with experimental material (see Fig. 9.48). The theoretical results are again in good agreement with the experimental data set. The deviations at the corner points are given by the fact that the experimentally adjusted Reynolds number is not infinite. However, the results correspond qualitatively very well with the continuous bending, also revealed by the theoretical model (compare also with the Reynolds shear stress presented in Fig. 9.48).

In Fig. 9.49, the average velocity profile in the boundary or overlap region is denoted by a "power law" or "logarithmic law." The logarithmic law of the wall is at present generally the accepted law for turbulent flows along a plane wall. However, in this review in Sect. 5.5 this law was questioned and criticized. In the next section, it is demonstrated for this kind of flows why in this book a deficit power law is proposed as a replacement for the (universal) logarithmic law of the wall.

## 9.5 "Wall Turbulent" Flows

In the last two sections, we developed the theories of plane turbulent Couette and Poiseuille flows in the regions far from the walls, which is the domain called core region. It was not necessary to derive the solutions closer to the wall, because there, both show the same phenomenon, called "wall turbulence." This "wall turbulent" flow is the flow of a turbulent fluid flowing parallel along a plane wall, and is the theme of this new section.

When a turbulent flow over a plate or a wing (of an aircraft) is experimentally investigated, it is difficult to visualize the characteristics of the flow. It cannot be seen by the naked eye and also visualization methods have their limits (Wu and Moin 2009). However, in the 1930s, it was discovered that coherent structures are present also in turbulent flows over walls (see Green 1995). There was always an interest to separate a complex turbulent flow field into coherent and noncoherent domains providing the possibility to find elementary "objects," which can be classified and idealized so that also simpler mathematical methods apply. Such a case is, e.g., the Biot–Savart law for (small) vortex structures and vorticity (see, e.g., Mathur et al. 2007), e.g., shown in Fig. 5.3. Such structures show temporal coherence, i.e., they persist in an only slowly changing form giving the possibility to apply statistical methods for their investigation. The difficulty to experimentally observe a coherent structure was somehow overcome by the development of DNS data visualization, which today makes coherent structures easily visible. Therefore, today the experiments are primarily used to validate the observations made by high-power numerical simulations, e.g., to prove the existence of a numerically discovered new "structure" in a real flow. In "wall turbulent" flows a characteristic vortex structure was discovered, called hairpin vortices (see, e.g., Adrian et al. 2001; Adrian 2007) (Fig. 9.52). Their hairpin-like loops led to their naming. Hairpin vortices are created because at different distances from the wall there are differences in the relative upward flowing velocities. Figure 9.52 shows that these hairpin vortices have a preferential mean orientation, and it is observed that they grow in a self-similar manner in the downstream direction (Wu and Moin 2009). Furthermore, they can combine, leading to annihilation, production, amplification, etc. They show eruptive behavior, transferring fluid spots from the wall to wall-distant regions and by this are

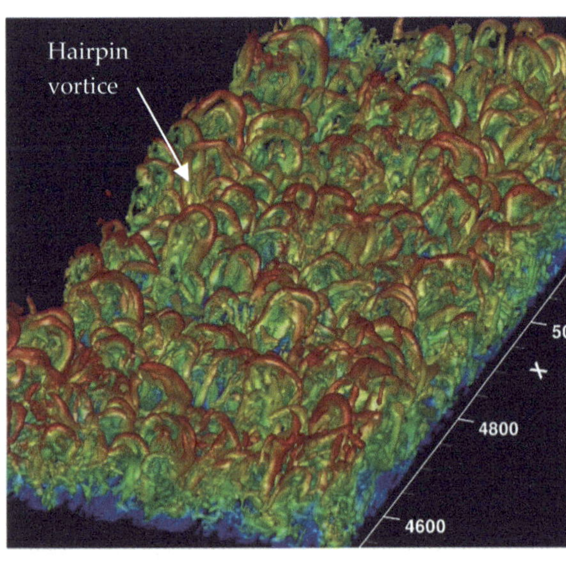


**Fig. 9.52** Inside a turbulent flow over a plate semiregular structures, called hairpin vortices, are observed. This image was extracted from a numerical simulation of a flow in a flat plate boundary layer, which grows from laminar through transition to its turbulent states. Courtesy Wu and Moin (2009) © J. Fluid Mech., reproduced with permission by Cambridge University Press

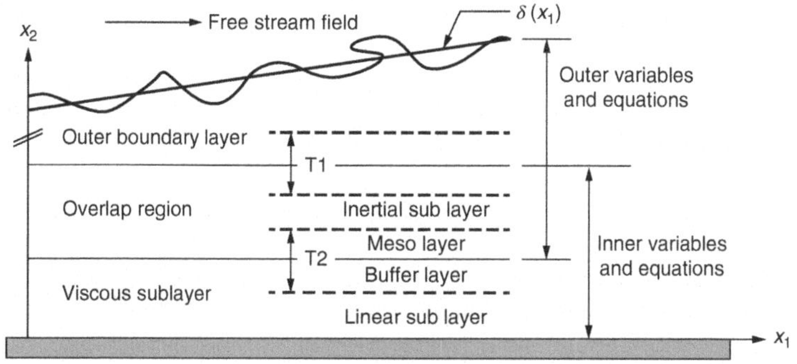


**Fig. 9.53** In a turbulent flow along a wall a manifold of different flow domains are distinguished. Four regions occur, namely from the wall to its far distance: the viscous sublayer, the overlap region, the outer boundary layer and the free stream field. In plane turbulent Couette and Poiseuille flows the outer boundary layer is symmetric to the center line between the two plates and is called core region. Courtesy Egolf (2009) © Int. J. Refrigeration, reproduced with permission by Elsevier Science & Technology

significant for an intensification of mass, heat, or/and particle transport vertically to the wall. A further, more detailed, discussion of coherent structures and their behavior is continued in Chap. 11.

A "wall-turbulent" flow is extremely complex. Such a flow can be partitioned into several different domains as shown by George and Castillo (1997) (see also the reproduced Fig. 9.53).

In this review chapter, only investigations of the overlap region are presented. The content of this section is mainly based on an article by Egolf (2009). To describe turbulent flows along a wall, also in this case the continuity and momentum equations are transformed to Reynolds averaged equations (Hinze 1975). Then, they are simplified by using the standard Prandtl boundary layer approximation technique. To eliminate the pressure gradient a functional relation, derived from the Bernoulli equation for the free stream field far from the wall with velocity $U(x_1)$, is considered

$$P + \frac{1}{2}\rho\, U^2 = \text{const} \quad \Rightarrow \quad \frac{1}{\rho}\frac{\mathrm{d}P}{\mathrm{d}x_1} = -U\frac{\mathrm{d}U}{\mathrm{d}x_1}. \tag{892a, b}$$

Equation (892b) is applied to replace the pressure term in the turbulent momentum equation (Reynolds equation). This procedure leads to the following system of coupled nonlinear partial differential equations (see Hinze 1975 or Schlichting 1979):

$$\begin{aligned} &\frac{\partial \bar{u}_1}{\partial x_1} + \frac{\partial \bar{u}_2}{\partial x_2} = 0, \\ &\bar{u}_1\frac{\partial \bar{u}_1}{\partial x_1} + \bar{u}_2\frac{\partial \bar{u}_1}{\partial x_2} - U\frac{\partial U}{\partial x_1} + \frac{\partial \overline{u_2' u_1'}}{\partial x_2} - \nu\frac{\partial^2 \bar{u}_1}{\partial {x_2}^2} = 0. \end{aligned} \tag{893a, b}$$

The free stream velocity $U$ obeys the scaling law

$$U(x_1) = U_0\left(\frac{x_1}{l}\right)^{p_1} \quad \Rightarrow \quad U_0 = U(l). \tag{894a, b}$$

For the average velocities dimensionless coordinates, called outer variables (see Fig. 9.53), are introduced

$$\overline{u}_\beta = U_0\left(\frac{x_1}{l}\right)^{p_\beta} f_\beta(\eta), \quad \beta \in \{1, 2\}. \tag{895a, b}$$

This similarity representation is equally applied to the Reynolds shear stress

$$\overline{u_2' u_1'} = -{U_0}^2\left(\frac{x_1}{l}\right)^{p_{21}} f_{21}(\eta). \tag{896}$$

The dimensionless spanwise coordinate is chosen according to

$$\eta = \frac{x_2}{\delta}, \quad \delta = \frac{1}{\omega}\left(\frac{x_1}{l}\right)^{p_0} l. \tag{897a, b}$$

In the above relations, $l$ is a characteristic length, e.g., the length of the plate, under consideration.

The scaling laws (894a) to (897b) are now substituted into the continuity Eq. (893a). This leads to the alternative equation

$$\left(\frac{x_1}{l}\right)^{p_1-1}\left(p_1 f_1 - p_0\eta\frac{\mathrm{d}f_1}{\mathrm{d}\eta}\right) + \omega\left(\frac{x_1}{l}\right)^{-p_0+p_2}\frac{\mathrm{d}f_2}{\mathrm{d}\eta} = 0 \tag{898}$$

or

$$p_1 f_1 - p_0\eta\frac{\mathrm{d}f_1}{\mathrm{d}\eta} + \omega\left(\frac{x_1}{l}\right)^{-p_0-p_1+p_2+1}\frac{\mathrm{d}f_2}{\mathrm{d}\eta} = 0, \tag{899}$$

respectively,

$$p_1 f_1 - p_0\eta\frac{\mathrm{d}f_1}{\mathrm{d}\eta} + \omega\frac{\mathrm{d}f_2}{\mathrm{d}\eta} = 0, \tag{900}$$

where self-similarity ($x_1$-independence), just as in previous sections, demands that the following equation relating exponents must hold:

$$p_0 + p_1 - p_2 - 1 = 0. \tag{901}$$

Notice that self-similarity in wall-bounded shear flows is expected to be exactly valid only in the infinite Reynolds number limit (George and Castillo 1997).

The same self-similarity procedure is also applied to the NSEs (893b); it yields

$$\left(\frac{x_1}{l}\right)^{2p_1-1}\left[p_1 f_1{}^2 - p_0\eta f_1\frac{\mathrm{d}f_1}{\mathrm{d}\eta}\right] + \omega\left(\frac{x_1}{l}\right)^{-p_0+p_1+p_2} f_2\frac{\mathrm{d}f_1}{\mathrm{d}\eta} - p_1\left(\frac{x_1}{l}\right)^{2p_1-1}$$
$$-\omega\left(\frac{x_1}{l}\right)^{p_{21}-p_0}\frac{\mathrm{d}f_{21}}{\mathrm{d}\eta} - \frac{1}{\mathrm{Re}_0}\omega^2\left(\frac{x_1}{l}\right)^{-2p_0+p_1}\frac{\mathrm{d}^2 f_1}{\mathrm{d}\eta^2} = 0, \qquad (902)$$

where the Reynolds number,

$$\mathrm{Re}_0 = \frac{U_0\, l}{\nu} \qquad (903)$$

was introduced; it is defined as the Reynolds number of the overall flow (compare, e.g., with Eq. (694)). Obvious steps transform Eq. (902) to

$$p_1 f_1{}^2 - p_0\eta f_1\frac{\mathrm{d}f_1}{\mathrm{d}\eta} + \omega\left(\frac{x_1}{l}\right)^{-p_0-p_1+p_2+1} f_2\frac{\mathrm{d}f_1}{\mathrm{d}\eta} - p_1$$
$$-\omega\left(\frac{x_1}{l}\right)^{-p_0-2p_1+p_{21}+1}\frac{\mathrm{d}f_{21}}{\mathrm{d}\eta} - \frac{1}{\mathrm{Re}_0}\omega^2\left(\frac{x_1}{l}\right)^{-2p_0-p_1+1}\frac{\mathrm{d}^2 f_1}{\mathrm{d}\eta^2} = 0, \qquad (904)$$

respectively,

$$p_1 f_1{}^2 - p_0\eta f_1\frac{\mathrm{d}f_1}{\mathrm{d}\eta} + \omega f_2\frac{\mathrm{d}f_1}{\mathrm{d}\eta} - p_1 - \omega\frac{\mathrm{d}f_{21}}{\mathrm{d}\eta} - \frac{1}{\mathrm{Re}_0}\omega^2\frac{\mathrm{d}^2 f_1}{\mathrm{d}\eta^2} = 0, \qquad (905)$$

provided that

$$p_0 + 2p_1 - p_{21} - 1 = 0, \qquad p_1 + 2p_0 - 1 = 0. \qquad (906a, b)$$

However, Eq. (906b), that is related to low Reynolds number viscous dissipation, should not be applied (see below). Integrating by parts the dimensionless continuity Eq. (900) and applying $f_2(0) = 0$ yields

$$f_2 = \frac{p_0}{\omega}\eta f_1 - \frac{p_0 + p_1}{\omega}\int_0^{\eta} f_1(\xi)\mathrm{d}\xi. \qquad (907)$$

Next, let us apply the DQTM. The generalized "mixing length" is equal to the distance from the wall

$$\chi_2 = x_2. \qquad (908)$$

The largest important distance occurring in the model is the average width of the turbulent domain, which is a function of $x_1$ (see boarder line of turbulent region in Fig. 9.53)

$$x_{2\,\max} = \delta(x_1). \tag{909}$$

The quantity $\delta(x_1)$ is chosen as the wall distance in the $x_2$ direction, where, over a specific cross section perpendicular to the wall, the average velocity is at its maximum

$$\overline{u}_{1\,\max} = \overline{u}_1(x_1, \delta(x_1)) = U(x_1). \tag{910a, b}$$

This velocity is equal to the free flow velocity. The minimum velocity occurs at the wall and is zero,

$$\overline{u}_{1\,\min} = 0. \tag{911}$$

Furthermore, one immediately sees that

$$b = \delta(x_1), \quad \sigma = \frac{\mathrm{d}\delta(x_1)}{\mathrm{d}x_1}. \tag{912a, b}$$

In total, the DQTM (see Eq. 307b) thus leads to

$$\overline{u_2' u_1'} = -\alpha \frac{\mathrm{d}\delta}{\mathrm{d}x_1} x_2\, \overline{u}_1 \frac{U(x_1) - \overline{u}_1(x_2)}{\delta(x_1) - x_2}, \tag{913}$$

where the quantities given by Eqs. (908) to (912b) have been substituted into (307b). In the above scaling a constant $\alpha$ was also introduced, which will be determined below. Because it stands in front of the Reynolds shear stress, it is the quantity that describes the excitation and strength of the turbulence intensity in this fluid dynamic system. Therefore, as in plane Couette and plane Poiseuille flows, it determines the stress parameter of the physical system. To introduce the DQTM, dimensionless Eqs. (894a) to (897b), with $\beta = 1$ (see 895a), are substituted into (913), leading to

$$f_{21} = \alpha \frac{p_0}{\omega} \frac{\eta}{1 - \eta} \left(f_1 - f_1{}^2\right). \tag{914}$$

Moreover, the same equation for the exponents (906a) applies, which was harvested from the NSE. Therefore, we notice that the three remaining equations (901), (906a, b) for the quantities $p_0$, $p_1$, $p_2$, and $p_{21}$ are not sufficient to solve the system of three equations with four variables. Just as in wake flows, where an additional equation by a momentum consideration was determined, here one additional equation must be formulated. So far, this has not been achieved. However, still numerous new and valuable results may be revealed (see below).

By inserting Eqs. (907) and (914) into (905), a single nonlinear integro-differential equation for $f_1$ can be derived; it reads

$$p_1 f_1{}^2 - (p_0 + p_1)\frac{\mathrm{d}f_1}{\mathrm{d}\eta}\int\limits_0^\eta f_1 \mathrm{d}\xi - p_1 - \alpha\, p_0 \frac{\mathrm{d}}{\mathrm{d}\eta}\left[\frac{\eta}{1-\eta}\left(f_1 - f_1{}^2\right)\right] - \frac{\omega^2}{\mathrm{Re}_0}\frac{\mathrm{d}^2 f_1}{\mathrm{d}\eta^2} = 0. \tag{915}$$

This equation fails to exhibit self-similarity because $\mathrm{Re}_0$ is $x$ dependent (see Hinze 1975). The last term describes viscous effects. We, therefore, restrict ourselves to the overlap region, i.e., the region for which the following two approximations

$$\mathrm{Re}_0 = \frac{U_0 l}{\nu} \gg 1 \quad \text{and} \quad \eta \ll 1 \tag{916a, b}$$

apply. So, by omitting the small terms, accordingly the main differential equation of the wall problem reduces to an equation for only $f_1$,

$$\begin{aligned} &p_1 f_1^2 - (p_0 + p_1)\frac{\mathrm{d}f_1}{\mathrm{d}\eta}\int\limits_0^\eta f_1 \mathrm{d}\xi - p_1 \\ &-\alpha\, p_0 \left(f_1 - f_1^2 + \eta\frac{\mathrm{d}f_1}{\mathrm{d}\eta} - 2\eta f_1 \frac{\mathrm{d}f_1}{\mathrm{d}\eta}\right) = 0. \end{aligned} \tag{917}$$

This leads us to

**Proposition 5** *The flow behavior in the overlap domain of wall-bounded turbulent flows is described by an ordinary nonlinear integro-differential equation that inherently contains a cooperative phenomenon* [see below].

The phase change character of the present theory will be demonstrated in the following sections.

As already discussed in detail in Sect. 5.6, starting in 1917 by Prandtl and 1921 by von Kármán, for the overlap region an average velocity profile described by a power law with exponent 1/7 was proposed, and some years later with Prandtl's mixing length theory the "logarithmic law of the wall," with two empirical constants, was derived and became well established. Then, over the years authors "proved" the superiority of one law over the other, and so, alternatively, either a power law or the logarithmic law were favored, with a slight advantage for the logarithmic law. Now, having in hands the main differential equation of "wall-turbulent" flows, a first trial was to solve it by a logarithmic Ansatz. However, this attempt failed miserably as the interested reader may immediately retrace by himself. Because of this first negative experience, it is reasonable to try the alternative power law

$$f_1(\eta) = A\eta^k + B. \tag{918}$$

If this proposal is substituted into (917), an equation constituted by terms with the three powers 0, $k$, and $2k$ occurs. Their corresponding three coefficient equations take the forms

$$\begin{aligned}
\eta^0 &: p_1 B^2 - p_1 - \alpha\, p_0 B + \alpha\, p_0 B^2 = 0,\\
\eta^k &: 2p_1 B - p_1 kB - p_0 kB - \alpha\, p_0 + 2\alpha\, p_0 B - \alpha\, p_0 k + 2\alpha\, p_0 kB = 0, \quad (919\text{a} - \text{c})\\
\eta^{2k} &: p_1 - p_0 k + 3\alpha\, p_0 k + \alpha\, p_0 + 2\alpha\, p_0 k^2 = 0.
\end{aligned}$$

The trial solution (918) solves neither of the equations (905) nor (915) exactly! It is, however, convenient that in all equations the constant $A$ cancels out. This presents the possibility of patching Eq. (918) to the boundary conditions defined by the average velocities of the adjacent domains. Note that

$$B = 1 \tag{920}$$

solves Eq. (919a). This $B$-value is introduced into (919b); the emerging equation is then solved, viz.,

$$p_1 = \frac{p_0 k - \alpha\, p_0(1 + k)}{2 - k}. \tag{921}$$

A second relation for $p_1$ is obtained from Eq. (919c), namely

$$p_1 = p_0\left(k - 3\alpha k - \alpha - 2\alpha\, k^2\right). \tag{922}$$

Equating (921) and (922) then implies

$$\alpha = \frac{k(k - 1)}{2k^3 - k^2 - 4k - 1}. \tag{923}$$

The quantity $\alpha$ represents the parameter in front of the Reynolds shear stress (see Eq. (913)) and, therefore, it is a natural stress parameter of the system, analogous to the Reynolds number. It is expected to be positive and to increase monotonically with the Reynolds number. The function $\alpha(k)$ is displayed in Fig. 9.54. The pole $k_P$ at which $\alpha$ tends to infinity is determined by a vanishing denominator of the term on the right-hand side of Eq. (923). Of the three real number solutions of the cubic equation that of significance is

$$k_P = \frac{1}{4}\left(3 - \sqrt{17}\right) \quad \Rightarrow \quad k_P = -0.2808. \tag{924a, b}$$

We know, at the critical Reynolds number, $\mathrm{Re}_c$, the first instability of laminar "wall flow" occurs. It marks the critical stress parameter, respectively, the critical driving parameter of the flow system, $\alpha_c$, occurring in this theory. Therefore, in our model the critical quantity $k_c$ corresponds to the critical Reynolds number, $\mathrm{Re}_{c,}$ in conventional fluid dynamic theories. It is determined by taking in Eq. (923), the derivative of $\mathrm{d}\alpha(k)/\mathrm{d}k$ and setting it equal to zero. The dependent variable to the stress

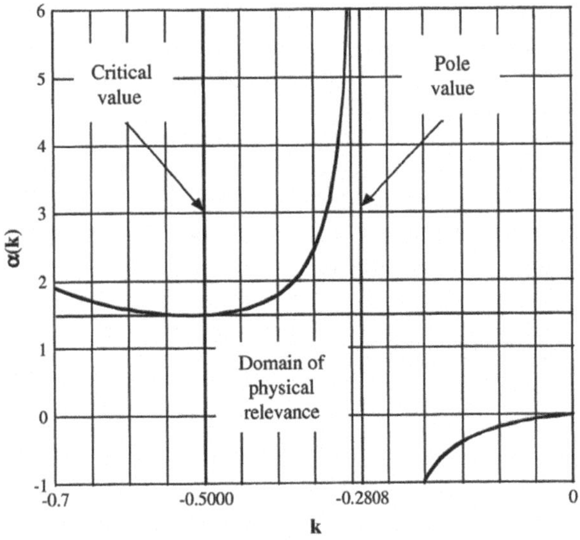


**Fig. 9.54** The abscissa quantity *k* is the negative of the exponent of the new power law (see below). Only the monotonically increasing part of the positive branch is of physical relevance and describes a critical phenomenon (compare with later Fig. 9.55). Courtesy of Egolf (2009) © Int. J. Refrigeration, reproduced with permission by Elsevier Science & Technology

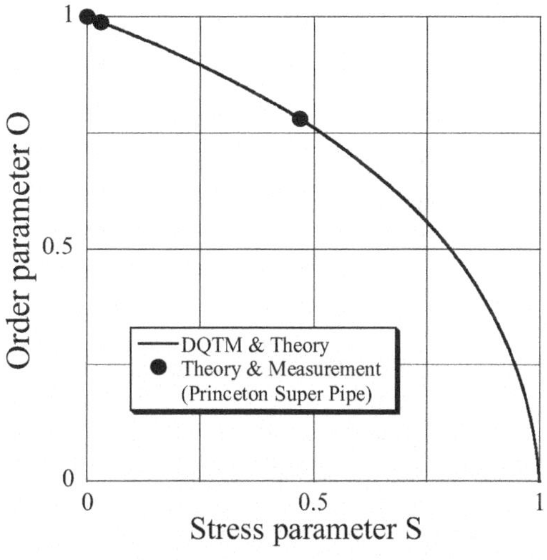


**Fig. 9.55** The order parameter *O* of "wall turbulent" flows as a function of the stress parameter or control parameter *S*, respectively [Interesting is that this interpretation leads to a larger order for a smaller inverse Reynolds number]. Therefore, highest order occurs for an infinite Reynolds number, describing fully turbulent flows. Courtesy of Egolf (2009) © Int. J. Refrigeration, reproduced with permission by Elsevier Science & Technology

parameter $\alpha$ is the order parameter $-k$ (see below). Its critical value is obtained by substituting $\alpha_c$ into Eq. (923) (see also in Fig. 9.54)

$$(\alpha_c, -k_c) = \left(\frac{3}{2}, \frac{1}{2}\right). \tag{925}$$

The theory reveals a critical point and a macroscopic view of the transition to turbulence (for more information see Chap. 11). However, our macroscopic model cannot reveal all the details of the instabilities and (temporal) transition processes that occur in this fluid dynamic system. A tremendous amount of literature can be found dedicated to this topic (theoretical work, e.g., Chandrasekhar 1981; Lin 1955, or also numerical work, e.g., Kleiser and Zhang 1991; Obrist and Schmid 2011).

Notice that even without explicit knowledge of the power law exponent, the system of equations decoupled and led directly to the solution $\alpha(k)$. This functional relation connects the stress parameter with the order parameter (see below) of the fluid dynamic system and will become important in the description of turbulent features at various Reynolds numbers.

In Fig. 9.53, it is shown that for the overlap region, inner variables or outer variables may be applied. Therefore, we now switch from the outer variables $(x_2, \overline{u}_1)$ to inner variables $(y^+, u^+)$; they are related by

$$y^+ = \frac{u^* x_2}{\nu}, \quad u^+ = \frac{\overline{u}_1}{u^*}, \tag{926a, b}$$

where $u^*$ is chosen to be the friction velocity derived from the wall shear stress defined at the bounding wall $u^* = \sqrt{\tau_0/\rho}$ (see also Eq. (136)).

By solving Eqs. (897a) and (926a), both for $x_2$, and then setting them to be identical, one obtains for $\eta$,

$$\eta = \frac{\nu y^+}{u^* \delta} = \frac{1}{\mathrm{Re}^+} y^+ \quad \text{or} \quad \mathrm{Re}^+ = \frac{u^* \delta}{\nu} =: \delta^+. \tag{927a – d}$$

Equation (927d) follows from (926a), provided that $x_2 = \delta$. Next, combining Eqs. (926b), (894a), and (895a) yields

$$u^+ = \frac{\overline{u}_1}{U} \frac{U}{u^*} = \frac{U}{u^*} f_1, \tag{928a, b}$$

so that with Eq. (918) one concludes that

$$f_1 = \frac{u^*}{U} u^+ = A\eta^k + 1, \tag{929a, b}$$

in which (920) was implicitly used. A conceptually significant result is obtained, if Eqs. (929a) and (929b) are combined and $\eta$ in the emerging expression is replaced by Eq. (927b). This leads to

$$u^+ = \frac{U}{u^*} \left[ 1 + A \left( \frac{1}{\mathrm{Re}^+} y^+ \right)^k \right]. \tag{930}$$

Because $u^+ \leq U/u^*$, we write

$$u^+ = \frac{U}{u^*}\left[1 - C\frac{1}{(y^+)^m}\right], \tag{931}$$

which yields by comparison with Eq. (930)

$$C = -(\mathrm{Re}^+)^m A \geq 0, \quad m = -k \geq 0. \tag{932a – d}$$

Equations (931) and (932a–d) represent the new alternative "power law of the wall." The exponent $m$ is related to the stress parameter $\alpha$ by Eq. (923) via (932c). This positive quantity is ideal to define the order parameter of "wall-turbulent" flows (see Chap. 11). The actual power law term or kernel in the new law describes the mean velocity deficit compared to the corresponding average velocity of the free flow at distance $\delta$. In the turbulence literature, this kind of law is called "deficit" or "defect power law." Therefore, we state:

**Theorem 7** *For plane "wall-turbulent" flows the DQTM reveals a deficit or defect power law, respectively. The power law exponent m varies from ½ in the laminar case down to 0.2808 for fully turbulent flows at infinite Reynolds number.*

In the scientific literature on turbulence a serious discussion, not clarified to present, is whether the solution of "wall turbulent" flows is a limiting average velocity profile that differs from a constant or a rectangular profile $f_1 \equiv 1$. The modeling with the DQTM leads to a satisfying insight for this limiting case. It states that an asymptotic pole value exponent proves the contrary, namely the occurrence of a monotonously bended function in this limit of highest degree turbulence. After George and Castillo (1997), a constant profile $\overline{u}_1/U \equiv 1$ is mused not to be unsatisfying, but at least interesting to be discussed. These authors, in their article, state that it may look plausible when the free stream velocity is increased or the viscosity decreased, but it is less reasonable, if the infinite Reynolds number limit is approached by increasing the streamwise distance proceeding along the plate. Finally, also in this limit the DQTM provides a satisfying solution in

**Theorem 8** *The infinite Reynolds number solution for the average velocity profile given by the DQTM is a monotonously curved limiting power law function and not a constant profile.*

Now, by applying material extracted from experimental work with the Princeton super pipe (see Zagarola et al. (1997a, b, 2018a, b) and Fig. 9.56) the new power law is tested. This device was specially constructed to obtain highest Reynolds numbers in a laboratory. The measured data are, e.g., available on the web and can be downloaded (Zagarola et al. 2018a, b). For the overlap region, distant from the center of the pipe, these data may be taken as reference also for wall turbulence, even if the experiment was performed in a cylindrical device with an internal diameter of 129.3 mm. A smooth-wall finish of a roughness <0.15 micrometer was achieved.

The conclusions of the scientists working with the super pipe support very much the present theoretical work (Zagarola et al. 2018b): *Results have led to a new set of*

**Fig. 9.56** High Reynolds number Test Facility (HRTF). "The Super Tunnel" at Princeton University, where the measurements for a comparison with our theory of "wall turbulence" were extracted. From Princeton (2019)

*scaling laws for turbulent pipe flows showing that the almost universally accepted logarithmic behavior does not hold in the near-wall region. There, a power law holds instead, and a logarithmic behavior appears only at Reynolds numbers that are considerably greater than those commonly achieved in the laboratory.*

Table 9.4 (in the upper two parts of it), contains the experimental super pipe data. Besides $Re_{0wall}$ and $Re^+$, all these data were given by the experimentalists (Zagarola et al. 2018a, b). $Re_{0wall}$ is obtained by dividing $Re_{0channel}$ by a factor of 2. That follows because the characteristic diameter of a pipe or rectangular channel is its diameter or width $w$, and $\delta$ of a corresponding wall flow is $\delta = w/2$. $Re^+$ is obtained by dividing $Re_{0wall}$ by $U_0/u^*$, a ratio of measured velocities that can be found in the last column of Table 9.4.

In the third part of the table, the quantity in the third column, $x_{2max}/\delta$, has been chosen in such a manner that criterion (916b) is fulfilled by a factor 1/10. With increasing Reynolds number, fewer points fall into the overlap region defined by this criterion.[1]

Now the quantity $y^+_{max}$ in the second column is obtained by combining Eqs. (897a), (926a), and (927c), which yields

$$\eta_{max} = \frac{x_{2\,max}}{\delta} = \frac{y^+_{max}}{Re^+} \cong \frac{1}{10} \quad \Rightarrow \quad y^+_{max} \cong \frac{Re^+}{10}. \qquad (933a-d)$$

These maximum values are observed in Fig. 9.57 ($y^+_{max} = 72$), Fig. 9.58 ($y^+_{max} = 1655$), and Fig. 9.59 $\left(y^+_{max} = 44,552\right)$. The methods how the next values $\alpha$ and $p_1/p_0$ and $\alpha\ p_0/p_1$ were determined are explained below.

In the fourth and last table section, the quantities $C$ and $m$ are listed. They were determined by a two-dimensional fitting process by applying the deficit power law (931) (with its two free constants) to the measured data. The results of this procedure are shown in Fig. 9.57 for the lowest, in Fig. 9.58 for a medium and in Fig. 9.59 for the highest Reynolds number data set. All 26 measurements of the mean downstream

[1]This can be seen by counting the measurements in Figs. 9.57, 9.58, 9.59. They are at a Reynolds number of Re = 15,789 twenty points, at Re = 511,900 twelve points and at Re = 17,629,500 only ten points, which fulfil the above requirement.

**Table 9.4** From a series of 26 measurements at different Reynolds numbers in the Princeton super pipe (Zagarola et al. 2018a, b)

| No. | $Re_{0channel}$ | $Re_{0wall}$ | $Re^+$ |
|---|---|---|---|
| 1 | 31,577 | 15,789 | 684 |
| 2 | 1,023,800 | 511,900 | 16,894 |
| 3 | 35,259,000 | 17,629,500 | 467,626 |
| – | – | $\infty$ | $\infty$ |
| No. | $U_0$ | $u^*$ | $U_0/u^*$ |
| 1 | 4.821 | 0.2089 | 23.1 |
| 2 | 9.820 | 0.3245 | 30.3 |
| 3 | 33.09 | 0.8787 | 37.7 |
| – | – | – | – |

| No. | $y^+_{max}$ | $x_{2max}/\delta$ | $\alpha$ | $p_1/p_0$ |
|---|---|---|---|---|
| 1 | 71.562 | 0.1046 | 3.202 | −1.064 |
| 2 | 1655.4 | 0.0980 | 50.68 | −16.03 |
| 3 | 44,452 | 0.0951 | 5140 | −1620 |
| – | – | – | $\infty$ | $-\infty$ |
| No. | $\alpha p_0/p_1$ | $C$ | $m$ | Method |
| 1 | −3.010 | 1.259 | 0.3290 | Exp. |
| 2 | −3.162 | 1.891 | 0.2832 | Exp. |
| 3 | −3.171 | 3.832 | 0.2808 | Exp. |
| – | −3.171 | – | 0.2808 | Theor. |

Here for the presentation the first (No. 1), the thirteenths (No. 2), and the twenty-sixth (No. 3) data set were applied. This corresponds to the sets belonging to the lowest, a medium, and the highest Reynolds number. In a pipe it follows that $Re = Re_0$ because $U = U_0$ for all downstream locations

velocity in the overlap region show an agreement as good as the three profiles chosen for presentation here, which is excellent. However, a critical reader may claim that it is the result of a fitting process with two free parameters and that this undermines these good results. Such critics must be accepted. Yet, a much finer confirmation of the model follows beneath.

Why can the quantities at present only be determined by a fitting procedure? The first quantity, namely $C$, is determined by the boundary condition. This theory is only valid for the overlap region at some distance from the wall. $C$ cannot be mathematically determined until the law for the average downstream velocity closer to the wall is at one's disposal. Here, eventually also a transition (see Fig. 9.51) should be taken into consideration. This enrichment of the present model does not seem so difficult to be performed, but it has not been worked out so far.

That the second quantity, the exponent $m$, must also be determined by a fitting process has its origin in the early introduction of the Reynolds number as a stress parameter of this kind of turbulent flow by Reynolds more than hundred sixty years ago. The present theory in a natural manner proposes an alternative stress parameter, $\alpha$, that is reversely proportional to a monotonic growing function of the Reynolds number $Re_0$. In plane Couette flows, for instance, we have experienced a strict proportionality of $\alpha$ with the inverse Reynolds number $1/Re_0$. However, for "wall-

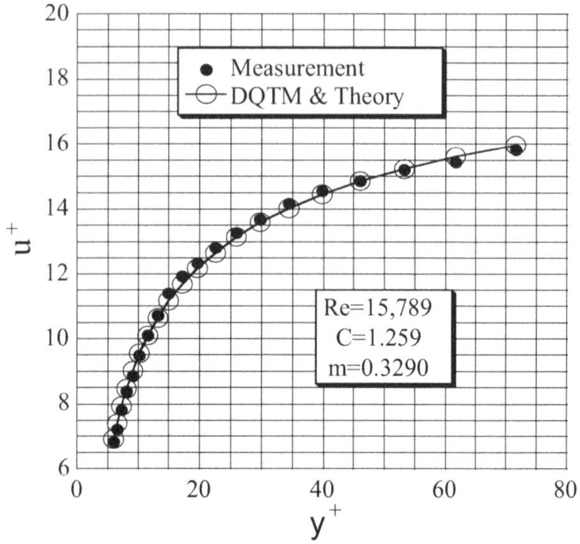


**Fig. 9.57** Analytically calculated average velocity profile for Re = 15,789 compared with super pipe data. Courtesy Egolf (2009) © Int. J. Refrigeration, reproduced with permission by Elsevier Science & Technology

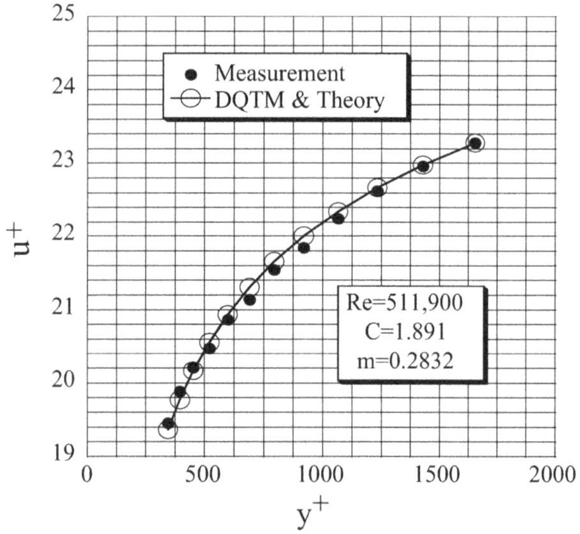


**Fig. 9.58** Analytically calculated average velocity profile for Re = 511,900 compared with super pipe data. Courtesy of Egolf (2009) © Int. J. Refrigeration, reproduced with permission by Elsevier Science & Technology

turbulent" flows the theoretical relation $\alpha(\mathrm{Re}_0)$ has not been derived. Today, there is no other possibility to relate $\alpha$ to $\mathrm{Re}_0$ than by experimental means. Therefore, for an experiment with known Reynolds number by fitting the average velocity profile in the overlap region with the deficit power law (931) determines the related exponent $m$. By combining Eq. (923) with (932c), it follows

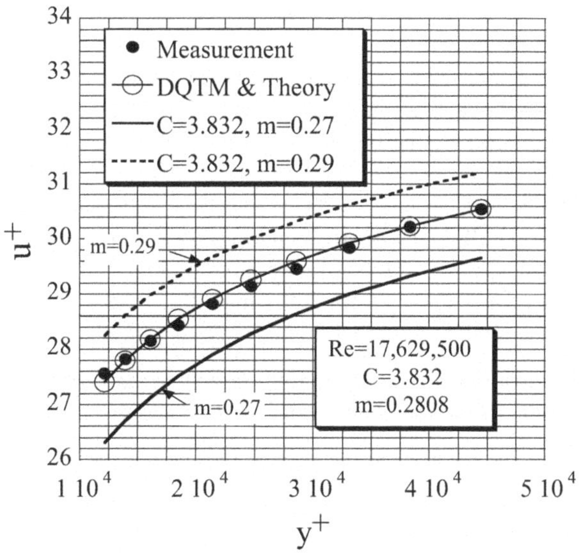


**Fig. 9.59** Analytically calculated average velocity profile for Re = 17,629,500 and three different values for $m$ compared with super pipe data. Courtesy of Egolf (2009) © Int. J. Refrigeration, reproduced with permission by Elsevier Science & Technology

$$\alpha = -\frac{m(m+1)}{2m^3 + m^2 - 4m + 1}. \tag{934}$$

If the experimentally determined value of $m$ is substituted into (934), the corresponding $\alpha$ value is obtained. So for a Reynolds number $\mathrm{Re}_0$, the related stress parameter $\alpha$ is determined. If this process is repeated for numerous Reynolds numbers, the relation $\alpha(\mathrm{Re}_0)$ can be established.

Fitting the theoretical deficit power law with its two free parameters to the twenty experimental data pairs $(y^+, u^+)$ works well for the lowest Reynolds number. For the medium Reynolds number, the eight pairs with the lowest values of $y^+$ cannot be brought to fit the theory and for the highest Reynolds number there are even ten such pairs. Nevertheless, it is very satisfying that criterion (933d) excludes exactly those pairs that were found to be problematic to fit the deficit power law.

Up till now, we have experienced good agreement between experiment and model calculations. However, the final and most convincing argument is still missing and shall now be outlined. There is a set of points where the relation $\alpha(\mathrm{Re}_0)$ is fully determined, namely at infinite Reynolds number. In this limit, a theoretical one-to-one relation is valid, namely

$$\alpha \to 0 \quad \Leftrightarrow \quad \mathrm{Re}_0 \to \infty. \tag{935a, b}$$

For this case, the theory has predicted the exponent

$$\mathrm{Re}_0 \to \infty \qquad m = 0.2808, \tag{936a, b}$$

with an accuracy of four digits after the comma (see Table 9.4). If we now assume that the highest Reynolds number achieved in the super pipe is so high that it can be represented by the infinite Reynolds number, an independent fitting should lead to result (936b). As it is shown in Table 9.4 this is indeed the case. The experimental fitting result coincides with the experimentally gained value with an absolute error of $\varepsilon = \pm 10^{-4}$. The corresponding relative error is only 0.04%. Egolf and Hutter 2016a, applying a safety factor larger than ten, claimed that the DQTM predicts this exponent with an error of less than 1%. How much superior is the quality of this model prediction to that of Sect. 5.5! We highlight this as:

**Proposition 6** *The DQTM prediction of the exponent of the deficit power law for infinite Reynolds number flow shows a relative error smaller than 0.1%.*

In Table 9.4, $p_1/p_0$ in the fifth column of the third part follows from Eqs. (921) and (932c)

$$\frac{p_1}{p_0} = -\frac{m + \alpha(1 - m)}{2 + m}. \tag{937}$$

By introducing Eq. (934) into (937) and simplifying the expression, a direct determination only using $m$ is also possible

$$\frac{p_1}{p_0} = -\frac{2m^2}{2m^2 + 3m - 1}. \tag{938}$$

Moreover, Table 9.4 (second column of the fourth part) contains $\alpha p_0/p_1$. It is obtained by dividing $\alpha$ by $p_1/p_0$ . Using Eqs. (934) and (937) this yields

$$\frac{\alpha\, p_0}{p_1} = \frac{(m + 1)}{2m(m - 1)}. \tag{939}$$

From Eq. (921) with (932c), for $\alpha \to \infty$, it follows that

$$\lim_{\alpha\to\infty} \frac{\alpha\, p_0}{p_1} = \frac{m + 2}{m - 1}. \tag{940}$$

By substituting the pole value, $m_P$, given by Eq. (924a) and again applying (932c), one obtains

$$\lim_{\alpha\to\infty} \frac{\alpha\, p_0}{p_1} = \frac{\sqrt{17} + 5}{\sqrt{17} - 7} = -3.171. \tag{941a, b}$$

Also, this result is in perfect agreement with the value obtained by evaluation from experiments (see Table 9.4). In Table 9.4, looking at the values of $\alpha p_0/p_1$, one could conclude that this may be a universal constant of the "wall turbulent" problem. Now, we calculate its value at the critical point (index $c$) by introducing in Eq. (939) $m_c = 0.5$, which leads to (942a) and the pole value (index $p$) (942b)

$$\left.\frac{\alpha\, p_0}{p_1}\right|_c = -3.000, \quad \left.\frac{\alpha\, p_0}{p_1}\right|_p = -3.171. \qquad (942\text{a, b})$$

In Table 9.4 for the lowest Reynolds number, the quantity $\alpha\, p_0/p_1$ is $-3.010$, only slightly smaller than $-3.000$ and in harmony with Eq. (942a). Furthermore, it is very stable and insensitive to changes of the stress parameter $\alpha$. From the critical value $\alpha = 3/2$ up to $\alpha \to \infty$ it only changes by 5.7%. After the critical value it converges to its asymptotic value. Therefore, the quantity $\alpha p_0/p_1$ may be taken as a kind of universal constant for all $\alpha$, or $\mathrm{Re}_0$, respectively.

The agreement between experimental and theoretical results of the average downstream velocity profiles is in all cases excellent. This statement is further confirmed, because the results are sensitive to changes in the exponent $m$. In Fig. 9.59, the three curves for $m = 0.27$, $m = 0.2808$, and $m = 0.29$ demonstrate that for a constant $C$, large changes of the mean velocity profile occur, even if the exponent $m$ is only slightly changed.

A comparison of the experimental super pipe data with the logarithmic law is shown in several articles, see, e.g., Barenblatt and Chorin (1998). It must be stressed that these comparisons also look very convincing. Nevertheless, we make the following conclusion:

*There is no theory that predicts the logarithmic law with its correct coefficients nearly as well as the DQTM predicts the deficit power law. Therefore, we claim that (with our present knowledge) the deficit power law is superior to the logarithmic law of the wall (see Table 5.3).*

For the velocity in the direction perpendicular to the wall, one introduces the definition

$$v^+ = \frac{\overline{u}_2}{u^*}. \qquad (943)$$

Furthermore, in agreement with Eq. (895b), one has

$$\overline{u}_2 = Vf_2 = U_0\left(\frac{x_1}{l}\right)^{p_2} f_2(\eta). \qquad (944\text{a, b})$$

Our starting point is Eq. (907). In a preparatory calculation the integral is determined. Furthermore, the deficit power law (918) with $B = 1$ is substituted into the integral to yield

$$\int_0^\eta f_1(\xi)\mathrm{d}\xi = \int_0^\eta \left(A\xi^k + 1\right)\mathrm{d}\xi = \eta\left(\frac{A}{k+1}\eta^k + 1\right) = \eta\left(\frac{A}{1-m}\eta^{-m} + 1\right). \tag{945a – c}$$

However, this kind of calculation is approximate. To obtain a good quality of the profile from the wall to the outer layer, the corresponding function of each layer must always be inserted into the integral (945a). For the viscous sublayer, this would be a linear function, adjacent possibly to a transition function and then for the overlap region the deficit power law, not being valid for the total domain as applied above. Because a complete theory with different wall functions is still to be derived, we restrict our considerations to this partial solution. A complete theory for all domains parallel to the wall is left for future elaboration.

Now, Eq. (907) leads to

$$f_2 = \frac{p_0}{\omega}\eta\left(1 + A\eta^k\right) - \frac{p_0 + p_1}{\omega}\left(1 + \frac{A}{1+k}\eta^k\right), \tag{946}$$

which can be transformed to

$$f_2 = \frac{p_0}{\omega}\eta\left[\left(1 + A\eta^k\right) - \left(1 + \frac{p_1}{p_0}\right) \times\frac{k + \left(1 + A\eta^k\right)}{k+1}\right]. \tag{947}$$

This representation was chosen, because now, in Eq. (947) $1 + A\eta^k$ can be expressed by $f_1$ (compare with Eqs. (918) and (920)) as

$$f_2 = \frac{p_0}{\omega}\eta\left[f_1 - \left(1 + \frac{p_1}{p_0}\right)\eta\frac{k + f_1}{k+1}\right]. \tag{948}$$

Next, this is transformed to inner variables by using Eqs. (927b), (929a), (932c), (943) and (944a)

$$v^+ = \frac{1}{\mathrm{Re}^+}\frac{V}{u^*}\frac{p_0}{\omega}y^+\left[\frac{u^*}{U}u^+ - \left(1 + \frac{p_1}{p_0}\right)\frac{\frac{u^*}{U}u^+ - m}{1-m}\right]. \tag{949}$$

Substituting a rearranged Eq. (931) leads to

$$v^{+} = K_1\, y^{+} \left\{ \left[ 1 - \frac{C}{(y^{+})^{m}} \right] - \left( 1 + \frac{p_1}{p_0} \right) \left[ 1 - \frac{1}{1-m} \frac{C}{(y^{+})^{m}} \right] \right\}, \tag{950}$$

with the constant

$$K_1 = \frac{1}{\mathrm{Re}^{+}} \frac{V}{u^{*}} \frac{p_0}{\omega}. \tag{951}$$

In Fig. 9.60 the function, describing the averaged transverse velocity, is shown for the highest Reynolds number. It was explained above that the solution (950) with (951) is best fulfilled for high Reynolds numbers, where the viscous sublayer becomes negligibly small. The constant $K_1$ can approximately be determined (with the respective numerical quantities in Table 9.4). However, $V$, for the moment, remains unknown, because the power law exponent $p_2$ in Eq. (944b) arises in that equation, but is still not explicitly determined (among all $p$-exponents). This unknown relation is the reason why in Fig. 9.60 $(1/K_1)v^{+}$ is used as ordinate.

It is also possible to substitute (938) into (950) to eliminate $p_0$ and $p_1$ in favor of $m$; the result is

$$v^{+} = K_1\, y^{+} \left\{ \left[ 1 - \frac{C}{(y^{+})^{m}} \right] - \left( \frac{4m^2 + 3m - 1}{2m^2 + 3m - 1} \right) \left[ 1 - \frac{1}{1-m} \frac{C}{(y^{+})^{m}} \right] \right\}. \tag{952}$$

The Reynolds shear stress, which is given by Eq. (913), is normalized by $u^{*2}$ to obtain the expression

$$\overline{u_2' u_1'}^{+} = \frac{\overline{u_2' u_1'}}{u^{*2}}. \tag{953}$$

Then it follows from Eq. (913) that

$$\overline{u_2' u_1'}^{+} = -\alpha \frac{\mathrm{d}\delta}{\mathrm{d}x_1} x_2 \frac{\overline{u}_1}{u^{*}} \frac{\dfrac{U}{u^{*}} - \dfrac{\overline{u}_1}{u^{*}}}{\delta - x_2}. \tag{954}$$

Applying here again the approximation (916b) leads to

$$\overline{u_2' u_1'}^{+} = -\alpha \frac{1}{\delta} \frac{\mathrm{d}\delta}{\mathrm{d}x_1} x_2 \frac{\overline{u}_1}{u^{*}} \left( \frac{U}{u^{*}} - \frac{\overline{u}_1}{u^{*}} \right). \tag{955}$$

The inner variables (926a,b) and $\mathrm{Re}^{+}$, given by Eq. (927c), yield

$$\frac{x_2}{\delta} = \frac{y^{+}}{\mathrm{Re}^{+}}, \tag{956}$$

which, inserted into (955), leads to the form

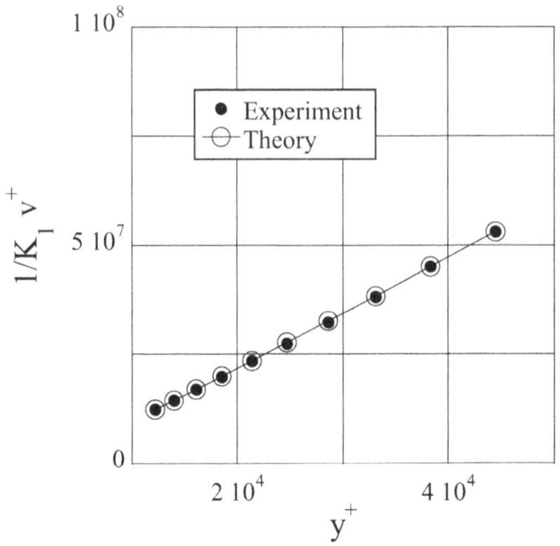


**Fig. 9.60** Average transverse velocity in the overlap region of "wall turbulent" flows. The measurements are not directly determined values. They are merely the transformed measured black points in Fig. 9.59 by having applied Eq. (952)

$$\overline{u_2'u_1'}^+ = K_2 y^+ u^+ (U^+ - u^+), \tag{957}$$

with the constant

$$K_2 = -\frac{\alpha}{\mathrm{Re}^+}\frac{\mathrm{d}\delta}{\mathrm{d}x_1}. \tag{958}$$

The determination of the constant $K_2$ is fraught with similar indeterminacies as $K_1$.

Equations (957) and (958) determine the Reynolds shear stress in inner variables; in Fig. 9.61 this dependence is shown. To be clear with all abbreviations carrying a plus sign (+), the following formulas, in agreement with definition (894a), are stated here

$$U^+ = \frac{U}{u^*} = U_0^+\left(\frac{x_1}{l}\right)^{p_1}, \quad U_0^+ = \frac{U_0}{u^*}. \tag{959a – c}$$

Now, the deficit power law (931) is substituted into (957). This leads to

$$\overline{u_2'u_1'}^+ = K_2\left(\frac{U}{u^*}\right)^2 y^+\left[1 - C\frac{1}{(y^+)^m}\right]\left\{1 - \left[1 - C\frac{1}{(y^+)^m}\right]\right\}, \tag{960}$$

which simplifies to the following final expression for the Reynolds shear stress

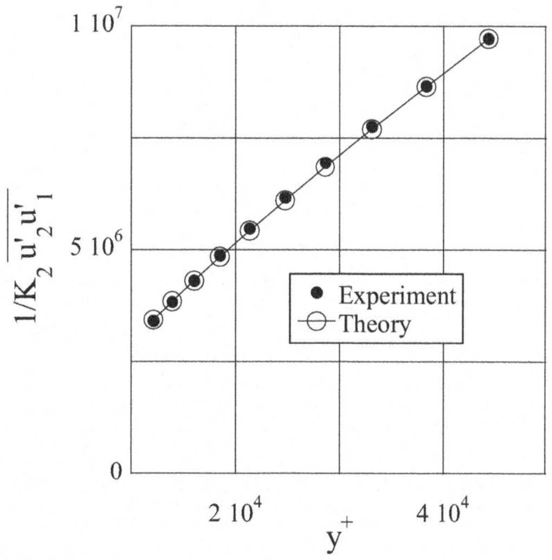


**Fig. 9.61** The normalized and scaled Reynolds shear stress in the overlap region of "wall turbulent" flows. The measurements are not directly determined values. They are just the transformed measured black points in Fig. 9.59 by having applied Eq. (957)

$$\overline{u_2'u_1'}^+ = K_3(y^+)^{1-m}\left[1-\frac{C}{(y^+)^m}\right], \tag{961}$$

with the constant $K_2$ given by Eq. (958) and

$$K_3 = C\left(\frac{U}{u^*}\right)^2 K_2 = -C\frac{\alpha}{\mathrm{Re}^*}\frac{\mathrm{d}\delta}{\mathrm{d}x_1}\left(\frac{U}{u^*}\right)^2. \tag{962a, b}$$

In the physically relevant domain of Fig. 9.54 one finds the order parameter curve of "wall turbulent" flows in the overlap region. The flow along the wall reveals a critical phenomenon. To show this, in Fig. 9.54 the panel is turned by 90° in the mathematical positive (counter clockwise) sense. Then, one observes a stress and order parameter relation of the function with physical relevance.

For the normalization of the stress parameter $S$ we propose

$$S = \frac{\alpha_c}{\alpha}, \tag{963}$$

a quantity varying from 0 to 1.

On the other hand, the order parameter $O$ will be defined by

$$O = \frac{m - m_c}{m_p - m_c}, \tag{964}$$

with the power law exponents at criticality (index $c$) and at the pole (index $p$). In Fig. 9.54, these two scaling procedures of the quantities lead directly to a suitable order parameter curve showing analogies to other phase transitions (see Goldenfeld 1992 and Chap. 11).

It is valuable to return to the study of the asymptotic value of the power law exponent. The pole value (see Eqs. (924a) and (932c)),

$$m_P = \frac{1}{4}\left(\sqrt{17} - 3\right) = 0.2808, \tag{965a, b}$$

is obviously an important number of "wall turbulence" research; and it is of a special mathematical type, namely

$$\frac{a + b\sqrt{n}}{c}, \quad a, b, c \neq 0, \quad n \in N^+, \quad n > 1, \tag{966}$$

and, furthermore, $n$ must be free of a quadratic. Such special numbers are called "quadratic irrationals" (Gellert et al. 1977).

The most famous of these numbers is a number with numerous names, e.g., *golden mean*, *golden section*, *golden number*, and *divine proportion*,

$$g = \frac{1 + \sqrt{5}}{2} = 1.6180\ldots. \tag{967a, b}$$

Such numbers are given by continued fractions, viz,

$$a_1 + \cfrac{1}{a_2 + \cfrac{1}{a_3 + \cfrac{1}{a_4 + \ldots}}}, \tag{968}$$

with integer numbers $a_1, a_2, a_3, \ldots$ (Penrose 2007). If one chooses $a_1 = a_2 = a_3 = \ldots = 1$, the Fibonacci sequence is obtained, discovered in ancient Indian mathematics,

$$g = 1 + \cfrac{1}{1 + \cfrac{1}{1 + \cfrac{1}{1 + \ldots}}}. \tag{969}$$

The result of this infinite continued fraction is the golden mean $g$. Over the centuries, artists and architects had a preference to apply the golden mean of height $h$ to width $w$ of paintings, buildings, windows, etc. to be $h/w = g$.
With the sequence 0,3,1,1,0,3,1,1... we have

$$m_p = 0 + \cfrac{1}{3 + \cfrac{1}{1 + \cfrac{1}{1 + \cfrac{1}{0 + \cfrac{1}{3 + \ldots}}}}} = 0.2808\ldots, \tag{970a, b}$$

**Fig. 9.62** The circular sitting area contains a red pebble stone inner circle and a larger granite area enclosed by a single row of red pebble stones. The diameter of the inner to the outer circle presents the "marvellous proportion" $m_p$ (new naming by author P.W. Egolf)

which is the "wall turbulence" pole value $m_p$. The principal author (P.W.E) applied this number in garden architecture and calls it the *marvellous proportion* (see Fig. 9.62).

# References


Abramovich, G.N.: The Theory of Turbulent Jets. MIT Press, Cambridge, MA (1963). LCC-CN: 63-21743

Abramowitz, M., Stegun, I.A.: Pocketbook of Mathematical Functions. Verlag Harri Deutsch, Thun (1984). ISBN 3-87144-818-4

Adrian, R.J.: Hairpin vortex organization in wall turbulence. Phys. Fluids. **19**, 041301 (2007)

Adrian, R.J., Balachandar, S., Lin, Z.C.: Spanwise growth of vortex structure in wall turbulence. KSME Int. J. **15**(12), 1741 (2001)

Aydin, E.M., Leutheusser, H.J.: Plane Couette flow between smooth and rough walls. Exp. Fluids Eng. **11**, 302 (1991)

Barenblatt, G.I., Chorin, A.J.: Comment on: log laws or power laws: the scaling in the overlap region. Phys. Fluids Comments. **10**(4), 1043 (1998)

Bech, K.H., Tillmark, N., Alfredsson, P.H.: An investigation of turbulent plane Couette flow at low Reynolds numbers. J. Fluid Mech. **286**, 291 (1995)

Brauckmann, H.J., Eckhardt, B.: Direct numerical simulation of local and global torque in Taylor–Couette flow up to Re=30000. J. Fluid Mech. **718**, 398 (2013)

Burgers, J.M., Mitchner, M.: On homogeneous non-isotropic turbulence connected with a mean motion having a constant velocity gradient. Proc. Kon. Ned. Akad. V. Wet. **B56**, 228 and 343 (1953)

Busse, F.: Bounds for turbulent shear flows. J. Fluid Mech. **41**(1), 219 (1970)

Chandrasekhar, S.: Hydrodynamic and Hydromagnetic Stability. Dover Press, Mineola, NY (1981). ISBN 0-486-64071-X. 978-0-521-19868

Corrsin, S.: Investigation of Flow in an Axially Symmetrical Heated Jet of Air. National Advisory Committee for Aeronautics, NACA wartime report No. 94, USA (1943, unpublished)

Corrsin, S., Uberoi, M.S.: Further Experiments on the Flow and Heat Transfer in a Heated Turbulent Air Jet. National Advisory Committee for Aeronautics, NACA report No. 1865, Washington, USA (1949)
Deissler, R.O.: Effect of initial condition on weak homogeneous turbulence with uniform shear. Phys. Fluids. **13**, 1868 (1970)
Douce, L.N., Egolf, P.W., Busse, F.: Nonlocal Modeling of Taylor-Couette Turbulent Flows. Bachelor diploma work at the University of Applied Sciences of Western Switzerland in collaboration with Bayreuth University, Germany (2016)
Egolf, P.W.: Die mittlere Geschwindigkeit längs der Achse eines turbulenten, isothermen Freistrahls. Gesund. Ing. **113**(4), 198 (1992)
Egolf, P.W.: Difference-quotient turbulence model: a generalization of Prandtl's mixing-length theory. Phys. Rev. E. **49**(2), 1260 (1994)
Egolf, P.W.: An attempt to prove the difference-quotient turbulence model. Helv. Phys. Acta. **71**(1), 7 (1998)
Egolf, P.W.: Lévy statistics and beta model: a new solution of "wall" turbulence with a critical phenomenon. Int. J. Refrig. **32**, 1815 (2009)
Egolf, P.W., Hutter, K.: A nonlocal zero-equation turbulence model and a deficit-power law of the wall with a dynamical critical phenomenon. PAMM. Appl. Math. Mech. **16**, 889 (2016a)
Egolf, P.W., Hutter, K.: Turbulent shear flow described by the algebraic difference-quotient turbulence model. In: Peintke, J., et al. (eds.) Progress in Turbulence VI, Springer Proceedings in Physics, vol. 165. Springer, Heidelberg (2016b). ISBN 978-3-319-29129-1
Egolf, P.W., Weiss, D.A.: Model for plane turbulent Couette flow. Phys. Rev. Lett. **75**(16), 2956 (1995a)
Egolf, P.W., Weiss, D.A.: Elementary geometric figures in fully developed turbulent plane Poiseuille flow. Helv. Phys. Acta. **68**, 219 (1995b)
Egolf, P.W., Weiss, D.A.: Difference-quotient turbulence model: the axi-symmetric isothermal jet. Phys. Rev. E. **58**(1), 459 (1998)
Egolf, P.W., Weiss, D.A.: Difference-quotient turbulence model: analytical solutions for the core region of plane Poiseuille flow. Phys. Rev E. **62**(1-A), 553 (2000)
El Telbany, M.M.M., Reynolds, A.J.: The structure of turbulent plane Couette flow. J. Fluids Eng. **104**, 367 (1982)
Fenstermacher, P.R., Swinney, H.L., Gollub, J.P.: Dynamical instabilities and the transition to chaotic Taylor vortex flow. J. Fluid Mech. **94**, 103 (1979)
Frisch, U.: Turbulence – The Legacy of A.N. Kolmogorov, 1st edn. Cambridge University Press, Cambridge, UK (1995). ISBN 0-85-403-441-2
Gelfand, I.M., Schilov, G.E.: Verallgemeinerte Funktionen. VEB Deutscher Verlag der Wissenschaften, Berlin (1962). (in German)
Gellert, W., Küstner, H., Hellwich, M., Kästner, H.: Kleine Enzyklopädie der Mathematik, 2nd edn. Harri Deutsch, Thun and Frankfurt a. M. (1977). (in German)
George, W.K., Castillo, L.: Zero-pressure-gradient turbulent boundary layer. Appl. Mech. Rev. **50** (11), 689 (1997)
Glauert, M.B.: The wall jet. J. Fluid Mech. **1**, 625 (1956)
Goldenfeld, N.: Lectures on Phase Transitions and the Renormalization Group, Frontiers in Physics, 5th edn. Adison-Wesley, Reading, PA (1992). ISBN 13-978-0-201-55408-3
Gollub, J.P., Swinney, H.L.: Onset of turbulence in a rotating fluid. Phys. Rev. Lett. **35**, 927 (1975)
Green, S.I.: Fluid Vortices: Fluid Mechanics and its Applications. Kluwer Academic, Dordrecht (1995). ISBN 0-7923-33-76-4
Hamiliton, J.M., Kim, J., Waleffe, F.: Regeneration mechanisms of near-wall turbulence structures. J. Fluid Mech. **287**, 317 (1995)
Heisenberg, W.: Über Stabilität und Turbulenz von Flüssigkeitsströmen. Ann. Phys. **379**(15), 577 (1924). (in German)
Hinze, J.O.: Turbulence, 2nd edn. McGraw-Hill, New York (1975). ISBN 0-07-029037-7

Hutter, K., Jöhnk, K.: Continuum Methods of Physical Modeling. Springer, Berlin (2004). ISBN 13-978-364-205-831-8

Hutter, K., Wang, Y.: Fluid and thermodynamics. In: Advanced Fluid Mechanics and Thermodynamic Fundamentals, vol. 2. Springer, Berlin (2016). ISBN 978-3-319-33635-0

Kampé de Fériet: Sur le spectre de la turbulence homogène. J. Acad. Sci. Paris. **208**, 772 (1939). (in French)

Katz, P.: Luftströmung im Raum. In: "Lehrbuch der Klimatechnik" Arbeitskreis der Dozenten für Klimatechnik, Band 1, "Grundlagen". Verlag C.F. Müller, Karlsruhe (1974). ISBN 3-7880-7052-8 (in German)

Khinchin, A.Y.: Continued Fractions. Dover, Mineola, USA (1992). ISBN 13-978-0-486-69630-0

Kleiser, L., Zhang, T.A.: Numerical simulation of transition in wall-bounded shear flows. Annu. Rev. Fluid Mech. **23**, 495 (1991)

Landau, L.D., Lifshitz, E.M.: Lehrbuch der Theoretischen Physik, Hydrodynamik, vol. VI. Akademie–Verlag, Berlin (1974). (in German)

Laufer, J.: The Structure of Turbulence in Fully Developed Pipe Flow. National Advisory Committee for Aeronautics, USA (1954). Technical Report No. 1174

Lee, M.J., Kim, J.: The structure of turbulence in a simulated plane Couette flow. In: Proceedings of the Eighth Symposium on Turbulent Shear Flows. Technical University Munich (1991)

Libby, P.A.: Introduction to Turbulence Combustion, An International Series. Taylor & Francis, Abingdon, UK (1996). ISBN 1-56032-100-8

Lin, C.C.: The Theory of Hydrodynamic Stability. Cambridge University Press, Cambridge, UK (1955)

Mathur, M., Haller, G., Peacock, T., Ruppert-Felsot, J., Swinney, H.: Uncovering the Lagrangian skeleton of turbulence. Phys. Rev. Lett. **98**(14), 144502 (2007)

Monchaux, R..: http://wwwy.ensta.fr/~monchaux/couette_flows.html (2019a)

Monchaux R..: http://wwwy.ensta.fr/~monchaux/PhD%20Plane%20Couette%20Flow%20FR.pdf (2019b)

Obrist, D., Schmid, P.J.: Algebraically diverging modes upstream of a swept bluff body. J. Fluid Mech. **683**, 346 (2011)

Pai, S.I.: On turbulent flow between parallel plates. J. Appl. Mech. **20**, 109 (1953)

Panchapakesan, N.R., Lumley, J.L.: Turbulence measurements in axisymmetric jets of air and helium. Part 1. Air jet. J. Fluid Mech. **247**, 197 (1993)

Pearson, J.R.A.: The effect of uniform distortion on weak homogeneous turbulence. J. Fluid Mech. **5**, 274 (1959)

Penrose, R.: The Road to Reality. Vintage Books, New York (2007). ISBN 978-0-679-77631-4

Princeton: https://www.princeton.edu/~gasdyn/HRTF.html (2019)

Regenscheit, B.: Die Berechnung von radial strömenden Frei- und Wandstrahlen, sowie Rechteckstrahlen. Gesundheitsingenieur. **92**, 193 (1971). (in German)

Reichardt, H.: Vollständige Darstellung der turbulenten Geschwindigkeitsverteilung in glatten Leitungen. ZAMM. **31**(7), 208 (1951). (in German)

Reichardt, H.: Über die Geschwindigkeitsverteilung in einer geradlinigen turbulenten Couetteströmung. ZAMM. **36**, 26 (1956). (in German)

Reichardt, H.: Gesetzmässigkeiten der geradlinigen turbulenten Couette-Strömung, Mitteilungen aus dem Max-Planck-Institut für Strömungsforschung und der aerodynamischen Versuchsanstalt, Göttingen (1959). (in German)

Resonance.: https://resonance.is/fractal-dissipation-energy-turbulent-flows-fluid-dynamics-underlies-wide-range-physical-phenomenal/ (2019)

Reynolds, A.J., Wieghardt, K.: Another look at unidirectional turbulent flow. J. Fluid Mech. **287**, 75 (1995)

Ricou, F.P., Spalding, D.B.: Measurements of entrainment by axisymmetrical turbulent jets. J. Fluid Mech. **11**, 21 (1961)

Robertson, J.M., Johnson, H.F.: Turbulence structure in plane Couette flow. J. Eng. Mech. Div. **6**, 1171 (1970)

Ruelle, D., Takens, F.: On the nature of turbulence. Commun. Math. Phys. **20**, 167 (1971a)
Ruelle, D., Takens, F.: On the nature of turbulence. Commun. Math. Phys. **23**(4), 343 (1971b)
Schlichting, H.: Boundary-Layer Theory. McGraw-Hill, New York (1979). ISBN 0-07-055-334-3
Schuster, H.G.: Deterministic Chaos: An Introduction, 2nd edn. Verlagsgesellschaft mbH (VGH), Weinheim (1988). ISBN 3-527-26862-6
Sommerfeld, A.: Ein Beitrag zur hydrodyamischen Erklärung der turbulenten Flüssigkeitsbewegung. In: Proceedings of the 4th International Congress of Mathematicians, vol. 3, p. 116, Rome (1908). (in German)
Stanišić, M.M.: The Mathematical Theory of Turbulence, 2nd edn. Springer, Berlin (1988). ISBN 0-387-96685-4
Tennekes, H., Lumley, J.L.: A First Course in Turbulence. MIT Press, Cambridge, MA (1972). ISBN 0-262-200-19-8
Townsend, A.A.: Measurements in the turbulent wake of a cylinder. Proc. R. Soc. Lond. **190A**, 551 (1947)
Townsend, A.A.: The fully developed wake of a circular cylinder. Aust, J. Res. **A2**, 451 (1949)
Townsend, A.A.: Structure of Turbulent Shear Flow. Cambridge University Press, Cambridge, UK (1976). ISBN 0521-20710-X
van Dike, M.: An Album of Fluid Motion. Parabolic Press, Stanford, CA (1982). ISBN 0-915760-02-9
Weisshaar, E., Busse, F.H., Nagata, M.: Twist vortices and their instabilities in the Taylor–Couette system. J. Fluid Mech. **226**, 549 (1991)
Wu, X., Moin, P.: Direct numerical simulation of turbulence in a nominally zero-pressure-gradient flat-plate boundary layer. J. Fluid Mech. **630**, 5 (2009)
Wygnanski, I., Fiedler, J.: Some measurements in the self-preserving jet. J. Fluid Mech. **38**, 577 (1969)
Yangate: http://www.yang.gatech.edu/projects/Liquid%20Rocket%20Injector%20Dynamics...%20(Nan)/index.htm (2019)
Youtube: https://www.youtube.com/watch?v=LOmQ-QLejCU (2019)
Zagarola, M.V.: Electronic Letter to First Author dated on 3rd of March (1998)
Zagarola, M.V., Perry, A.E., Smits, A.J.: Log laws or power laws: The scaling in the overlap region. Phys. Fluids. **9**(7), 2094 (1997a)
Zagarola, M.V., Smits, A.J.: Scaling of the mean velocity profile for turbulent pipe flow. Phys. Rev. Lett. **78**(2), 239 (1997b)
Zagarola, M.V., Perry, A.E., Smits, A.J.: http://www.princeton.edu/~gasdyn/#superpipedata (2018a)
Zagarola, M.V., Perry, A.E., Smits, A.J.: http://www.princeton.edu/~gasdyn/#superpipe_data (2018b)

# Chapter 10
# Thermodynamics of Turbulence

## 10.1 Introduction

### *10.1.1 Microscopic and Macroscopic Theories*

Fact in turbulence research is that there are mainly two limiting cases to describe turbulent flows, a microscopic and a macroscopic description.

The first one is based on the entire ensemble of atoms or molecules in the flow field. The characteristic amount of the considered entities, respectively, the number of degrees of freedom of such a large-scale system, e.g., an atmospheric turbulent flow, is larger than the order of the value of the Avogadro number ($6.022 \times 10^{23}$), which is very large. Castaign (1989, 1996) pointed out that controlling each of these single degrees is impossible and a futile program. The related problem is so large that it can at best be handled by molecular dynamics and supercomputing.

On the macroscopic side, it is the description of fluid behavior by the NSEs, which reduce the problem to only four macroscopic variables, namely three velocity components and the pressure (see Chap. 2). Naively, one could now assume that by this enormous reduction of variables, which is roughly a factor $10^{23}$, the problem has been sufficiently simplified, so that it could be easily solved. However, the nonlinear terms in the NSEs lead to another quasi-unsolvable problem, which has amply been discussed in Chap. 2. Applying Kolmogorov's microscales, it is easy to show that the number of grid points $N$ demanded to solve the NSEs by DNS is $N \propto \mathrm{Re}^{9/4}$ (see, e.g., Frisch 1995). This leads to an enormous demand of Central Processing Unit (CPU) time to calculate a turbulent field at high Reynolds number.

That the two most usual physical concepts show such a high resistance to a solution makes it attractive to search for alternative mesoscopic or macroscopic models. Such models are semi-microscopic models, employing the Langevin Equation (LE) and the Fokker–Planck Equation (FPE). Furthermore, also thermodynamic models reduce the degrees of freedom enormously. Because turbulent flows are generically irreversible and nonequilibrium processes, we will further concentrate

---

The original version of this chapter was revised. A correction to this chapter can be found at https://doi.org/10.1007/978-3-030-26033-0_14



P. W. Egolf, K. Hutter, *Nonlinear, Nonlocal and Fractional Turbulence*,
https://doi.org/10.1007/978-3-030-26033-0_10

efforts on nonequilibrium thermodynamic concepts. Returning to the considerations of Castaign (1996), we cite his conclusion that "*a thermodynamic approach could be advantageous, where only a few important parameters allow a complete description of the large fluid dynamic system.*"

### *10.1.2 Langevin and Fokker–Planck Equations*

We have seen that a successful method is to start to model turbulence by considering the mechanism of Brownian motion. Originally, Langevin (1908) designed his first version of the Langevin equation (LE) (Kawasaki 1973) to describe Brownian motion (Reif 1975; Rogers and Williams 2000). This equation is a stochastic differential equation that describes only a subset of the entire degrees of freedom. The remaining variables are typically collective and macroscopic variables (Haken 1983, 1987). They change only slowly compared to the faster microscopic variables, which are compacted in a stochastic force term of the Langevin differential equation. With knowledge of this review, it can be foreseen that an analogous version of a Fractional LE (FLE) (Mainardi and Pironi 1996; Kobelev and Romanov 1999; Lutz 2001; Camargo et al. 2009), describing anomalous diffusion processes, could play a role in future descriptions of turbulent flows.

The Fokker–Planck Equation (FPE) (Fokker 1914; Planck 1917) shows a similar bundling of a high number of degrees of freedom in a time-dependent probability distribution of the stochastic variables. Kolmogorov (1931) independently derived the equation. Applied to particle distributions the equation is called Smoluchovski equation (see, e.g., Dhont 1996), and if diffusion is missing, in statistical physics, it is called Liouville equation. This partial differential equation describes the temporal evolution of the probability density of the velocities of the particles under the influence of regular forces, but also a random force given by the stochastic properties of microscopic variables, e.g., also obeying Brownian movements. Research on a Fractional FPE (FFPE), describing subdiffusion, was published, e.g., by Barkai (2001).

### *10.1.3 Reduction of the Degrees of Freedom by Scaling*

The main development of the DQTM with Lévy statistics as outlined in this review, borrows its ideas from the microscopic world and applies them to the equally exorbitant high number of eddies in a high Reynolds number flow. However, a main difference is that the largest eddies are of truly macroscopic dimensions. *This special type of approach, which was initiated by Prandtl, could be called "microscopic-macroscopic modelling." In this framework, the degrees of freedom are a priori not small, and a reduction of its number is given by scaling laws.*

Let us come back to Castaign's insight, which is in full agreement with that of Goldenfeld (1992), who in winter 1983–1984 realized that certain nonequilibrium phenomena [like 2-d turbulence] could be analyzed by using renormalization group techniques, which at the time was mainly employed in solid-state physics for field theoretical approaches and critical phenomena, among them also phase transitions to "integrate out" some degrees of freedom. With our conviction that turbulence *is* a critical phenomenon, from today's point of view, this is like closing a circle. Based on such ideas, it becomes evident that thermodynamic models may also be successful to reduce the complexity. Especially if the features of nonlinear systems are coupled with nonequilibrium thermodynamic concepts, important new results may be expected to be revealed.

### 10.1.4 Different Thermodynamic Concepts

Before diving into the details of the peculiarities of the entropy concept, apt for the thermodynamic description of turbulence, it should be explained that essentially two classes of thermodynamic formulations or axiomatisms are in use in different branches of modern natural sciences. Both classes start with Boltzmann–Gibbs (BG) statistics from which equilibrium thermodynamics emerged with its entropy $S = -k \log_e p$. In the second half of the nineteenth century and beyond, starting with Clausius (1850), followed by Gibbs (1902), Duhem (1895, 1911), Natanson (1896), Eckart (1940), Meixner (1943), Onsager (1931), Casimir (1945) and others (see, e.g., de Groot and Mazur (1984) for a summary of these early contributions). In short, a structure of a field theory was introduced from which a theoretical procedure was developed that allowed a clear definition of thermostatic equilibrium states of physical systems and small deviations from them, characterizing the irreversible, local, and linear description of dissipation. Because of the smallness of these deviations, a tendency of the basic understanding had developed that led to the claims of thermodynamic understanding that (1) entropy was only defined in equilibrium and (2), nonlinear processes, with possibly large deviations from equilibrium, could not be based on a rational description of the first and second law. Both claims turned out to be incorrect, but the resolutions of the above items (1) and (2) led to two different classes of thermodynamic conceptual extensions, which may be coined rational and extended or nonextensive thermodynamics[1], respectively.

"Rational Thermodynamics" is an axiomatic formulation using the first and second laws together with the remaining balance laws of a physical system (balances of mass, momenta, energy, etc.). Constitutive equations in these formulations are functional statements of those physical quantities arising in these balance laws,

[1]There exists also a form of this type of thermodynamics, called "extended thermodynamics," which has nothing in common with what we shall describe below, see, e.g., Müller and Ruggeri (1998).

which must be formulated in terms of the remaining variables such that the emerging functional (differential) equations can uniquely be integrated, at least in principle. Processes, which produce no entropy are called thermodynamic equilibria. They define via a Gibbs-like relation the independent variables of the equilibrium entropy (and other equilibrium variables, e.g., pressure, absolute temperature, and other canonical energies). The reduced entropy inequality (often called the dissipation inequality), which remains after the equilibrium properties are separated from the original entropy inequality, is then used to determine the irreversible parts of the constitutive variables. There exist several sub-models of rational thermodynamics, which exploit the entropy inequality, either as part of a closed or open system of field equations. Typical reviews of these procedures are Truesdell (1984), Müller and Ruggeri (1998), Jou et al. (1988) and Hutter (1977) or Hutter and Wang (2016b).

"Nonextensive Thermodynamics" focusses on postulations of nonequilibrium entropies. Their devotees have developed concepts of generalized entropies, such that in thermodynamic equilibrium the BG–entropy emerges. As far as we know, no reference to the second law of thermodynamics as a fundamental physical law (expressed, e.g., in form of the entropy inequality) is made or used, when generalizing this extended entropy concept. However, it is argued that this kind of generalized entropy is compulsory to be able to describe nonlinear processes. The attitude in this type of thermodynamics, which surely still has major aspects to be revealed, is to start with classical BG statistical mechanics and to seek an extension of the classical statistically averaged models by nonlinearly adjusting the asymptotic tail of very large fluctuations of the micro-variables arising in the models. We shall make the details of this approach explicit in the remainder of this review, and stress here, that rational thermodynamics will not explicitly play a role from now on.

It is, however, apt to present an additional side view on activities in which effects described by cooperative dynamics and ordering have been brought into thermodynamic concepts. A milestone in this area was the phenomenological approach by Prigogine (1947), which gave rise to a new way of studying irreversible thermodynamic systems, already taking nonlinear effects and some of their consequences into consideration. Haken (1983, 1987) developed and summarized knowledge on cooperative dynamics in complex physical systems with hierarchical structures (see also Takayama 1988). They report on the slaving principle, where a single mode can reach a superior importance compared to others, and by this presents itself as an ideal order parameter of a complex system. This is a procedure that opens the door for statistical thermodynamic descriptions of nonlinear systems, also showing cooperative phenomena, including phase transitions (see also Chap. 11).

Clothing fluid dynamic systems by a statistical suit started with Gibbs 1902; Boltzmann 1909; Onsager (1949) (see Chapman and Cowling 1970; Truesdell and Muncaster 1980). Today, mainly two approaches are available (i) The Euler flow system is developed by a finite number of point vortices (see, e.g., Novikov 1976; Pointin and Lundgren 1976; Saffman and Baker 1979; Fröhlich and Ruelle 1982; Montgomery 1985; Robert and Sommaria 1991 and (ii) the energy and vorticity are described by a truncated set of (finite) Fourier components (Lee 1952; Kraichnan

1975; Kraichnan and Montgomery 1980). Explicitly, a Hamiltonian (Miller 1990) can be formulated that contains two constants of the motion, which in a 2-d approach are the kinetic energy and the square of the vorticity (or enstrophy) and in a 3-d formulation the kinetic energy and the helicity. This method is the basis for our theoretical considerations of this section. The loss of information by the truncation process was tried to be overcome by constructing the Gibbs free energy for the full Euler system (see, e.g., Albeverio and Cruzerio 1990; Boldrighini and Frigio 1980). Robert (1990, 1991), to obtain the Gibbs potential, goes a similar way as we will also go in this section; namely he works with Young measures and the Kullback–Leibler entropy (see below). Sect. 10.4 will follow the work of Tsallis (2009) and Egolf and Hutter (2018).

## 10.2 A Brief Review of Some Essentials of Boltzmann–Gibbs Thermodynamics

In this section, we start with standard or equilibrium thermodynamics often referred to as statistical thermodynamics of the Boltzmann–Gibbs (BG) type (see, e.g., Reif 1975; Becker 1978; Le Bellac et al. 2006). This formulation was applied with success to numerous physical systems in thermal equilibrium, e.g., dilute gases, Boson gases, photon radiation, phonons in solids, fermions, and ferromagnetic magnons at low temperatures (see, e.g., Becker 1978 and Brenig 1992).

The systems, studied in this section, exhibit order/disorder phenomena. High order is described by low entropy and vice versa. We support the viewpoint that applying the right entropic functional is essential for physical modeling (see Tsallis 2009). The entropy associated to a BG physical system was proposed by Boltzmann and refined by Gibbs for general systems by the entropy

$$S_{\mathrm{BG}} = -k \sum_{i=1}^{N} p_i \log_{\mathrm{e}} p_i, \tag{971}$$

where the $p_i$'s denote the microscopic probabilities of the system occupying a partial region $\Omega_i$ $(1 \leq i \leq N)$ of the total phase space $\Omega$ (see Sect. 6.1). The quantity $k$ is a constant.[2] These probabilities obey the required normalization

[2]In the case of standard BG thermodynamics, this is the Boltzmann constant $k = k_B$. Here, we denote it by $k$ to emphasize reference to the generalizations of nonlinear dynamical systems, where it shows other values.

$$\sum_{i=1}^{N} p_i = 1. \tag{972}$$

For equal probabilities $p_i = 1/N$ one derives

$$S_{\mathrm{BG}} = k \log_{\mathrm{e}} N. \tag{973}$$

When two probabilistic independent subsystems $A$ and $B$ with numbers of states $N_A$ and $N_B$ are put in contact, the joint probabilities obey the special relation

$$p_{ij}^{A+B} = p_i^A p_j^B, \quad \forall i, j. \tag{974}$$

In this case, the entropy functional (compare with Eq. (971)) is

$$\begin{aligned} S_{\mathrm{BG}}(A+B) &= -k \sum_{i=1}^{N_A} \sum_{j=1}^{N_B} p_{ij}^{A+B} \log_{\mathrm{e}} p_{ij}^{A+B} \\ &= -k \sum_{i=1}^{N_A} \sum_{j=1}^{N_B} p_i^A p_j^B \log_{\mathrm{e}} \left( p_i^A p_j^B \right) \\ &= -k \sum_{i=1}^{N_A} \sum_{j=1}^{N_B} p_i^A p_j^B \left( \log_{\mathrm{e}} p_i^A + \log_{\mathrm{e}} p_j^B \right) \\ &= -k \left( \sum_{j=1}^{N_B} p_j^B \right) \sum_{i=1}^{NA} p_i^A \log_{\mathrm{e}} p_i^A - k \left( \sum_{i=1}^{N_A} p_i^A \right) \sum_{j=1}^{N_B} p_j^B \log_{\mathrm{e}} p_j^B. \end{aligned} \tag{975a – d}$$

By applying Eq. (972) to the two parentheses in Eq. (975d), this simplifies to

$$S_{\mathrm{BG}}(A+B) = -k \sum_{i=1}^{N_A} p_i^A \log_{\mathrm{e}} p_i^A - k \sum_{j=1}^{N_B} p_j^B \log_{\mathrm{e}} p_j^B. \tag{976}$$

With the help of Eq. (971), it follows that the Boltzmann–Gibbs entropy of a joint system $A + B$ with independent subsystems $A$ and $B$ is additive, viz,

$$S_{\mathrm{BG}}(A+B) = S_{\mathrm{BG}}(A) + S_{\mathrm{BG}}(B). \tag{977}$$

Furthermore, the BG entropy is at its maximum at equal probabilities and shows expansibility to new states and concavity (for proofs of these features see, e.g., Tsallis 2009).

Consider a system, with phase space $\Omega$, in thermal equilibrium. For a microscopic or at least a very small part of this system, with phase space element $\Omega_i$, and with some simplifying assumptions, the probabilities are demonstrated to be

$$p_i = \frac{e^{-\beta E_i}}{Z_{\mathrm{BG}}} \tag{978}$$

(see, e.g., Reif 1975). $E_i$ is the energy related to the domain $\Omega_i$ of the system. This expression is called Boltzmann factor. Furthermore, the cited derivation yields

$$\beta = \frac{1}{kT}, \tag{979}$$

with the Boltzmann constant $k$, the Kelvin temperature $T$ and the partition function of Boltzmann–Gibbs type, given by

$$Z_{\mathrm{BG}} = \sum_{i=1}^{N} e^{-\beta E_i}. \tag{980}$$

## 10.3 Kraichnan's BG Equilibrium Thermodynamics of 2-d and 3-d Turbulent Fields

Now, the question is: Can these thermodynamic laws describe turbulence? Kraichnan saw a kind of analogy between a weakly coupled Boson gas below the Bose–Einstein condensation temperature threshold in thermal equilibrium and turbulence.

He stated (see Frisch 2011): "*There is a fairly close dynamical analogy in which the number density and the kinetic energy of the Bosons play the roles of kinetic energy and squared vorticity* [which is the specific enstrophy of a two-dimensional turbulent field (see, e.g., Hutter and Wang 2016a)]." We state that this example is one of the most instructive ones to show how thermodynamics principally applies to turbulence.

Again, we assume the fluid to be incompressible. Let us start with the divergence of the momentum equation for such a fluid (see also Foias et al. 2001)

$$\operatorname{div}\left[\frac{\partial \vec{u}}{\partial t}+\left(\vec{u}\cdot\nabla\right)\vec{u}-\nu\,\Delta\vec{u}+\frac{1}{\rho}\nabla p-\vec{f}\right]=0. \tag{981}$$

This transforms to

$$\frac{\partial}{\partial t}\operatorname{div}\vec{u}+\operatorname{div}\left[\left(\vec{u}\cdot\nabla\right)\vec{u}\right]-\nu\,\Delta\operatorname{div}\vec{u}+\frac{1}{\rho}\Delta p-\operatorname{div}\vec{f}=0, \tag{982}$$

which, owing to incompressibility, simplifies to the relation

$$\operatorname{div}\left[\left(\vec{u}\cdot\nabla\right)\vec{u}\right]+\frac{1}{\rho}\Delta p-\operatorname{div}\vec{f}=0. \tag{983}$$

For a divergence-free-specific force field $\vec{f}$, i.e., if it is a vorticity field, it follows that

$$\frac{1}{\rho}\Delta p=-\operatorname{div}\left[\left(\vec{u}\cdot\nabla\right)\vec{u}\right]. \tag{984}$$

Furthermore, it is evident that

$$\begin{aligned}\operatorname{div}\left[\left(\vec{u}\cdot\nabla\right)\vec{u}\right]&=\sum_{i,j}^{3}\frac{\partial}{\partial x_i}\left(u_j\frac{\partial u_i}{\partial x_j}\right)=\sum_{i,j=1}^{3}\left(\frac{\partial u_j}{\partial x_i}\frac{\partial u_i}{\partial x_j}+u_j\underbrace{\frac{\partial^2 u_i}{\partial x_j\partial x_i}}_{=\,0}\right)\\&=\sum_{i,j=1}^{3}\frac{\partial u_j}{\partial x_i}\frac{\partial u_i}{\partial x_j}.\end{aligned} \tag{985a – c}$$

Now, Eq. (984) becomes

$$\frac{1}{\rho}\Delta p=-\sum_{i,j=1}^{3}\frac{\partial u_j}{\partial x_i}\frac{\partial u_i}{\partial x_j}. \tag{986}$$

Because the Laplace operator of the pressure is expressible as $\int_\Gamma \Delta p\,\mathrm{d}V=\int_\Gamma \operatorname{div}\left(\nabla p\right)\mathrm{d}V$, the integral over a domain $\Gamma$ with boundary $\partial\Gamma$, for smooth pressure $p$, can be transformed by the divergence theorem to

$$\int_{\Gamma} \Delta p \, \mathrm{d}V = \int_{\Gamma} \mathrm{div}(\nabla p) \, \mathrm{d}V = \int_{\partial\Gamma} \frac{\partial p}{\partial \vec{n}} \, \mathrm{d}S. \qquad (987\mathrm{a}, \mathrm{b})$$

in which $\vec{n}$ is the unit normal vector on $\partial\Gamma$ pointing to the outside of the domain $\Gamma$.

For a Neumann problem, for which

$$\Delta p = -\rho \sum_{i,j=1}^{\infty} \frac{\partial u_j}{\partial x_i} \frac{\partial u_i}{\partial x_j}, \qquad (988)$$

is solved in $\Gamma$ subject to the flux boundary condition

$$\frac{\partial p}{\partial \vec{n}} = \nabla p \cdot \vec{n}, \qquad (989)$$

prescribed on $\partial\Gamma$, we may regard the right-hand side of Eq. (988) as prescribed in $\Gamma$. We know that this so-called Neumann problem for the Poisson equation is mathematically well posed and solvable. This view point allows us to infer that the pressure $p$ is a quadratic form (functional) of the fluid velocity field $\vec{u}$.

Therefore, we have learnt that in an incompressible fluid the pressure $p$ at each instant of time $t$ is fully determined by the total velocity field $\vec{u}$ in the (entire) fluid domain and its initial pressure distribution $p_0$ at time $t_0$, where the velocity field takes the form $\vec{u}_0$. Then, it follows that

$$\frac{p}{\rho} = \psi\left(\vec{u}\right), \qquad \frac{p_0}{\rho} = \psi\left(\vec{u}_0\right), \qquad (990\mathrm{a}, \mathrm{b})$$

in which $\psi$ is a differentiable function. Thus, we can define the quadratic operator

$$\mathrm{B}\left(\vec{u}, \vec{u}\right) = -\left(\vec{u} \cdot \nabla\right) \vec{u} - \nabla\psi\left(\vec{u}\right), \qquad (991)$$

so that the momentum equation can be written as

$$\frac{\partial \vec{u}}{\partial t} - \mathrm{B}\left(\vec{u}, \vec{u}\right) - \nu \, \Delta\vec{u} = \vec{f}. \qquad (992)$$

Foias et al. (2001) wrote: "*This equation is not the usual form of the Navier-Stokes equation, but it puts in evidence the fact that those equations may be regarded as evolution equations for the velocity* $\vec{u}$."

Kraichnan (1967) (see also Frisch 2011) studied the force-free Euler equation

$$\frac{\partial \vec{u}}{\partial t} = \mathrm{B}\left(\vec{u}, \vec{u}\right), \tag{993}$$

exactly in this sense. However, he additionally applied a spectral Ritz–Galerkin truncation with a filtering operator that in the following shall be outlined in detail.

Consider the NSEs with periodic boundary conditions, e.g., the solutions of the equation in the square bracket of Eq. (981) equal to zero ($[\ldots] = 0$) with boundary conditions that are spatially periodic and for simplicity assume that they have the maximum period $L$ in each of the three directions. Then a Fourier series expansion of the velocity is

$$\vec{u}^{(\kappa')}\left(\vec{x}\right) = \sum_{n_1=1, n_2=1, n_3=1}^{+\infty} \hat{\vec{u}}_{\vec{n}}^{(\kappa')} e^{\frac{2\pi i}{L} \vec{n} \cdot \vec{x}} \tag{994}$$

with the dimensionless integer wave number vector $\vec{n} = (n_1, n_2, n_3)$ and the physical wave number vector $\vec{k} = \kappa'(n_1, n_2, n_3)$, in which $\kappa'$ has dimension 1/length. The hat in $\hat{\vec{u}}_{\vec{n}}^{(\kappa')}$ denotes the Fourier component.

In particular, the following conditions hold

$$\hat{\vec{u}}_{\vec{k}} \cdot \vec{k} = 0, \quad \hat{\vec{u}}_{-\vec{k}} = \overline{\hat{\vec{u}}}_{\vec{k}}. \tag{995a, b}$$

The lowest component of the wave number is defined by

$$\kappa' = \frac{2\pi}{L}. \tag{996}$$

If the spectrum is also cut at a large wave number (small wave length), defined by

$$\kappa'' = \frac{2\pi}{L} N, \tag{997}$$

the twofold truncated velocity vector is written by

$$\vec{u}^{(\kappa', \kappa'')}\left(\vec{x}\right) = \sum_{n_1=1, n_2=1, n_3=1}^{N} \hat{\vec{u}}_{\vec{n}}^{(\kappa', \kappa'')} e^{i\kappa' \vec{n} \cdot \vec{x}}. \tag{998}$$

[For more details on the truncation procedure, cutting away eddies of larger wave numbers than that of a predefined threshold value $\kappa''$ ($n_i > N$, $i = 1, 2, 3$), which is identical to setting their Fourier components equal to zero, see also Kraichnan (1967), Herring and Kraichnan (1972), and Eyink et al. (2011)]. This is tantamount to eliminating all Fourier components with wave numbers larger than $\kappa''$, a fact that is

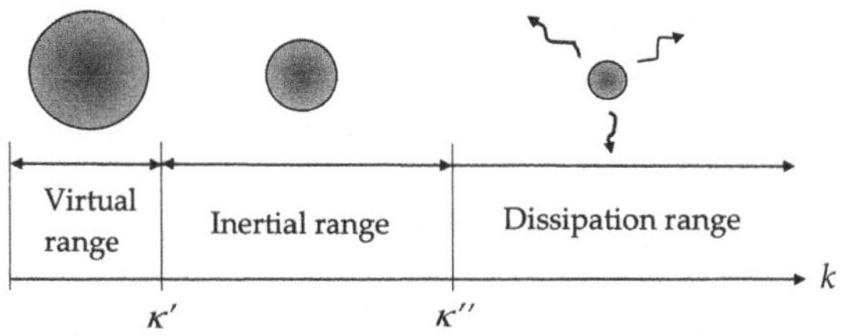


**Fig. 10.1** Three domains are distinguished: A virtual range that belongs to virtual eddies of diameter larger than the fluid basin (range to the left), the so-called inertial range with the full ensemble of eddies with a diameter $L$ down to $l_K$ and the dissipation range with eddies that would have an eddy diameter smaller than the Kolmogorov dissipation scale if they would exist. The circular objects characterize eddies of typical sizes of their domains. The standard Ritz-Galerkin truncation neglects wave numbers larger than the bounding wave number value $\kappa''$

reminiscent of introducing the Kolmogorov dissipation length $l_K$, if we assume that $\kappa''$ is the wave number closest or identical to this inverse dissipation length

$$\kappa'' \cong \frac{1}{l_K}. \tag{999}$$

Galerkin–Ritz spectral procedures project a vector $\vec{u}$ from the entire vector space onto a space spanned by the $N$ first eigenvectors $\vec{w}_n$ of the Stokes operator (for a definition using duality, see Foias et al. 2001), which is orthonormal

$$\left(\vec{w}_n \cdot \vec{w}_m\right) = \delta_{nm}, \quad i \in \{1,2,3\}. \tag{1000}$$

In a spectral method, a velocity vector $\vec{u}$ can be written by means of the vectors of an orthonormal basis $\vec{w}_n$ and the Fourier components $\widehat{u}_n$, viz,

$$\vec{u} = \sum_{n=1}^{\infty} \widehat{u}_n \vec{w}_n. \tag{1001}$$

As described above, a Ritz–Galerkin procedure cuts away the highest modes. In a modification, we demand that it removes both, lowest and largest modes (see Fig. 10.1). A projection is then described by

$$P_N \vec{u} = \vec{\overline{u}} = \sum_{n=1}^{N} \left(\vec{\overline{u}} \cdot \vec{w}_n\right) \vec{w}_n = \sum_{n=1}^{N} \widehat{u}_n \vec{w}_n, \quad \widehat{u}_n = \left(\vec{\overline{u}} \cdot \vec{w}_n\right). \tag{1002a – d}$$

A Ritz–Galerkin truncation can be interpreted as a projection method to convert a continuous operator problem, e.g., an infinite set of differential equations into a discrete problem, in our case a finite set of evolution equations describing a dynamical system (see Cuillière 2016).

Then, a quadratic series development is introduced (compare similarity with definition in Kraichnan (1967))

$$\widetilde{u}_i^2 = \dot{\widetilde{x}}_i^2 = \sum_{m,n=1}^{N} A_{imn}\widetilde{x}_m\widetilde{x}_n, \quad i \in \{1,2,3\}, \tag{1003a, b}$$

where the $\widetilde{x}_i$'s and $\widetilde{u}_i$'s represent independent Cartesian replicas of the space and velocity fields and the $A_{ijk}$'s are spectral coupling coefficients. In this projection operation the constraint

$$\frac{\partial \widetilde{u}_i}{\partial \widetilde{x}_i} = 0, \tag{1004}$$

remains valid because the Ritz–Galerkin truncation operator $P_N$ commutes with derivatives (this is a modified interpretation of Kraichnan's method to account for the fluid incompressibility).

It follows that Kraichnan's 2-d and 3-d systems conserve each a subspace, namely (1) in the two-dimensional problem defined by the 2-d kinetic energy per unit of surface mass $\rho'$ (kg m$^{-2}$)

$$E_\varepsilon = \frac{1}{2}\rho' \int_\Gamma \vec{u}^2 \, \mathrm{d}A = \text{const}, \tag{1005a, b}$$

and the enstrophy $\Omega_\varepsilon$, defined by

$$\Omega_\varepsilon = \varepsilon = \int_\Gamma \vec{\omega}^2 \mathrm{d}A = \text{const}, \quad \vec{\omega} = \nabla \times \vec{u}. \tag{1006a – d}$$

(2) in the three-dimensional case, it is the 3-d kinetic energy

$$E_H = \frac{1}{2}\rho \int_\Gamma \vec{u}^2 \, \mathrm{d}V = \text{const}, \tag{1007a, b}$$

and the helicity, defined by

$$\Omega_H = H = \int_\Gamma \vec{u} \cdot \vec{\omega} \, \mathrm{d}V = \text{const}. \tag{1008a – c}$$

For reasons of brevity, we will employ a different notation, namely $\Omega_\varepsilon{=}\varepsilon$, $\Omega_H = H$, $E_\varepsilon$ and $E_H$.

It is clear that, if $\kappa''$ tends to infinity, a transformation from the discrete to the continuous case can be performed by replacing sums by corresponding integrals.

**Remark** *The enstrophy definition, which in Eq. (1006b,d) has been introduced, is a special case where incompressibility is assumed to hold. The more general definition of enstrophy is*

$$\Omega_\varepsilon = \int_\Gamma \left|\frac{\partial u_i}{\partial x_j}\right|^2 \mathrm{d}A, \tag{1009}$$

*where* $|\ldots|$ *denotes the Frobenius norm, which is the sum of the absolute values squared of all the occurring Jacobian matrix elements* $J_{i,\,j} = \partial u_i \,/\, \mathrm{d}x_j$, $i \in \{1, 2, 3\}$.

Now, following Kraichnan and going back to the discrete description, one distinguishes:

(i) *Energy conservation*:

$$\widetilde{E}_\chi \propto \sum_{i=1}^{d} \widetilde{u}_i^2, \quad \chi \in \{\varepsilon, H\}, \quad d \in \{2, 3\}, \tag{1010a – d}$$

with $\chi = \varepsilon$ for $d = 2$ and $\chi = H$ for $d = 3$ (we write $\propto$, because we neglect the constants $\rho' / 2$ and $\rho \,/\, 2$ in this and the following equations).

With Eqs. (1000) and (1002c) one derives

$$\begin{aligned}\widetilde{E}_\chi &\propto \sum_{i=1}^{d}\sum_{m=1}^{N}\widehat{u}_m w_{i,m}\sum_{n=1}^{N}\widehat{u}_n w_{i,n} = \sum_{i=1}^{d}\sum_{m,n=1}^{N}\widehat{u}_m\widehat{u}_n w_{i,m} w_{i,n} = \sum_{m,n=1}^{N}\widehat{u}_m\widehat{u}_n \\ &\times \sum_{i=1}^{d} w_{i,m} w_{i,n} = \sum_{m,n=1}^{N}\widehat{u}_m\widehat{u}_n\left(\vec{w}_m\cdot\vec{w}_n\right) = \sum_{m,n=1}^{N}\widehat{u}_m\widehat{u}_n\delta_{mn} = \sum_{n=1}^{N}\widehat{u}_n^2,\end{aligned} \tag{1011a – f}$$

where the commas in $w_{i,m}$ and $w_{i,n}$ do not denote derivatives; they separate the denotations (indices) of the coordinates and the number of the replicas of the Ritz–Galerkin truncation procedure.

Similarly, Kraichnan defined

(ii) *Enstrophy conservation*:

$$\widetilde{\Omega}_{\varepsilon} = \sum_{i=1}^{2} \widetilde{\Omega}_{\varepsilon i} = \sum_{i,j=1}^{2} \left| \frac{\partial \widetilde{u}_i}{\partial x_j} \right|^2, \tag{1012a, b}$$

again using the Frobenius norm. Here, we have also used the fact that the operations "taking the derivatives" and "applying the Ritz–Galerkin projection" commute.

We start with Eq. (994) and in the following auxiliary calculation for brevity reasons omit the superscript in the velocity components

$$\begin{aligned} \frac{\partial u_i}{\partial x_j} &= \frac{\partial \left( \sum\limits_{n_1=1, n_2=1, n_3=1}^{\infty} \widehat{u}_{i,\vec{n}} \, e^{i\,\kappa'\,\vec{n}\cdot\vec{x}} \right)}{\partial x_j} \\ &= i\kappa' n_j \sum_{n_1=1, n_2=1, n_3=1}^{\infty} \widehat{u}_{i,\vec{n}} \, e^{i\kappa'\vec{n}\cdot\vec{x}} = i\kappa' n_j u_i. \end{aligned} \tag{1013a – c}$$

Furthermore, denoting conjugate complex numbers by overbars, it follows with summation over $i$ and $j$,

$$\left| \frac{\partial u_i}{\partial x_j} \right|^2 = \frac{\partial u_i}{\partial x_j} \overline{\frac{\partial u_i}{\partial x_j}} = \left( i\kappa' n_j u_i \right) \left( -i\kappa' n_j u_i \right) = \kappa'^2 n_i n_i u_j u_j. \tag{1014a – c}$$

With the discrete version of Eqs. (1009) and (1014c), it is concluded that

$$\begin{aligned} \Omega_{\varepsilon} &= \kappa'^2 \left( n_1^2 \widetilde{u}_1^2 + n_1^2 \widetilde{u}_2^2 + n_2^2 \widetilde{u}_1^2 + n_2^2 \widetilde{u}_2^2 \right) = \kappa'^2 \left( n_1^2 + n_2^2 \right) \left( \widetilde{u}_1^2 + \widetilde{u}_2^2 \right) \\ &= \kappa'^2 \vec{n}^2 \cdot \left| \vec{\widetilde{u}} \right|^2, \end{aligned} \tag{1015a – c}$$

or

$$\Omega_{\varepsilon} = \left| \vec{k} \right|^2 \cdot \left| \vec{\widetilde{u}} \right|^2. \tag{1016}$$

Next, we again apply the Ritz–Galerkin truncation and Eqs. (1014c) to (1016) and obtain

$$\widetilde{\Omega}_\varepsilon \propto \left|\frac{\partial P_N u_i^{(\kappa')}}{\partial x_j}\right|^2 = \frac{\partial P_N u_i^{(\kappa')}}{\partial x_j} \overline{\frac{\partial P_N u_i^{(\kappa')}}{\partial x_j}} = \vec{k}^2 \left(P_N u_i^{(\kappa')}\right)^2$$
$$= \left|\vec{k}\right|^2 \sum_{i=1}^{d} \widetilde{u}_i^2. \qquad (1017\text{a} - \text{d})$$

With the analogous procedure as described by Eqs. (1010) and (1011) for the enstrophy, we can derive the final representation

$$\widetilde{\Omega}_\varepsilon = k^2 \left|\widehat{\vec{u}}\right|^2 = k^2 \sum_{n=1}^{N} \widehat{u}_n^2, \qquad (1018\text{a, b})$$

where $k$ denotes the modulus of the wave number $\vec{k}$.

(iii) *Helicity conservation*:

The helicity is defined by

$$\widetilde{\Omega}_\text{H} = k \left|\vec{\widetilde{u}}\right|^2 = k \sum_{n=1}^{N} \widehat{u}_n^2. \qquad (1019\text{a, b})$$

Because in this chapter, we are mainly concerned with 2-d turbulence, we shall not prove representations (1019a,b). The 3-d case shall be treated in future work elsewhere.

Notice that all these conserved quantities are quadratic forms of the velocity vector. However, only for the Euler equation they are strictly conserved. Then, for the discretized form, Kraichnan introduced a set of probability densities by

$$p_{\chi n} = \frac{e^{-\left(\alpha \widetilde{E}_{\chi n} + \beta\, \widetilde{\Omega}_{\chi n}\right)}}{Z_{\text{BG}\chi}}, \quad Z_{\text{BG}\chi} = \sum_{n=1}^{N} e^{-\left(\alpha \widetilde{E}_{\chi n} + \beta\, \widetilde{\Omega}_{\chi n}\right)}, \quad \chi \in \{\varepsilon, H\} \qquad (1020\text{a} - \text{d})$$

with the partition function $Z_{\text{BG}\chi}$. The generalized energy $\breve{E}_\chi$, and its average quantity $\langle \breve{E}_\chi \rangle$, is now defined as

$$\breve{E}_\chi = \alpha\,\widetilde{E}_\chi + \beta\,\widetilde{\Omega}_\chi, \quad \langle \breve{E}_\chi \rangle = \sum_{n=1}^{N} p_{\chi n}\,\breve{E}_{\chi n} = \frac{\sum_{n=1}^{N} \breve{E}_{\chi n}\, e^{-\breve{E}_{\chi n}}}{Z_{\mathrm{BG}\chi}}. \qquad (1021\mathrm{a}-\mathrm{d})$$

Because in Eqs. (1010a, 1016, 1018a,b, 1019a,b), $\widetilde{E}_\chi$ and $\widetilde{\Omega}_\chi$ are quadratic functions of $\widetilde{u}_i$, we know that this system of evolution equations, constituting a conservative dynamical system, is obeying Gaussian statistics.

A Fourier transform of $\breve{E}_\varepsilon$ for 2-d turbulence leads to the following generalized energy-enstrophy spectrum (see Refs. Frisch 1995; Kraichnan 1967 and for its derivation the Appendix E, Eq. (E-34)):

$$\text{2-d}: \qquad \widehat{\breve{\mathrm{E}}}_\varepsilon(k) = \frac{1}{2}\,\frac{1}{\alpha + \beta\,k^2}, \qquad (1022)$$

where $\alpha$ and $\beta$ are constants. On the other hand, a Fourier transform of $\breve{E}_H$ for 3-d turbulence leads to the following energy spectrum (see also Frisch 1995 and Kraichnan 1967):

$$\text{3-d}: \qquad \widehat{\breve{\mathrm{E}}}_H(k) \propto k^2, \qquad (1023)$$

in which $\widehat{\breve{\mathrm{E}}}_H(k)$ is the Fourier transform of $\breve{\mathrm{E}}_H$. Kraichnan calls expressions in Eqs. (1022) and (1023) the "mean modal intensity spectra." Then, the isotropic energy and enstrophy spectrum for 2-d turbulence follow by the formulae[3]

$$S_\chi(k) = \frac{\mathrm{d}\widehat{\breve{E}}_\chi}{\mathrm{d}k} \propto \frac{1}{k}\widehat{\breve{E}}_\chi, \quad \chi \in \{\varepsilon, H\}, \qquad (1024\mathrm{a}-\mathrm{d})$$

In the 2-d situation Eq. (1022) implies

$$k \ll \sqrt{\frac{\alpha}{\beta}} \Rightarrow \widehat{\breve{\mathrm{E}}}_\varepsilon(k) = \frac{1}{2\alpha} = \text{const}, \quad k \gg \sqrt{\frac{\alpha}{\beta}} \Rightarrow \widehat{\breve{\mathrm{E}}}_\varepsilon(k) = \frac{1}{2\beta}\,\frac{1}{k^2}. \quad (1025\mathrm{a}-\mathrm{c})$$

By applying Eq. (1024a), the energy and enstrophy spectrum for the low and high wave number regimes are

$$k \ll \sqrt{\frac{\alpha}{\beta}} \quad \Rightarrow \quad \mathrm{S}_\varepsilon(k) \equiv 0, \quad k \gg \sqrt{\frac{\alpha}{\beta}} \quad \Rightarrow \quad \mathrm{S}_\varepsilon(k) \propto \frac{1}{k^3}. \qquad (1026\mathrm{a,b})$$

[3]Equation (1024b) is only valid, if we are dealing with power laws (see later).

Whereas the energy spectrum for $S_\varepsilon(k)$ shows the useless value zero, the enstrophy spectrum is in excellent agreement with computer experiments (see Fig. 10.6). This suggests that at high wave numbers, the occurring small eddies are close to or even in quasi-thermodynamic equilibrium.

However, the experimenters state that their simulation results may be slightly erroneous because of aliasing and slow convergence of their calculations. This is a motivation for us to also generalize the results for the high wave number regime. Like this in future, if even higher precision experiments would be available, the power law exponent of this spectrum could be adjusted with a small energy leakage contribution. In any case $q$ values very close to "1" (see Sect. 10.4) and power law exponents close to -3, are expected to occur. Kraichnan called the high wave number regime "inverse cascade of two-dimensional turbulence" (Frisch 2011; Siggia and Aref 1981; Frisch and Sulem 1984).

Now, we ask ourselves: "What is the essential conclusion of this instructive spectral example derived by Kraichnan 50 years ago?" The answer to this question is

**Proposition 7** *By a Ritz–Galerkin truncation of the 2-d and 3-d Euler equations, these equations can be transformed to a finite system of evolution equations with two conserved quadratic forms that obey Gaussian statistics.*

Furthermore, we have corroborated and now state:

**Proposition 8** *Truncating the momentum equation and applying Boltzmann–Gibbs equilibrium statistical mechanics (thermodynamics) leads to a loss of essential features of small wave number turbulence. It is a too bold approach of modeling turbulent fields.*

Today, it is known that the entropy functional and the partition function, respectively, define universality classes showing identical critical exponents (see Egolf and Hutter 2017b). This means that a system close to its criticality exhibits a kind of universal behavior that is not determined by the specific microscopic interactions of neighboring particles. It was shown, for instance, that a liquid–gas transition and a ferromagnetic system (lattice gas and Ising model), in the region just below criticality, possess with high precision the same scaling exponents because they belong to the same universality class (Ma 1982; Stanley 1971). This agrees with the statement of Goldenfeld (1992) that phenomena with the same set of critical exponents are said to form a universality class. Therefore, one ought to distinguish between exact equivalences, where two models are analogous to one another (with a unique mapping from one to the other) and approximate equivalences, in which the partition functions may be (slightly) different, but nevertheless comprise two systems that show the same or similar behavior near criticality. According to Goldenfeld (1992), members of the same universality class have the following three properties in common:

1. The symmetry group of the Hamiltonian
2. The dimensionality of the physical problem
3. The range of the forces (short or long).

With these facts, we assert that the Boltzmann–Gibbs thermodynamics is inadequate to describe turbulence sufficiently accurately. In nonlinear dynamics (see Beck and Schlögl 1993) different approaches have been worked out, which are also presented in Tsallis (2009). They involve different entropy functionals, e.g., the Normalized entropy, Shannon entropy, Rényi entropy, Kolmogorov-Sinai entropy, and Escort entropy. Till date, not all questions have been answered concerning what kind of generalized entropy functional is optimal and which one best applies to a specific complex system.

## 10.4 An Introduction to the Nonextensive Thermodynamics of Tsallis

Below we demonstrate that yet a further entropy functional, called Tsallis entropy, is particularly optimal to describe turbulence. This entropy can be directly related to fractal geometry and Lévy flight statistics. Furthermore, the above class of entropies have many features in common and the Tsallis entropy enjoys simple relations with other entropies listed above.

Let us introduce the Tsallis entropy in a metaphoric manner (see Tsallis 2009). To this end, he aims for the derivation of Eq. (973) by considering the linear differential equation

$$\frac{\mathrm{d}N}{\mathrm{d}S_{\mathrm{BG}}} = \frac{1}{k}N, \quad N(0) = 1, \tag{1027a, b}$$

in which, for simplicity, we concentrate on the case of a single probability, i.e., $N = 1$; for this case

$$p = \frac{1}{N} \quad \Leftrightarrow \quad S_{\mathrm{BG}} = \langle S_{\mathrm{BG}} \rangle. \tag{1028a, b}$$

Writing Eq. (1027a) as

$$\frac{\mathrm{d}N}{N} = d(\log_{\mathrm{e}} N) = \frac{1}{k}\mathrm{d}S_{\mathrm{BG}}, \tag{1029a, b}$$

its integration yields

$$\log_{\mathrm{e}} N = \frac{1}{k}S_{\mathrm{BG}} + \log_{\mathrm{e}} C, \tag{1030}$$

in which $\log_{\mathrm{e}} C$ is the constant of integration. It follows that

$$N(S_{\mathrm{BG}}) = Ce^{\frac{1}{k}S_{\mathrm{BG}}}. \tag{1031}$$

With Eq. (1030) the constant $C$ is determined by expressing the statement of certainty (which is equivalent to $S_{\mathrm{BG}}(1) = 0$) as

$$S_{\mathrm{BG}}(1) = k \log_{\mathrm{e}}(1) - k \log_{\mathrm{e}}(C) = 0, \tag{1032a, b}$$

implying $C = 1$, so that, owing to Eq. (1030),

$$\langle S_{\mathrm{BG}} \rangle(N) = k \log_{\mathrm{e}} N, \tag{1033}$$

which is also the entropy of a system with elements having equal probabilities that was already anticipated by Eq. (973). With Eqs. (1028a) and (1033), this is generalized in the form

$$S_{\mathrm{BG}}(N) = p\langle S_{\mathrm{BG}} \rangle = -k\, p \log_{\mathrm{e}} p, \tag{1034a, b}$$

(compare with Eq. (971) for a single state $p$). Furthermore, Eq. (1031) reduces to

$$N(S_{\mathrm{BG}}) = e^{\frac{1}{k}S_{\mathrm{BG}}}. \tag{1035}$$

At this stage, the differential equation with its initial condition (1027a,b) is suggestive for a generalization of the Boltzmann–Gibbs entropy functional. Therefore, we now choose a nonlinear scaling exponent with value $q \in \mathbb{R}$ added to the linear differential Eq. (1027a), which is

$$\frac{\mathrm{d}N}{\mathrm{d}S_q} = \frac{1}{k} N^q, \quad N(0) = 1. \tag{1036a, b}$$

Writing (1029a) here alternatively as

$$\frac{\mathrm{d}N}{N^q} = \frac{1}{1-q} \mathrm{d}\left(N^{1-q}\right) = \frac{1}{k} \mathrm{d}S_q, \tag{1037a, b}$$

a straightforward integration yields the result

$$N^{1-q} = (1-q)\frac{1}{k} S_q + C, \tag{1038}$$

where $C$ is a constant of integration, which is determined to be $C = 1$, since for $N = 1$, $S_q = 0$. Thus, the general integral of (1036a) is

$$N(S_q) = \left[1 + (1-q)\frac{1}{k}S_q\right]^{1/(1-q)}. \tag{1039}$$

Now, we shall define a generalization of the exponential function called $q$-exponential function as suggested by Tsallis (2009) and given by Eq. (1039),

$$e_q^x := [1 + (1-q)\,x]^{1/(1-q)}, \tag{1040}$$

which is displayed in Fig. 10.2 for different values of $q$. In agreement with relation (1040) and $n = 1/(1-q)$ the well-known formula, defining the usual exponential function, emerges if the limit $q \to 1$ of Eq. (1040) is taken, viz,

$$e_1^x = \lim_{q\to 1}[1 + (1-q)x]^{1/(1-q)} = \lim_{n\to\infty}\left[1 + \frac{x}{n}\right]^n = e^x. \tag{1041a – c}$$

From Eqs. (1039) and (1041a–c), we draw the following inference:

$$N(S_1) = e^{\frac{1}{k}S_1} = e^{\frac{1}{k}S_{BG}} = N(S_{BG}). \tag{1042a – c}$$

For $q = 1$, the BG-entropy $S_{BG}$ and the Tsallis entropy $S_q$ are the same.
On the other hand, the inverse function of (1039) is

$$S_q(N) = k\frac{1}{1-q}\left(N^{1-q} - 1\right). \tag{1043}$$

This now suggests the definition of the $q$-logarithmic function

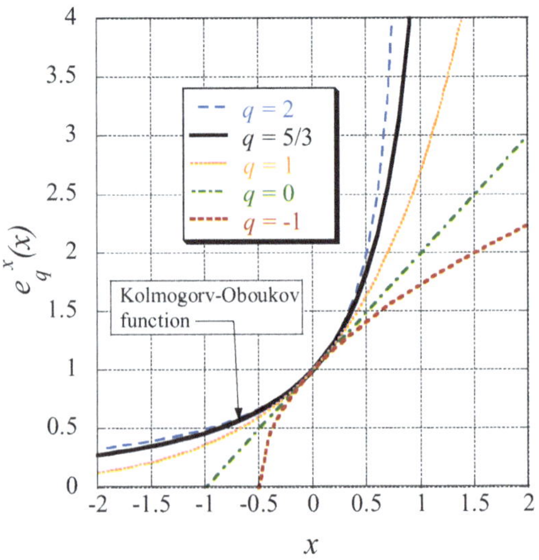


**Fig. 10.2** The $q$-exponential function $e_q^x$ for typical values of $q$. Courtesy of Tsallis (2009) © Springer Science and Business Media, reproduced with permission

$$\ln_q(x) := \frac{1}{1-q}\left(x^{1-q} - 1\right), \tag{1044}$$

where $\log_e x = \ln x$ is used to avoid a confusion between the $q$ exponent and the basis $b$ of the logarithmic function $\log_b$. The assignment (1044) also implies

$$\ln_q\left(\frac{1}{x}\right) := \frac{1}{1-q}\left(x^{q-1} - 1\right). \tag{1045}$$

Moreover, in agreement with Eq. (1044) and the assignment $n = 1/(1-q)$, the formula defining the usual logarithmic function is deduced from the following chain

$$\ln_1(x) = \lim_{q\to 1}\frac{1}{1-q}\left(x^{1-q} - 1\right) = \lim_{n\to\infty} n\left(x^{1/n} - 1\right) = \ln(x). \tag{1046a – c}$$

From the above presented metaphor and Eqs. (1043) and (1044), we postulate the following generalization of Eq. (973) for equal probabilities:

$$S_q(N) = k \ln_q(N). \tag{1047}$$

In Fig. 10.3, this function is displayed for $k = 1$ and different $q$-values.
With $p = 1/N$ Eq. (1047) is

$$S_q(p) = k \ln_q\left(\frac{1}{p}\right). \tag{1048}$$

Notice that it must be carefully checked which rules of the usual logarithm apply to its generalized form[4]; thus, we cannot write this equation in the form of Eq. (971) for a single probability. Therefore, the next step is to demand

$$S_q(N) = k\left[\ln_q\left(\frac{1}{p}\right)\right]. \tag{1049}$$

In this generalization process, arbitrary probabilities must be introduced as

$$S_q(N) = k\sum_{i=1}^{N} p_i \ln_q\left(\frac{1}{p_i}\right). \tag{1050}$$

In agreement with Eq. (1046a–c), we conclude that

[4]The relation $\ln_q(1/p)$ is not identical to $-\ln_q(p)$ unless $q = 1$.

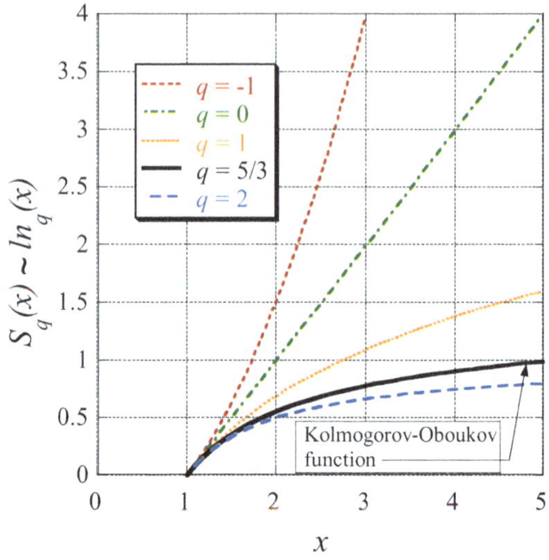


**Fig. 10.3** The equiprobability entropic functional $S_q(N)$ plotted against the number of states $N$ for $k = 1$ and different parameters $q$. Courtesy of Tsallis (2009) © Springer Science and Business Media, reproduced with permission

$$S_1 = S_{\mathrm{BG}} \tag{1051}$$

and, therefore, as already stated after Eq. (1042a–c), equilibrium thermodynamics of the Boltzmann–Gibbs type is a special case of the, thus, proposed Tsallis nonequilibrium thermodynamics.
Inserting definition (1045) of the $q$-logarithmic function into (1050) leads to

$$S_q(N) = \frac{k}{1-q} \sum_{i=1}^{N} p_i \left(p_i{}^{q-1} - 1\right), \tag{1052}$$

which, with the help of Eq. (972), is identical to

$$S_q(N) = k \, \frac{1 - \sum\limits_{i=1}^{N} p_i{}^q}{q-1}. \tag{1053}$$

This $q$-entropy or Tsallis entropy generalizes the entropy of an equilibrium to that of a nonequilibrium thermodynamic system, respectively, from systems with linear to those with nonlinear and complex behavior; the latter, e.g., suitable for turbulence. Notice that this entropy initially was introduced by ideas of fractal geometry.

It follows that

$$\begin{aligned} q < 1: \quad &\Leftrightarrow \quad p_i^q > p_i, \\ q = 1: \quad &\Leftrightarrow \quad p_i^q = p_i, \\ q > 1: \quad &\Leftrightarrow \quad p_i^q < p_i. \end{aligned} \tag{1054a – c}$$

It is helpful to introduce a new set of probabilities by defining

$$P_i = \frac{p_i^q}{Z_{\mathrm{ES}}}, \quad Z_{\mathrm{ES}} = \sum_{i=1}^{N} p_i^q; \tag{1055a, b}$$

the full set $\{P_i\}$ is called *escort probabilities*. They obviously also fulfill the normalization condition

$$\sum_{i=1}^{N} P_i = 1. \tag{1056}$$

The condition $q < 1$ makes the escort probabilities larger than the $p_i's$ and $q > 1$ makes them smaller. Therefore, $q < 1$ enhances the occurrence of rare or large-scale events and $q > 1$ enhances those of frequent or small-scale events. We experience here a connection with anomalous diffusion (treated in Sect. 7.2.4), where the case $q < 1$ corresponds to subdiffusivity and subadditivity, $q = 1$ describes the standard Brown diffusion, obeying Gaussian statistics, and the case $q > 1$ belongs to superdiffusivity (Table 10.1) and superadditivity.

We will identify more quantitative relations between anomalous diffusion, Lévy statistics, and extensive thermodynamics later in this section. When $q < 1$, it is important that summations exclude all probabilities of zero value, whereas for $q > 1$ the summation may include all occurring probabilities.

Next, let us investigate the additivity of the Tsallis entropy. Then, from Eq. (1044), we conclude that

$$(1-q)\ln_q(x_A x_B) = \left[(x_A x_B)^{1-q} - 1\right]. \tag{1057}$$

This is identical to the following extended mathematical expression:

$$\begin{aligned} &(x_A x_B)^{1-q} - 1 = \left(x_A{}^{1-q} - 1\right) + \left(x_B{}^{1-q} - 1\right) \\ &+(x_A x_B)^{1-q} - x_A{}^{1-q} - x_B{}^{1-q} + 1 = \left(x_A{}^{1-q} - 1\right) \\ &+\left(x_B{}^{1-q} - 1\right) + \left(x_A{}^{1-q} - 1\right)\left(x_B{}^{1-q} - 1\right), \end{aligned} \tag{1058a, b}$$

which is now rewritten to become

**Table 10.1** The three main types of diffusivity (sub, normal, and super diffusivity), listed with a number of application examples (Shlesinger et al. 1987)

| | |
|---|---|
| Subdiffusivity (dispersive diffusion) | Disordered media, fractal structures, trapping in condensed matter, sticking, glasses, etc. |
| Normal diffusivity (Brown's diffusivity) | Gases, molecules, tracers, heat transfer, etc. |
| Superdiffusivity | Chaotic dynamics, phase diffusion in chaotic regimes of Josephson junctions, turbulence, etc. |

$$(x_A x_B)^{1-q} - 1 = (1-q)\ln_q(x_A)$$
$$+(1-q)\ln_q(x_B) + (1-q)\ln_q(x_A)\,(1-q)\ln_q(x_B). \tag{1059}$$

With the help of Eq. (1044), these expressions are transformed back to Tsallis entropies. This yield

$$\frac{S_q(A+B)}{k} = \frac{S_q(A)}{k} + \frac{S_q(B)}{k} + (1-q)\frac{S_q(A)}{k}\frac{S_q(B)}{k}, \tag{1060}$$

which demonstrates nonadditivity of $S_q$ for $q \neq 1$. Depending on $q$, it follows that

$$\begin{aligned} q &< 1: \quad S_q(A+B) > S_q(A) + S_q(B), \\ q &= 1: \quad S_q(A+B) = S_q(A) + S_q(B), \\ q &> 1: \quad S_q(A+B) < S_q(A) + S_q(B). \end{aligned} \tag{1061a – c}$$

If $q < 0$ $S_q$ is convex whereas for equal probabilities, $p_1 = p = 1/2$ and $p_2 = 1-p = 1/2$, it takes a minimum. On the other hand, for $q > 0$ it is concave with a maximum for equal probabilities in this case (see Fig. 10.4). *Slowly, it becomes transparent that nonextensive thermodynamics is the appropriate framework to tackle anomalous diffusion* [and as we have demonstrated in Sect. 7.2.4 also turbulence] (see Budde et al. 2001).

## 10.5 Relation Between Lévy Statistics and Tsallis Nonextensive Thermodynamics

To make the entire situation more transparent, we now outline the connections between Lévy flight dynamics and the Tsallis nonextensive and nonadditive thermodynamics (see Table 10.2).

The mean square displacement of diffusive problems is generalized by setting (Zumhofen et al. 1989)

$$\langle x^2(t) \rangle \propto t^\alpha, \tag{1062}$$

with the time $t$ and the exponent $\alpha = 1$ describing Brown diffusion (Table 10.3).

After Alemany and Zanette (1994), the jump probability distribution can be obtained from Eq. (971) by maximizing the entropy $S_{BG}$, now written as a Lebèsque integral

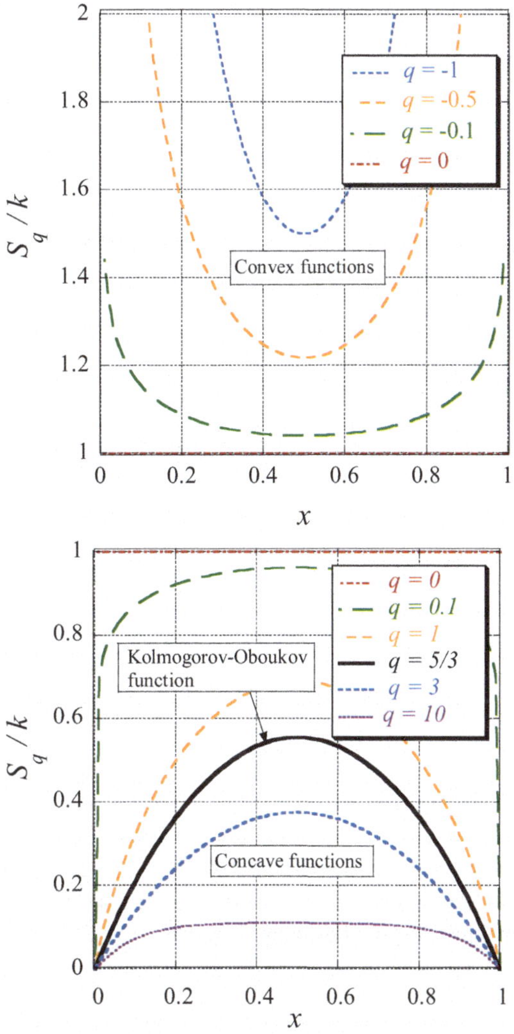


**Fig. 10.4** The $q$ entropy for $N = 2$ for negative and positive $q$ values. We stress that all $q$ entropies are strictly convex or concave, a feature that not all generalized entropies (e.g., Renyi and Escort) fulfill Courtesy of Tsallis (2009) © Springer Science and Business Media, reproduced with permission

**Table 10.2** Physical systems with conventional linear (column 2) and with nonlinear behavior (column 3)

| Scientific term | Linear | Nonlinear |
|---|---|---|
| Statistics | Gaussian | Non-Gaussian, Lévy distribution |
| Variance | Finite | Infinite |
| Scales | Single | Manifold |
| Diffusion | Brown | Lévy |
| Thermodynamics | Equilibrium | Nonequilibrium |
| Entropy | Additive | Nonadditive |

**Table 10.3** Characteristic values for diffusion domains

| Diffusion | $\alpha$ | $h$ | $\gamma_L$ | $q$ |
|---|---|---|---|---|
| Sub | $<1$ | $<1/2$ | $=2$ | $0 < q < 1$ |
| Normal | $=1$ | $=1/2$ | $=2$ | $=1$ |
| Super | $>1$ | $>1/2$ | $=2$ | $1 < q < 5/3$ |
| Super-K | $>1$ | $>1/2$ | $=2$ | $q = 5/3$ |
| Super-L | $>1$ | $>1/2$ | $<2$ | $5/3 < q < 3$ |

Notice that $h$ denotes the Hurst exponent, K the Kolmogorov, and L the Lévy flight regime

$$S_{\mathrm{BG}}[p\,(x)] = -k\int_{\Gamma} p(x)\log_{\mathrm{e}} p(x)\,\mathrm{d}x, \tag{1063}$$

subject to the two auxiliary constraints, namely

$$\int_{\Gamma} p(x)\,\mathrm{d}x - 1 = 0 \tag{1064}$$

and

$$\int_{\Gamma} x^2 p(x)\,\mathrm{d}x - \sigma^2 = 0. \tag{1065}$$

The quantity $\sigma$ plays the role of a kind of standard deviation. For simplicity, we have chosen the 1-d representation. For higher dimensions, the considerations are analogous (see also Alemany and Zanette 1994). The problem of maximizing the entropy (1063), subject to the constraints (1064) and (1065) is technically performed by maximizing the Lagrange function

$$\begin{aligned}\Phi[p\,(x)] = -k\int_{\Gamma} p(x)\log_{\mathrm{e}} p(x)\,\mathrm{d}x - \lambda_1\left[\int_{\Gamma} p(x)\,\mathrm{d}x - 1\right] \\ - \lambda_2\left[\int_{\Gamma} x^2 p(x)\,\mathrm{d}x - \sigma^2\right],\end{aligned} \tag{1066}$$

in which $\lambda_1$ and $\lambda_2$ are Lagrange parameters. The solution of this extremum principle determines these Lagrange parameters as well as the jump probability distribution, which takes the form

$$p(x) = A\ \exp\left(-\frac{x^2}{2\sigma^2}\right) \tag{1067}$$

where for details it is suggested to consult, e.g., the book on fluctuation theory by Montrol and Lebowitz (1976).

A Fourier transform for small wave number $k$ (large wave length $\lambda$) leads to the characteristic function (see Alemany and Zanette 1994) of the following form:

$$G(k) \approx 1 - \frac{1}{2}\sigma^2 k^2 + \ldots. \tag{1068}$$

With the characteristic function of Lévy flights, this is now generalized as

$$G(k) = \exp\left(-ck^{\gamma_L}\right) \approx 1 - ck^{\gamma_L} + \ldots \tag{1069a, b}$$

An inverse Fourier transform of this equation reveals the asymptotic Lévy distribution appropriate for large distances, with the scaling property

$$p(x) \propto x^{-\gamma_L - 1}. \tag{1070}$$

This is called the long tail of the Lévy probability distribution. To be precise, the approximate large distance (small $k$) Fourier transform of (1067) is (1068), a result of the linear equilibrium theory, which is then made nonlinear in (1069), and this approximation in Fourier space is assumed to be representative for the longtail behavior of the Lévy probability distribution expressed as Eq. (1070).

The same idea of applying an extremum principle and by taking into account the generalized constraints (1064) and (1065), namely

$$\begin{aligned} &\int_\Gamma p(x)^q dx - \lambda = 0, \\ &\int_\Gamma x^2 p(x)^q dx - \sigma_q^2 = 0, \end{aligned} \tag{1071a, b}$$

with the $q$-entropy (see Eq. (1053)), viz,

$$S_q[p\,(x)] = -k\frac{1 - \int_\Gamma p(x)^q\,dx}{1 - q}, \tag{1072}$$

leads to the result (Alemany and Zanette 1994)

$$p(x) = \left[\frac{kq}{\lambda_1(q-1)} + \frac{\lambda_2 q}{\lambda_1} x^2\right]^{1/(1-q)}, \quad (1073)$$

with the two Lagrange multipliers $\lambda_1$ and $\lambda_2$, which here are not important for our purposes and therefore not explicitly listed (for more information see Sect. 10.14 and Shlesinger et al. 1995).

For large distances $x$, setting equal the exponents of the formulas Eqs. (1070) and (1073) yields

$$\gamma_L = \frac{3-q}{q-1}, \quad q = \frac{3+\gamma_L}{1+\gamma_L}. \quad (1074\text{a, b})$$

These two relations directly connect Lévy flight statistics with the thermodynamic formalism of Tsallis. The corresponding characteristic parameters are presented in Table 10.3.

## 10.6 Escort Probability Distribution and Expectation Values

Shlesinger et al. (1987) introduced a Lévy flight distribution, where the jump size and jump time are related by a velocity in a more general manner than in our specific turbulence-related approach. However, interesting in the present context is that they prove that the Lévy flight probability distribution can be written in a formulation of the problem in which the dynamics and the physical observables are described by an integral equation in which the multiplicity of scales is explicitly considered. Furthermore, they also find that: "*The basic transport equation involves an integral equation with a coupled (scaled) memory nonlocal in space and time.*" Even though their general way of writing the theory is not presented here (the interested reader may consult the reference given above), it is evident that their approach is in agreement with our work presented in Sects. 7.2.4 and 7.3.

The entropy optimization method, as applied by Kraichnan (1967) for BG thermodynamics, shall now analogously be applied to the Tsallis thermodynamic theory because it is a prerequisite to perform thermodynamics with generalized entropic forms (see, e.g., Tsallis 2009). With small deviations, we follow mainly this reference.

Let us now, in analogy to Eq. (1055a,b), introduce the continuous escort probability distribution

$$P(x) = \frac{[p(x)]^q}{Z_{\mathrm{ES}}}, \quad Z_{\mathrm{ES}} = \int_0^\infty [p(x)]^q \mathrm{d}x, \tag{1075a, b}$$

in which the integral is introduced as a Lebèsque integral with continuous measure. The distribution is also normalized, viz,

$$\int_0^\infty P(x)\, \mathrm{d}x = 1. \tag{1076}$$

Furthermore, by defining the $q$-average value of a function $f$ as

$$\langle f(x) \rangle_q = \int_0^\infty f(x)P(x)\, \mathrm{d}x = F_q, \tag{1077a, b}$$

it follows for $f = x$, that

$$\langle x \rangle_q = \int_0^\infty xP(x)\, \mathrm{d}x = X_q, \tag{1078a, b}$$

which is the $q$-average value of $x$.

To optimize the entropy distribution $S_q[P]$ (adapted from Eq. 1066), we define the Lagrange function $\Phi$, containing the two constraints, Eqs. (1076) and (1077b), for the energy function being given by $f(x) = E(x)$. This Lagrangian is

$$\Phi[P] = S_q[P] - \lambda_1 \left[ \int_0^\infty P(x)\, \mathrm{d}x - 1 \right] - \lambda_2 \left[ \int_0^\infty P(x)E(x)\, \mathrm{d}x - E_q \right]. \tag{1079}$$

Now, we substitute the $q$ entropy, Eq. (1072),

$$\Phi[P] = k \frac{1 - \int_0^\infty P(x)\mathrm{d}x}{q-1} - \lambda_1 \left[ \int_0^\infty P(x)\mathrm{d}x - 1 \right] - \lambda_2 \left[ \int_0^\infty P(x)E(x)\, \mathrm{d}x - E_q \right], \tag{1080}$$

where the denominator $q-1$ is a quantitative measure of “Non-Boltzmann–Gibbs behavior.”

Maximization is obtained by performing the optimization $\mathrm{d}\Phi[P]/\mathrm{d}P = 0$; it leads to the optimum result $P_{\max}$, where, upon dropping the index, one obtains (see Tsallis 2009)

$$P[E(x)] = \frac{e_q^{-\beta_q\,[E(x)-E_q]}}{\int_0^\infty e_q^{-\beta_q\,[E(x)-E_q]}\mathrm{d}x}, \tag{1081}$$

in which $E_q$ is the average energy. Its definition will be given below (see Eq. (1085)).

The optimization process determines the generalized Boltzmann factor $P[E(x)]$, now called Tsallis factor. One can show that the Lagrange multiplier $\lambda_1$, due to Eq. (1076), factorizes out.

Now, a $q$-generalized partition function is defined by

$$Z_q := \int_0^\infty e_q^{-\beta_q\,[E(x)-E_q]}\mathrm{d}x. \tag{1082}$$

When applying definitions (1040) and (1082), (1081) is found to be given by

$$P[E(x)] = \frac{\left\{1 + (1-q)\beta_q\left[E(x) - E_q\right]\right\}^{1/(1-q)}}{Z_q}. \tag{1083}$$

Next, the averaged energy is calculated by

$$\langle E(x)\rangle_q = E_q = \int_0^\infty P[E(x)]\,E(x)\mathrm{d}x. \tag{1084a, b}$$

Substituting Eq. (1083) into (1084b) leads to

$$E_q = \frac{\int_0^\infty \left\{1 + (1-q)\beta_q\left[E(x) - E_q\right]\right\}^{1/(1-q)} E(x)\,\mathrm{d}x}{Z_q}. \tag{1085}$$

## 10.7 Generalized Thermodynamic Potentials

Tsallis (2009) shows that his nonequilibrium thermodynamics with natural $q$-generalizations allows the application of the Legendre transformation technique to derive different potentials, e.g., the free energy $F_q$, the Gibbs free energy $G_q$ and the Helmholtz free energy $H_q$. Therefore, it is concluded that the Legendre formalism is $q$-invariant. Furthermore, for the free energy, $F_q$, e.g., it follows, in analogy to a well-known classical BG relation (see, e.g., Reif (1975)) that

$$F_q = U_q - TS_q = -\frac{1}{\beta_q} \ln_q Z_q. \tag{1086a, b}$$

We recognize that well-known formulas of equilibrium thermodynamics keep their validity in the context of nonextensive irreversible thermodynamics, if the usual symbols are replaced by the $q$-quantities. An interesting example, which reveals the generalized temperature of a turbulent flow, $T_q$, as

$$\frac{1}{T_q} = \frac{\partial S_q}{\partial U_q}, \quad U_q = \int_0^\infty P(E)E(x)\,\mathrm{d}x, \tag{1087a, b}$$

is derived with the $q$-internal energy $U_q$. In Boltzmann–Gibbs systems, the temperature usually is the stress parameter of the system. Therefore, Eq. (1087a) may reveal a generalized temperature $T_q$, that can be related to the inverse overall Reynolds number, Re (see Chap. 11). Furthermore, the $q$-generalized specific heat is given by (for the BG formula and its derivation see, e.g., Reif 1975)

$$C_q = \frac{\partial U_q}{\partial T_q} = T_q \frac{\partial S_q}{\partial T_q} = -T_q \frac{\partial^2 F_q}{\mathrm{d}T_q^2}. \tag{1088a – c}$$

However, here we have now reached the boundary of the present knowledge of this specific section; we leave further nonequilibrium thermodynamic developments of the $q$-concept for future investigations (as far as they are not already to be found in Tsallis 2009).

## 10.8 Fractional Calculus: A Promising Future-Oriented Method to Describe Turbulence

In analogy to Kraichnan's calculus of the energy-enstrophy spectrum, to obtain the $q$-generalized versions of Eqs. (1022), (1025), and (1026), we will apply fractional calculus. To be able to generalize the steps between these equations, the Fourier transform of a derivative and its square must be determined. Therefore, we deal here in a general manner with fractional derivatives. The reason is that the common Riemannian derivative, in our case, insufficiently describes turbulent phenomena. The adequate new tools are fractional derivatives and integrals. Such were, e.g., also applied with success by the present authors to model turbulent shear flows (see, e.g., Egolf and Hutter 2016a, 2018).

In Fourier calculus, the following main principle holds:

$$k^m\left(i\frac{\mathrm{d}}{\mathrm{d}k}\right)^n\widehat{f}(k)=F\left[\left(-i\frac{\mathrm{d}}{\mathrm{d}x}\right)^m x^n f(x)\right], \tag{1089a, b}$$

where the quantities $\widehat{f}$ and $F$ denote the Fourier transforms of the argument function. Now, we study the special case $n = 0$, which yields:

$$k^m\widehat{f}(k)=F\left\{(-i)^m\frac{\mathrm{d}^m f(x)}{\mathrm{d}x^m}\right\}. \tag{1090a, b}$$

We observe that by a Fourier transform, a derivative of order $m$ of a function $f(x)$ is transformed to a power law of the wave number $k$, with exponent $m$, times the Fourier transformed function. This is the basic rule to generalize the usual derivative to the Fourier fractional derivative of order $m = q$:

$$F\left\{(-i)^q\frac{\mathrm{d}^q f(x)}{\mathrm{d}x^q}\right\}=k^q\widehat{f}(k),\quad (-i)^q=e^{-i\frac{\pi}{2}q}. \tag{1091a, b}$$

The rule for usual integer-order derivatives and power laws with integer exponents is generalized in fractional calculus to real and complex numbers (see, e.g., Herrmann 2011, 2014); in the context of this book real-order derivatives are sufficient. By applying the Fourier fractional derivative to a power law, one finds (see, e.g., Herrmann 2014):

$$\frac{\mathrm{d}^q}{\mathrm{d}x_q}x^\alpha=\frac{\Gamma(q+\alpha)}{\Gamma(\alpha)}x^{\alpha-q},\quad x>0, \tag{1092}$$

in which $\Gamma$ denotes the gamma function.
Now, we set $q = 1$, which yields:

$$\frac{\mathrm{d}}{\mathrm{d}x}x^\alpha=\frac{\Gamma(1+\alpha)}{\Gamma(\alpha)}x^{\alpha-1},\quad x>0. \tag{1093}$$

With the formula (see Bronstein and Semendjajew 1981)

$$\frac{\Gamma(1+\alpha)}{\Gamma(\alpha)}=\alpha, \tag{1094}$$

Equations (1093) and (1094) yield

$$\frac{\mathrm{d}}{\mathrm{d}x}x^\alpha=\alpha\, x^{\alpha-1},\quad x>0, \tag{1095}$$

which is the usual Riemannian differentiation rule for a power law with exponent $\alpha$. In fractional calculus, there is no consensus on the differentiation of a constant. There are essentially two definitions:

$$\frac{\mathrm{d}^q}{\mathrm{d}x^q}\mathrm{const} = 0 \quad \text{and} \quad \frac{\mathrm{d}^q}{\mathrm{d}x^q}\mathrm{const} = c\,x^{0-q} = c\,x^{-q},\ c = c(\Gamma(q)), \qquad (1096\mathrm{a}-\mathrm{d})$$

where definition (1096a) is due to Caputo's fractional derivative (see Herrmann 2014) and Eq. (1096c) is the solution for a Riemannian fractional derivative (Herrmann 2014) with a constant $c$ being a function of Gamma functions and is motivated by Eq. (1092) with $\alpha = 0$. As the first definition leads also to the useless result (1026a), derived with BG thermodynamics, the second one is very promising and, therefore, leads us with some pragmatism to give it preference (see Sect. 10.11).

## 10.9 Jackson's Fractional Derivative and the DQTM

Instead, we shall now show that a nonlocal derivative, named the Jackson derivative, is ideal to describe, on the one hand, the DQTM, and, on the other hand, also the Tsallis entropy. Its derivation inherently shows scale-invariance stemming from fractal geometrical and fractional dynamical properties.

More than 100 years ago, Jackson (1909, 1910) generalized the local differential quotient d/d$x$ by a nonlocal operator $\mathrm{D}_\alpha$ acting on a function $f$ as follows:

$$\mathrm{D}_\alpha f(x) = \frac{f(\alpha x) - f(x)}{\alpha x - x}. \qquad (1097)$$

At first, we step back and introduce the stretch and compress difference operator

$$\mathrm{C}_\alpha f(x) = f(\alpha x) - f(x). \qquad (1098)$$

Then it is clear that

$$\mathrm{D}_\alpha f(x) = \frac{\mathrm{C}_\alpha f(x)}{\alpha x - x}. \qquad (1099)$$

Now, in the DQTM (307b), we can write the average velocity difference

$$\begin{aligned}\overline{u}_1(x_2) - \overline{u}_{1\,\min} &= \overline{u}_1(x_2) - \overline{u}_1(x_{2\,\min}) = \overline{u}_1(x_2) - \overline{u}_1(\alpha x_2) = -[\overline{u}_1(\alpha x_2) - \overline{u}_1(x_2)] \\ &= -\mathrm{C}_\alpha[\overline{u}_1(x_2)],\end{aligned} \qquad (1100\mathrm{a}-\mathrm{d})$$

and the difference quotient by

$$\frac{\overline{u}_{1\,\max} - \overline{u}_1}{x_{2\,\max} - x_2} = \frac{\overline{u}_1(x_{2\,\max}) - \overline{u}_1(x_2)}{x_{2\,\max} - x_2} = \frac{\overline{u}_1(\beta\, x_2) - \overline{u}_1(x_2)}{\beta x_2 - x_2} = \mathrm{D}_\beta[\overline{u}_1(x_2)]. \tag{1101a – c}$$

We have replaced $f$ in (1099) by $\overline{u}_1$ and $x$ by $x_2$. Furthermore, the following variables were introduced:

$$\alpha = \frac{x_{2\,\min}}{x_2}, \quad \beta = \frac{x_{2\,\max}}{x_2}. \tag{1102a, b}$$

Equations (1100d) and (1101c) allow to write the DQTM in an elegant way, namely as

$$\tau_{21} = -\rho\sigma\chi_2\, \mathrm{C}_\alpha[\overline{u}_1(x_2)]\, \mathrm{D}_\beta[\overline{u}_1(x_2)]. \tag{1103}$$

Notice that $\chi_2$ is a nonlocal constant or variable. This comprehensive manner of writing the DQTM with two nonlocal and fractional operators would also suit well into Sect. 7.3.7.

Now, let us apply this nonlocal formalism to the Tsallis entropy. Abe (1997) noticed the following remarkable property of the Gibbs–Boltzmann entropic functional:

$$S_{\mathrm{BG}} = -k\frac{\mathrm{d}}{\mathrm{d}x}\sum\nolimits_{i=1}^{N} p_i^x\Big|_{x=1}. \tag{1104}$$

Below a more general treatment will also prove this relation.

Next, we write for the Tsallis entropy (1053)

$$\begin{aligned} S_q &= k\,\frac{1-\sum\limits_{i=1}^{N} p_i^q}{q-1} = k\,\frac{\sum\limits_{i=1}^{N} p_i - \sum\limits_{i=1}^{N} p_i^q}{q-1} = k\,\frac{\sum\limits_{i=1}^{N} p_i^x - \sum\limits_{i=1}^{N} p_i^{qx}}{qx-x}\Bigg|_{x=1} = -k\sum_{i=1}^{N}\frac{p_i^{qx}-p_i^x}{qx-x}\Bigg|_{x=1} \\ &= -k\sum_{i=1}^{N} \mathrm{D}_q\left[p_i^x\right]\big|_{x=1}, \end{aligned} \tag{1105a – e}$$

and since the operations $\sum\limits_{i=1}^{N}$ and $\mathrm{D}_q$ commute, we find the final result

$$S_q = -k\,\mathrm{D}_q\left(\sum_{i=1}^{N} p_i^x\big|_{x=1}\right), \tag{1106}$$

which for $q = 1$ also proves Eq. (1104). Tsallis (2009) writes: "*We consider this as an inspiring property, where the usual infinitesimal translational operation is replaced by a finite operation, namely, in this case, by that one which is basic for scale-invariance."*

At the beginning of this section, we experienced that Kraichnan (see Kraichnan and Montgomery (1980)) recognized that BG-thermodynamics is insufficient to describe the scaling laws in numerous vortex systems, involving turbulence. Similar insufficiencies were reported in stellar systems and black holes, which are not the topic of this review. Then, we demonstrated that Tsallis nonequilibrium thermodynamics is well suited to describe turbulent phenomena with a close relation to non-Gaussian statistics.

## 10.10 Beck-Tsallis Thermodynamics of Turbulence

Therefore, it is intriguing to follow a first attempt, where the Tsallis thermodynamic concept was applied to turbulence, an idea put forward by Beck et al. (2001). They assumed that the velocity difference

$$u(r) = u_1(x_1 + r) - u_1(x_1) \tag{1107}$$

in fully developed Taylor–Couette flows (see beginning of Sect. 9.3) extremizes the Tsallis generalized entropy $S_q$. This led these authors to formulate a stationary velocity difference probability distribution $p(u)$. This distribution was in the focus of an experimental validation method that showed excellent agreement between experiment and theory (see Fig. 10.5). The comparison of the theoretical with the

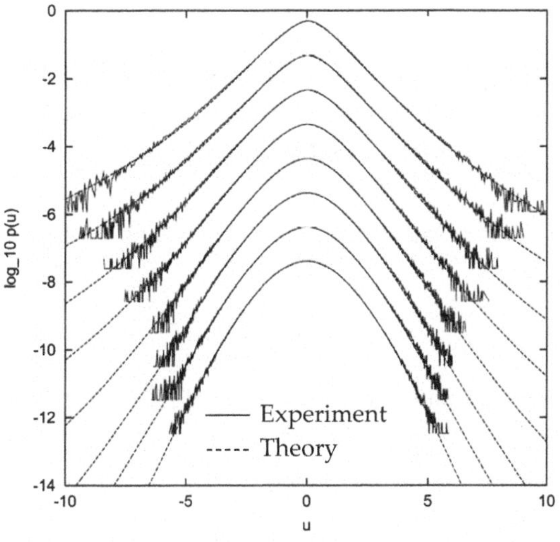


**Fig. 10.5** Probability density functions of the velocity difference for the Taylor–Couette flow experiment at Re = 540,000, compared with the nonextensive thermodynamic theory. The distances $r/\eta$ are from the top to the bottom 11.6, 23.1, 46.2, 92.5, 208, 399, 830, and 14.400, where $\eta$ is the Kolmogorov dissipation length. The $q$-values vary only very slightly from top to bottom, namely between 1.168 and 1.038. Courtesy Beck et al. (2001) © Phys. Rev. E, reproduced with permission by APS

experimental results leaves no doubt that this concept is just perfect for a description of turbulent flows. For more details, see Beck et al. (2001).

Slightly different than in previous approaches with a truncation of Fourier components, our generalized theory seems to be quite insensitive to the limitation of wave numbers of this method and reveals convincing energy spectra. Our view, based on Tsallis' nonextensive thermodynamics, is that turbulent flows, including the Kolmogorov–Oboukov description, are per se always nonequilibrium thermodynamic phenomena.

In many presentations of nonequilibrium thermodynamics, the linear laws for driving forces (e.g., Fick's law, Fourier's law, and Ohms law) are retained. Furthermore, many concepts concern linear responses close to equilibrium, e.g., the fluctuation dissipation theorem (e.g., Grenard et al. 2008), originally introduced by Nyquist. We know from our considerations in Chaps. 4 and 5 that linear laws as driving forces for fluxes, even with nonlinear modifications added to the pre-factor, and small linear expansions from equilibrium are no option to develop useful descriptions of turbulent flows. So, knowing about many valuable results of these alternative theories, it was necessary to again break with conventional ideas and methods based on linearization. Therefore, our decision was to cope with an innovative and new nonequilibrium thermodynamic formulation, specifically the nonextensive thermodynamics of Tsallis. Its well-defined comprised form with direct and clear relations to the features of our fluid dynamic systems (nonlinearity, non-Gaussian characteristics, Lévy statistics, fractal behavior, scale invariance, power laws, etc.) and the formal conservation of the thermodynamic formalism of the Boltzmann–Gibbs type was the reason to integrate this new concept into our work. Further, valuable results are expected to be revealed by research following in this and similar directions.

## 10.11 Fractional Generalization of Kraichnan's Energy-Enstrophy Spectrum and its Validation by Numerical Experiments

With such knowledge, Eq. (1024a) is now $q$-generalized in the following manner:

$$S_\varepsilon(k) = \mathrm{D}_k^{(q)} \widehat{\breve{E}}(k) = \frac{\mathrm{d}^q}{\mathrm{d}k^q} \widehat{\breve{E}}(k). \tag{1108a, b}$$

Now, a Fourier transform of $\breve{E}_\varepsilon$ for 2-d turbulence leads to the following generalized energy-enstrophy spectrum (its proof is outlined in Appendix E, Eq. (33))

$$\widehat{E}_\varepsilon(k) = \frac{1}{2}\frac{1}{Z_{\mathrm{ES}\varepsilon}}\frac{1}{q^2}\frac{1}{\alpha+\beta\,k^{2q}}. \tag{1109}$$

Two wave number regions emerge from this formula:

(A) The first is the small wave number regime, for which, from Eq. (1109), it follows that

$$\widehat{E}_\varepsilon(k) = \frac{1}{2\alpha}\frac{1}{Z_{\mathrm{ES}\varepsilon}}\frac{1}{q^2} = \mathrm{const.} \tag{1110a, b}$$

Now, by applying Eq. (1096c), one calculates the fractional derivative of this constant

$$S_\varepsilon(k) = \frac{1}{2\alpha}\frac{1}{Z_{\mathrm{ES}\varepsilon}}\frac{1}{q^2}c(\Gamma)\frac{1}{k^q} = \mathrm{const}\frac{1}{k^q}, \tag{1111a, b}$$

which is different from Eq. (1026a). Notice that in Eq. (1111b), with $q = 1$, a reciprocal energy spectrum does not result, because in this case $c = 0$ will guarantee agreement with the usual differentiation, so that (1026a) also occurs as special case of this new treatment. This would be a wave number spectrum related to the thermodynamic equilibrium of a turbulent flow field at small wave numbers, which does not exist!

However, as we can see in the third line of Table 10.4, there is another special case, namely that corresponding to super diffusion, which is characterized by $q = 5/3$, leading to the well verified Kolmogorov–Oboukov energy spectrum:

$$S_\varepsilon(k) = \mathrm{const}\,k^{-5/3}. \tag{1112}$$

This result is in best agreement with computer experiments for 2-d turbulence performed and reported by Lilly (1969) and reviewed by Kraichnan (1967). Let us summarize:

**Proposition 8** *Tsallis' nonextensive thermodynamics generalizes Kraichnan's constant turbulent energy by a power law energy spectrum with exponent $-q$, which relates this generalized spectrum directly with the Kolmogorov–Oboukov $-5/3$ law*

**Table 10.4** Characteristic values of the three diffusion domains, some of their main contributors and related thermodynamics

| $q$ | Main contributors | Thermodynamics |
|---|---|---|
| 1 | Kraichnan | Quasi-equilibrium |
| 5/3 | Kolmogorov–Oboukov | Nonequilibrium |
| >5/3 | Lévy–Frisch | Nonequilibrium |

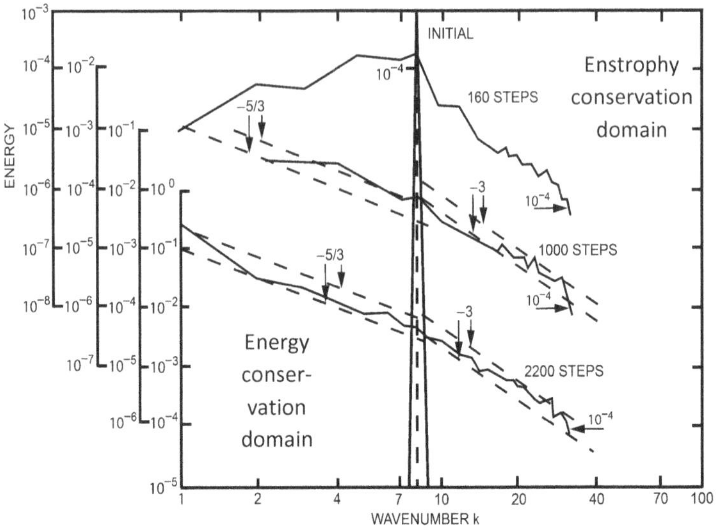


**Fig. 10.6** For comparisons of the power laws describing the spectra of the inertial energy and the enstrophy range, early computer simulations of Lilly (1969) have already been cited and interpreted by Kraichnan (1967). These numerical experiments show a convergence toward the predicted values of the power law exponents, which for the highest computation time (2200 time steps) are at low wave number close to –5/3 and at high wave number close to the equilibrium value −3. Reproduced from Lilly, 1969, Physics of Fluids 12, II-240 (1969), with the permission of AIP Publishing [This double regime spectrum was also confirmed by newer DNS results (see Boffetta 2007 and references therein)]

*and its fractal generalization, e.g., given by the fractal $\beta$-model (see, e.g., Frisch* 1995*), describing turbulent flow fields with intermittency* (see below).

With Kolmogorov's assumption that the energy spectrum depends only on the wave number modulus $k$ and the dissipation rate $\varepsilon$, the exponent −5/3 power law is obtained for two as well as for three-dimensional turbulence (see Hunt et al. 1991 and Fig. 10.6). Notice that this spectrum is valid only at low wave numbers up to an intersecting value $k_I$ after which a $k^{-3}$ power law (see also Fig. 10.6) is observed that dominates the behavior up to Kolmogorov's dissipation wave number (Kraichnan 1967; Hunt et al. 1991).

Furthermore, notice also that a Kolmogorov–Oboukov turbulent flow is thermodynamically equally a nonequilibrium process. For Re $\to \infty$, turbulent fields converge toward a state without intermittency. Tsallis (2009) wrote that chaotic systems with strong chaos (showing at least one positive Lyapunov exponent) approach [quasi-] thermodynamic equilibrium.

Weakly chaotic and turbulent systems show more complex behavior than infinite Reynolds number turbulent flows. These systems are closer to criticality. Table 10.3, in its last line, shows that turbulent systems usually are characterized by $q$ values between 5/3 and 3. Therefore, with (1111b) also the next refined energy spectrum follows

$$S(k) = \text{const}\, k^{-(5/3+\mu)}. \tag{1113}$$

The intermittency exponent $\mu$ is closely related to the structure function $\varsigma_p$ of order $p$ of the fluid velocity (see Frisch 1995). From the $q$-value range in the last line of Table 10.3, we also deduce that

$$q = \frac{5}{3} + \mu, \quad \mu = \frac{d - D}{3}, \quad 0 \le \mu \le \frac{4}{3}. \tag{1114a – c}$$

and conclude that Tsallis' $q$ factor copes directly with basic quantities of research on turbulence, which are known for many decades. Note that a power law exponent with a magnitude larger than 5/3 was a hypothesis of Kolmogorov (1962) and Obukhov (1962) for an intermittency correction (see also in Kraichnan 1975), which by the present analysis is more than just qualitatively confirmed. A comparison of Eq. (1114a) with the maximum value $q$ reveals Eq. (1114c). In this equation, $d$ denotes the spatial dimension and $D$ the Hausdorff–Besicovitch dimension, which is a fractal dimension. All this is in order as long as one works with Lévy flight statistics (see, e.g., Gotoh and Kraichnan 2004; Shlesinger et al. 1987, 1995), where a cascade of Lévy flight sizes directly relates to corresponding diameters of eddies in a way that basic kinematic laws of turning eddies with different angular velocities are fulfilled (see Egolf 2009). This is also in full agreement with results of the fractal $\beta$-model presented in detail by Frisch (1995).

On the other hand, Gotoh and Kraichnan (2004) critically comment on the application of Tsallis' nonequilibrium thermodynamics (with a constant $q$-factor) to turbulence. Arimitsu and Arimitsu (2004) make the link of Tsallis thermodynamics to the multifractal model of turbulence. In their model the intermittency and, thereby, its exponent $\mu$ are scale dependent. Notice that intermittency increases with decreasing scale. These considerations are in perfect agreement with the direct relation Eq. (1114a) between $\mu$ and $q$, where a scale-dependent intermittency exponent also calls for a scale-dependent $q$ factor. In the language of Tsallis thermodynamics, different wave number subsystems show different $q$ values.

(B) The second is the large wave number regime, for which, from Eq. (1109), it follows that

$$\widehat{E}_\varepsilon(k) = \frac{1}{2\beta} \frac{1}{Z_{\mathrm{ES}\varepsilon}} \frac{1}{q^2} \frac{1}{k^{2q}}. \tag{1115}$$

Now, by applying Eq. (1096c), the fractional derivative of this function is derived, which yields

$$S_\varepsilon(k) = \text{const} \frac{1}{k^{3q}}, \tag{1116}$$

which for $q = 1$ is identical with Kraichnan's former result (1026b), derived with BG equilibrium thermodynamics, leading to a "minus three power law exponent." Also, this result is in good agreement with the numerical experiments of Lilly (1969), presented in Fig. 10.6. The good coincidence between the equilibrium thermodynamic theory of Kraichnan,[5] being the thermodynamic special case with $q = 1$ of Tsallis, and experiments suggest that small eddies, which have smaller turnover times or life times, respectively, compared to large eddies (see Sect. 7.2.4.5), are close to or even exactly in thermal equilibrium. If in future experimentally determined small deviations to the power law exponent "−3" would be observed, then, with high probability they would be due to small nonequilibrium effects (see Egolf and Weiss 1996). In this context, note that Kraichnan (1967) proposed a small logarithmic correction to the enstrophy range of the spectrum. In the next section, more information on structure functions shall be given.

## 10.12 Velocity Structure Functions

The velocity structure function of order $p$ is defined by

$$S_p\left(\vec{r}\right) = \left\langle \left[\vec{u}\left(\vec{x} + \vec{r}\right) - \vec{u}\left(\vec{x}\right)\right]^p \right\rangle = \left\langle \delta\vec{u}\left(\vec{r}\right)^p \right\rangle. \tag{1117a, b}$$

It scales with

$$S_p(l) \propto u_0^p \left(\frac{l}{l_0}\right)^{\zeta_p}, \tag{1118}$$

with (see Frisch 1995)

$$\zeta_p = \frac{p}{3} + (d - D)\left(1 - \frac{p}{3}\right) = \frac{p}{3} + \frac{1}{3}(d - D)(3 - p), \tag{1119a, b}$$

in which the codimension $d - D$ appears again. By introducing the intermittency exponent (1114b), one can conclude that

$$\zeta_p = \frac{p}{3} + \mu\,(3 - p). \tag{1120}$$

Notice that $\mu$ also occurs in Eq. (366b), where it describes intermittency in the energy spectrum of turbulent flows. Rearranging Eq. (1120) gives

---

[5]Being slightly negligent we name quasi-equilibrium thermodynamic systems simply thermodynamic systems.

$$\mu = \frac{\zeta_p - \frac{p}{3}}{3 - p}. \tag{1121}$$

Of high importance is the second-order structure function $\zeta_2$, obtained by setting in Eq. (1121) $p = 2$ that leads to the following formula:

$$\mu = \zeta_2 - \frac{2}{3}. \tag{1122}$$

Notice that in experiments determining the intermittency exponent $\mu$, one must be certain that the inertial range is of high quality. Even a very small dissipation in this area could considerably change the exponent of the energy spectrum. This was pronounced by Egolf and Weiss (1996) in a note, where they proposed to add a term to the exponent of the structure function (1119a,b)

$$\zeta_p = \frac{p}{3} + \frac{1}{3}(d - p)\,(3 - D) - \frac{p}{3}\tilde{q}, \tag{1123}$$

where $\tilde{q}$ describes the effect of the energy leakage by dissipation in the energy cascade. Notice that so far, it has not yet been proven that a description of dissipative effects in the inertial range in the structure function is more than only an effective modeling concept. Frisch (1995) writes: "*Simulations of Borue and Orszag (*1995*) display a wider and steeper inertial range with an exponent of the energy spectrum m* = *1.85* ± *0.05. The discrepancy could be caused by a modification of the dissipation term, instead of an ordinary Laplacian. According to She and Lévêque (*1995*) this can significantly affect the value of the inertial-range exponents.*" A very sophisticated study on universal features of the turbulent energy cascade and its nonequilibrium thermodynamic processes was undertaken by Reinke et al. (2017).

Another refinement is to replace the intermittency exponent by bifractality. In the bifractal model the velocity has, close to a subset in physical space $S_1$, the scaling exponent $h_1 = \zeta_1$ and, close to a second subset $S_2$, the value $h_2 = \zeta_2$ (see Frisch 1995)

$$\frac{\delta u(r)}{u_0} \propto \begin{cases} \left(\frac{l}{l_0}\right)^{h_1}, r \in S_1, \quad \dim S_1 = D_1 \\ \left(\frac{l}{l_0}\right)^{h_2}, \; r \in S_2, \quad \dim S_2 = D_2. \end{cases} \tag{1124a, b}$$

Measurements of Van Atta and Park (1972) and Anselmet et al. (1984) show that $\zeta_p(p)$ beyond approximately $p = 8$ deviates from linearity. To take this effect fully into account the bifractal model is further generalized. In this procedure, a range of exponents in an interval $[h_{\min}, h_{\max}]$ is considered leading to the multifractal model with the first-order structure function

$$\frac{\delta u(r)}{u_0} \propto \int_I \mathrm{d}\mu(h) \left(\frac{l}{l_0}\right)^{h+3-D(h)}, \quad h \in [h_{\min}, h_{\max}], \tag{1125}$$

The measure $\mu(h)$ weighs the distribution of the different exponents. For more details of the multifractal model, see Frisch (1995).

For our low-order modeling concept, it was sufficient to take the $\beta$-fractal model into account. The only property that was important was that self-similarity leads to a simple large-scale description, a feature that is also inherently existing in its two generalized versions. Extensions to multifractal generalizations are left for future studies.

## 10.13 Justification of the Quadratic Form of the Energy as a Function of Space and Velocity Coordinates

In this section, we give a motivation of the quadratic form of the energy functional, which Kraichnan discovered. From the constancy of the real space and velocity variables (see below), viz

$$\left.\begin{array}{l} \vec{\tilde{x}} = \text{const} \\ \vec{\tilde{u}} = \text{const} \end{array}\right\} \Rightarrow \vec{\tilde{x}}^{\,2} + \vec{\tilde{u}}^{\,2} = \text{const}, \tag{1126a – c}$$

also the constancy of its sum follows. The quantity $\vec{\tilde{x}}$ is scaled to have the same dimension as the velocity (see text and formula just after Eq. (1148)). The statement of Kraichnan (1967) explains that the constant energy surfaces in phase space are hyperspheres.

To describe the eddy fluctuation motion, we start with a 2-d-coupled dynamical system

$$\dot{x}_1 = \omega\, x_2, \quad \dot{x}_2 = -\omega\, x_1, \quad \omega = \text{const}, \tag{1127a – c}$$

where for simplicity we omitted all the wavy overbars. Differentiation of Eq. (1127a) leads to

$$\ddot{x}_1 = \omega\, \dot{x}_2. \tag{1128}$$

Substituting Eq. (1127b) into (1128) then yields

$$\ddot{x}_1 = -\omega^2 x_1. \tag{1129}$$

By differentiating Eq. (1127b) and substituting (1127a) an additional second-order differential equation follows, so that we now face a system of two ordinary differential equations

$$\ddot{x}_1 + \omega^2 x_1 = 0, \quad \ddot{x}_2 + \omega^2\, x_2 = 0. \tag{1130a, b}$$

These are differential equations of harmonic oscillators for the two coordinates $x_1$ and $x_2$ with the solutions

$$x_1 = A \sin\left(\omega\, t + \phi\right), \quad x_2 = A \cos\left(\omega\, t + \phi\right). \tag{1131a, b}$$

The squared velocity is

$$\vec{u}^{\,2} = u_1^2 + u_2^2 = \dot{x}_1^2 + \dot{x}_2^2. \tag{1132a, b}$$

Substituting Eqs. (1127a,b), it follows that

$$\vec{u}^{\,2} = \omega^2\left(x_1^2 + x_2^2\right) = \left(\omega^2 r^2\right) = \left(\omega\, r\right)^2. \tag{1133a – c}$$

By a substitution of solution (1131a,b) into (1132b) and the application of the Pythagoras' law for the trigonometric functions the same result is obtained. This describes the energy of a rotating eddy, which is

$$E = \frac{1}{2}\vec{u}^{\,2} = \frac{1}{2}\omega^2 r^2 = \text{const.} \tag{1134a – c}$$

In a harmonic oscillator, the total energy $E$ is the sum of the kinetic energy $T$ and the potential energy $U$, which is constant. In the oscillation process, these two exchange their energies under the restriction that

$$E = T + U = \text{const.} \tag{1135a, b}$$

On the other hand, in a turbulent eddy there is an exchange between two kinetic energies of the two coupled oscillators

$$E = T_1 + T_2 = \frac{1}{2}u_1^2 + \frac{1}{2}u_2^2 = \frac{1}{2}\vec{u}^{\,2} = \text{const.} \tag{1136a – d}$$

Now, we take advantage of the fractal-$\beta$ model (see Frisch 1995). Thereby, we follow the presentation in Sect. 7.2.4 and Egolf (2009). In this model eddies are assumed to occur in classes with eddies of different sizes $l_n$ (see Fig. 7.5). The largest eddies show a diameter $d_0 = l_0$. Kraichnan in his work attributed the index 1 to the eddies of largest size, $d_1$, and this convention was also applied in an article of Egolf and Hutter (2018), dedicated to a nonequilibrium thermodynamic generalization of

Kraichnan's energy-enstrophy spectrum. However, throughout this book we shall keep to the standard denotation of turbulence research where largest diameters are denoted by $d_0$.

We now rewrite Eq. (1136d) in the notation of the fractal-$\beta$ model by renaming $r$ by $l/2$ and assigning the index 0 to several quantities

$$E_0 = \frac{1}{2}\vec{u}_0^{\,2} = \frac{1}{8}\omega_0{}^2 l_0^2 = \frac{1}{8}\omega_0{}^2\left(x_1^2 + x_2^2\right), \quad r = \frac{l_0}{2}. \qquad (1137\text{a} - \text{d})$$

Kraichnan stated that the total energy $E$ is constant

$$E = \sum_{n=0}^{N} E_n = \text{const.} \qquad (1138\text{a, b})$$

With the theory comprising Eqs. (339) to (358), it follows for the energies of eddies of different classes $n$ that

$$E_n = E_0 \left(\frac{l_n}{\mathrm{l}_0}\right)^{2/3+(d-D)/3}. \qquad (1139)$$

A comparison of this exponent with Eq. (1114a,b) reveals that

$$\frac{2}{3} + \frac{d-D}{3} = q - 1. \qquad (1140)$$

with the Tsallis parameter $q$. Now, Eq. (1139) is rewritten as

$$E_n = E_0 \left(\frac{l_n}{l_0}\right)^{q-1}. \qquad (1141)$$

With Eq. (346a) it follows from (1141) that

$$\begin{aligned} E_n &= E_0 \left[\left(\frac{1}{b}\right)^n\right]^{q-1} = E_0\left(b^{1-q}\right)^n \Rightarrow \\ \frac{E_{n+1}}{E_n} &= b^{1-q} < 1, \text{since } b > 1, q > 1. \end{aligned} \qquad (1142\text{a} - \text{c})$$

To obtain the total energy of the eddies of all classes, this equation is inserted into the sum of Eq. (1138a)

$$E = E_0 \sum_{n=0}^{N} \left(b^{1-q}\right)^n = \text{const.} \qquad (1143\text{a, b})$$

The final result is

$$E = \frac{1 - b^{(N+1)(1-q)}}{1 - b^{1-q}} E_0 = \kappa E_0. \quad (1144\text{a, b})$$

This result deserves the next formal:

**Proposition 9** *By scaling laws the total turbulent kinetic energy can be expressed as a function of the energy contained in the eddies of the largest size* $E_0$ *only. The turbulent kinetic energy of the eddies of the remaining classes* $E_1 \ldots E_N$ *is described by the multiplicative factor* $\kappa - 1$ *times* $E_0$*. This is a consequence of the self-similarity property of the eddy cascade.*

The special case $N = 0$ (only one class of eddies) leads to the correct solution of a total energy belonging to that of the single class containing only largest eddies, namely

$$E = E_0. \quad (1145)$$

The special case $q = 1$ yields a division "0/0." Therefore, to Eq. (1144a) the rule of "Bernoulli-de l'Hôpital" is applied by differentiating the numerator and denominator with respect to $b$, yielding

$$E = (N + 1)\, b^{N\,(1-q)} E_0, \quad (1146)$$

which leads to a direct consequence of the energy equipartition law between the eddies of different classes (see Eq. (1142c) with $q = 1$)

$$E = (N + 1)\, E_0. \quad (1147)$$

Notice that $N$+1 denotes the number of occurring classes.
For emphasis, the result Eq. (1142c) with $q = 1$ suggests

**Proposition 10** *Equilibrium BG-turbulent flows show energy equipartition in the eddies of all classes and, consequently, for infinite Reynolds number flow its turbulent kinetic energy diverges.*

Next, with Eqs. (1144b), (1137c) takes the form

$$E = \frac{1}{8} \kappa\, {\omega_0}^2 \left(x_1^2 + x_2^2\right). \quad (1148)$$

By modifying the coordinates by the stretching transformation $\widehat{x}_i = \sqrt{\kappa/8}\, \omega_0 x_i, i \in \{1, 2\}$—note that this space coordinate has the dimension of a velocity—one finds

$$u^2 = \left(\widehat{x}_1^2 + \widehat{x}_2^2\right), \quad (1149)$$

which is the result that is also confirmed in Appendix E by Eq. (E-8), q.e.d.

The same results, with a different constant $\kappa$, could also be obtained by taking into account in the derivation of Eq. (1144a) the occupation probability. This type of modeling includes the birth rates and life times of eddies of different classes (for details see Egolf 2009).

Finally, another very important result is written as

**Proposition 11** *By applying a Ritz–Galerkin truncation, the continuous spectrum of turbulent flows is discretized. It belongs to a system of 2N coupled oscillators with N pairs, each with its own fluctuation frequency. The large number of frequencies* $\omega_n$, $n \in \{1, \ldots N\}$ *scale in a self-similar manner.*

A description by periodic or quasi-periodic motions, as originally proposed by Landau (see in Landau and Lifshitz 1974) in an earlier stage of turbulence research, finally provides a somewhat crude model of turbulence. However, this does not throw any remarkable shadow on this kind of valuable description of isotropic turbulence. Rather, Kraichnan's model and its generalization in this book offers considerable insights into the physics of 2-d turbulent flows, as it will also become clear in the next section.

## 10.14 A Generalized Temperature of Turbulence

In Sect. 10.5, we were working with the Non-Gaussian Lévy flight probability distribution (1073). On the other hand, in Sect. 10.11 the proof of Eq. (1109) was performed with a weakly $q$-deformed Gaussian probability distribution (see insert Eq. (E-3) in Appendix E below). Therefore, the question could be raised, whether this might be an inconsistency in our modeling. However, in this section convincing explanations are given on these two distributions, and it is shown that they are so closely related that in the present physical modeling no claim is justified that the procedure is deprived by rationality.

We start by clarifying what we define as a weakly $q$-deformed Gaussian distribution and what was already earlier defined to be a strongly or just a $q$-deformed Gaussian distribution.

(A) Firstly, a weakly $q$-deformed Gaussian distribution is proposed by writing a usual Gaussian probability distribution as an escort probability distribution by combining Eq. (1067) with (1075)

$$P(x) = \frac{\left[\exp\left(-\frac{x^2}{2\sigma^2}\right)\right]^q}{Z_{\mathrm{ES}_\varepsilon}}, \quad Z_{\mathrm{ES}_\varepsilon} = \int_0^\infty \left[\exp\left(-\frac{x^2}{2\sigma^2}\right)\right]^q \mathrm{d}x. \qquad (1150\mathrm{a,b})$$

Then, it follows that

$$P(x) = \frac{\exp\left(-q\frac{x^2}{2\sigma^2}\right)}{Z_{\mathrm{ES}_\varepsilon}}. \tag{1151}$$

As demanded, the case $q = 1$ leads to the original classical Gaussian probability distribution (1067). Definition (1151) was applied in Appendix E, Eq. (E-4), and the motivation for this choice is given below.

(B) Secondly, according to, e.g., Abe (1997), Bercher (2018), a $q$-deformed Gaussian distribution is defined by applying Tsallis' generalized exponential function (1040)

$$P(x) = \frac{e_q^{-\left[\frac{x^2}{2(\sigma_q)^2}\right]}}{Z_q} = \frac{1}{Z_q}\left[1-(1-q)\frac{x^2}{2\sigma_q^2}\right]^{1/(1-q)}, \quad \sigma_1 = \sigma, \tag{1152a – c}$$

where also here, based on considerations in Sect. 10.5, for $q = 1$, the classical Gaussian distribution (1067) follows. The generalized standard deviation $\sigma_q$ is defined by Eq. (1071b).

Gotoh and Kraichnan (2004) provide the confirmation that the weakly deformed $q$-Gaussian probability distribution is the right choice in the context of our work. They state that with the Ritz–Galerkin description of the truncated dynamical system and the three constraints:

(a) $E_\varepsilon = E$ is the sum of energy contributions $E_n$: $E = \sum_{n=0}^{N} E_n$ (see Eq. (1011f))
(b) The dynamics conserves the turbulent kinetic energy, $E = \frac{1}{2}\widetilde{u}^2 = \text{const}$ (see Eq. (1005b))
(c) The Liouville property $\sum_{n=0}^{N}\frac{\partial\dot{\widetilde{x}}_n}{\partial\widetilde{x}_n} = 0$ holds (see Eq. (1004))

that in phase space, an energy form exists that is a hypersphere. Then also when $q \neq 1$, the energy shell shows a homogeneous probability density, and if additionally, $N \to \infty$, the probability distribution converges toward the Gaussian distribution

$$p_n(\widetilde{x}_n) = \frac{1}{\widetilde{u}}\sqrt{\frac{N+1}{2\pi}}\exp\left[-(N+1)\frac{1}{2}\frac{\widetilde{x}_n^2}{\widetilde{u}^2}\right]. \tag{1153}$$

Note, we assume that our finite dimensional dynamical system contains a very high number ($2N$) of binary oscillators, so that in equations following (1153), also for finite $N$, we will write instead of the symbol "≈" the sign "=". Furthermore, we know that in turbulence dissipation occurs, an effect that destroys the Liouville property (c). On the other hand, in a delicate equilibrium between a random forcing and dissipation, which, e.g., can be described by a fractional Langevin equation

(FLE), Tsallis thermodynamics can be extended to weakly $q$-deformed Gaussian distributions as applied in our proof in Appendix E, which, incidentally, has been motivated by Gotoh and Kraichnan's arguments above.

Next, with the statement (b) above and with the help of (1075a,b) we generalize the distribution (1153) as

$$P(x) = \left(\frac{1}{u}\sqrt{\frac{N+1}{2\pi}}\right)^{q} \exp\left[-\frac{q}{2}(N+1)\frac{x^2}{u^2}\right]. \tag{1154}$$

in which tildes have been dropped and where we have written $x^2$ for $\left|\vec{x}\right|^2$. Our approach is based on the escort generalized Boltzmann probability distribution and yields

$$P(x) = \frac{1}{Z_{\mathrm{ES}_\varepsilon}}\left[\exp\left(-\beta_q\, E\right)\right]^{q} = \frac{1}{Z_{\mathrm{ES}_\varepsilon}} \exp\left(-q\beta_q\, E\right). \tag{1155a, b}$$

By inserting the energy (1021a) (see also Appendix E, Eq. (E-9)), it follows that

$$P(x) = \frac{1}{Z_{\mathrm{ES}_\varepsilon}} \exp\left[-q\beta_q\left(\alpha + \beta\, k^{2q}\right)x^2\right]. \tag{1156}$$

Comparing the prefactors of Eqs. (1154) and (1156) yields

$$Z_{\mathrm{ES}_\varepsilon} = \left(u\sqrt{\frac{2\pi}{N+1}}\right)^{q}, \tag{1157}$$

and by comparing the arguments of the exponential functions, and applying $u^2 = 2E$, we get

$$E = \frac{1}{4}\frac{N+1}{\beta_q}\frac{1}{\alpha + \beta\, k^{2q}}, \tag{1158}$$

which, with $Z_{\mathrm{ES}\varepsilon} = 2\beta_q/[q^2(N+1)]$, is consistent with Eq. (1109). Notice that the constant in Eq. (1158) depends on the space dilatation or contraction given after Eq. (1148)! Next, Eq. (1154) is rewritten as

$$\begin{aligned} P(x) &= \left(\frac{1}{4\pi}\frac{N+1}{E}\right)^{q/2} \exp\left(-\frac{1}{4}q\,\frac{N+1}{E}x^2\right) = \\ &\left(\frac{1}{4\pi}\frac{N+1}{E}\right)^{q/2}\left\{\exp\left(-\frac{1}{4}\,(1-q)\frac{N+1}{E}x^2\right)\right\}^{q/(1-q)}. \end{aligned} \tag{1159a, b}$$

Assuming that the argument in the exponential function in (1159b) is an order of magnitude smaller than unity, a linear Taylor series expansion is applied to yield

$$P(x) = \left(\frac{1}{4\pi}\frac{N+1}{E}\right)^{q/2}\left(1+\frac{1}{4}(q-1)\frac{N+1}{E}x^2\right)^{q/(1-q)}, \quad Z_{\mathrm{ES}\varepsilon} = \left(4\pi\frac{E}{N+1}\right)^{q/2}. \tag{1160a, b}$$

This expression is equivalent to

$$P(x) = \left[\left(\frac{1}{4\pi}\frac{N+1}{E}\right)^{(1-q)/2} + \frac{1}{4}(q-1) \times \left(\frac{1}{4\pi}\right)^{(1-q)/2}\left(\frac{N+1}{E}\right)^{(1-q)/2+1} x^2\right]^{q/(1-q)}. \tag{1161}$$

The first term in the braces can be compared with the first term of Eq. (1073) to the power $q$. This suggests the identification

$$\lambda_1 = \frac{kq}{q-1}\left(\frac{1}{4\pi}\frac{N+1}{E}\right)^{(q-1)/2} \quad \Rightarrow \quad \lambda_1 = \frac{kq}{q-1} Z_{\mathrm{ES}\varepsilon}{}^{(1-q)/q} \tag{1162a, b}$$

In an analogous procedure now the second term of Eq. (1073) is raised to the power $q$ and compared to the second term of (1161). The resulting expression is then solved for $\lambda_2$, which leads to

$$\lambda_2 = \frac{1}{4}\frac{q-1}{q}\left(\frac{1}{4\pi}\right)^{(1-q)/2}\left(\frac{N+1}{E}\right)^{(1-q)/2+1}\lambda_1. \tag{1163}$$

Inserting (1162a) into this equation finally yields

$$\lambda_2 = \frac{1}{4}k\frac{N+1}{E}. \tag{1164}$$

Next, we consider the BG special case. Inserting $q = 1$ into Eq. (1162b) delivers

$$\lambda_1 = \lim_{q\to 1}\frac{k}{(q-1)} \to \infty, \tag{1165}$$

which corresponds to a diverging first Lagrange parameter. Now, when we apply Eqs. (1147) to (1164), $\lambda_2$ is modified to

$$\lambda_2 = \frac{1}{4}\frac{k}{E_0}. \tag{1166}$$

Next, we introduce a generalized temperature of turbulent flows $T_q$ and immediately omit its index. We further assume that eddies of class $n$ with generalized energy $E_n$ are characterized by their generalized temperature $T_n$. Thermodynamics (for an object with $m$ degrees of freedom) suggests that the relation between energy and generalized temperature is

$$E_n = \frac{m}{2} k\, T_n, \quad n \in \{ 0, 1, 2, \ldots, N\}. \tag{1167}$$

$T_n$ may be called the generalized turbulent Kelvin temperature of an eddy of class $n$. Applying this relation with $m = 4$ and $n = 0$ yields

$$E_0 = 2k\, T_0. \tag{1168}$$

When this is substituted into (1166),

$$\lambda_2 = \frac{1}{8T_0} \quad \Rightarrow \quad T_0 = \frac{1}{8\lambda_2} \tag{1169a, b}$$

is obtained and this result is stated as

**Proposition 12** *In turbulent BG flows the inverse second Lagrange parameter $\lambda_2$ defines the generalized temperature of the eddies of class zero.*[6]

Next, we apply the energy equipartition law of equilibrium turbulent flows (see Eq. (1142c) with $q = 1$), viz,

$$E_0 = E_1 = \ldots = E_{N-1} = E_N. \tag{1170}$$

Then, we conclude with the help of Eqs. (1142c) and (1168) for the energy-enstrophy spectrum that

$$T_0 > T_1 > \ldots > T_{I-1} > T_I = T_{I+1} \ldots = T_{N-1} = T_N. \tag{1171}$$

This new result is now stated as

**Proposition 13** *In 2-d turbulence eddies in the energy transfer range belonging to the classes $n = 0, 1, \ldots, I$ ($I$ = intersection between the two ranges characterized by the wavenumber $k_I$) show decreasing generalized temperatures: $T_0 > T_1 > \ldots > T_{I-1} > T_I$, whereas the eddies in the enstrophy transfer range belonging to classes $I$+1, $I$+2,…, $N-1$, $N$, show generalized temperatures that are equal: $T_{I+1} = T_{I+2} = \ldots = T_{N-1} = T_N$. This is a consequence that in the energy range the eddies are not in equilibrium, whereas in the enstrophy range they are.*

These findings are consistent with Kraichnan's statement that in the enstrophy range, where eddies belonging to different wave number intervals are in thermal equilibrium; there is no energy transfer (see Kraichnan 1959, 1967; Lilly 1969).

[6]Temperatures of turbulence were also proposed by Chorin (1994) and Castaign (1996).

Finally, it is stated that the probability distribution (1073) and the escort generalized weak Gaussian probability distribution (1154), applied in our model, do not lead to an inconsistency, because Eq. (1073) can be interpreted to be a first-order Taylor expansion of Eq. (1154). Equally satisfying is that comparisons between the two probability distributions lead to important correct statements on the equilibrium and nonequilibrium thermodynamics of eddies of different classes.

## 10.15 Final Discussion on Nonextensive Thermodynamics of Turbulence

Chapter 10 is mainly based on results of Kraichnan (1959, 1967), Tsallis (1995, 2009) and Tsallis et al. (1996), and our extensions, first published in Egolf and Hutter (2018). A further important contribution, namely the relation between Tsallis thermodynamics and Lévy statistics is due to Alemany and Zanette (1994). Our derivations incorporate and combine these works, but also provide a majority of new results; some remained unanswered and remained unresolved for decades as open ends. In this memoir, a unique modern theory of 2-d turbulence was created, e.g., by implementing new ideas of fractional calculus, so that the new theory now presents itself with completeness and shows close agreement with experiments. We have also found a physically clearly defined generalized temperature of turbulent flows. Its variation acts as a measure of deviation of a turbulent system from equilibrium.

Kraichnan applied equilibrium BG thermodynamics to 2-d and 3-d turbulence. For 2-d turbulence two partial regimes are observed, a low wave number regime, where the energy is constant, and a large wave number regime in which the enstrophy is constant (Kraichnan 1971). The generalization of the 3-d turbulence energy spectrum with Tsallis nonextensive formalism and fractional calculus seems promising, but has not yet been derived and is proposed for further investigation.

On the other hand, Tsallis has been a forerunner of generalizing thermodynamics to nonequilibrium processes. Turbulence at moderate Reynolds number is a highly nonequilibrium physical phenomenon and demands new nonlocal concepts that are still in their infancy. A variety of different entropic forms were proposed, which luckily have some mutual connections, so that it is often more a matter of taste than correctness to apply one or another. The Tsallis entropic form works with a new formalism that generalizes the significant classical statistical-based terms by a $q$-factor and, thus, leaves well-established formulas of equilibrium thermodynamics preserved and in a generalization to nonequilibrium thermodynamics practically unaltered, if the basic functions are $q$-generalized (e.g., by the $q$-exponential function, the $q$-logarithmic function, or the $q$-deformed fractional derivative (Weberszpil and Chen 2017)). We introduced the weakly $q$-deformed Escort-Gaussian probability distribution. It is the right choice for a Ritz–Galerkin truncated system of the NSE that develops to a high-degree dynamical coupled system of $2N$ oscillators or

$N$ eddies, respectively. For a large number $N$ of eddy classes, the Gaussian distribution is a good approximation and in the limit as $N \rightarrow \infty$ it is even the correct one. We found that the Non-Gaussian Lévy probability distribution can be sought to be a Taylor expansion to this limiting case. However, it is well known that the most complex behavior occurs above criticality and at medium Reynolds numbers and simplifies in the infinite Reynolds number limit.

According to a study of Alemany and Zanette (1994), Tsallis' extended thermodynamics is directly related to Lévy walks and flights. And Lévy statistics had been applied to successfully describe turbulence (see, e.g., Frisch 1995; Egolf 2009)! A direct conclusion is that Tsallis' contributions to statistical physics yield a favorite tool to build a solid foundation of the thermodynamics of turbulence. However, Gotoh and Kraichnan (2004) offered critical remarks concerning the application of Tsallis thermodynamics with a single $q$-value to turbulence. They also discussed the application of a wave number-dependent $q$ value to overcome missing precision. Such a dependence emerges naturally in our calculus in the context of modeling turbulent fields with intermittency (compare with Eq. (1114a–c)).

Furthermore, it is known that Lévy random walks and flights can be related to fractional Langevin and Fokker–Planck equations that are based on fractional calculus. It is evident that the dynamics belonging to Richardson's and Mandelbrot's fractal geometry is fractional dynamics and calculus. Egolf and Hutter (2016a, 2017a) introduced it to turbulence modeling. Therefore, it is a direct consequence that Tsallis' thermodynamics can be formulated with the help of fractional calculus. Without this tool it is impossible to generate by a differentiation of a constant energy the Kolmogorov–Oboukov $k^{-5/3}$ energy intensity spectrum. Moreover, a fractional generalization of the enstrophy is of importance to generalize Kraichnan's energy-enstrophy spectrum.

With all these new developments, it was tempting for us to apply Tsallis' thermodynamics to generalize Kraichnan's important work on the Ritz–Galerkin truncated turbulent energy and enstrophy spectrum of 2-d turbulence. The new findings, related to the energy and enstrophy spectra, are consistent with three well-known types of descriptions of turbulence with an increasing level of complexity (1) the BG thermodynamic results of Kraichnan; (2) the spectrum of Richardson, Oboukov and Kolmogorov; and (3) newer results of Lévy and Frisch, etc., describing turbulence including intermittency effects.

In the context of the proof of the generalized 2-d turbulence spectrum of Kraichnan, it has become clear that the usual definition of enstrophy with a square of the first derivative of the velocity relates intimately to Boltzmann–Gibbs ($q = 1$) equilibrium thermodynamics and requires a generalization, e.g., with the nonextensive ($q \neq 1$) Tsallis thermodynamics including fractional derivatives. In future, a generalization of the definition of enstrophy in articles and standard textbooks may become useful.

All these new studies show that nonequilibrium thermodynamics and fractional calculus, involving nonlocality, are likely to play a crucial role in modern turbulence research and computation. Whereas such theoretical approaches have started to actively bloom, time seems now also ready for fluid dynamic computational experts to introduce into their numerical algorithms and codes new fractional concepts (see Chap. 12). Modern zero-equation turbulence models, e.g., the DQTM (Egolf and Hutter 2016b), are more than just a compensation of the eliminated wave number terms (Kraichnan 1971), e.g., by a Ritz-Galerkin truncation method. They contain scaling laws and couple, in a statistical sense, small with large wave number eddies in a nonlocal manner. Therefore, we conjecture and predict that in future low-order fractional turbulence modeling will be superior and more accurate as compared to the past and present results of higher-order conventional calculus computations, based on linear and local concepts, and will decrease the demanded computation power (CPU time) likely by at least an order of magnitude.

## References


Abe, S.: A note on the $q$-deformation-theoretic aspect of the generalized entropies in nonextensive physics. Phys. Rev. Lett. A. **224**, 326 (1997)

Albeverio, S., Cruzerio, A.: Global flows with invariant Gibbs measures for Euler and Navier-Stokes two dimensional fluids. Commun. Math. Phys. **129**, 431 (1990)

Alemany, P.A., Zanette, D.H.: Fractal random walks from a variational formalism for Tsallis entropies. Phys. Rev. E. **49**(2), R956 (1994)

Anselmet, F., Gagne, Y., Hopfinger, J., Antonio, R.A.: High-order velocity structure functions in turbulent shear flows. J Fluid Mech. **140**, 63 (1984)

Arimitsu, T., Arimitsu, N.: Multifractal analysis of fluid particle accelerations in turbulence. Phys. D. **193**, 218 (2004)

Barkai, E.: Fokker-Planck equation, solution, and application. Phys. Rev. E. **63**, 046118 (2001)

Beck, C., Schlögl, F.: Thermodynamics of Chaotic Systems – An Introduction, Cambridge Nonlinear Science Series 4. Cambridge University Press, Cambridge, UK (1993). ISBN 13-978-0-521-43367-9

Beck, C., Lewis, G.S., Swinney, H.L.: Measuring nonextensitivity parameters in a turbulent Couette-Taylor flow. Phys. Rev. E. **63**(3), 035303 (2001)

Becker, R.: Theorie der Wärme. Springer, Berlin (1978). ISBN 3-540-08988-8 (in German)

Bercher, J.F.: On escort distributions, q-gaussians and Fisher information. http://www.lss.supelec.fr/MaxEnt2010/paper/066.pdf (2018)

Boffetta, G.: Energy and enstrophy fluxes in the double cascade of 2d turbulence. J. Fluid Mech. **589**, 253 (2007)

Boldrighini, C., Frigio, S.: Equilibrium states for a plane incompressible perfect fluid. Commun. Math. Phys. **72**, 55 (1980)

Boltzmann, L.: In: Hasenöhrl, F., Barth, J.A. (eds.) Wissenschaftliche Abhandlungen, vol. I–III, New York (1909). Chelsea, 1969

Borue, V., Orszag, S.A.: Forced three-dimensional homogeneous turbulence with hyperviscosity. Europhys. Lett. **29**, 687 (1995)

Brenig, W.: Statistische Theorie der Wärme – Gleichgewichtsphänomene. Springer, Berlin (1992). ISBN 3-540-55723-7-3. (in German)

Bronstein, I.N., Semendjajew, K.A.: Taschenbuch der Mathematik. Edition Harri Deutsch, Thun (1981). ISBN 3-87-144-492-8 (in German)

Budde, C., Prato, D., Ré, M.: Superdiffusion in decoupled continuous time random walks. Phys. Lett. A. **283**, 309 (2001)

Camargo, R.F., Chiacchio, A.O., Capelas de Oliveira, E.: On anomalous diffusion and the fractional generalized Langevin equation for a harmonic oscillator. J. Math. Phys. **50**, 123518 (2009)

Casimir, H.B.G.: On Onsager's principle of microscopic reversibility. Rev. Mod. Phys. **17**, 343 (1945)

Castaign, B.J.: Consequences d'un principe d'extremum en turbulence. J. Phys. France. **50**, 147 (1989). (in French)

Castaign, B.J.: The temperature of turbulent flows. J. Phys. II. **6**, 105 (1996)

Chapman, S., Cowling, T.G.: The Mathematical Theory of Non-Uniform Gases: An Account of the Kinetic Theory of Viscosity, Thermal Conduction and Diffusion in Gases. Cambridge University Press, Cambridge, UK (1970). ISBN 10-0521075777

Chorin, A.J., Vorticity and Turbulence. Applied Mathematical Sciences 103. Springer edition, New York, Inc. (1994), ISBN 0-387-94197-5

Clausius, R.: Über die bewegende Kraft der Wärme und die Gesetze, welche sich daraus für die Wärmelehre selbst ableiten lassen. Ann. Phys. und Chemie. **155**, 368 (1850). (in German)

Cuillière, J.-C.: Introduction à la méthode des éléments finis, 2nd edn. Dunod, Paris (2016). ISBN 978-2100-74-26-22 (in French)

de Groot, S.R., Mazur, P.: Non-equilibrium Thermodynamics. Dover, New York (1984). ISBN 0-486-64741-2

Dhont, J.K.G.: An Introduction to Dynamics of Colloids. Elsevier, Amsterdam (1996). ISBN 0-08-053507-0

Duhem, P.: Revue des deux mondes, T. **129**, 869 and T **130**, 380 and 851 (1895). (in French)

Duhem, P.: Energétique, vol. 2. Gauthier-Villars, Paris (1911). (in French)

Eckart, C.: The thermodynamics of irreversible processes. I. The simple fluid. Phys. Rev. **58**, 267 (1940)

Egolf, P.W.: Lévy flights and beta model: a new solution of "wall" turbulence with a critical phenomenon. Int. J. Refrig. **32**, 1815 (2009)

Egolf, P.W., Hutter, K.: From linear and local to nonlinear and nonlocal zero equation turbulence models. In: Proceeding of IMA Conference for Turbulence, Waves and Mixing, Kings College, pp. 71–74, Cambridge, UK (2016a). 6–8 July

Egolf, P.W., Hutter, K.: Turbulent shear flow described by the algebraic difference-quotient turbulence model. In: Peintke, J., et al. (eds.) Progress in Turbulence VI, Springer Proceedings in Physics, vol. 165. Springer, Heidelberg (2016b). ISBN 978-3-319-29129-1

Egolf, P.W., Hutter, K.: Fractional turbulence models. In: Peintke, J., et al. (eds.) Progress in Turbulence VII, Springer Proceedings in Physics, vol. 165. Springer, Heidelberg (2017a). ISBN 978-3-319-57933-7

Egolf, P.W., Hutter, K.: The mean field theories of magnetism and turbulence. J. Entropy. **19**, 589 (2017b)

Egolf, P.W., Hutter, K.: Tsallis extended thermodynamics applied to 2-d turbulence: Lévy statistics and q-fractional generalized Kraichnanian energy and enstrophy spectra. J. Entropy. **20**, 109 (2018)

Egolf, P.W., Weiss, D.A.: Turbulent energy cascade model including intermittency and non-equilibrium energy production and dissipation. Helv. Phys. Acta. **69**(1), 11 (1996)

Eyink, G.L., Frisch, U., Robert, H.: In: Davidson, P.A., Kaneda, Y., Moffat, K., Sreenivasan, K.R. (eds.) Voyage Through Turbulence. Cambridge University Press, Cambridge, UK (2011). ISBN 978-0-521-19868-4

Foias, C., Manley, O., Rosa, R., Temam, R.: Navier-Stokes equations and turbulence. In: Encyclopedia of Mathematics and its Applications. Cambridge University Press, Cambridge, UK (2001). ISBN 0-52136032-3

Fokker, A.D.: Die mittlere Energie rotierender elektrischer Dipole im Strahlungsfeld. Ann. Phys. **348**, 810 (1914). (in German)

Frisch, U.: Turbulence – The Legacy of A.N. Kolmogorov, 1st edn. Cambridge University Press, Cambridge, UK (1995). ISBN 0-85-403-441-2

Frisch, U.: Robert Harry Kraichnan (1928–2008), Lecture at the Conference "Turbulence – the Historical Perspective", Warsaw (2011). https://www.youtube.com/watch?v=UPgVuoO6RWY&list=PL0jAiXLfdUJBAoYmevhgcao62CBX9fIlF&index=11

Frisch, U., Sulem, P.L.: Numerical simulation of the inverse cascade in two-dimensional turbulence. Phys. Fluids. **27**, 1921 (1984)

Fröhlich, J., Ruelle, D.: Statistical mechanics of vortices in an inviscid two-dimensional fluid. Commun. Math. Phys. **87**, 1 (1982)

Gibbs, J.W.: Elementary Principles in Statistical Mechanics, Developed with Especial Reference to the Rational Foundation of Thermodynamics. Charles Scribner's and Sons, New York (1902)

Goldenfeld, N.: Lectures on Phase Transitions and the Renormalization Group, Frontiers in Physics, 5th edn. Adison-Wesley, Reading, PA (1992). ISBN 13-978-0-201-55408-3

Gotoh, T., Kraichnan, R.H.: Turbulence and Tsallis statistics. Phys. D. **193**, 231 (2004)

Grenard, V., Garnier, N.B., Naert, A.: The fluctuation–dissipation relation on a Melde string in a turbulent flow; considerations on a 'dynamical temperature'. Lett. J. Stat. Mech. Theory and Exp. **9**, L09003 (2008)

Haken, H.: Synergetik: Eine Einführung. Springer, Berlin (1983). ISBN 3-540-12597-3 (in German)

Haken, H.: Advanced Synergetics: Instability, Hierarchies of Self-Organizing Systems and Devices. Springer, Berlin (1987). ISBN 3-540-12162-5

Herring, J.R., Kraichnan, R.H.: Comparison of some approximations for isotropic turbulence. In: Statistical Models and Turbulence, Lecture Notes in Physics, vol. 12. Springer, Berlin (1972). ISBN 978-3-540-37093-2

Herrmann, R.: Fractional Calculus. World Scientific, New Jersey (2011). ISBN 13-978-981-4340-24-3

Herrmann, R.: Fraktionale Infinitesimalrechnung, 2nd edn. Books on Demand, Norderstedt (2014). ISBN 978-3-7357-4109-7 (in German)

Hunt, J.C.R., Phillips, O.M., Williams, D.: Turbulence and Stochastic Processes: Kolmogorov's Ideas 50 Years on, The Royal Society, London. University Press, Cambridge, UK (1991). ISBN 0-854-03-441b-2

Hutter, K.: The foundations of thermodynamics, its basic postulates and implications. A review of modern thermodynamics. Acta Mech. **27**, 1 (1977)

Hutter, K., Wang, Y.: Fluid and thermodynamics. In: Basic Fluid Mechanics, vol. 1. Springer, Berlin (2016a). ISBN 978-3-319-33632-9

Hutter, K., Wang, Y.: Fluid and thermodynamics. In: Advanced Fluid Mechanics and Thermodynamic Fundamentals, vol. 2. Springer, Berlin (2016b). ISBN 978-3-319-33635-0

Jackson, F.: Generalization of the differential operative symbol with an extended form of Boole's equation. Mess. Math. **38**, 57 (1909)

Jackson, F.: On q-definite integrals. Quart. J. Pure Appl. Math. **41**, 193 (1910)

Jou, D., Casas-Vázquez, J., Lebon, G.: Extended irreversible thermodynamics. Rep. Prog. Phys. **51**, 1105 (1988)

Kawasaki, K.: Simple derivations of generalized linear and nonlinear Langevin equations. J. Phys. A, Math Nucl. Gen. **6**, 1289 (1973)

Kobelev, V., Romanov, E.: Fractional Langevin equation to describe anomalous diffusion. In: 4th International Summer School/Conference "Let's Face Chaos through Nonlinear Dynamics" (1999). 28 June–11 July, Maribor

Kolmogorov, A.N.: Über die analytischen Methoden in der Wahrscheinlichkeitsrechnung. Math. Annalen. **104**, 415 (1931). (in German)

Kolmogorov, A.N.: A refinement of previous hypotheses concerning the local structure of turbulence in a viscous incompressible fluid at high Reynolds number. J. Fluid Mech. **13**, 82 (1962)

Kraichnan, R.H.: The structure of isotropic turbulence at very high Reynolds numbers. J. Fluid Mech. **5**(4), 495 (1959)

Kraichnan, R.H.: Inertial ranges in two-dimensional turbulence. Phys. Fluids. **10**(7), 1417 (1967)

Kraichnan, R.H.: Inertial-range transfer in two-and three-dimensional turbulence. J. Fluid Mech. **47** (3), 525 (1971)

Kraichnan, R.H.: Statistical dynamics of two-dimensional flow. J. Fluid Mech. **67**, 155 (1975)

Kraichnan, R.H., Montgomery, M.T.: Two-dimensional turbulence. Rep. Prog. Phys. **43**, 547 (1980)

Landau, L.D., Lifshitz, E.M.: Lehrbuch der Theoretischen Physik, Hydrodynamik, vol. VI. Akademie–Verlag, Berlin (1974). (in German)

Langevin, P.: Sur la théorie du mouvement brownien. CR Acad. Sci. Paris. **146**, 530 (1908). (in French)

Le Bellac, M., Mortessagne, F., Batrouni, G.G.: Equilibrium and Non-Equilibrium Statistical Thermodynamics. Cambridge University Press, Cambridge, UK (2006). ISBN 978-0-521-82143-8

Lee, T.D.: On some statistical properties of hydrodynamical and magneto-hydrodynamical fields. Q. Appl. Math. **10**, 69 (1952)

Lilly, D.K.: Numerical simulation of two-dimensional turbulence. Phys. Fluids. **12**(suppl. S2), 240 (1969)

Lutz, E.: Fractional Langevin equation. Phys. Rev. E. **64**, 051106 (2001)

Ma, S.-K.: Modern Theory of Critical Phenomena, Frontiers in Physics. The Benjamin Cummings, Reading, MA (1982). ISBN 0-8053-6670-9

Mainardi, F., Pironi, P.: The fractional Langevin equation: Brownian motion revisited. Extracta Math. **10**(1), 140 (1996)

Meixner, J.: Zur Thermodynamik der irreversiblen Prozesse in Gasen mit chemisch reagierenden, dissoziierenden und anregbaren Komponenten. Ann. Phys. **435**(4), 244 (1943). (in German)

Miller, J.: Statistical mechanics of Euler equations in two dimensions. Phys. Rev. Lett. **65**(17), 2137 (1990)

Montgomery, D.: Maximal entropy in fluid and plasma turbulence. In: Ray Smith, C., Grandy, W.T. (eds.) Maximum Entropy and Bayesian Methods in Inverse Problems. Reidel, Doordrecht (1985). ISBN 978-90-481-8418-7

Montrol, E.W., Lebowitz, J.L.: Fluctuation Phenomena. North Holland Personal Library (1976). ISBN 0-444-87038-5

Müller, I., Ruggeri, T.: Rational Extended Thermodynamics. Springer, Heidelberg (1998). ISBN 978-1-4612-2210-1

Natanson, L.Z.: Ueber die Gesetze nicht umkehrbarer Vorgänge. Phys. Chem. **21**, 193 (1896). (in German)

Novikov, E.A.: Dynamics and statistics of a system of vortices. Sov. Phys. JETP. **41**, 937 (1976)

Obukhov, A.M.: Some specific features of atmospheric tubulence. J. Fluid Mech. **13**, 77 (1962)

Onsager, L.: Reciprocal relations in irreversible processes. I. Phys. Rev. **37**, 405 (1931)

Onsager, L.: Statistical hydrodynamics. Neuvo Cimento. **6**(2), 279 (1949)

Planck, M.: Über einen Satz der statistischen Dynamik und seine Erweiterung in der Quantentheorie, vol. 24. Sitzungsbericht Preuss. Akad. Wiss (1917). (in German)

Pointin, Y.B., Lundgren, T.S.: Statistical mechanics of two-dimensional vortices in a bounded container. Phys. Fluids. **10**, 1459 (1976)

Prigogine, I.: Etude thermodynamique des phénomènes irréversibles. Dunod, Paris (1947). (in French)

Reif, F.: Grundlagen der physikalischen Statistik und der Physik der Wärme. Walter de Gruyter, Berlin (1975). ISBN 3-11-004103-0 (in German)

Reinke, N., Nickelson, D., Peinke, J.: On Universal Features of the Turbulent Cascade and its Non-equilibrium Thermodynamic Process, Unpublished Manuscript (2017)

Robert, R.: Etats d'équilibre statistique pour l'écoulement bidimensionnel d'un fluide parfait. C.R. Acad. Sci. Paris. **311**(Série 1), 575 (1990). (in French)

Robert, R.: A maximum-entropy principle for two-dimensional perfect fluid dynamics. J. Stat. Phys. **65**(3/4), 531 (1991)

Robert, R., Sommaria, J.: Statistical equilibrium states for two-dimensional flows. J. Fluid Mech. **229**, 291 (1991)

Rogers, L.C.G., Williams, D.: Diffusions, Markov Processes and Martingales, Cambridge mathematical library, 2nd edn. Cambridge University Press, Cambridge, UK (2000). ISBN 13-978-0521775946

Saffman, P.G., Baker, G.R.: Vortex interactions. Annu. Rev. Fluid Mech. **11**, 95 (1979)

She, Z.S., Lévêque, E.: Quantized energy cascade and log-Poisson statistics in fully developed turbulence. Phys. Rev. Lett. **72**, 336 (1995)

Shlesinger, M.F., West, B.J., Klafter, J.: Lévy dynamics of enhanced diffusion: Application to turbulence. Phys. Rev. Lett. **58**(11), 1100 (1987)

Shlesinger, M.F., Zaslafsky, G.M., Frisch, U.: Lévy Flights and Related Topics in Physics. Springer, Berlin (1995). ISBN 978-3-662-140-48-2

Siggia, E.D., Aref, H.: Point-vortex simulation of the inverse energy cascade in two-dimensional turbulence. Phys. Fluids. **24**, 171 (1981)

Stanley, H.E.: Introduction to Phase Transitions and Critical Phenomena, International Series of Monographs on Physics. Oxford Science, Oxford, UK (1971). ISBN 0-19-5053-8

Takayama, H.: Cooperative dynamics in complex physical systems. In: Proceedings of the Second Yukawa International Symposium, Kyoto (1988). 24–27 August, Reprint Edition Springer, Berlin. ISBN 3-540-508-65-1

Truesdell, C.: Rational Thermodynamics. Springer, New York (1984). ISBN 978-1-4612-9737-6

Truesdell, C.A., Muncaster, R.G.: Fundamentals of Maxwell's Kinetic Theory of a Simple Monoatomic Gas: Treated as a Branch of Rational Mechanics. Academic, Cambridge, MA (1980). ISBN 0-12-701350-4

Tsallis, C.: Comment on "Thermodynamic stability conditions for the Tsallis and Rényi entropies" by J.D. Ramshaw. Phys. Lett. A. **206**(5–6), 389 (1995)

Tsallis, C.: Introduction to Nonextensive Statistical Mechanics. Springer, Cham (2009). ISBN 978-0-387-85358-1

Tsallis, C., Levy, S.V.F., Souza, A.M.C., Maynard, R.: Statistical-mechanical foundation of the ubiquity of the Lévy distribution in nature, Phys. Rev. Lett. **77**(27), 5442 (1996)

van Atta, C.W., Park, J.: Statistical self-similarity and inertial subrange turbulence. In: Statistical Models and Turbulence, Lecture Notes in Physics, vol. 12, p. 402. Springer, Berlin (1972)

Weberszpil, J., Chen, W.: Generalized Maxwell relations in thermodynamics with metric derivatives. J. Entropy. **19**, 407 (2017). https://doi.org/10.3390/e19080407

Zumhofen, G., Blumen, A., Klafter, J., Shlesinger, M.F.: Lévy walks for turbulence: a numerical study. J. Stat. Phys. **54**(5/6), 1519 (1989)

# Chapter 11
# Turbulence: A Cooperative Phenomenon

## 11.1 Introduction

Turbulence is an order–disorder phenomenon belonging to the field of statistical physics (see, e.g., Itsykson and Drouffe 1989a, b). Ordinary people classify chaos and turbulence to be inherently linked with disorder. However, the scientific terminology is inverse. In this book, the smooth laminar flow defines the phase in a fluid being in disorder, but the more it is externally excited and becomes turbulent, by symmetry breakings, the more order appears. It will be shown that order in turbulent flows is defined by the fluctuation intensity or its turbulent kinetic energy. Therefore, the constant velocity of laminar flows (stress parameter) is not a result of symmetry breaking as is the occurrence of a well-defined magnetization (order parameter) in a magnetic system is. As a result, we can say that a laminar flow, with the turbulence intensity as order parameter, shows full translational and rotational symmetries. Usually for large Reynolds numbers it is believed that a flow approaches homogeneous and isotropic turbulence, which is statistically without structure and therefore, without any symmetries. However, microscopically this is not the case. If in a model consideration small vortices are assumed to occur, in the infinite Reynolds number limit their number tends to infinity and their diameter to zero. Thus, to conclude that they disappear is a wrong inference. Finally, the result is that usually the creation of an ensemble of different structure scales, especially small ones, leads to higher symmetries. On the other hand, in macroscopic models of turbulence such assumptions are allowed to be made and, thereby, a virtual higher symmetry may be assumed that leads to simpler results in this limiting case. As a result of all these considerations, a corresponding generalized entropy (see, e.g., Tsallis 2009) for increasing Reynolds number decreases. An increasing stress of a fluid dynamic system is described by an increasing characteristic velocity of the physical system,

The original version of this chapter was revised: Table 11.2 has been updated. The correction to this chapter is available at https://doi.org/10.1007/978-3-030-26033-0_13

The original version of this chapter was revised. A correction to this chapter can be found at https://doi.org/10.1007/978-3-030-26033-0_14



P. W. Egolf, K. Hutter, *Nonlinear, Nonlocal and Fractional Turbulence*,
https://doi.org/10.1007/978-3-030-26033-0_11

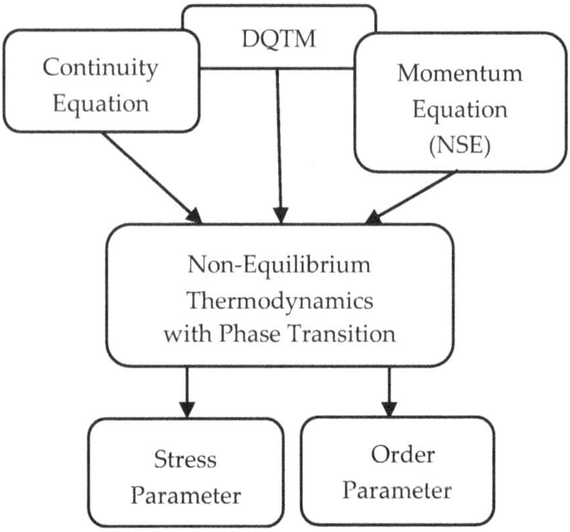


**Fig. 11.1** The dynamical theory of CE (continuity equation), NSE, and DQTM reveal a critical phenomenon with a continuous phase transition. The two phases are laminar streaks and turbulent patches (coherent structures). Courtesy Egolf and Hutter (2017b) © J. Entropy, reproduced with permission by Springer Science+Business Media

or in a dimensionless number presentation, by an increasing overall Reynolds number, $\mathrm{Re}_0$ (Hinze 1975). Therefore, flows with the highest turbulence intensity are occurring when its Reynolds number is infinite. Order in such systems has to do with cooperative behavior, and it usually occurs when a critical value of the external stress is exceeded; in our case this is the critical overall Reynolds number $\mathrm{Re}_{0c}$.

Other physical systems showing cooperative or critical behavior are magnetic systems, where magnetic moments or spins align (one may think of the aligned hairs in a crew cut of a soldier), defining order in a very obvious manner. Disorder occurs here above a critical temperature, called the Curie temperature, $T_c$, and the order of the system increases if the temperature $T$ is decreased below $T_c$, reaching its maximum at $T = 0$ K.

As demonstrated in Sects. 9.3–9.5 Egolf et al. solved analytically plane Couette (Egolf and Weiss 1995a), plane Poiseuille (Egolf and Weiss 1995b, 2000), and turbulent "wall" flows (Egolf 2009) by applying a nonlocal and fractional turbulence model (Egolf and Hutter 2016a, b, c), the Difference-Quotient Turbulence Model (DQTM). If the continuity and the Navier–Stokes equation (see Hutter and Wang 2016a, b) are in a self-similar manner combined with the DQTM) (see Chap. 8), in all these cases a critical phenomenon with a continuous phase transition is revealed (see Fig. 11.1).

Confirming statements of order-disorder in the work of Egolf and Weiss (see Refs. above), the stress parameter (for a definition see below) occurs inversely, namely as $1/\mathrm{Re}_0$. Thus, one may state that, in analogy to magnetism, the Reynolds number should have been defined inversely or that thermodynamics should be consequently performed by using as its stress parameter the coldness $1/T$, instead of the temperature $T$.

It is beyond the scope of this memoir to review all the articles on turbulence, where their authors have discovered critical phenomena and the phase change character of turbulence, usually without actively recognizing this phenomenon. However, such a review of the present authors is found in Egolf and Hutter (Egolf and Hutter 2017b). The entire Chap. 11 mainly follows this review. Briefly, we may state that most articles on cooperative phenomena are experimentally motivated (see, e.g., Ravelet et al. 2008; Wester et al. 2017). The related fluid dynamic experiments show a criticality and their authors are now aware that there is a turbulence quantity that serves as an order parameter. Cortet et al. (2011) used time series of stroboscopic particle image velocimetry data to study the response of von Kármán swirling flows between Re = 100 and 1,000,000. The flows can be characterized by a scalar, the modulus of the global angular momentum. Their response is linear with a slope depending on Re and shows a divergence at a critical Reynolds number. This divergence coincides with a spontaneous symmetry breaking, whereas the statistics transforms from a Gaussian to a Non-Gaussian distribution with metastable and nonsymmetrical states. Time intermittencies between metastable states are observed. The authors write in their final sentence of their abstract "*We show that these observations can be interpreted in terms of divergence of the susceptibility to symmetry breaking, revealing the existence of a phase transition. An analogy with the ferromagnetic-paramagnetic transition in solid-state physics is presented.*"

More than 30 years ago, Pomeau (1986) described the laminar–turbulent transition of a fluid by coupled oscillators. This was a kind of preliminary stage for a statistical description of this transition and paved the way for applications of percolation ideas performed by Allhoff and Eckhardt (2012), Kreilos and Eckhardt (2014), Lemoult et al. (2016) and Wester et al. (2017). We further discuss results of the research division of the last listed authors. These scientists apply knowledge from percolation analysis in order to temporally and spatially resolve a boundary layer transition in a channel flow. The percolation theory allows them to describe a complex phase transition with only three critical exponents. Particle Image Velocimetry (PIV) experiments yield the basis for these investigations. In percolation theory, the data need to be binarized with the help of a threshold value, which in their case is the magnitude of the fluid velocity. In the percolation theory, a cell is either laminar or turbulent. It is evident that this concept is in excellent agreement with the phase change concept of turbulent flows, where the fluid field is separated also into two states or phases, respectively, namely into laminar streaks and turbulent patches. Wester et al. (2017) validate critical exponents (see below) of the directed percolation theory by experimental techniques with good accuracy. Also, other authors realized that there could exist an analogy to magnetic systems (e.g., Nelkin 1973; Tabeling and Willaime 2002) and generalized temperatures were introduced as stress parameters of turbulent systems (e.g., Robert and Sommaria 1991; Brown 1982). Furthermore, many authors claimed about a lack of existence of clear theories and models, describing turbulent phase transitions (Tabeling and Willaime 2002; Castaign 1989). The review article of Egolf and Hutter (2017b) with new results presents a first attempt to improve this incompleteness in the field of turbulence. It hopefully leads to a standardization in the sense that stress and order parameters are not arbitrarily chosen, with definitions that vary from paper to paper.

In this book, we have two main objectives:

(A) Firstly, the right analogy between magnetism and turbulence is introduced in the context of critical phenomena exhibiting continuous phase transitions. For a good understanding of the analogies, it will be necessary to outline the field theory of magnetism to a large extent. Having knowledge of this, the analogy helps to understand turbulence with knowledge of magnetism and vice versa.

(B) Secondly, with this analogy, it is possible to transform thermodynamic theories for magnetic systems to systems describing turbulent flows. In a first attempt, this has now been performed for the simplest model, the Mean Field Theory (MFT). This procedure already reveals two main results. First, in analogy to the Curie law in magnetism, a new law—called by us Curie law of turbulence—was discovered. Second, strict mathematical derivations based on this law lead to the right response function and energy of turbulence. We take these results as a first validation of this new discovered law. A collection of other new laws, where some of them show a divergence at the critical Reynolds number (compare with experimental observations in the article of Cortet et al. 2011), are outlined, and this occurs again in analogy to corresponding formulas in magnetism.

## 11.2 Cooperative Phenomena

In this chapter, we introduce the basic concepts of cooperative phenomena and of phase transitions in an ad hoc manner. A verification of these results and a presentation of a firmer basis, at least on the level of a macroscopic model, will follow in Sects. 11.3 and 11.4. One should be aware that models like the MFT are generally called ad hoc solution models (see Stanley 1971).

### *11.2.1 What is a Critical or a Cooperative Phenomenon?*

For readers, not so familiar with critical phenomena, we start our explanations with a solid–liquid phase transition. Lowering at constant pressure $p_0$ the temperature $T$ of a sample of liquid, e.g., water, from above its critical temperature, $T_c = T_W = 0\ °C$ at normal pressure (1 atm = 101.325 Pa), below this critical value, leads to a sudden change from its liquid to its solid state. Each of the two phases is characterized by its own specific volume $v$, specific energy $e$, entropy $s$, etc. At criticality some of these quantities jump, or, as we say in a more mathematical manner, they are discontinuous. Therefore, such changes in the material state are called discontinuous phase transitions. These phase transitions are alternatively also called first-order phase transitions (see Fig. 11.2a). If we now add a certain percentage of a freezing suppressing additive, e.g., alcohol or glycol, to the water, then, by decreasing the temperature, the solid phase is continuously produced. The reason is that the freezing process is practically reduced to the water/ice transition. Therefore, in the shrinking water content the additive concentration rises and, thus, shifts the transition temperature of transforming water to ice continuously to lower temperatures. So, this process yields a continuous phase transition (Fig. 11.2b). The substance then

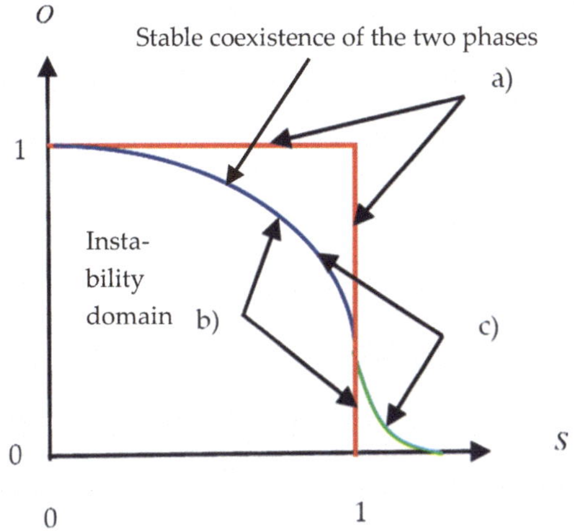


**Fig. 11.2** A discontinuous phase transition (**a**) and two continuous phase transitions (**b**) and (**c**). The latter shows scaling behavior above and below criticality. This occurs, e.g., in a magnetic system with a nonzero magnetic field. Courtesy Egolf and Hutter (2017b) © J. Entropy, reproduced with permission by Springer Science+Business Media

transforms from the liquid to the solid region through a mushy two-phase region. Such melting–freezing processes are successfully modeled, e.g., by the Continuous-Properties Model (CPM) of Egolf and Manz (1994). In this approach, melting and freezing are calculated by nonlinear diffusion. These authors showed that, like in solutions of the Burgers equation, the temperature profiles show a steepening effect. Egolf and Manz observed theoretically steepening of the profiles to the front or to the back and both. For water, the mushy substance is called ice slurry, a binary fluid that is applied in refrigeration technologies to transport efficiently the cold. Ice slurry may, for instance, be modeled as a Bingham fluid, which has higher flow resistance than pure water, but, because of the very high latent heat, is still energetically favorable for the transport of cold energy (see, e.g., Frei and Egolf 2000; Hansen et al. 2001; Egolf and Kauffeld 2005; Kauffeld et al. 2005).

Intriguing is that a variety of systems in different areas of scientific domains exhibit analogous critical phenomena. They are observed in fields where statistical physics applies, just as it is also the case in turbulence. Such systems are, as we have just discussed, liquid/solid transitions, but they also exist as gas/liquid transitions, magnetic systems with spontaneous magnetization in solid state probes with small internal magnetic field changes (see, e.g., Ma 1982; Thomson 1988), systems with spin-ordering in Ising ferromagnets or Ising antiferromagnets and spin glasses (Goldenfeld 1992), $He^4$ at the critical lambda-point (Krey 1980), etc.

Another feature occurring in phase transitions is symmetry breaking. In liquid/solid transitions this phenomenon occurs, because a regular crystal microscopically has a higher symmetry than the irregularly located atoms or molecules in the liquid phase. On the other hand, a gas/liquid transition has no such change of symmetry and, therefore, also no symmetry breaking. An example of a magnetic system with symmetry breaking will also be given below.

### 11.2.2 Stress and Order Parameter

In a description of materials showing a phase change, it is essential to identify the main external stress parameter of a system, $s$, which by division with its critical value, $s_c$, at which eventually symmetry breaking occurs, becomes dimensionless,

$$S = \frac{s}{s_c} \quad \Rightarrow \quad S \in (0, \infty). \tag{1172}$$

This definition was already used as Eq. (963). However, here it is expressed in more general symbols. The parameter $S$ characterizes the external forcing of a system, and therefore, is also called control parameter. A nonequilibrium thermodynamic system is externally forced away from equilibrium. This led scientists of nonlinear dynamics to prefer the designation stress parameter, as its monotonic numerical amplification increases the stress on a dynamic system.

In a liquid/solid transition system the stress parameter is the temperature $T$ or, alternatively, its coldness $1/T$. Related to the imposed impact on a system, it reacts in its specific manner. Nonequilibrium systems can even spontaneously organize their internal structure and raise the order, which is a process that is today well known in many scientific areas; it is called *self-organization process* (see, e.g., Haken 1983, 1987). It is initiated by instabilities and bifurcations, which are related to the critical stress parameters. The occurrence of a von Kármán eddy distribution behind a cylinder is an impressive example of such a process (see Sect. 11.2.3). The internal organization and, simultaneously, the order of the system is described by an order parameter $O$. This parameter is generally set to zero above criticality; below this, for lower temperatures, it is (monotonically) increasing toward "1," indicating the route to highest order, and lowest entropy, respectively (see Fig. 11.2b). It appears evident that such an order parameter curve $O(S)$ essentially characterizes a physical system.

Let us now define the order parameter. With knowledge that can be acquired by studying Egolf and Weiss (1995a, 2000), Egolf (2009), and Egolf and Hutter (2017b), we set the order parameter, in agreement to its special case Eq. (964), in a more general manner as

$$O = \frac{o - o_c}{o_p - o_c}, \quad \Rightarrow \quad O \in [0, 1], \tag{1173}$$

where $o$ denotes an appropriate changing property, which may be constant above criticality (Fig. 11.2b), but not as displayed in Fig. 11.2c, and increases monotonically below it toward smaller values of $S$. Furthermore, the quantity $O$ is its dimensionless counterpart. The index $c$ denotes here the critical and $p$ a pole value, which indicate the properties of the lowest and the highest order phase, respectively. Often, the critical parameter $o$ is so defined that its value at criticality is $o_c = 0$ and Eq. (1173) simplifies to

$$O = \frac{o}{o_{\mathrm{p}}}. \tag{1174}$$

Note that order parameters may be numerous mathematical objects as scalars, pseudo-scalars, vectors, tensors, elements of symmetry groups, etc. (see Krey 1980; Goldenfeld 1992).

Returning to the static fluid system, its density has the properties which an order parameter must possess, namely e.g., a monotonic increase of its value toward lower temperatures. Therefore, in agreement with Eq. (1174), we may write

$$O = \frac{\rho - \rho_{\mathrm{liquid}}}{\rho_{\mathrm{solid}} - \rho_{\mathrm{liquid}}} \quad \Rightarrow \quad O \in [0, 1]. \tag{1175}$$

In a model, it is often assumed that the pure liquid and solid phases show temperature-independent physical properties. Then, we have at and above criticality a pure liquid phase $\rho = \rho_{\mathrm{liquid}}$ and the order parameter is at its lowest value ($O = 0$). On the other hand, at $T = 0$ K there exists only the pure solid phase $\rho = \rho_{\mathrm{solid}}$, and here the order is at its maximum ($O = 1$) (see Fig. 11.2a).

Not all systems show homogenously dispersed mushy regions. However, it could be that ice blocks are floating in equilibrium with water, as it can be observed in the arctic sea. In such cases, the property $o$ (in the present case the density $\rho$) would have to be an integral measure, respectively, a spatial average of the material containing domain.

The relation of the order parameter to the inherent order of the system can be easily explored in a paramagnetic–ferromagnetic phase transition (see Fig. 11.3). In a paramagnetic sample, above a critical temperature, $T_c$, all spins of the system are homogenously disordered. The order defined by the spins is characterized by their magnetization $m$. A spin-up counts $s^{\uparrow} = +1$ (or + ½ in the case of an electronic spin) and a spin down $s^{\downarrow} = -1$ (or −½), respectively. Therefore, above $T_c$, statistically the magnetization

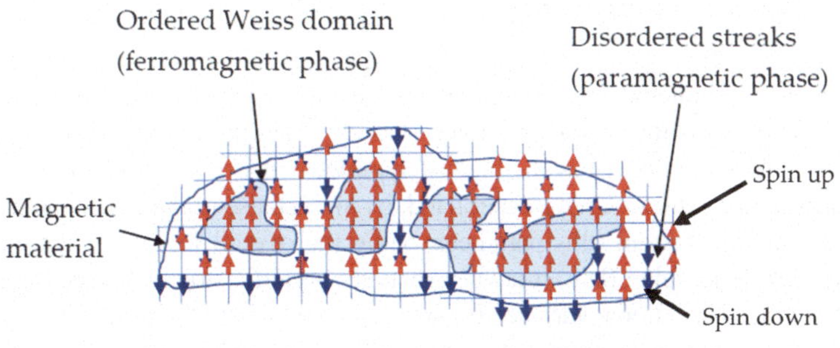


**Fig. 11.3** Weiss domains show spins only in the upward direction (red color) and, thus, have a maximum magnetic moment. The disordered streaks, surrounding the Weiss domains, show upward (red) and downward directed spins (blue), and, in the spatial mean, have no magnetization and, therefore, the magnetic moment in this intermediate zones is statistically equal to zero. Courtesy Egolf and Hutter (2017b) © J. Entropy, reproduced with permission by Springer Science+Business Media

$$M = \frac{m}{m_\mathrm{p}} \cong \frac{\sum s^{\uparrow} + \sum s^{\downarrow}}{\sum s}, \quad \sum s = \sum s^{\uparrow} + \sum |s^{\downarrow}| \qquad (1176\mathrm{a} - \mathrm{c})$$

is zero. In Eq. (1176a), we have assumed that the spins are very small and numerous, so that the magnetization $M$ can be regarded as a continuous variable. Because we only consider spins in a predefined and its opposite direction, we have written the otherwise vectorial quantity "magnetization" as a scalar. When lowering the temperature $T$ below the critical stress parameter $S_\mathrm{c} = T_\mathrm{c}$, the spins begin to order and by this the magnetization increases. If all spins are directed upward, their number is equal to the total sum of all absolute spin values, $\sum s$, and then the normalized magnetization $M$ is equal to "1." So, the magnetization can serve as an ideal order parameter of this magnetic system. In magnetism, such orderings may occur in patches (islands of aligned magnetic moments), where complete order occurs, whereas in the remaining local areas the elementary magnets are still in a fully disordered state. In terms of phase transitions, the ordered patches are one phase and the disordered ones define the second phase. The ordered domains are called Weiss domains (Fig. 11.3). At criticality the spins are disordered. Decreasing the temperature birth is given to small Weiss domains, which, with lower temperature, occur more numerously and start to grow, till at $T = 0$ K the entire area is a single fully ordered Weiss-domain. This picture is analogous to the ice blocks in the arctic sea, and, also here one requires the introduction of an integral quantity as a suitable order parameter

$$\vec{\mathcal{M}} = \int_{\Gamma} \vec{M}(\vec{x})\, \mathrm{d}V, \qquad (1177)$$

a global quantity, that is an effective magnetization, known as magnetic moment.

### 11.2.3 Symmetry Breaking

We already encountered two examples of physical systems showing symmetry breaking. The first is the von Kármán vortex street. Laminar flow in the downstream direction behind a cylinder has the highest symmetry; it is insensitive to all kinds of translations and rotations. On the other hand, by an increase of the stress parameter of the fluid dynamic system (this parameter being the overall Reynolds number $\mathrm{Re}_0$) above its threshold value $\mathrm{Re}_{0\mathrm{c}}$, it creates spatially periodic structures (see Fig. 11.4). These limit the translational variance for size-scaled eddies to the distance of periodic length $2\lambda$. Such a restriction is called symmetry breaking and is related to a decrease of the (generalized) entropy (see, e.g., Tsallis 2009; Beck and Schlögl 1993). At this point a subtlety must be explained. In the simple thermodynamic modeling of this exposure the first instability defines the critical point and that's it! This is the bifurcation point where quasi-steady behavior transforms to temporal behavior. In

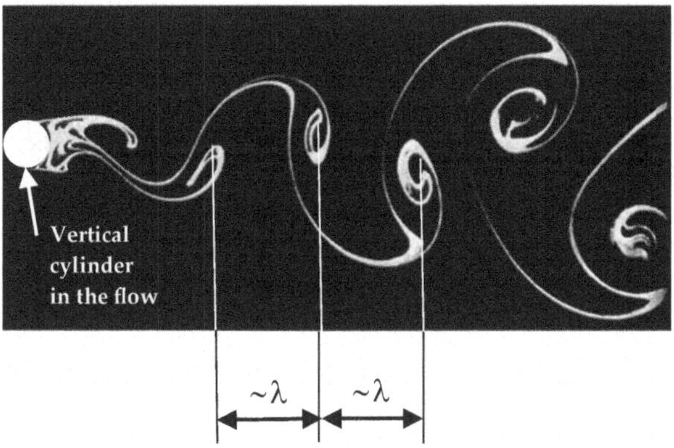


**Fig. 11.4** A von Kármán vortex street shows an increase of its width in the streamwise direction. This flow shows clockwise turning eddies in the upper half and anticlockwise turning eddies in the lower half plane showing (on the average) a periodic distribution. A higher order of this flow structure compared to laminar flow seems evident, whereas this is not so evident for chaotic and turbulent flows. Background open source. From van Dyke (1982), reproduced with changes

the MFT of a von Kármán flow behind a cylinder, it is also the critical Reynolds number where periodic von Kármán eddies appear. These structures are not turbulent; only after a further increase of the overall Reynolds number a transition to quasi-periodic structures, chaos, soft, and strong turbulence occurs. Different fluid dynamic systems show different scenarios of transitions to turbulence (see, e.g., Haken 1983, 1987). It is beyond the scope of this chapter to discuss these different types of transitions. This lack is also justified, because the MFT, presented in this treatise, does not describe the transitions in all its details. An important subtlety is that our model describes the transition from laminar to pulsating or fluctuating behavior, including also the turbulent states in a single step.

The second example is the magnetic physical system just discussed above. The stress parameter is the temperature, which by a decrease below its critical value, the Curie temperature $T_c$, initiates an ordering of spins. By this the homogeneity is lost and the system, in a natural manner, creates a preferential direction. Rotational symmetry is then only preserved around the axis pointing in this direction. By alignments of elementary magnetic moments, respectively spins, the energy of the system decreases to become a minimum at zero absolute temperature. The disordered phase, occurring in the interval $T_c \leq T \leq \infty$, is called paramagnetic phase and the phase with some order, observed between $0 \leq T < T_c$, is denoted ferromagnetic phase. In a magnetic system, the alignment of magnetic moments is also remarkably influenced by a second stress parameter, namely the internal magnetic field $H$.

### 11.2.4 Response Functions and Critical Exponents

Now, let us assume that there are two stress parameters $s_1$ and $s_2$ of a system and a single order parameter $O(s_1, s_2)$. Following Eq. (1173), we study the dependence of $O$ on $s_1$ by setting $s_2 = 0$,

$$O(s_1, 0) = f\left(\frac{s_1 - s_{1c}}{s_{1p} - s_{1c}}\right), \quad s_1 \le s_{1c}, \qquad (1178a, b)$$

where the exact form of the function $f$ is sought. For criticality, we request (see Fig. 11.2b)

$$f(s_{1c}) = f(0) = 0, \qquad (1179a, b)$$

and for the pole value

$$f(s_{1p}) = f(1) = 1. \qquad (1180a, b)$$

A power law with a positive (possibly fractional) exponent $\beta'$ fulfills these two requirements (references with tables and quantitative values are given below and in Table 11.2)

$$O(s_1, 0) = \left(\frac{s_1 - s_{1c}}{s_{1p} - s_{1c}}\right)^{\beta'}, \quad s_1 \le s_{1c}, \quad \beta' > 0. \qquad (1181a-c)$$

If $O(s_{1c}, 0) \neq 0$, then we redefine the order parameter as the difference $\Delta O = O(s_1, 0) - O(s_{1c}, 0)$. This is the case for the liquid–solid transition, but not for the magnetic transition, as we will see.

Such a power law description may or may not be approximate when applied to the entire domain [0,1] of $O$, however, close to criticality, it yields experimentally a fairly accurate formula. Therefore, from now on we mainly will apply this equation only in the limit as $s_1$ approaches criticality and define

$$\beta' = \lim_{\varepsilon \to 0} \frac{\log_e O(\varepsilon)}{\log_e \varepsilon}, \quad \varepsilon = \frac{s_1 - s_{1c}}{s_{1p} - s_{1c}}, \quad s_1 \le s_{1c}. \qquad (1182a-c)$$

The exponent $\beta'$ is called critical exponent. Finally, the order parameter is recalled to be

$$O(s_1, 0) = \left(\frac{s_1 - s_{1c}}{s_{1p} - s_{1c}}\right)^{\beta'}, \quad s_1 \le s_{1c}, \quad \beta' > 0 \qquad (1183a-c)$$

For consistency reasons, we denote this exponent by $\beta'$ just as in the main literature (see, e.g., Ma 1982; Krey 1980; Stanley 1971; Goldenfeld 1992). In Fig. 11.2 it is seen that we can choose $s_{1\mathrm{p}} = 0$, thereby transforming Eq. (1183a–c) to

$$O(s_1, 0) \propto \left(1 - \frac{s_1}{s_{1\mathrm{c}}}\right)^{\beta'}, \quad s_1 \leq s_{1\mathrm{c}}, \quad \beta' > 0. \qquad (1184\mathrm{a}-\mathrm{c})$$

Next, the order parameter $O$ will be studied as a function of the second stress parameter $s_2$ at criticality $s_{2\mathrm{c}}$ and slightly above at fixed $s_1 = s_{1\mathrm{c}}$, i.e., along an $s_1$–isoline,

$$O(s_{1\mathrm{c}}, s_2) \propto \left(\frac{s_2 - s_{2\mathrm{c}}}{s_{2\mathrm{c}}}\right)^{1/\delta}, \quad \delta > 0. \qquad (1185\mathrm{a,b})$$

Here, the critical exponent $\beta'$ is written for reasons of tradition as the constant $1/\delta$. The reaction of the property of the bulk material and of the order parameter $o$ to a change of the stress parameter $s_2$ at constant $s_1$ can be estimated by the quantity

$$\psi := \left.\frac{\partial O}{\partial s_2}\right|_{s_1}. \qquad (1186)$$

Assume that $O(s_{1\mathrm{c}}, s_2)$ describes the order parameter with a form as shown in Fig. 11.2c. Then, scaling, given by a power law, is present on both sides of the critical point. Now, one may write, e.g., for the exponent $\gamma'$ if $s_1 < s_{1\mathrm{c}}$ and $\gamma$ if $s_1 > s_{1\mathrm{c}}$. The quantity $\psi$ is discontinuous at criticality, as is clearly seen from

$$\psi \propto \begin{cases} \left(1 - \frac{s_1}{s_{1\mathrm{c}}}\right)^{-\gamma'}, & s_1 < s_{1\mathrm{c}}, \quad \gamma' > 0, \\ \left(\frac{s_1}{s_{1\mathrm{c}}} - 1\right)^{-\gamma}, & s_1 > s_{1\mathrm{c}}, \quad \gamma > 0. \end{cases} \qquad (1187\mathrm{a}-\mathrm{f})$$

The related critical exponents are usually identical; i.e., $\gamma' = \gamma$ (see, e.g., Ma 1982). The above twofold values occur for most critical exponents as discussed below. However, because of brevity, in the following sections we always merely formulate a single version (Eq. 1187a or 1187b). The negative signs in the exponents of these equations are introduced to have positive values for $\gamma$ and $\gamma'$. A further advantage is that a negative sign visibly signalizes that this physical quantity diverges when criticality is approached.

Still a further quantity, that diverges as $s_{1\mathrm{c}}$ is approached, may exist; if so, it is described as

$$\phi \propto \left(1 - \frac{s_1}{s_{1\mathrm{c}}}\right)^{-\alpha'}, \quad s_1 \leq s_{1\mathrm{c}}, \quad \alpha' > 0, \qquad (1188\mathrm{a}-\mathrm{c})$$

with the critical exponent $\alpha'$.

The special case of a liquid–solid transition is obtained by identifying the first stress parameter with the temperature, $s_1 = T$, and the second with $s_2 = p_c$. The order parameter was already introduced to be the density $\rho$ of the fluid. With these identifications Eq. (1184) becomes

$$\Delta\rho = \rho(T, p_c) - \rho(T_c, p_c) \propto \left(1 - \frac{T}{T_c}\right)^{\beta'}, \quad T \leq T_c, \quad \beta' > 0, \qquad (1189a-d)$$

provided, of course, $\rho$ only depends on $T$ and $\rho_c$. Correspondingly, Eq. (1185) transforms to

$$\Delta\rho = \rho(T_c, p) - \rho(T_c, p_c) \propto \left(\frac{p - p_c}{p_c}\right)^{1/\delta}, \quad \delta > 0. \qquad (1190a-c)$$

Next, let us describe the reaction of a static fluid system, subjected to a change of its order parameter (being the volume $O = V$) by an altering pressure field $s_2 = p$ at constant temperature $s_1 = T$. It is given by Eq. (1186), which is adapted to the symbols of the physical properties of a static fluid, and where the negative sign guarantees a positive value of the compressibility, viz,

$$\kappa_T = -\frac{1}{V}\frac{\partial V}{\partial p}\bigg|_T. \qquad (1191)$$

$\kappa_T$ is the *compressibility at constant temperature* and must be distinguished from the adiabatic *compressibility at constant entropy* $\kappa_S$. If the volume change, caused by a unit of pressure change, is large, the system is said to be highly compressible, and if this quantity is zero, the fluid is incompressible. Following Eq. (1187a–c), the compressibility diverges toward criticality from below (exponent $\gamma'$) and above (exponent $\gamma$). Then, e.g., the low temperature case is described by the formula

$$\kappa_T \propto \left(1 - \frac{T}{T_c}\right)^{-\gamma'}, \quad T \leq T_c, \quad \gamma' > 0. \qquad (1192a-c)$$

Another diverging quantity of the type of Eq. (1188) of a fluid system is the specific heat at constant volume, $C_V$, characterized by the critical exponent $\alpha'$,

$$C_V \propto \left(1 - \frac{T}{T_c}\right)^{-\alpha'}, \quad T \leq T_c, \quad \alpha' > 0, \qquad (1193a-c)$$

which must be distinguished from the specific heat at constant pressure, $C_p$.

In analogy to the liquid–solid transitions, the special case of a magnetic system is obtained by substituting for the first stress parameter the temperature, $s_1 = T$, and for the second one, the external magnetic field $s_2 = H_0$. The dimensionless order parameter was already introduced to be the magnetization $M$. Thus, Eq. (1184) becomes

$$M(T,0) \propto \left(1-\frac{T}{T_c}\right)^{\beta'}, \quad T \leq T_c, \quad \beta' > 0. \qquad (1194a-c)$$

Furthermore, Eq. (1185) transforms to the following order parameter/stress parameter relation

$$M(T_c, H_0) \propto (H_0 - H_{0c})^{1/\delta} = {H_0}^{1/\delta}, \quad \delta > 0, \qquad (1195a-c)$$

where $H_{0c} = 0$ is usually suppressed. Next, consider the reaction of a magnetic system, described by a change of its order parameter (the magnetization $O = M$) to an altering magnetic field $s_2 = H_0$, at constant temperature $s_1 = T$. It is given by Eq. (1186), which is adapted to the magnetic symbols; this yields

$$\chi_T = \left.\frac{\partial M}{\partial H_0}\right|_T. \qquad (1196)$$

This derivative of $M$ with respect to $H_0$ at fixed $T$ is called the *differential magnetic susceptibility* at constant temperature and is denoted by $\chi_T$. Just as it was the case for the compressibility, also the susceptibility diverges as criticality is approached (from below and above). The first (low temperature) case is described by the formula

$$\chi \propto \left(1-\frac{T}{T_c}\right)^{-\gamma'}, \quad T \leq T_c, \quad \gamma' > 0. \qquad (1197a-c)$$

In a magnetic system, its ability to react to opposed stress is quantified by its susceptibility $\chi$. If the magnetization change is large, caused by a unit of magnetic field change, the system is said to be easily magnetizable, and if this quantity is zero, the material is non-magnetizable. Quantities, such as the compressibility of a fluid and the susceptibility of a magnetic system are called analogous "response functions."

Furthermore, following Eq. (1188a–c), another diverging quantity of a paramagnetic to ferromagnetic phase transition system is the specific heat at constant magnetic field,[1] $C_{H_0}$, formally identical to Eq. (1187a–c), which, on the other hand, is characterized by its exponent $\alpha'$,

[1]In strict analogy, it would have to be $C_M$, the specific heat at constant magnetization $M$, instead of $C_{H_0}$, the specific heat at constant external magnetic field $H_0$.

**Table 11.1** Analogy between a static fluid phase transition and a magnetic phase transition

| Quantity | Fluid | Magnet |
|---|---|---|
| Stress parameter 1 | $T$ | $T$ |
| Stress parameter 2 | $p$ | $H_0$ |
| Order parameter | $v(\rho)$ | $M$ |
| Specific heat | $C_V$ | $C_M$ |
| Response function | $\kappa_T$ | $\chi_T$ |

Recall that the specific volume is related to the density of the fluid by $v = 1/\rho$

$$C_{H_0} \propto \left(1 - \frac{T}{T_c}\right)^{-\alpha'}, \quad T \leq T_c, \quad \alpha' > 0. \tag{1198a – c}$$

These thermomagnetic quantities are reviewed by Egolf et al. (see Kitanovski and Egolf 2006; Egolf et al. 2015).

The alert reader must have recognized that there is a perfect analogy between the discussed fluid and magnetic systems. The corresponding quantities are shown in Table 11.1.

In the discussion of the most important critical exponents two additional ones are important, namely the pair correlation and the correlation length.

### *11.2.5 Pair Correlation Function and Correlation Length*

To derive the pair correlation, let us study the particle number density $n(\vec{r})$ of a physical system with $N$ particles [in this paragraph we follow the presentation in Stanley (1971)]

$$n(\vec{r}) = \sum_{k=1}^{N} \delta(\vec{r} - \vec{r}_k). \tag{1199}$$

Defining $\langle(\ldots)\rangle$ as the ensemble average of $(\ldots)$, its "pair correlation function," defined as

$$\varsigma_2 = \langle n(\vec{r})n(\vec{r}')\rangle, \tag{1200}$$

is proportional to the conditional probability of meeting a particle at position $\vec{r}$ if there is another particle at position $\vec{r}'$. Closely related to this quantity is the pair or density–density correlation

$$G\left(\vec{r}, \vec{r}'\right) = \left\langle \left[n\left(\vec{r}\right) - \left\langle n\left(\vec{r}\right)\right\rangle\right] \cdot \left[n\left(\vec{r}'\right) - \left\langle n\left(\vec{r}'\right)\right\rangle\right] \right\rangle, \tag{1201}$$

which is a measure of the correlations of the fluctuations of the particle density.

In the special case that a system is spatially uniform (translationally invariant), it follows that

$$G\left(\vec{r}, \vec{r}'\right) = G\left(\vec{r} - \vec{r}'\right). \tag{1202}$$

Employing the notation,

$$n \equiv \left\langle n\left(\vec{r}\right)\right\rangle = \left\langle n\left(\vec{r}'\right)\right\rangle, \tag{1203a, b}$$

Equations (1201) and (1203a,b) yield

$$\begin{aligned} G\left(\vec{r} - \vec{r}'\right) &= \left\langle \left[n\left(\vec{r}\right) - n\right] \cdot \left[n\left(\vec{r}'\right) - n\right] \right\rangle \\ &= \left\langle \left[n\left(\vec{r}\right)n\left(\vec{r}'\right) - n\,n\left(\vec{r}\right) - n\,n\left(\vec{r}'\right) + n^2\right] \right\rangle \\ &= \left\langle n\left(\vec{r}\right)n\left(\vec{r}'\right)\right\rangle - 2n\left\langle n\left(\vec{r}\right)\right\rangle + n^2. \end{aligned} \tag{1204a – c}$$

Applying again (1203a), yields

$$G\left(\vec{r} - \vec{r}'\right) = \left\langle n\left(\vec{r}\right)n\left(\vec{r}'\right)\right\rangle - n^2. \tag{1205}$$

For $\left|\vec{r} - \vec{r}'\right| \to \infty$ it may be assumed that the probability of finding a particle at position $\vec{r}'$ is independent of the presence of a particle at $\vec{r}$. However, if the densities are uncorrelated, it follows that

$$\left\langle n\left(\vec{r}\right)n\left(\vec{r}'\right)\right\rangle = \left\langle n\left(\vec{r}\right)\right\rangle\left\langle n\left(\vec{r}'\right)\right\rangle = n^2. \tag{1206a, b}$$

Hence, Eq. (1205) shows that, in the limit of large distances between $\vec{r}$ and $\vec{r}'$, the pair correlation function $G$ vanishes,

$$G\left(\vec{r} - \vec{r}'\right) \to 0, \quad \left|\vec{r} - \vec{r}'\right| \to \infty. \tag{1207}$$

In linear and in thermal equilibrium systems, this behavior is often described by an exponential decay at $T_c$ and $p_c$ (see, e.g., Goldenfeld 1992; Yeomans 1992)

$$G(r) \propto \frac{1}{r^{d-2+\eta}} \exp\left(-\frac{r}{\xi}\right). \tag{1208}$$

with $\eta$ being the critical exponent of the pair correlation function $G$. In this formula, $d$ denotes the Euclidean dimension of the system and $\xi$ is the correlation length of the density fluctuations. It is a measure giving the distance over which cooperative behavior is perceptible and corresponds to characteristic sizes of the already discussed patches, e.g., Weiss domains. Toward criticality, the correlation length scales also with a power law

$$\xi \propto \left(1 - \frac{T}{T_c}\right)^{-\nu'}, \quad T \leq T_c, \quad \nu' > 0. \tag{1209a – c}$$

and diverges with the critical exponent $\nu'$.

Stanley (1971) demonstrates by a clear and brief computation, using the Boltzmann factor of a grand canonical ensemble, that

$$\frac{\kappa_T}{\kappa_{T,0}} = \frac{1}{n} \int_\Gamma G(\vec{r})\, \mathrm{d}V, \tag{1210}$$

where $\kappa_{T,0}$ is the isothermal compressibility of an ideal gas. This law is the analogue to the fluctuation-dissipation theorem of a static fluid. It proves that an increase of the compressibility is related to an increase of the density fluctuations (divergence) and to the range of the density–density correlation function.

### *11.2.6 Universality: Yes or No ?*

We have now introduced the main critical exponents $\alpha'$, $\beta'$, $\delta$, $\gamma'$, $\eta$, and $\nu'$ and the counterparts $\alpha$ and $\gamma$, of which some experimentally determined values are listed in Table 11.2.

Why is in the literature and in this section so much attention given to the limit toward criticality, respectively to critical exponents, if the complete functions contain much more information? The answer is that experimentally it was observed that different systems, to a very high experimental accuracy, show the same values of the exponents. This can be seen, e.g., in Table 11.2, where the critical exponents of the order parameter of the fluids $CO_2$ and Xe are 0.34 and 0.35, respectively. Many more such fluids show these values (see, e.g., Ma 1982). Yet, not only this, also the magnetic system, e.g., composed by EuS, shows a critical exponent $\beta' = 0.33$. Furthermore, e.g., the 3-$d$ Heisenberg model (see, e.g., Papon et al. 2002; Le Bellac et al. 2006) predicts a value exactly in this range. Therefore, some decades ago, scientists were convinced that critical exponents are a manifestation of a kind of

**Table 11.2** Quantities of critical exponents given by models and experiments for a selection of different physical systems for the range $T < T_c$. Data are from Ma (1982) and Stanley (1971)

| Quantity | $\alpha'$ | $\beta'$ | $\delta$ | $\gamma'$ | $\nu'$ |
|---|---|---|---|---|---|
| *Fluids* | --- | --- | --- | --- | --- |
| $CO_2$ | ~ 0.1 | 0.34 | 4.2 | ~ 1.0 | - |
| Xe | < 0.2 | 0.35 | 4.4 | ~ 1.2 | 0.57 |
| *Magnets* | --- | --- | | --- | --- |
| Ni | -0.3 | 0.42 | 4.22 | 1.32 | - |
| EuS | -0.15 | 0.33 | - | - | - |
| $CrBr_3$ | - | 0.368 | 4.3 | - | - |
| *Models* | --- | --- | | --- | --- |
| Classical | 0 | 1/2 | 3 | 1 | 1/2 |
| Spherical 3-*d* | - | 1/2 | 5 | - | - |
| Ising 2-*d* | 0 | 1/8 | ~15 | ~7/4 | 1 |
| Ising 3-*d* | ~ 1/8 | ~ 5/16 | ~5 | ~5/4 | - |
| Heisenberg 3-*d* | - | ~ 0.35 | ~5 | - | - |

universal behavior of systems showing phase transitions. For example, in the three-dimensional Ising antiferromagnetic material $DyAlO_3$, the exponent $\beta'$ was, for $T \rightarrow T_c$ (from below), experimentally determined to be $\beta' = 0.311 \pm 0.005$. Goldenfeld (1992) writes that the observed values of $\beta'$ for a liquid–solid transition and a para-ferromagnetic phase transition system, within the accuracy of the performed experiments, were determined to be the same. So, early researchers believed that the critical exponents of all order parameter curves were 1/3. However, Ho and Lister (1969) (see also, e.g., Stanley 1971) demonstrated unequivocally that for the insulating $CBr_3$ material $\beta' = 0.368 \pm 0.005$ (see Table 11.2), disproving such an assumption.

Even if such deviations and imperfections are generally accepted today, in the scientific community the consensus is that toward criticality, where correlation and long-range order measures increase, the nature of the short-range interactions may become less significant. This explains that systems in different areas of physics reveal so similar or even identical critical laws.

Basic interaction models, e.g., listed in Table 11.2, reveal equations and inequalities that relate the different critical exponents to each other. These dependencies yield the possibility that if, e.g., two critical exponents are known that a third critical exponent can be calculated. In Table 11.2, one can find the critical exponents for static fluids and magnetic systems and the theoretical values given by different magnetic interaction models. The classical values correspond to the values derived with the MFT (see Table 11.2). The "spherical model" is also a simple theory to describe ferromagnetism. It was solved by Berlin and Kac (1952). It is a model that can be analytically solved in the presence of an external field. The Ising model (see eg., Papon et al. (2002)) was formulated by Wilhelm Lenz in 1920 and solved by his student Ernst Ising. It describes magnetic dipole moments also as next neighbor entities in a regular lattice configuration. The one-dimensional Ising model does not show a phase transition, whereas the higher dimensional Ising models do. The Heisenberg model (see Le Bellac et al. 2006) is more sophisticated than the models discussed above and serves for the study of critical behavior and phase transitions of quantum mechanical systems. More information on all these models and their derivations are found in the following Refs. Ma (1982), Krey (1980), Stanley (1971), and Goldenfeld (1992). If a complete description of critical exponents is envisaged, the above mentioned relations between the critical exponents are very important. However, because investigations on phase transitions in turbulence are still in their infancy, we at present do not already have need for them. For a reader, who wants to explore more on this topic, an excellent survey is given by Stanley (1971).

In this review, by discussing later the MFT, we only touch the surface of standard and well-accepted theories on phase transitions, including the theoretical discoveries made possible by the application of the DQTM to solve elementary turbulent shear flows. Our intention is to highlight and explore analogies occurring between turbulent flows and other physical systems known for decades to reveal phase transitions. So, let us now switch from the liquid and magnetic model systems back to turbulence.

### *11.2.7 Turbulent Phase Transition with its Two Phases*

If turbulence is stated to be a cooperative or critical phenomenon exhibiting phase transitions, the first question that arises is what the two phases are and what makes their difference. By thinking in analogies this question can be easily answered. In a liquid system, the low-order phase is the pure liquid and the high-order phase is the pure solid phase. In turbulence, the lowest order phase, with the highest symmetry properties, is laminar flow and the highest order phase, in this case, would be an infinite Reynolds number turbulent flow. Therefore, e.g., in a medium Reynolds number flow, we should be able to distinguish subdomains of calm laminar streaks from regions of high turbulent activity, showing a high production rate of kinetic fluctuation energy, of vorticity and enstrophy, etc. This picture fits very well with

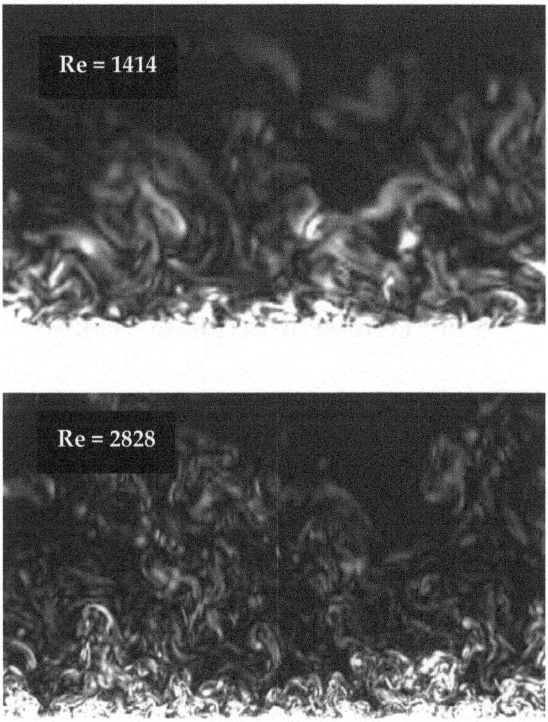


**Fig. 11.5** DNS vorticity magnitude contours of a three-dimensional boundary layer flow at two Reynolds numbers. The upper turbulent flow was evaluated at a low Reynolds number and the lower flow at twice this value, From Eprints (2019), reproduced with changes

Leonardo da Vinci's view, demonstrated in numerous drawings of turbulent water flows. In a description Leonardo even wrote "*Observe the motion of the surface of the water, ..., which has two motions, ..., one part of which is due to the principal current, the other to the random and reverse motion* (see Fig. 5.1 and Gad-el-Hak 2000)." It is intriguing how his observations support the two-phase picture of turbulence, which is strongly advertised in this presentation! Therefore, we summarize that coherent structures are separated from rather inactive flow regions. Furthermore, we claim that the vorticity-rich regions are low-entropy regions, and we classify them to be the analogous patterns, called Weiss domains in magnetic phase transition systems. Today, the study of coherent structures is a main activity in turbulence research, where some part of the studies is experimental and the larger part is performed by direct numerical simulations (DNS). Because of reasons of length, it is beyond the scope to review this research domain in this chapter. However, a concise review of the authors on laminar streaks and coherent structures, e.g., can be found in Holmes et al. (1996), Hussain (1983), and Egolf and Hutter (2018).

To give some impression of the two distinct phase patterns, Fig. 11.5 is presented. It shows DNS data in form of vorticity contour plots for two Reynolds numbers given in the figure inserts. In both figures high vorticity domains are shown as white regions, which in a simplification are related to whorls and eddies. These domains are distinguished from the dark laminar streaks, where the flow is of lower specific turbulent energy, e.g., because it lacks rotational energy. At lower Reynolds number, the whirling coherent structures show coarse-grained patterns and at higher Re values a finer distribution. According to the proposed phase change concept, it is assumed that the finer dispersed patterns occurring at a higher Reynolds number are more space filling, a fact that can be clearly observed by comparing the flow structures of the higher located figure to those in the lower positioned one.

## 11.3 Mean Field Theory of a Paramagnetic to Ferromagnetic Phase Transition

We start with an old, but very successful model of phase transitions, namely the Mean Field Theory (MFT) and introduce it by following mainly arguments of Ma (1982) . *The method to advance is to review its simplest introduction for a ferromagnetic–paramagnetic phase transition, and then, in the next section, in analogy immediately apply it to turbulence.*

Let us assume that a ferromagnetic body is brought into an external magnetic field $H_0$. An electron in the solid material is in a local internal magnetic field $H$, that is established by the external magnetic field $H_0$ plus an average magnetic or magnetization field, $m$, that originates from neighboring spins, which have the same direction as or the opposite one from the external field. Thus, the following theory is again written only in scalar variables. In the MFT, it is assumed that the field $m$ is a function of the average of all neighbouring spins. Therefore, its value may slightly deviate from the actual magnetization of all spins that is strictly fulfilling the relation

$$H = H_0 + M. \tag{1211}$$

For small $m$'s, we may assume that the external field is linear in this averaged magnetization

$$H = H_0 + a\,m, \tag{1212}$$

with the relation

$$M \cong a\,m \cong m, \tag{1213a, b}$$

assuming a value of the constant $a$ close to "1." The quantity $m$ follows the Curie law, i.e.,

$$m = c\,\frac{H}{T}, \tag{1214}$$

where $c$ is another constant. By solving for $m$ and eliminating the internal field $H$, in a combination of Eqs. (1212) and (1214), it follows that

$$m = \frac{cH_0}{T - ac}. \tag{1215}$$

The product $ac$ must be a characteristic temperature of the problem. The only suggestive choice is the critical temperature,

$$T_{\mathrm{c}} = ac. \tag{1216}$$

Then, Eq. (1215) takes its final form

$$m = \frac{cH_0}{T - T_{\mathrm{c}}}, \quad T > T_{\mathrm{c}}. \tag{1217a, b}$$

Next, with the assumption in Eq. (1213b), the differential susceptibility can be derived from (1217) to yield

$$\chi_T \cong \left.\frac{\partial M}{\partial H_0}\right|_T = \frac{c}{T - T_{\mathrm{c}}}, \quad T > T_{\mathrm{c}}. \tag{1218a − c}$$

If the temperature converges toward the critical temperature, the magnetization and the susceptibility diverge, the latter with exponent $\gamma = 1$. These results are in agreement with experiments (see, e.g., Ma 1982).

This most simple approach delivers good results above criticality; however, it fails below the Curie temperature $T_{\mathrm{c}}$.

To obtain reasonable results for the full domain, one must develop approximation Eq. (1212) to a higher order

$$H = H_0 + a\,m + b\,m^3 + O(m^5), \tag{1219}$$

where the odd power of $m$ guarantees that the oppositely directed spins (e.g., those positioned in the downward direction) will contribute negatively to the internal magnetic field $H$. In this equation $b$ is a negative constant (see below). Now, combining Eqs. (1214) and (1219), it follows that

$$bc\,m^3 + ac\,m - T\,m + cH_0 = 0. \tag{1220}$$

Equation (1216) remains valid and changes Eq. (1220) to

$$bc\,m^3 + (T_{\mathrm{c}} - T)\,m + cH_0 = 0. \tag{1221}$$

Because the quadratic term is missing in this polynomial function of degree three, this is a reduced cubic equation of $m$ that can be analytically solved, depending on the numerical values of the constants, by either the Casus Cardani or the Casus irreducibilis method. Here, it is sufficient to study only some special cases, as outlined below.

(A) Firstly, we look at a domain of vanishing external magnetic field $H_0 = 0$ and temperatures below criticality, $T < T_c$. Then from Eq. (1221), there follows (apart from the trivial solution $m = 0$) a quadratic equation with the solution

$$m\,(T,0) = \left[\frac{T - T_c}{bc}\right]^{1/2}, \quad T < T_c, \quad b < 0, \quad c > 0, \qquad (1222\text{a} - \text{d})$$

which is only meaningful for $b < 0$ (sic!). Moreover, by applying Eq. (1213b) and a comparison with Eq. (1194), the critical exponent is extracted to be

$$\beta' = \frac{1}{2}. \qquad (1223)$$

To derive the susceptibility, the easiest way is to differentiate Eq. (1221) with respect to $H_0$ at constant temperature $T$,

$$3bc\,m^2 \frac{\partial m}{\partial H_0}\bigg|_T + (T_c - T)\frac{\partial m}{\partial H_0}\bigg|_T + c = 0, \quad T < T_c, b < 0, c > 0, \quad (1224\text{a} - \text{d})$$

With Eqs. (1213b) and (1218a) this is identical to

$$\left[\,3bc\,m^2 + (T_c - T)\,\right]\chi_T + c = 0, \quad T < T_c, b < 0, c > 0. \qquad (1225\text{a} - \text{d})$$

Now, Eq. (1222a) is inserted for $m$ and Eq. (1225a) is solved for $\chi_T$, which leads to

$$\chi_T = \frac{c}{2}\frac{1}{(T_c - T)}, \quad T < T_c, \quad c > 0, \qquad (1226\text{a} - \text{c})$$

Comparing this equation with Eq. (1197) yields a further critical exponent, namely

$$\gamma' = 1. \qquad (1227)$$

(B) Secondly, we assume that $H_0 \neq 0$ and study the behavior of the magnetic system at the critical temperature, $T = T_c$. In this case, Eq. (1221) simplifies to

$$b\,m^3 + H_0 = 0, \quad T = T_c, \quad b < 0. \qquad (1228\text{a} - \text{c})$$

revealing, in the following formula, a further critical exponent, $1/\delta$,

$$m(T_c, H_0) = \left(-\frac{H_0}{b}\right)^{1/3}, \quad b < 0. \qquad (1229\text{a, b})$$

alerting us that the constant $b$ must be negative. Comparison with Eqs. (1195b) and (1213b) leads to

$$\delta = 3. \tag{1230}$$

(C) Thirdly, we look at the domain of vanishing external magnetic field $H_0 = 0$ and temperatures above criticality, $T > T_c$. For this region, the results are practically the same as for $T < T_c$, and we obtain

$$m(T,0) = \left(\frac{T - T_c}{bc}\right)^{1/2}, \quad H_0 = 0, \quad T > T_c, b > 0, c > 0. \tag{1231a – e}$$

where $b$ now must be positive. We conclude that

$$\beta = \beta' = \frac{1}{2}. \tag{1232a, b}$$

As in the first case, but with $T - T_c$ instead of $T_c - T$ (see Eq. (1226a)), the susceptibility is derived to be analogous to Eq. (1226)

$$\chi_T = \frac{c}{2} \frac{1}{(T - T_c)}, \quad T < T_c, c > 0. \tag{1233a – c}$$

resulting in the equivalence of the critical exponents of these quantities,

$$\gamma = \gamma' = 1. \tag{1234a, b}$$

With the approximation Eq. (1213b), the magnetic energy for a material with linear magnetization can be estimated by the formulas (see, e.g., Kitanovski and Egolf 2006)

$$E = -\mu_0 \mu_r HM \cong -\mu_0 \mu_r Hm. \tag{1235a, b}$$

The constant $\mu_0 = 4\pi \cdot 10^{-7} \mathrm{NA}^{-2}$ denotes the magnetic permeability and $\mu_r$ the dimensionless relative magnetic permeability. We will only study the case of a vanishing external magnetic field $H_0 = 0$.

(A) In the first case for ferromagnetic materials at temperatures below criticality, $T < T_c$, with Eq. (1212), it follows that

$$E \cong -\mu_0 \mu_r \left(H_0 + a\, m\right) m \cong -\mu_0 \mu_r\, m^2. \tag{1236a, b}$$

In this case, the external magnetic field is zero, however, the internal magnetic field, due to spontaneous magnetization, may be nonzero. Inserting Eqs. (1222a) into (1236b), it leads to

$$E \cong -\frac{\mu_0 \mu_r}{bc}(T - T_c), \quad H_0 = 0, \quad T < T_c. \tag{1237a – c}$$

Furthermore, the specific heat $C_{H_0}$ at constant magnetic field $H_0$ is defined by the derivative

$$C_{H_0} = \left.\frac{\partial E}{\partial T}\right|_{H_0}, \quad H_0 = 0, \tag{1238a, b}$$

which with Eq. (1237a) leads to

$$C_{H_0} \cong -\frac{\mu_0 \mu_r}{bc}, \quad H_0 = 0, \qquad T < T_c, \tag{1239a – c}$$

a specific heat that is positive because $b$ is negative. Comparison with Eq. (1198) delivers a next critical exponent, namely

$$\alpha' = 0, \tag{1240}$$

because Eq. (1239a) is temperature independent. □

(B) The second case, where the magnetic system is exactly at its critical temperature, $T = T_c$, reveals a discontinuity in the specific heat. This is shown by proving that the specific heat above criticality is also constant (see third case), but shows a different value from that below criticality (see Eq. 1239a). Thus, with the help of the third case, it follows that the specific heat shows a discontinuity. □

(C) In the third case at temperatures above criticality, $T > T_c$, there is no spontaneous magnetization, and it follows that

$$E \equiv 0, \quad T > T_c. \tag{1241a, b}$$

This implies that

$$C_{H_0} \equiv 0, \quad T > T_c. \tag{1242a, b}$$

and

$$\alpha = \alpha' = 0, \quad T > T_c. \tag{1243a – c}$$

Comparing Eqs. (1239a) and (1242), it is evident that the specific heat is discontinuous at criticality. □

Summarizing, in the simplest manner, we have developed the MFT for a ferromagnetic–paramagnetic phase transition and have derived some classical critical exponents shown in Table 11.2. They are also called mean field exponents. Ma (1982) writes: "*They do not agree very well with the [measured] values. However, in view of how little we put in, the theory is remarkably successful. It shows that the field provided by neighboring spins is responsible for generating a nonzero magnetisation below* $T_c$." In Table 11.2, the reader may compare the classical critical exponents with the corresponding measured values and notice that they correspond to one another (even if some only very approximately). The theory also predicts a divergent susceptibility and exponents that are independent of any details. Because the ideas of the MFT, in strict analogy, can be applied to antiferromagnetic materials, liquid–gas, binary alloys, and other critical systems, it is tempting to assume that the models in these alternative fields must reveal the same critical exponents. This led to the idea of the earlier formulated universality.

In the following sections, our objective is to test whether the classical and successful MFT, up-to-present mainly applied in magnetism, could have an application to turbulence and might deliver new insights and eventually even new physical results. We will experience that this is indeed the case.

## 11.4 Mean Field Theory of Turbulence

Notice that in this section, the MFT is developed for turbulence in an analogous manner to magnetism, described in Sect. 11.3. Therefore, for better understanding, it is valuable to have Table 11.3 at one's disposal, and in case of a doubt to check the analogous model derivations in the magnetism part, which are slightly more extensive than those in this section on turbulence.

Following Egolf and Weiss's discovery (see Egolf and Weiss 1995a, b) that a generalized temperature, $T$, of plane Couette flow is inverse to the overall Reynolds number $\mathrm{Re}_0$, that is, proportional to the characteristic velocity $u_0$, we assume that the following analogy holds between magnetism and turbulence

$$S_1 = T \quad \leftrightarrow \quad S_1 = 1/u_0. \tag{1244a, b}$$

This and further analogies are listed in Table 11.3. Furthermore, we remark that an external magnetic field $H_0$ may be a constant or slowly varying field and the up and down flipping spins, leading to the magnetization $M$, have the character of a fluctuating quantity. Therefore, we propose the analogy of the basic (scalar) relation of magnetism with the decomposed first velocity component $u$ of a fluid,

$$H = H_0 + M \quad \leftrightarrow \quad u = \overline{u} + u', \tag{1245a, b}$$

which is composed of the one-dimensional mean velocity, $\overline{u}$, and fluctuation velocity component, $u'$. Furthermore, as second stress parameter (in a positive form), we assume the correspondence

$$S_2 = H_0 \quad \leftrightarrow \quad S_2 = |\overline{u}|. \tag{1246a, b}$$

Moreover, the positive order parameter in the analogous formulation is proposed to be the rms-fluctuation velocity

**Table 11.3** Analogous quantities of a magnetic (second column) and a fluid dynamic system exhibiting turbulence (third column)

| | | |
|---|---|---|
| Stress parameter $S_1$ | Temperature $T$ | Inverse characteristic velocity $1/u_0$ |
| Stress parameter $S_2$ | External magnetic field $H_0$ | Absolute mean velocity $\lvert\overline{u}\rvert$ |
| Hermaphrodite parameter | Internal magnetic field $H$ | Absolute velocity $\lvert u\rvert$ |
| Order parameter $O$ | Magnetisation $M$ | RMS fluctuation quantity $\sqrt{\overline{u'^2}}$ |
| Response function | Isothermal susceptibility $\chi_T$ | Vorticibility at constant characteristic velocity $\tau_{u_0}$ |
| Energy | Magnetisation energy $E$ | Turbulent kinetic energy $E$ |
| Specific thermal capacity | Specific heat at constant external magnetic field $C_{H_0}$ | "Specific heat" at constant absolute mean velocity field $C_{\lvert\overline{u}\rvert}$ |

In light gray, one finds the basic stress and order parameters and in dark gray the quantities derived with the MFT in this chapter

$$O = M \quad \leftrightarrow \quad O = \sqrt{\overline{u'^2}}. \tag{1247a, b}$$

One could introduce, e.g., the quantity $\big||u| - |u|_c\big|$ as order parameter, a quantity that is zero or positive and Galilei invariant. However, the fact that it is not identical to zero in the entire laminar domain is a strong argument against such a choice. Such argumentation also annihilates other similar postulations.

Next, we assert that fluctuations are favorably initiated in neighborhoods of already fluctuating domains. More specifically, in analogy to magnetism, where an averaging is performed only over neighboring spins, here an averaging of fluctuations is also performed only over neighboring cells. This does not seem to be an unrealistic assumption. Furthermore, a next neighbor approximation of the fluctuation quantity by an averaging only over neighboring domains is required (which we assume without a concrete definition) with a linear dependence

$$u = \bar{u} + av', \quad \overline{v'} = 0, \tag{1248a, b}$$

assuming here the value of $a$ to be also close to "1," so that $v' \approx u'$. There is a small, however, important difference between a magnetic and a fluid dynamic system. Above criticality, the magnetic system has always a magnetization given by its spins. However, in the statistical mean it may disappear. This is different in the fluid dynamic system, where above criticality there are never any fluctuations present. Notice that in the MFT it is only important that above criticality toward higher values of the stress parameter the order parameter (which is an averaged quantity) decreases or is identical to zero.

Now, by comparing Eqs. (1245b) with (1248a), we write in analogy to Eq. (1213a,b)

$$\sqrt{\overline{u'^2}} \cong a\sqrt{\overline{v'^2}} \cong \sqrt{\overline{v'^2}}. \tag{1249a, b}$$

Next, we conjecture (again in analogy to magnetism, with its Curie law : $M = cH/T$) that a fluid dynamic system also follows such a law, which we call "Curie law of turbulence,"[2]

$$\sqrt{\overline{v'^2}} = cu_0\big||u| - |u|_c\big|. \tag{1250}$$

The correctness of the law Eq. (1250) is not so easy to recognize. However, let us be pragmatic and see whether eventually ensuing results will come out to be more evident and, thereby, could give support to the validity of this new formula.

Imposing a linear approximation of Eq. (1251) (see below), different from that in magnetism, we experience that in modeling turbulence the theory fails below as well as above criticality. The reason is the strict absence of fluctuations in the high-order or laminar phase.

Therefore, here we immediately introduce an approximation up to third order in the approximate rms-fluctuation velocity and, for consistency reasons, also introduce the critical value of the absolute flow velocity, $|\bar{u}|_c$. This quantity is obtained by adjusting the fluid dynamic system at the overall critical Reynolds number and by

[2]In magnetism one has: $H_{0c} = 0$ and in turbulence $|\bar{u}|_c \neq 0$. Therefore, the generalization of introducing the critical value in the turbulence case extends the applicability of the model without leading to any difficulties.

measuring a velocity component at the field location of interest, averaging it and then finally applying the operation "absolute value." This yields

$$\left||u| - |u|_c\right| = |\overline{u}| - |\overline{u}|_c + a\sqrt{\overline{v'^2}} + b\sqrt{\overline{v'^2}}^3 + O\left(\sqrt{\overline{v'^2}}^5\right), \qquad (1251)$$

with coefficients $a$ and $b$ analogous to those in Eq. (1219). Combining Eqs. (1250) and (1251), eliminating $|\overline{u}| - |\overline{u}|_c$ and setting $ac = 1/u_{0c}$, in analogy to Eq. (1221) yields

$$bc\sqrt{\overline{v'^2}}^3 + \left(\frac{1}{u_{0c}} - \frac{1}{u_0}\right)\sqrt{\overline{v'^2}} + c\left||\overline{u}| - |\overline{u}|_c\right| = 0, \qquad (1252)$$

which is a cubic equation for $\sqrt{\overline{v'^2}}$.
We could obtain analytical solutions of Eq. (1252) for the three roots and then construct inferences for these. However, for the moment it is sufficient to solve some simple special cases.

Unfortunately, without a driving field $\left||\overline{u}| - |\overline{u}|_c\right|$, no spontaneously created turbulent fluctuations exist. However, we may study a case, not by demanding that $\left||\overline{u}| - |\overline{u}|_c\right| = 0$, but instead, e.g.,

$$\sqrt{\overline{v'^2}} \gg c\frac{\left||\overline{u}| - |\overline{u}|_c\right|}{\left(\dfrac{1}{u_{0c}} - \dfrac{1}{u_0}\right)}, \quad \text{and} \quad \sqrt{\overline{v'^2}} \gg \left(\frac{\left||\overline{u}| - |\overline{u}|_c\right|}{|b|}\right)^{1/3}. \qquad (1253a, b)$$

This is the case when the fluctuations are very large and the driving mean velocity field small. It is not expected that such realizations are often occurring in a usual fluid of a geophysical or technical flow[3] with a Reynolds number above criticality. On the other hand, one might find some applications in a superfluid that shows vanishing viscosity (see, e.g., Putterman 1974; Tilley and Tilley 1986; Egolf 1990). What we assume here is a superfluid applied as a fluid with practically negligible viscosity and no internal entropy (wave) generation. This is a model prototype of a low-viscosity fluid.

Under the above restrictions, the last term in Eq. (1252) may be neglected, so that

[3]Except in special flow realizations, e.g., a fluid in a container with vibrating walls.

$$bc\left(\sqrt{\overline{v'^2}}\right)^2+\left(\frac{1}{u_{0c}}-\frac{1}{u_0}\right)=0,\quad u_0>u_{0c}, b<0, c>0. \qquad (1254a-d)$$

Solving for the full and approximate rms-fluctuation intensity leads us to the results

$$\sqrt{\overline{u'^2}}\cong\sqrt{\overline{v'^2}}=\left[\frac{1}{bc}\left(\frac{1}{u_0}-\frac{1}{u_{0c}}\right)\right]^{1/2},\quad u_0>u_{0c}, b<0, c>0, \qquad (1255a-e)$$

which are analogous to Eq. (1222a–d), equally leading again to the critical exponent $\beta'=1/2$.
Also in analogy to Eq. (1221), we differentiate Eq. (1252), with respect to $|\overline{u}|$ at constant inverse characteristic velocity $1/u_0$, to obtain

$$3bc\sqrt{\overline{v'^2}}^{\,2}\left.\frac{\partial\sqrt{\overline{v'^2}}}{\partial|\overline{u}|}\right|_{u_0}+\left(\frac{1}{u_{0c}}-\frac{1}{u_0}\right)\left.\frac{\partial\sqrt{\overline{v'^2}}}{\partial|\overline{u}|}\right|_{u_0}+c=0,\ u_0>u_{0c}, b<0, c>0. \quad (1256a-d)$$

In analogy to Eq. (1218a), we introduce the “differential response function of turbulence”

$$\tau_{u_0}:=\left.\frac{\partial\sqrt{\overline{u'^2}}}{\partial|\overline{u}|}\right|_{u_0}\cong\left.\frac{\partial\sqrt{\overline{v'^2}}}{\partial|\overline{u}|}\right|_{u_0}, \qquad (1257a, b)$$

which we call “vorticibility,” which in a linearization is identical with the turbulence intensity Eq. (1258a) and, for isotropic turbulence, also with the turbulence degree Eq. (1258b)

$$\tau_{u_0}=\left.\frac{\sqrt{\overline{u'^2}}}{|\overline{u}|}\right|_{u_0},\quad \mathrm{Tu}=\left.\frac{\sqrt{\overline{u'^2}+\overline{v'^2}+\overline{w'^2}}}{\sqrt{\overline{u}^2+\overline{v}^2+\overline{w}^2}}\right|_{|\vec{u}_0|}. \qquad (1258a, b)$$

The differential vorticibility is substituted into Eq. (1256) and (in analogy to Eq. 1225a) yields

$$\left[3bc\sqrt{\overline{v'^2}}^{\,2}+\left(\frac{1}{u_{0c}}-\frac{1}{u_0}\right)\right]\tau_{u_0}+c=0,\quad u_0>u_{0c}, b<0, c>0. \quad (1259a-d)$$

Now, Eq. (1255b) is inserted to estimate $\sqrt{\overline{v'^2}}$ to obtain, in an analogy to Eq. (1226a), the following result

$$\tau_{u_0} = \frac{c}{2} \frac{1}{\left(\frac{1}{u_{0c}} - \frac{1}{u_0}\right)}, \quad u_0 > u_{0c}, c > 0, \qquad (1260a - c)$$

which allows identification of the critical exponent $\gamma' = 1$. These properties are summarized by

**Proposition 14** *The vorticibility is a response function of a turbulent flow. The higher the increase of the turbulence intensity is, that is, created by an increase of a unit of the absolute mean velocity, the larger will be the response of the turbulent system. A fluid with zero vorticibility is not able to produce per se any instabilities and fluctuations and in analogy corresponds to an incompressible fluid or non-magnetizable material. In a superfluid the "vorticibility" as a function of the inverse characteristic velocity approaches criticality from below with a divergence, described by a power law with critical exponent $\gamma' = 1$.*

More generally valid is the description of the turbulent system just at the inverse critical characteristic velocity $1/u_{0c}$. Then, Eq. (1252) simplifies to

$$b\sqrt{\overline{v'^2}}^3 + \left||\overline{u}| - |\overline{u}|_c\right| = 0, \quad \text{at} \quad u = u_{0c}, \qquad (1261a, b)$$

revealing a further critical exponent

$$\sqrt{\overline{u'^2}}\left(\frac{1}{u_{0c}}, |\overline{u}| - |\overline{u}|_c\right) \cong -\left(\frac{|\overline{u}| - |\overline{u}|_c}{b}\right)^{1/3}, \quad \text{at} \quad u = u_{0c}, b < 0. \qquad (1262a - c)$$

**Proposition 15** *The fluctuation intensity at constant (critical) inverse characteristic velocity varies as a function of the absolute mean fluid velocity difference $\left||\overline{u}| - |\overline{u}|_c\right|$ by a power law with the critical exponent 1/δ, where $\delta = 3$.*

The turbulent kinetic energy $E$ of the turbulent system is also the product of a stress and an order parameter, which in analogy to magnetism, where $E = -\mu_0\mu_r HM$, is given by

$$E = d\left||u| - |u|_c\right|\sqrt{\overline{u'^2}} \cong d\left||u| - |u|_c\right|\sqrt{\overline{v'^2}}, \qquad (1263a, b)$$

with a constant $d > 0$ that will later be identified. We simply concentrate on the part describing the turbulent kinetic energy. Then, by inserting the linear term in $\sqrt{\overline{v'^2}}$ of Eq. (1249a) and setting $a = 1$, it follows that

$$E \cong \frac{\rho}{2}\overline{v'^2} \cong \frac{\rho}{2}\overline{u'^2}, \qquad (1264a, b)$$

and the constant was identified to be $d = \rho/2$. We notice that the energy term of magnetism, in its analogy to turbulence, leads to the expected correct expression. However, it is rightly the positive turbulent kinetic energy of the flow, which is here presented in a one-dimensional approach. A generalization to the *3-d* form would be straightforward.

Finally, the analogy between magnetism and turbulence is different by the sign in their energies. Whereas a magnetic system lowers its energy in a cooling process from zero to negative values with its minimum at the absolute zero point, a turbulent system also shows zero turbulent intensity at criticality, but with decreasing inverse overall Reynolds number, it shows an increasing turbulence intensity. This is taken into consideration by setting different signs in Eqs. (1235a,b), (1263a,b), and (1264a,b).

Next, the restricted form Eq. (1255b) is substituted into Eq. (1264a), which yields

$$E \cong \frac{\rho}{2bc}\left(\frac{1}{u_0} - \frac{1}{u_{0c}}\right), \quad u_0 > u_{0c}, b < 0, c > 0. \tag{1265a – d}$$

This equation shows that at the critical point the turbulent kinetic energy is zero. Because above criticality, in a domain called the laminar regime, no fluctuations are present, one finds that

$$E \equiv 0, \quad u_0 < u_{0c}. \tag{1266a, b}$$

The specific heat at constant mean velocity $|\overline{u}|$ is

$$C_{|u|} = \left.\frac{\partial E}{\partial\left(\frac{1}{u_0}\right)}\right|_{|\overline{u}|}. \tag{1267}$$

For the domain below criticality, it follows with Eqs. (1265a–d) and (1267) that

$$C_{|u|} = \frac{\rho}{2bc} < 0, \;\; \text{since} \;\; b < 0. \tag{1268a, b}$$

In classical equilibrium thermodynamics, a condition of stability requires $C_{|u|}$ to be positive, when regarded as a function of the Reynolds number. However, here we have the inverse Reynolds number as independent variable. Therefore, when $C_{|u|} \equiv C_{|u|}(1/\mathrm{Re}) = C_{|u|}(\mathrm{Re})\, \mathrm{d}(1/\mathrm{Re})/\mathrm{d}\,\mathrm{Re} = -1/\mathrm{Re}^2 C_{|u|}(\mathrm{Re})$, which demonstrates that, in the present 1/Re-dependence of $C_{|u|}$, the sign of $C_{|u|}$ changes and, thus, in this notation stability prevails if Eq. (1268a,b) is valid. Hence,

**Proposition 16** *The "specific heat of turbulence" in a superfluid is approximately the turbulent kinetic energy introduced to the turbulent flow field occurring by an increase of the stress parameter of the system, which is the characteristic velocity or the overall Reynolds number, respectively.*

In the domain above criticality, one observes no fluctuations and, therefore, $E \equiv 0$ everywhere. Then, from Eq. (1267), it follows that

$$C_{|u|} \equiv 0. \tag{1269}$$

Comparing the specific heat below criticality (see Eq. 1268a) and above (Eq. (1269), one recognizes that:

**Proposition 17** *The specific heat of a turbulent superfluid at the critical Reynolds number is discontinuous. The critical exponents of the specific heat are $\alpha = \alpha' = 0$.*

Exponentially decaying pair correlation functions, e.g., Eq. (1208) may not be the adequate tool to describe complex nonlinear and turbulent systems. Large eddies are responsible for the fact that fluid lumps rotationally approach from very distant locations leading to nonlocal behavior, which is related to only weakly decaying or even constant correlations. This results in long correlation lengths that are almost identical to the characteristic sizes of the laminar streaks. Then, the phase change concept proposes at criticality correlation lengths of the size of the fluid domain (in a strict sense of infinite length). Some approximate results in this direction are found in Egolf and Hutter (2017b).

## 11.5 First Experiments for a Qualitative Comparison

The new approach to turbulence by critical phenomena demands the planning of new measurements for its evaluation. For instance, the inverse overall Reynolds number stress parameter calls for a different discretization of a corresponding stress parameter measuring interval. This leads to a high importance of experiments close to criticality. In the fluid dynamic literature, experiments were mostly performed at Reynolds numbers largely above the critical value. In a figure, the transformation of measured points to inverse abscissa values maps them close to the origin and, thus, makes it rather useless for a comparison of theoretical with experimental results. Furthermore, it is well known that experiments of high quality, targeted to be just slightly below criticality, are extremely difficult to be performed. Here, a future demand exists to seriously evaluate these "vortisation" curves over the entire inverse Reynolds number domain from zero to criticality.

Even if it was difficult to find accurate measurements for comparisons of experimental and theoretical "vortisation" curves, we found in a literature search an adequate experimental work. Ravelet et al. (2008) studied an inertially driven von Kármán flow between two counter-rotating large impellers with curved blades. They focused on the transition from laminar flow to fully developed turbulence over a wide range of Reynolds numbers, namely from Re = $10^2$–$10^6$. The azimuthal shear layer becomes unstable as a result of a Kelvin–Helmholtz instability (see Helmholtz 1868; Kelvin 1871). The flow creates traveling/drifting waves, called Kelvin–Helmholtz waves, which are modulated travelling waves, then temporal and spatiotemporal chaos sets in before turbulence.

The authors state that the transition to turbulence is globally supercritical. Their main conclusion is that the energy of the velocity fluctuations can be considered as an order parameter, characterizing the dynamics from the first occurring time dependence to fully developed turbulence (see Fig. 11.6). This conclusion corresponds fully with our ideas about the order parameter drawn in the context of developing the MFTT (MFT of turbulence). However, there could be an alternative to introduce as order parameter, e.g., the kinetic energy of the fluctuations or the squared value of the vorticity, integrated over the flow domain. With the information given by the DQTM that in turbulent flows the stress parameter is an inverse function of the Reynolds number, we redraw Fig. 11.6 to obtain Fig. 11.7. The above described higher sensitivity near criticality to details in the inverse Reynolds number presentation also becomes evident here. Whereas in Fig. 11.6, the impression is that there is a precisely determined and experimentally observed critical Reynolds number, our preferred presentation in Fig. 11.7 reveals the enormous difficulties of determining the fluid dynamic criticality with certainty. If the real order parameter, which is the root of the ordinate value of Fig. 11.7, is chosen, the shape of the curve will come a little bit closer to a usual convex shape of a magnetization curve.

In Fig. 11.8, the proposed analogy between magnetism and turbulent systems is visually demonstrated. The two pictures on the left show the two occurring phases of each of these two systems. In the top image, a magnetic state of a probe of EuO is shown. Clearly, the white Weiss domains are seen. The black domains correspond to

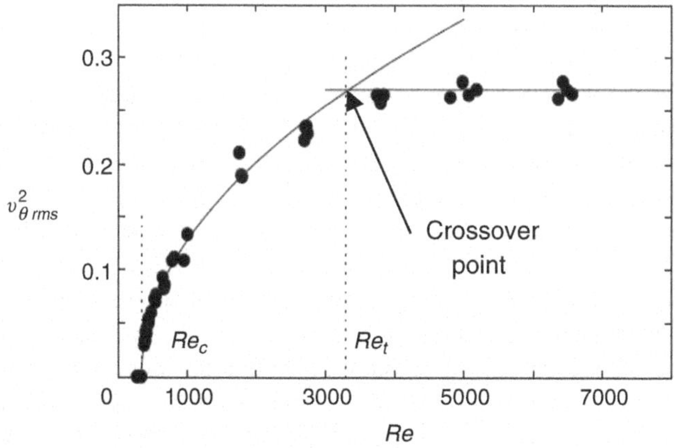


**Fig. 11.6** The non-normalized order parameter as a function of a non-normalized stress parameter of a von Kármán flow between two impellers. $Re_c$ characterizes the onset of the time behavior. $Re_t$ denotes the crossover Reynolds number, where the "turbulent kinetic energy" saturates. Courtesy Ravelet et al. (2008) © J. Fluid Mech., reproduced with permission by Cambridge University Press

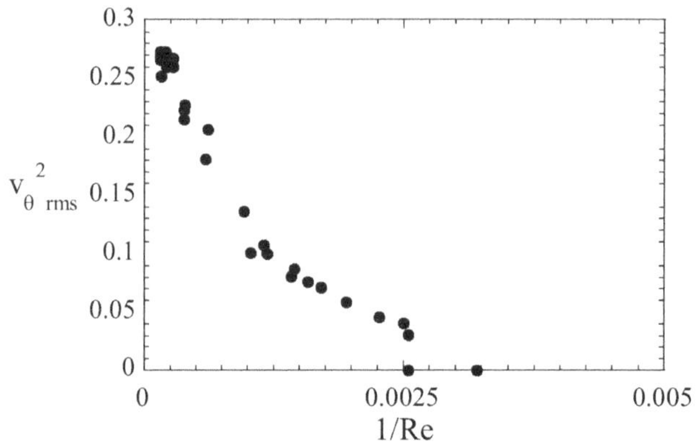


**Fig. 11.7** In this figure, the same experimental results as in Fig. 11.6 are presented. However, the abscissa is the inverse Reynolds number. Furthermore, the stress and order parameter have neither been normalized. Be aware that a normalization does not change the character of the curve. Here the square of the order parameter, given by the MFT, was chosen. Courtesy Egolf and Hutter (2017b) © J. Entropy, reproduced with permission by Springer Science+Business Media

domains with disordered and the white with ordered spins. With decreasing temperature, the white domains grow until at $T = 0$ K the entire picture becomes white. On the other hand, an increase of the temperature up to $T_c$ lets the disordered black patches grow until at criticality the entire picture is black. Identical behavior is demonstrated in the image at the bottom on the left. Here, black laminar streaks grow until at the inverse critical Reynolds number the entire picture presents itself in black color. The similarity of the ordered patches in magnetic and turbulent systems is very impressive.

In the image at the top right-hand side more quantitative results, namely the measured magnetization is compared with a theoretical curve obtained by a spin wave model leading to perfect agreement (see Krey 1980).

In analogy, the order parameter of the theory of "wall turbulence" as a function of its stress parameter, as presented in Sect. 9.5, is shown in the image at the bottom to the right. Qualitatively, the order parameter curve is the same; however, quantitatively it is a little flatter than its magnetic counterpart. For visualization also the three selected measurements gained in the Princeton super pipe have been transformed into this graph. One observes that even the lowest Reynolds number data set was performed in a state were the present concept already shows an order of approximately 95%.

Chorin (1994) presents a near-equilibrium theory of turbulence. His approach, using advanced similarity methods, shows also that the well-known von Kármán-Prandtl logarithmic law of the wall must be abandoned in favor of a power law. Furthermore, he explains very well why equilibrium thermodynamics may be applied especially to the smaller scales of turbulent flows and introduces a "non-touchable temperature" (turbulent instead of a Kelvin temperature) as stress parameter, which is of the form $1/\log_e(\mathrm{Re})$. Therefore, his considerations and conclusions have much in common with our work, especially also with our results in Sect. 10.14.

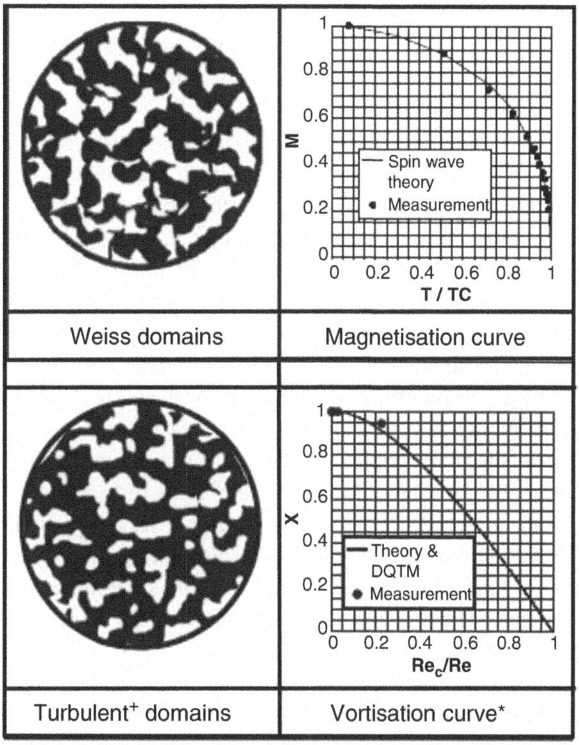


**Fig. 11.8** Analogy between the magnetic and the turbulent system with an image of two phases each and figures showing magnetization and "vortisation". The symbol * denotes a new name introduced by Egolf and Hutter (2016c) and + denotes that in analogy to "Weiss" a scientist's name of turbulence research again in analogy could eventually be more accurate than the word "turbulent". Courtesy Egolf and Hutter (2016c) © reproduced with permission by Springer International Publishing

A continuation of these ideas to even more complexity raises the question, if such an order concept also occurs in a generalized manner in a turbulent flowing ferrofluid (see, e.g., Rosensweig 1997). Interactions between turbulent patches and magnetic ordered islands are expected to occur. However, it is very important to understand turbulence in usual Newtonian fluids before a complete theory of more complex fluids, e.g., an interaction of fluid dynamics and electromagnetism (e.g., solar wind, earth's interior electromagnetic flows), may be established.

## 11.6 Discussion of Results

In the above sections, we described a perfect analogy between magnetism and turbulence by a transformation of the well-established MFT of magnetism to the analogous theory of turbulence, which is designated as MFTT. This new model reveals the response function of turbulence, which—in the context of critical phenomena—is proposed to be called "vorticibility." This quantity is known in turbulence research as the relative turbulence intensity and plays a key role in the description of turbulent flow fields since Reynolds in the late 1880s proposed a splitting of the velocity field into the averaged and the fluctuation parts. The differential form of the "vorticibility" is not widely used in the literature on turbulence, but our findings could likely support a more frequent use and evaluation in future studies of turbulence.

A second novel discovery is the "Curie law of turbulence." It states a proportionality of the rms-fluctuation velocity to the inverse stress parameter of the system multiplied with the difference of the absolute velocity from its critical value (see Eq. 1250). This is at least correct in the sense that the fluctuations are absent at criticality and strongly increase with increasing Reynolds number. This law is new and, in a first step, the authors followed a pragmatic way and just assumed it to be taken for granted. Some support for the law then emerged by employing further derived results that are extremely reasonable (e.g., the above mentioned definition of the response function "vorticibility" or the correct energy of the turbulent system). It is intriguing that this analogy provides us with correct new physical terms in an interdisciplinary context. Furthermore, an experimental validation of the new discovered "Curie law of turbulence" seems to us to be essential.

One may say "*Of course, anyone is free to postulate analogies and relations between physical variables. The question is, are they scientifically correct and do they actually describe or model real physical phenomena?*" Such criticism sounds plausible. However, it does not apply to our work, as we will explain in the next paragraph.

It is very impressive that the NSEs, combined with the DQTM, for turbulent flows along plane walls (e.g., plane Couette flows, plane Poiseuille flows and "wall turbulence") in each case reveals a complete theory of a critical phenomenon. Furthermore, it is clear that these theoretical discoveries are much more sophisticated than ad hoc model developments of phase transitions. Having been in the favorable position of possessing a good theory of the critical phenomenon of turbulence (be aware that the theories for the different flows are similar), we asked ourselves, whether simple ad hoc theories, as successfully developed in other fields of physics, in analogy would also apply to turbulence. So, with the knowledge of the DQTM critical phenomenon theory, we courageously proceeded and developed the MFTT. Therefore, this happened with much helpful background information. For example, the idea of an inverse Reynolds number stress parameter is a result from the DQTM-based considerations that could be incorporated into the development of an ad hoc MFTT.

In so doing, one has to be aware that with a wrong choice of analogous quantities neither the correct response function nor the right energy of a turbulent flow could be derived. The only alternative choice, that, in a first attempt, could generate some interest would be to take the overall velocity $u_0$ instead of its inverse quantity $1/u_0$ as main stress parameter. However, in this case the "Curie law of turbulence" would predict a rms-fluctuation quantity that is approximately constant with increasing Reynolds number. Therefore, in a useful description of turbulent phenomena, also this proposal cannot serve as a serious alternative. These arguments provide confidence in the usefulness of the presented analogy between MFTs in other scientific areas than turbulence and the new MFTT.

Furthermore, a critical reader must have realized that no definition of entropy was introduced, although significant characterizations of this work are order and disorder phenomena. We have not reached the level of introducing all the thermodynamic potentials. In this chapter, our model just relates to equilibrium thermodynamics and is, therefore, likely described by the Gibbs–Boltzmann thermodynamics, where the entropy of two subsystems in thermal contact is additive. The entropy of a turbulent system would follow again in analogy to magnetism, from a corresponding Gibbs potential. Because this treatment, with some new results, is a first attempt of applying existing phase change concepts to turbulence, we decided to dispose important extensions, which have not already been worked out and discussed in Chap. 10, to future work.

Finally, it is impressive that an old equilibrium theory, such as the MFT, already applies reasonably well to a quasi-steady turbulent flow field. In Table 11.3, the analogous quantities of magnetism and turbulence will be a help to successfully transform other more sophisticated thermodynamic models of magnetism to turbulence, which should lead to further physical insights of near-to-critical phenomena of turbulence. By this, a step-by-step approach to more sophisticated and accurate final thermodynamic models of turbulence, which in the end will be more complex and accurate, will be obtained. A convergence of such more sophisticated models toward the already discovered DQTM-based critical phenomenon theory is expected to occur.

## References


Allhoff, K.T., Eckhardt, B.: Directed percolation model for turbulence transition in shear flows. Fluid Dyn. Res. **44**(3), 031201 (2012)

Beck, C., Schlögl, F.: Thermodynamics of Chaotic Systems – An Introduction, Cambridge Nonlinear Science Series 4. Cambridge University Press, Cambridge, UK (1993). ISBN 13-978-0-521-43367-9

Berlin, T.H., Kac, M.: The spherical model of a ferromagnet. Phys. Rev. **86**(Ser. 2), 821 (1952)

Brown, T.M.: Information theory and the spectrum of isotropic turbulence. J. Phys. A Math. Gen. **15**(7), 2285 (1982)

Castaign, B.J.: Consequences d'un principe d'extremum en turbulence. J. Phys. France. **50**, 147 (1989). (in French)

Chorin A.J.: Vorticity and Turbulence. Springer, New York (1994). ISBN 0-387-94197-5

Cortet, P.P., Herbert, E., Chiffaudel, A., Daviaud, F., Dubrulle, B., Padilla, V.: Susceptibility divergence, phase transition and multistability of a highly turbulent closed flow. J. Stat. Mech. Theory and Exp. 1742-5468/11/P07012+21$33.00. (2011)

Egolf, P.W.: Selbstorganisation, Wellenturbulenz und kritische Phänomene in suprafl üssigem Helium II mit einer freien Oberfläche, Dissertation (PhD thesis), Swiss Federal Institute of Technology, ETHZ, No. 9290 (in German) (1990)

Egolf, P.W.: Lévy flights and beta model: a new solution of "wall" turbulence with a critical phenomenon. Int. J. Refrig. **32**, 1815 (2009)

Egolf, P.W., Hutter, K.: A nonlocal zero-equation turbulence model and a deficit-power law of the wall with a dynamical critical phenomenon. PAMM – Proc. Appl. Math. Mech. **16**, 889 (2016a)

Egolf, P.W., Hutter, K., 2016b, "From linear and local to nonlinear and nonlocal zero equation turbulence models," (Proceeding of IMA Conference for Turbulence, Waves and Mixing, Kings College, Cambridge, UK. 6–8 July, pp. 71–74

Egolf, P.W., Hutter, K.: Turbulent shear flow described by the algebraic difference-quotient turbulence model. In: Peintke, J., et al. (eds.) Progress in Turbulence VI, Springer Proceedings in Physics, vol. 165. Springer, Heidelberg (2016c). ISBN 978-3-319-29129-1

Egolf, P.W., Hutter, K.: Fractional turbulence models. In: Peintke, J., et al. (eds.) Progress in Turbulence VII, Springer Proceedings in Physics, vol. 165. Springer, Heidelberg (2017a). ISBN 978-3-319-57933-7

Egolf, P.W., Hutter, K.: The mean field theories of magnetism and turbulence. J. Entropy. **19**, 589 (2017b)

Egolf, P.W., Hutter, K.: Tsallis extended thermodynamics applied to 2-d turbulence: Lévy statistics and q-fractional generalized Kraichnanian energy and enstrophy spectra. J. Entropy. **20**, 109 (2018)

Egolf, P.W., Kauffeld, M.: Special Issue on Ice Slurrries of the Int. J. Refig. **28**, 4 (2005)

Egolf, P.W., Manz, H.: Theory and modeling of phase change materials with and without mushy regions. Int. J. Heat and Mass Transf. **31**(18), 2917 (1994)

Egolf, P.W., Weiss, D.A.: Model for plane turbulent Couette flow. Phys. Rev. Lett. **75**(16), 2956 (1995a)

Egolf, P.W., Weiss, D.A.: Elementary geometric figures in fully developed turbulent plane Poiseuille flow. Helv. Phys. Acta. **68**, 219 (1995b)

Egolf, P.W., Weiss, D.A.: Difference-quotient turbulence model: analytical solutions for the core region of plane Poiseuille flow. Phys. Rev E. **62**(1-A), 553 (2000)

Egolf, P.W., Gama, S., Vuarnoz, D.: Thermomagnetism with external and internal magnetic field quantities. J. Appl. Phys. **118**(2), 023903-1 (2015)

Eprints.: https://eprints.soton.ac.uk/66182/1/A_primer_on_DNS.pdf (2019)

Frei, B. Egolf, P.W. 2000, "Viscometry Applied to the Bingham Substance Ice Slurry," Second Workshop on Ice Slurries of the International Institute of Refrigeration, 25–26 May, Paris, pp. 48–60

Gad-el-Hak, M.: Flow Control: Passive, Active and Research Flow Management. Cambridge University Press, Cambridge, UK (2000). ISBN 13-978-0521036719

Goldenfeld, N.: Lectures on Phase Transitions and the Renormalization Group, Frontiers in Physics, 5th edn. Adison-Wesley, Reading, PA (1992). ISBN 13-978-0-201-55408-3

Haken, H.: Synergetik: Eine Einführung. Springer, Berlin (1983). ISBN 3-540-12597-3 (in German)

Haken, H.: Advanced Synergetics: Instability, Hierarchies of Self-Organizing Systems and Devices. Springer, Berlin (1987). ISBN 3-540-12162-5

Hansen, T.M., Kauffeld, M., Sari, O., Egolf, P.W., Pasche, F.: Research, development and applications of ice slurry in Europe - from ancient Rome to modern technologies. In: Proceedings of the Fourth Workshop on Ice Slurries, pp. 1–12 (2001). 12–13 November, Osaka

Helmholtz, H.L.F.: Über diskontinuierliche Flüssigkeits-Bewegungen. Monthly Reports of the Royal Prussian Academy of Philosophy in Berlin, vol. 23, p. 215 (1868) (in German)

Hinze, J.O.: Turbulence, 2nd edn. McGraw-Hill, New York (1975). ISBN 0-07-029037-7

Ho, J.T., Lister, J.D.: Magnetic equation of state of $CrBr_3$ near the critical point. Phys. Rev. Lett. **22**, 603 (1969)

Holmes, P., Lumley, J.L., Berkooz, G.A.L.: Turbulence, Coherent Structures, Dynamical Systems and Symmetry. Cambridge University Press, Cambridge, UK (1996). ISBN 0 521 55142 0

Hussain, A.K.M.F.: Coherent structures - reality and myth. Phys. Fluids. **26**, 2816 (1983). https://doi.org/10.1063/1.864048

Hutter, K., Wang, Y.: Fluid and thermodynamics. In: Basic Fluid Mechanics, vol. 1. Springer, Berlin (2016a). ISBN 978-3-319-33632-9

Hutter, K., Wang, Y.: Fluid and thermodynamics. In: Advanced Fluid Mechanics and Thermodynamic Fundamentals, vol. 2. Springer, Berlin (2016b). ISBN 978-3-319-33635-0

Itsykson, C., Drouffe, J.-M.: Statistical field theory. In: From Brownian motion to renormalization and lattice gauge theory, vol. 1. Cambridge Monographs on Mathematical Physics, Cambridge, UK (1989a). ISBN 0-521-34058-6

Itsykson, C., Drouffe, J.-M.: Statistical field theory. In: Strong Coupling, Monte Carlo Methods, Conformal Field Theory, and Random Systems, vol. 2. Cambridge Monographs on Mathematical Physics, Cambridge, UK (1989b). ISBN 0-521-37012-4

Kauffeld, M., Kawaji, M., Egolf, P.W.: Handbook on Ice Slurries. International Institute of Refrigeration, IIF/IIR, Paris (2005)

Kelvin Lord (Thomson W.): Hydrokinetic solutions and observations. Phil. Mag. **42**, 362 (1871)

Kitanovski, A., Egolf, P.W.: Thermodynamics of magnetic refrigeration. Int. J. Refrig. **29**, 3 (2006)

Kreilos, T., Eckhardt, B.: Increasing lifetimes and the growing saddles of shear flow turbulence. Phys. Rev. Lett. **112**(4), 044503 (2014)

Krey, G.: Phasenübergänge und kritische Phänomene. Friedrich Vieweg & Sohn, Braunschweig/ Wiesbaden (1980). (in German). ISBN 3-528-08422-7

Le Bellac, M., Mortessagne, F., Batrouni, G.G.: Equilibrium and Non-Equilibrium Statistical Thermodynamics. Cambridge University Press, Cambridge, UK (2006). ISBN 978-0-521-82143-8

Lemoult, G., Shi, L., Avila, K., Jalikop, S.V., Avila, M., Hof, B.: Directed percolation phase transition to sustained turbulence in Couette flow. Nat. Phys. **12**, 254 (2016)

Ma, S.-K.: Modern Theory of Critical Phenomena, Frontiers in Physics. The Benjamin Cummings, Reading, MA (1982). ISBN 0-8053-6670-9

Nelkin, M.: Intermittency in fully developed turbulence as a consequence of the Navier-Stokes equations. Phys. Rev. Lett. **30**, 1029 (1973)

Papon, P., Leblond, J., Meijer, P.H.E.: The Physics of Phase Transitions. Springer, Heidelberg (2002). ISBN 3-540-43236-1

Pomeau, Y.: Front motion, metastability and subcritical bifurcations in hydrodynamics. Physica D: Nonlinear Phenomena. **23**(1), 3 (1986)

Putterman, S.J.: Superfluid Hydrodynamics. Elsevier, Amsterdam (1974). ISBN 10-0444106812

Ravelet, F., Chiffaudel, A., Daviaud, F.J.: Supercritical transition to turbulence in an inertially driven von Kármán closed flow. J. Fluid Mech. **601**, 339 (2008)

Robert, R., Sommaria, J.: Statistical equilibrium states for two-dimensional flows. J. Fluid Mech. **229**, 291 (1991)

Rosensweig, R.E.: Ferrohydrodynamics. Dover, Mineola (1997). ISBN 0-486-67834-2

Stanley, H.E.: Introduction to Phase Transitions and Critical Phenomena, International Series of Monographs on Physics. Oxford Science, Oxford, UK (1971). ISBN 0-19-5053-8

Tabeling, P., Willaime, H.: Transition at dissipative scales in large-Reynolds-number turbulence. Phys. Rev. E. **65**, 066301 (2002)

Thomson, J.C.: Classical Equilibrium Statistical Mechanics. Clarendon Press, Oxford, UK (1988). ISBN 0-19-851984-2

Tilley, D.R., Tilley, J.: Superfluidity and Superconductivity, Graduate Student Series in Physics. Adam Hilger, Bristol (1986)

Tsallis, C.: Introduction to Nonextensive Statistical Mechanics. Springer, Cham (2009). ISBN 978-0-387-85358-1

van Dyke, M.: An Album of Fluid Motion. The Parabolic Press, Stanford, CA (1982). ISBN 0-915760-02-9

Waiwhakaiho: http://waiwhakaiho.org.nz/?p=222 (2016)

Wester, T., Traphan, D., Gülker, G., Peintke, J.: Percolation: statistical description of a spatial and temporal highly resolved boundary layer transition. In: Örlü, R., et al. (eds.) Progress in Turbulence VII, Springer Proceedings in Physics. Springer, Heidelberg (2017). ISBN 978-3-319-57933-7

Yeomans, J.M.: Statistical Mechanics and Phase Transitions. Oxford Science, Oxford, UK (1992). ISBN 019-158-970-5

# Chapter 12
# Conclusions and Outlook

There is no doubt that there is no complete and accurate theoretical understanding of the physics of turbulence, e.g., as it is the case of the fluid dynamics of laminar flows, of solid body mechanics, elasticity, and electromagnetism. According to George (2013), the worldwide costs of this ignorance is tremendous. Inadequate weather forecasts, inaccurate design, and calculations of mobiles (cars, ships, planes, etc.), of hydraulic and fluid-thermal systems lead to the introduction of excessively high safety factors. The result of this ignorance is that these engineering systems are operating distant from their optimal operation characteristics. This causes a tremendous waste of primary energy and creates a corresponding elevated global warming potential (GWP). George writes: *"The problem is of course, that the cost of our ignorance is* [not focused on a single area or technical object, but is] *spread across the entire economic spectrum of human existence."* We may be driven by an insatiable scientific theoretical curiosity to understand nature or may have ecological or/and economic goals as our motivation; in the end we believe that one day a complete picture will be constructed by a large amount of different ideas, but the solution of the millennium prize problem (Clay 2000) is just a single (however important) part of the puzzle of solving the enigma of turbulence. There will be rigorous mathematical-physical results, but also more approximate dynamical physical models, e.g., random walk diffusion models, extended thermodynamic models, and fractional models, which will play a significant role in a final understanding, description, and handling of turbulent flows.

This memoir is a large presentation of turbulent diffusion with practically all its features and many of its modern aspects. It has a unique character, because of its novelty in many of its ideas and physical models that cannot be found in other classical textbooks on turbulence. The possibility to prepare such an extensive new work was only possible for us, because of the discovery of the DQTM.

Stimulated by Prandtl's work, we were guided to embed the DQTM into its right position within the existing broad landscape of a huge amount of different theories and models of turbulence. The positive experience is that, as usual with valuable new physical theories, the new results are in complete harmony, and they perfectly fit into

The original version of this chapter was revised. A correction to this chapter can be found at https://doi.org/10.1007/978-3-030-26033-0_14



P. W. Egolf, K. Hutter, *Nonlinear, Nonlocal and Fractional Turbulence*,
https://doi.org/10.1007/978-3-030-26033-0_12

already existing accepted concepts. However, the derivation of the DQTM, e.g., by Lévy flight modeling and the fractal-$\beta$ model application, does not only perfectly cope with existing knowledge on turbulence; it additionally also leads to numerous new insights and results. These are all in perfect agreement with today's accepted standard techniques and theories of turbulence. Such considerations are the basis of this memoir containing also new nonlocal and fractional mathematical-physical modeling techniques of turbulence. Our developments and findings will be briefly summarized in the following sections.

Because of the conceptual novelty of this book, it seemed to be useful to start by criticizing older approaches, especially based on linearity and locality. The first step for an improvement is usually a critique and the finding of an inadequacy. Therefore, quite a lot of energy and time were invested to demonstrate deficiencies and fallacies of existing models and laws. Advantageous is that this is supported by existing critiques of other authors in the turbulence literature (see, e.g., Tsinober 2009, 2014 and Spalart 2015). The most famous law, at present still in debate, is the logarithmic law of the wall (see, e.g., George 2007). Even though experimental values are close to predicted results of this law, a satisfying theoretical derivation of it is still missing. By contrast, the Navier–Stokes equations combined with the DQTM reveals a deficit power law to describe "wall turbulence" with theoretical results that show an improved accuracy in comparison with experimental data. On the basis of serious critics, the new laws were carefully introduced, discussed, and established. The most intriguing and beautiful new results are shown in Fig. 12.1 and discussed in the next paragraph.

We start our explanations of the results (see Fig. 12.1) close to the top on the right-hand side and will turn around in the counter clock-wise direction. Application of mass and momentum considerations of *plane turbulent wake flows*, based on Newton's second law, without any empirics, leads to a closure of nonlocal nature. This important result should be considered as a milestone in the area of turbulent closure methods. For this special case, any correct closure proposition must produce this physically rigorous result. Furthermore, and the other way round, a natural generalization of this closure, which considers, e.g., Galilei invariance, directly leads to the new nonlocal closure equation, called *DQTM*. It is presented in the center of Fig. 12.1.

With this new closure many analytical solutions of elementary turbulent flow problems were derived. They all show excellent agreement with experiments. In this context, for a *round turbulent jet*, a highly nonlinear ordinary differential equation is obtained, which is analytically solved by the *Gaussian mean velocity profile* that for more than a hundred years is known to fit experiments with good agreement. This result is presented in Fig. 12.1 on the top to the right. Furthermore, the DQTM helps to solve the problem of "wall turbulence" and, in the violent debate of the logarithmic law versus the power law, gives high preference and superiority to the *deficit or defect power law*. For the infinite Reynolds number solution the power law shows a special exponent, being a quadratic irrational number (continued fraction), as the golden mean or golden proportion is, which in analogy, we call the *marvelous proportion* (see Fig. 12.1 on the top to the left). Another solution of great beauty

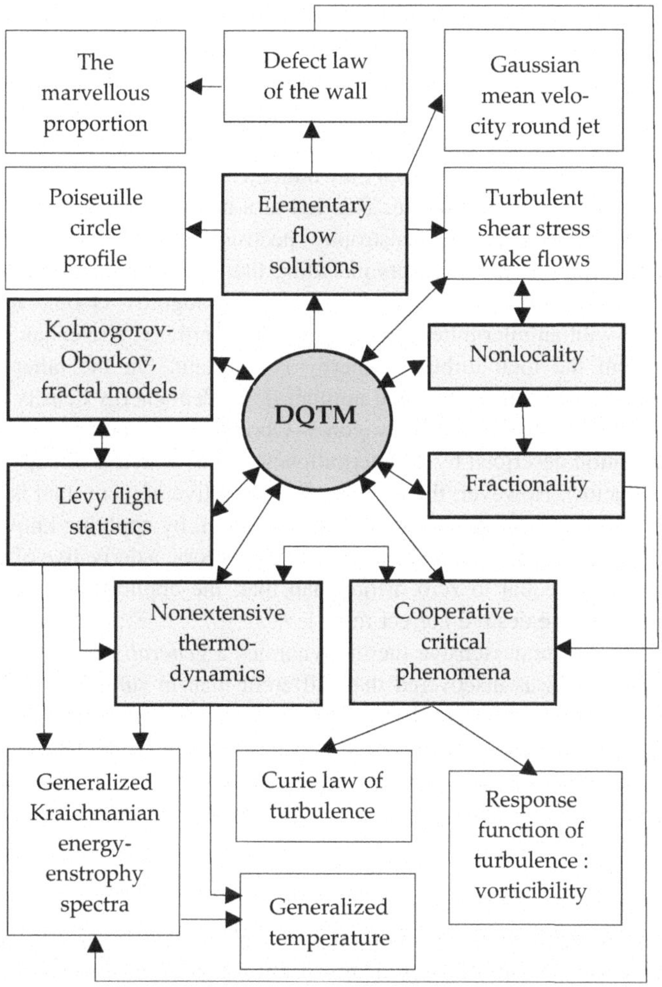


**Fig. 12.1** The DQTM with its main derivations, relations and a selection of especially important new results. All the occurring theories and models are tightly related. However, the presented arrows show the strongest connections, as for example between the DQTM and nonlocality, etc.

is the mean velocity profile of *plane turbulent Poiseuille flows* described by generalized real-number order Bessel functions that in the infinite Reynolds number limit simplify to a *circular profile*.

The fractal beta model, containing as a special case the *Kolmogorov–Oboukov self-similar power law*, combined with *Lévy flight statistics* serves as a useful and ideal tool to motivate the DQTM. On the other hand, Lévy flight statistics is intimately connected with nonextensive thermodynamics, e.g., the *nonextensive or q-thermodynamics of Tsallis*.

*Kraichnan* applied Boltzmann–Gibbs equilibrium thermodynamics to obtain *the energy-enstrophy spectrum of 2-d turbulence.* For small wave numbers, his results are very approximate, whereas for high wave numbers he obtained excellent agreement with experiments. The conclusion is that high wave number eddies are in thermal equilibrium, whereas such with a wave number below a critical value are not. Therefore, it was tempting to apply the nonextensive thermodynamics of Tsallis to low wave-number 2-*d* turbulence. This led in a natural derivation to a generalization of the Kraichnanian energy-enstrophy spectrum of 2-*d* turbulence, as it occurs in oceans and atmospheres of planets including the earth (see on the bottom to the left). For low wave numbers, the results are the Kolmogorov–Oboukov law or its generalization with an intermittency correction. Furthermore, Kraichnan stated that in this problem the total turbulent energy is constant. On the other hand, its derivative must lead to the right wave-number-dependent energy density spectrum, according to the just mentioned Kolmogorov–Oboukov $-5/3$ power law (with ev. a slight modification described by an intermittency factor, which is directly related to the Tsallis $q$ factor). However, the Riemannian derivative of a constant is zero. The only way out of this trap and solution of this problem, by applying known mathematical tools, is given by fractional calculus. In this theory, a derivative of a constant is not imperatively equal to zero. More than that: the application of a fractional derivative leads to the desired correct result!

In the context of nonextensive thermodynamics a *generalized absolute temperature of turbulence* was discovered that, different than in similar work (see, e.g., Chorin 1994), is not negative (see Fig. 12.1 to the bottom in the center). It is identical to a Lagrange parameter of a thermodynamic variational principle. Our temperatures vary in different eddy classes of characteristic sizes above a threshold value. On the other hand, the small wavelength eddies, with shorter turn around and relaxation times, are in thermal equilibrium, which correctly is characterized by identical generalized temperatures; this result is also revealed by our extensive theoretical considerations (Egolf and Hutter 2018).

Nonextensive thermodynamics is tightly related to *cooperative or critical phenomena of turbulence*, as it arises in the solutions of all close-to-wall turbulent elementary flows, solved with the help of the DQTM. A very beautiful and convincing critical phenomenon solution occurs in a tight entanglement with the flow theory of the defect law of "wall-turbulent flows." Even if turbulence reveals dynamical phase transition characteristics, including critical slowing down at the critical Reynolds number (see Hohenberg and Halperin 1977; Egolf and Di Nardo 1991; Egolf et al. 1993), it reveals an impressive analogy with static magnetic systems showing a ferro-paramagnetic phase transition. This analogy led to the discovery of the "*Curie law of turbulence*" and as a deduced result the *response function of turbulence*, which we call "*vorticibility*." It describes the ability of creation of turbulent fluctuations, respectively, changing the turbulent intensity (order parameter) as a result of a change of the inverse mean velocity (stress parameter).

We remark that in future it might be possible to supercompute the flight of each molecule in a fluid; by this more insight to the fluid dynamics of highly excited fluids will be gained. However, it will also be possible to develop new practical theories and this in a shorter period of time, which will also allow to gain a deeper physical understanding of the mechanisms of turbulence. Finally, such theories will also be very helpful to gain improved computation of turbulence, e.g., by an implementation of semi-analytical models into numerical programs (see, e.g., Egolf 2014), which lead to reduced computation time and increased numerical stability.

Finally, let us finish this textbook with a few additional and more general remarks concerning the future of turbulence. A very interesting question is posed and answered in Bradshaw (1994) "*Will turbulence modeling ever become obsolete?*"

Bradshaw's answer is

*"I'm sure that Keith Stewartson (1925–1983), an applied mathematician, regarded the whole phenomenon of turbulence as being 'unrigorous' and probably invented by the Devil on the seventh day of Creation (when the Good Lord wasn't looking)."*

Release of such hopeless situations usually comes by a "paradigm change" that opens a new door and changes the situation drastically. Surely the development of nonlinear dynamics and the introduction of fractal geometry led to such a paradigm change with a remarkable impact on the theory of turbulence. "*The considerations in this memoir show that in the field of fractional dynamics (dynamical systems on fractal sets) nonlocal and fractional operators yield new tools for better descriptions and investigations of turbulence. In this context, the difference-quotient mathematics and especially the DQTM 'closure' will, as a special case, play a crucial role.*"

In the context of such a paradigm change, we have to be ready to leave or even abandon theories, which eventually accompanied us during our entire university or industrial career. It is not easy for a lecturer who taught engineering students the logarithmic law with its universal von Kármán constant during decades to suddenly give preference to another law; and it is neither easy for a computer scientist, who applied $k-\varepsilon$ models in numerous projects and has become an expert in this kind of computation, to look out for and apply new approaches, e.g., with fractional derivatives.

Recently, there appeared first signs, mainly given by the papers referenced below, which announce that the newest developments also in numerical simulations in future will be based on nonlocal and fractional calculus. After a certain paving by theoretical works on fractional calculus in the field of turbulence by Chen (2006), Egolf and Hutter (2017), etc., Churbanov and Vabishchevic (2015) applied a fractional Laplacian (see also Lischke et al. 2018) to numerically solve steady-state duct flows. Epps and Cushman-Roisin (2018) also base their work on a fractional Laplacian operator, containing the parameter $\alpha$, which is directly related to the Lévy $\alpha$-stable distribution, which for $\alpha = 2$ reduces it to the classical Maxwell–Boltzmann distribution. A fractional Laplacian may be tuned to recover Kolmogorov's inertial subrange by applying an $\alpha = 2/3$-fractional derivative (Chen 2006). This can easily be demonstrated by applying a Fourier transform. In such a consideration for nonhomogeneous flows, the effect of the advection terms must be incorporated. Song and Karniadakis (2018) applied a Caputo fractional derivative or left-sided Riemann–Liouville derivative, respectively, to model the Reynolds shear stress for pipe flows. They assume the order of the fractional derivative of the Reynolds stress to be a function of the wall distance $\alpha(y)$ building into their model a wall-dependent nonlocality. Inspiring is a plot $\alpha(y^+)$ in the viscous wall unit for friction Reynolds numbers $\mathrm{Re}^+$ between 180 and 5200, because this results in a single universal curve. Sousa (2013) applied the Riesz fractional derivative to numerically model turbulent diffusion. Samiee et al. (2019) propose fractional-order modeling of subgrid-scale stresses in Large Eddy Simulations (LES) of turbulent flows that respect non-Gaussian statistics. Especially in LES models, the main challenge is to incorporate the cumulative statistical effects of small-scale fluctuations on the motion of large-scale vortices, an idea historically going back to Heisenberg's modeling of the turbulent energy spectrum (Heisenberg 1948). As Samiee and collaborators outline, a nonlocal extension of LES with its filtering procedure reveals some new challenging problems to be solved. Generally, one has to be aware that memory behavior and nonlocality imply a full matrix numerical simulation procedure which is extremely computing-time consuming. On the other hand, the DQTM as a special case, with its dependence only on spatially separated extremities of the mean velocity field and their distances, without any direct dependence on each single location in the intermediate zones, suggests a substantial reduction of the CPU time and, thereby, an economic numerical treatment of complex turbulent flows, if efficient algorithms to find the minima and maxima will be applied. In future research activities such and similar investigations are expected to enrich and improve numerical methods and simulations of turbulence.

## References

Bradshaw, P.: Turbulence: the chief outstanding difficulty of our subject. Exp. Fluids. **16**(3–4), 203 (1994)

Chen, W.: A speculative study of 2/3-order fractional Laplacian modeling of turbulence: Some thoughts and conjectures. Chaos. **16**, 023126 (2006)

Chorin A.J.: Vorticity and Turbulence. Springer, New York (1994). ISBN 0-387-94197-5

Churbanov, A.G., Vabishchevic, P.N.: Numerical investigation of a space-fractional model of turbulent flow in rectangular ducts. arXiv:1505.0519vl [cs.NA] (2015)

Clay: http://www.claymath.org/millennium-problems/millennium-prize-problems (2000)

Egolf, P.W.: Application of numerical calculation methods. In: Lecture Notes University of Applied Sciences of Western Switzerland. Yverdon-les-Bains (2014)

Egolf, P.W., Di Nardo, S.: Critical slowing down of surface waves on superfluid helium. Proceedings of the 19th International Conference on Low Temperature Physics, Brighton, UK. Physica B **169**, 517 (1991)

Egolf, P.W., Hutter, K.: Fractional turbulence models. In: R. Örlu et al. (eds.). Progress in Turbulence VII, Springer Proceedings in Physics, p. 196. Springer, Heidelberg (2017). ISBN 978-3-319-57933-7, https://doi.org/10.1007/978-3-319-57934-4_18

Egolf, P.W., Hutter, K.: Tsallis extended thermodynamics applied to 2-d turbulence: Lévy statistics and q-fractional generalized Kraichnanian energy and enstrophy spectra. J. Entropy. **20**, 109 (2018)

Egolf, P.W., Weiss, D.A., Di Nardo, S.: The appearance of oscillating second sound induced patterns on the free surface of He II. A nonequilibrium phase transition with critical slowing down. J. Low Temp. Phys. **90**(3/4), 269 (1993)

Epps, B.P. Cushman-Roisin, B.: Turbulence modeling via the fractional Laplacian. arXiv: 1803.05286 (2018)

George, W.K.: Is there a universal log law for turbulent wall-bounded flows? Phil. Trans. R. Soc. A. **365**, 789 (2007)

George, W.K.: Lectures in Turbulence for the 21st Century. www.turbulence-online.com (2013)

Heisenberg, W.: Zur statistischen Theorie der Turbulenz. Zeitschrift für Physik. **124**, 628 (1948). (in German)

Hohenberg, P.C., Halperin, B.I.: Theory of dynamic critical phenomena. Rev. Mod. Phys. **49**(3), 436 (1977)

Lischke, A., Pang, G., Gulian, M., Song, F., Glusa, Ch., Zheng, X., Mao, Z., Cai, W., Meerschart, M.M., Ainsworth, M., Karniadakis, G.E.: What is a fractional Laplacian?. arXiv:1801.09767v2 [math.NA] (2018)

Samiee M., Akhavan-Safaei, A., Zayernouri, M.: A fractional subgrid-scale model for turbulent flows: theoretical formulation and a priori study. arXiv: submit/2854703 [cs.CE] (2019)

Song, F., Karniadakis, G.E.: A universal fractional model of wall turbulence. arXiv: 1808.10276v1 [physics.flu-dyn] (2018)

Sousa, E.: Numerical solution of a model for turbulent diffusion. Int. J. of Bifurcation and Chaos. **23**(10), 1 (2013). https://doi.org/10.1142/SO218127413501666

Spalart, P.R.: Philosophies and fallacies in turbulence modeling. Prog. Aerosp. Sci. **74**, 1–15 (2015)

Tsinober, A.: An Informal Conceptual Introduction to Turbulence. Springer, Cham (2009). ISBN 978-90-481-3174-7

Tsinober, A.: The Essence of Turbulence as a Physical Phenomenon with Emphasis on Issues of Paradigmatic Nature. Springer, Dordrecht (2014). ISBN 978-94-007-7179-6

# Correction to: Turbulence: A Cooperative Phenomenon

**Correction to:**
**Chapter 11 in: P. W. Egolf, K. Hutter,**
***Nonlinear, Nonlocal and Fractional Turbulence*,**
**https://doi.org/10.1007/978-3-030-26033-0_11**

The original version of the book was inadvertently published with an error in Table 11.2. Owing to an oversight Table 7.2 was mistakenly duplicated as Table 11.2. The chapter has now been corrected and approved by the author. The correct presentation is included now.

---

The updated online version of the chapter can be found at
https://doi.org/10.1007/978-3-030-26033-0_11

 
P. W. Egolf, K. Hutter, *Nonlinear, Nonlocal and Fractional Turbulence*,
https://doi.org/10.1007/978-3-030-26033-0_13

**Table 11.2** Quantities of critical exponents given by models and experiments for a selection of different physical systems for the range $T < T_c$. Data are from Ma (1982) and Stanley (1971)

| Quantity | $\alpha'$ | $\beta'$ | $\delta$ | $\gamma'$ | $\nu'$ |
|---|---|---|---|---|---|
| *Fluids* | --- | --- | --- | --- | --- |
| $CO_2$ | ~ 0.1 | 0.34 | 4.2 | ~ 1.0 | - |
| Xe | < 0.2 | 0.35 | 4.4 | ~ 1.2 | 0.57 |
| *Magnets* | --- | --- | | --- | --- |
| Ni | -0.3 | 0.42 | 4.22 | 1.32 | - |
| EuS | -0.15 | 0.33 | - | - | - |
| $CrBr_3$ | - | 0.368 | 4.3 | - | - |
| *Models* | --- | --- | | --- | --- |
| Classical | 0 | 1/2 | 3 | 1 | 1/2 |
| Spherical 3-*d* | - | 1/2 | 5 | - | - |
| Ising 2-*d* | 0 | 1/8 | ~15 | ~7/4 | 1 |
| Ising 3-*d* | ~ 1/8 | ~ 5/16 | ~5 | ~5/4 | - |
| Heisenberg 3-*d* | - | ~ 0.35 | ~5 | - | - |

# Correction to: Nonlinear, Nonlocal and Fractional Turbulence

## Correction to: P. W. Egolf, K. Hutter, *Nonlinear, Nonlocal and Fractional Turbulence*, https://doi.org/10.1007/978-3-030-26033-0

Below corrections are updated in book.

**Text book cover image,**
**third line from bottom:** ...with permission. ©2019 American Institute of Physics

**Page v, 2nd line:** ...of irregularity, with all its facets, is a main topic...

**5th line:** ..., the main ingredients in life, art, music, etc.

**7th line** ... order and disorder, with different

**8th line** ...tempi and intensities, gives our...

**11th line** ... (energy input), it gives...

**20th line** ...(from its big bang), but even the creation...

**Page vi, 17th line** He, at ...

---

The updated online versions of the chapters can be found at
https://doi.org/10.1007/978-3-030-26033-0_1
https://doi.org/10.1007/978-3-030-26033-0_2
https://doi.org/10.1007/978-3-030-26033-0_3
https://doi.org/10.1007/978-3-030-26033-0_5
https://doi.org/10.1007/978-3-030-26033-0_6
https://doi.org/10.1007/978-3-030-26033-0_7
https://doi.org/10.1007/978-3-030-26033-0_8
https://doi.org/10.1007/978-3-030-26033-0_9
https://doi.org/10.1007/978-3-030-26033-0_10
https://doi.org/10.1007/978-3-030-26033-0_11
https://doi.org/10.1007/978-3-030-26033-0_12
https://doi.org/10.1007/978-3-030-26033-0



P. W. Egolf, K. Hutter, *Nonlinear, Nonlocal and Fractional Turbulence*,
https://doi.org/10.1007/978-3-030-26033-0_14

**18th line** …his time, did not distinguish between art and science.

**Page vii, Pic. 3, Tab. Capt.:** In June 1889, the famous impressionist artist Vincent van Gogh painted the picture "starry night" by oil on canvas.

**1st line:** In the literature on turbulence, Leonardo da Vinci's drawings…

**Page x, 3rd line:** This is all well known, as is the fact…

**16th line:** … a flux law, in analogy to Fick's law…

**Page xi, 5th line:** In April 1985, one of us (Peter William Egolf – PWE), deriving the *Difference-Quotient Turbulence Model* (DQTM), got automatically involved into the problem of nonlocality and started to tackle it.

**11th line:** … engineering experts, applying often stronger…

**16th line:** …and, thereby, describes a *hereditary*…

**25th to 27th line:** …approach, because in the mathematical formulation the positions in between these three distinct points are of no importance.

**34th line:** Based on these new ideas, the first author…

**Page xii, 10 th and 11th line:** …the entire spectrum and, for large wave numbers, contains Kraichnan's correct part as a special case.

**Page xiii, 33rd line:** …"constitutive equation", is reviewed and…

**40th to 42nd line:** Many zero-equation turbulence models of Prandtl and his contemporary scientists and engineers are reviewed, and their advantages and disadvantages are discussed.

**Page xiv, 13th line:** After its development in the middle of the 1980s and its first publication in …

**40th line:** … constant in the thermal quasi-equilibrium enstrophy range.

**Page xv, 21 st & 22nd line:** … their care and positive influence on our work-life balances, which is…

**26th line:** Good support of production was also given by Shivakumar Krithika.

**Page xviii and page 188:** 9.2.1 Jets in Quiescent Surroundings

**and page 220:** 9.2.2 Jets in Parallel Co-flows

**Pages xix and 338:** Justification of the Quadratic Form of the Energy as a Function of Space and Velocity Coordinates

**Page xix and page 358:** 11.2.1 What is a Critical…

**and page 372:** 11.2.7 Turbulent Phase Transition with its Two Phases…

**Page xx and page 413:** Appendix E: Fourier Transform for $q$-Generalized Energy...

**Page xxi:** $\overline{a}(\overline{t})$ Mean acceleration of a fluid lump on the axis of a two-dimensional turbulent jet

$\overline{\mathbf{A}}_2$ Averaged second Rivlin-Ericksen tensor

**Page xxii:** $b$ Half-width of turbulent shear flows, e.g., axi-symmetric jets

$c$ Constant in characteristic function of Lévy flight distribution

**Page xxiii:** $da$ Differential surface element

$dv$ Differential volume element

${}_L\mathrm{D}^{\alpha}_{+}$ Left-handed Liouville fractional derivative of order $\alpha$

${}_L\mathrm{D}^{\alpha}_{\pm}$ Left- and right-handed Liouville fractional derivative of order $\alpha$

$\hat{\check{e}}(k)$ Energy density in Fourier space of $\check{e}$ at wave number $k$

$E_\epsilon$ Turbulent kinetic energy of a two-dimensional flow

$E_H$ Turbulent kinetic energy of a three-dimensional flow

$\check{E}_\epsilon$ Generalized energy of a two-dimensional flow

$\hat{\check{E}}_\varepsilon(k)$ Fourier transform of the generalized energy $\check{E}_\epsilon$

$\check{E}_H$ Generalized energy of a three-dimensional flow

**Page xxiv:** $g(x,x')$ Nonlocality function, response function
$g_H(x,x')$ Hamba's response function
$g_K(x,x')$ Kraichnan's response function
$G_{21}$ Shear modulus of plane shear flows

**Page xxv: H** Hessian tensor of the pressure field

*k-l* Two-equation turbulence model for the transport of turbulent kinetic energy $k$ and Kolmogorov length $l$

*k-ω* Two-equation turbulence model for the transport of turbulent kinetic energy $k$ and vorticity $\omega$

$k'$ Lowest occurring wave number modulus

$k''$ Highest occurring wave number modulus

**K** Bachelor's diffusivity tensor ("coefficient")

**Page xxvi:** $m_p$ Scalar magnetization

$m_0$ Mass flux of axi-symmetric turbulent jet in the orifice

$n_{\min}$ Eddy density of the eddies on the lower side of $x_2^*$; see Fig. 7.5

$n_{\max}$ Eddy density of the eddies on the upper side of $x_2^*$; see Fig. 7.5

**Page xxvii:** $\overline{p}_{max}$ ... of the smaller scale...

$\vec{p}_B$ Momentum of particle B on trajectory $x_B$

$q$ Index to identify Tsallis's thermodynamic quantities

**Page xxviii**: $s$ Step size in a Lévy flight distribution

$\widehat{s}_{\pm}$ Positive and negative shift operator

**Page xxix**: $\bar{t}$ ... the orifice and moving on the jet axis

$\bar{u}$ Downstream velocity of a fluid lump moving on the axis of a turbulent axisymmetric jet

**Page xxx**: $\widehat{\vec{u}}_{\vec{n}}^{(\kappa',\kappa'')}$ ... and highest wave number $\vec{k}'' = \kappa''(n_1, n_2, n_3)$

$\widehat{\vec{u}}_{\vec{k}'}$ Fourier...

**Page xxxi:** $\bar{x}$ Downstream position of a fluid lump moving on the axis of a turbulent axisymmetric jet

$x_0$ Core length of a turbulent cylindrical jet

$\bar{x}_0$ Streamwise position of a fluid lump moving on the axis of a turbulent axisymmetric jet at the end of the theoretical core

$\bar{x}_1$ Streamwise position of a fluid lump moving on the axis of a turbulent axisymmetric jet

**Page xxxii :** $\alpha$ Order of Liouville fractional derivative

$\alpha$ Exponent of a power law describing the workman's sizes

$\alpha$ Critical exponent for a quantity that approaches criticality from above

$\alpha'$ Critical exponent for a quantity that approaches criticality from below

$\beta$ Factor for normal turbulence intensity.

$\beta*$ Exponent....to $\mathrm{Re}_{\mathrm{crit}}$

$\gamma$ Critical exponent for a quantity that approaches criticality from above

$\gamma$ Reduction factor for normal turbulence intensities...

$\gamma$ Scale factor in orthotropic turbulence intensity para..

**Page xxxiii:** $\gamma'$ Critical exponent for a quantity that approaches criticality from below

$\epsilon_m$ Eddy viscosity of turbulent momentum transfer

**Page xxxiv :** $\nu$ Critical exponent for a quantity that approaches criticality from above

$\nu'$ Critical exponent for a quantity that approaches criticality from below

$\xi$ Integration variable in Liouville fractional derivative

**Page xxxv:** $\sigma_q$ Standard deviation of $q$-Gaussian distribution

$\tau_0$ Viscous shear stress of flow at the wall

**Page 3, 6th line:** … boundary, as it was derived with Prandtl's mixing-length…

**11th line:** … working lifetime prior to 1988, was the most important…

**12th line:** Turbulence, by its field quantities, …

**14th and 15th line:** Chaos and turbulence, in all its facets, fascinates…

**Page 7:** Works of additional scientists and engineers, who studied fluid flows theoretically, numerically or at laboratory scale and contributed significantly to the field of turbulence, will be cited below.

**Page 8:** Clay,: http… prize-problems (2000)

**Page 16, 17th line:** The two terms describing the divergence of $\left\langle g'\vec{u}' \right\rangle$ and of $\left\langle \vec{g}' \otimes \vec{u}' \right\rangle$…

**Page 17, 16th line:** …contains the substantial derivative…

**Page 18, 10th line:** Therefore, six unknown components of this tensor must….

**Page 19, 17th line:** $\langle \text{mom. eq.} \rangle \cdot \left\langle \vec{u} \right\rangle = 0 \neq \left\langle \text{mom. eq.} \cdot \vec{u} \right\rangle = 0$ (20a-c)

**Page 20, 6th line:** … more convenient than Eq. (21), but mathematically…

**Page 21, 28th line:** $\langle mom.\ eq. \rangle \otimes \ldots$ (29)

**Page 31, 21st line:** 4. Mass of constituents.

**Page 33, 21st line:** $\tau_{21} = -\rho\left(\mathbf{T}_R\right)_{21} = \ldots$ (48a-c)

**Page 35, last line:** …. effects, from Eq. (16b) it follows that

**Page 37, 3rd line:** $\tau_{21} = \ -\rho\ldots$ (65a,b)

**23rd line:** …interpreted as a derivative somewhere between…

**Page 38, last line:** …boundary regions, because there the eddy…

**Page 39, 11th line:** Turbulence is rotational and, therefore, flows are…

**Page 42, 16th line:** This was already noticed by Taylor, who found…

**Page 47, 16th line:** $\ldots \frac{1}{\upsilon_{T'}}$ (106a,b)

**Page 49, 5th & 6th line:** …Prandtl number is defined as the ratio…

**Page 52, 10th & 11th line:** ...is applied to linear and nonlinear eddy viscosity models.

**Page 54, 20th line:** $\overline{u_2'u_1'} \neq \propto \frac{d\overline{u}_1}{dx_2}, \quad \alpha < \infty, \quad \mathrm{Re} \to \infty$ (131a,b)

**25th line:** ... eg., Eqs. 53a,b,69,71, and 89a,b). *It immediately...*

**Page 55, 21st line:** 3. The fallacies of turbulence modeling.

**25th line:** ...the role of higher-order moments...

**26th line:** ...influence of higher-order moments...

**26th line:** ..., then higher-order turbulence modeling...

**27th line:** ...superior to lower-order methods.

**28th line:** ..., applied to higher-order moment closures,...

**Page 56, 8th line:** ... 1960's and 1970's, because of the ambition to treat complex geometries, was mainly...

**Page 57, 17th line:** ...in the Reynolds stresses a higher-order term by products of...

**Page 62:** $u^+(y^+)|_{y^+=0} = 0$ (147)

**Page 63, 8th line:** ...with decreasing thickness as a function of increasing Reynolds number.

$B = A\left[\log_e\left(\frac{4}{A}\right) - 1\right], \quad A = \frac{1}{\kappa}$ (158a,b)

**Page 65:** Table 5.3 : ...applying Lie group methods, shows that for turbulent shear flows any function and, thus, any scaling law is admissible.
... dependent coefficients (incomplete similarity) is obtained.

**Page 66:** Toward the wall, approximately the same laws are valid for plane (plate or channel) and cylindrical (pipe) geometries.

**Page 67 :** $\lim_{\mathrm{Re}\to\infty} \frac{\overline{u}_1(\mathrm{Re}, x_2)}{U} = 1$ (162)

**Page 68 :** ..., using Lie group analysis, he demonstrated that any function is fully admissible.

**Page 71:** Eisner, F., Reibungswiderstand...(1932) (in German)

**Page 72:** Kevlahan, N.K.R., Farge....

**Page 76, 38th line:** ....trajectory of the real particle, contributing to fluctuations and..

**Page 79, 23rd line:** ...length of the overall flow field $L$ and its related turnover time $T$?

**Page 80, 16th line:** A first class of nonlocal models...

**22nd line:** ... (173a-c)

**Page 84, 20th & 21st line:** After multiplication of both sides of this equation with $dx_2$ and introducing the differences…

**Page 91, 2nd line:** …which is the Hessian tensor of the pressure…

**6th line:** …, it follows from Eq. (213a) that

**Page 92, 5th line:** Note the formal analogy with the one-dimensional version given by Eq. (207).

**Page 94, 20th line:** …where the D/D$t$ is the substantial time derivative. The…

**Page 95, 8th line :** …been neglected, because in turbulence…

**Page 98, 10th line:** By integrating Eq. (238), it follows that

**Page 99, 1st line:** $$\widehat{u}_{1a} - \widehat{u}_{1b} = -\int_{\tau}^{0} \frac{\partial \widehat{p}}{\partial \widehat{x}}(s)ds + \frac{1}{Re}\int_{\tau}^{0} \Delta\widehat{u}_1 ds, \tag{239}$$

**4th line:** … quantities yields the non-normalized

**Page 102, 28th line:** …the local strain rate (~ local mean velocity gradient) is…

**Page 104:** Liepmann, H.W.: Mécanique…

**Page 107, 5th line:** …will be determined for a wake flow behind a cylinder…

**Page 108:** At this stage, in April 1985, Egolf was somewhat sceptical to take this particular result for granted and to accept it as a basis for further investigation.

**Page 112:** $$\frac{d\delta_{21}}{dt} = -\lim_{\substack{\Delta t \to 0 \\ \Delta x_2 \to 0}} \frac{\arctan\left(\frac{\delta x_1}{\Delta x_2}\right)}{\Delta t} \tag{272}$$

**Page 117**: Throughout this page, the averaged quantities $\overline{\Delta x_1}$ and $\overline{\Delta x_2}$ must be replaced by $\Delta x_1$ and $\Delta x_2$.

$$\frac{d\delta_{T21}}{dt} = -\lim_{\substack{\Delta t \to 0 \\ \overline{\Delta x_2} \to \max}} \frac{\arctan\left(\frac{\delta \bar{x}_1}{\Delta x_2}\right)}{\Delta t} \tag{290}$$

Because the fluid motion fluctuates, the average deformation $\delta\bar{x}_1$ of a fluid element is considered.

**Page 117, 19th line**: …spatial difference $\Delta x_2$ and the mean velocity difference…

**Page 121, 16th line:** $\bar{u}_2^* = \text{const} \times \bar{u}_1^*$. (298)

**Page 125, Table 7.1, ninth line, second row:** Density of molecules $n$

**Table 7.1, ninth line, third row:** Distribution of eddy density $n$

**Page 126, 2nd line:** Consider a random walk with steps of…

**Page 129, 21st line:** In the static picture, describing…

**22nd line:** … describing probability distributions, we used…

**Page 130, 2nd line:** …such as turbulent momentum or particle concentration, are also …

**Page 135**: The size of eddies of this class is defined by Kolmogorov's dissipation length scale (see Sect. 7.2.4.11).

**Page 141**: The special case of Kolmogorov/Obukhov, for infinite Reynolds number flow, ($a$=$b$=2, $d$=$D$=3) leads to

**Page 144**: $\widehat{p}_n = \rho q_0 \left(\frac{a}{b^{(2+d-D)/3}}\right)^n u_0 \left(\frac{1}{b^n}\right)^{(1+D-d)/3} = \ldots$ (398a,b)

**Page 151**: $\kappa_{k,\infty} = \frac{\widehat{P}_{k,\infty}}{\widehat{P}_0} = \ldots, \quad \kappa_{m,\infty} = \frac{\widehat{P}_{m,\infty}}{\widehat{P}_0} = \ldots.$ (429a-d)

**Page 152, 8th line:** This is the DQTM which—at the time of its derivation in 1985—was developed by different reasoning.

**Page 154, 20th line:** … gorov microscales vanish.

**Page 160, 6th line:** … and consider the Liouville weight $\omega(\lambda, \tau) = \tau^{\lambda - 1}$ $(0 < \lambda < 1)$, …

**Page 163-165:** By replacing the two derivatives in Prandtl's mixing-length model (Eq. 53b) by the left- and right-handed Liouville fractional derivative each, …

In equations (468b), (470), (473) and (474) the spatial derivatives are the left- and right-handed Liouville fractional derivative ${}_LD^{\alpha}_{\pm}$.

**Page 165, second last line:** … leads to the same result, makes it…

**Pages 165, 167 and 168:** In equations (477), (481) and (482) the derivatives are $\frac{\partial \overline{u}_1}{\partial x_2'}$.

**Pages 168, 8th line:** *We thus have proven that the DQTM….*

**Page 171:** Einstein, A.: Zur…Ann. Phys. **19**, 371 (1906). (in German)

**Page 172:** Osterwalder, K.: Methoden…Mathematics and Physics (1980)…

**Page 175, last line:** … and the dimensionless Reynolds shear stress and the normal turbulence intensities are

**Page 179:** $\overline{u}_1 = U_G - \overline{u}_1^*, \qquad \overline{u}_1^* \geq 0, \qquad \overline{u}_2^* = \overline{u}_2,$ (500a-c)

**Page 181, 11th line:** $\frac{\partial}{\partial x_1}\left(\frac{\overline{u}_1^*}{U_G}\right) - \frac{\partial}{\partial x_2}\left(\frac{\overline{u_2' u_1'}}{U_G^2}\right) = 0$ (503a,b)

**Page 182, 17th line:** Inserting Eqs. (491a) and…

**Page 186:** … to obtain with $\sigma$=d$b$/d$x_1$ and Eq. (500a),…

**Page 186, 2nd line:** By substituting (522a,d, and e), it follows that

**18th line:** ... to obtain, with $\sigma$=d$b$/d$x_1$ and Eq. (500a), ...

**Page 190:** ..., where a decomposition of the downstream velocity $\overline{u}_1$ into a horizontal flow...

**Page 191, 1st line:** $p_1 f_1 - p_0 \eta \frac{df_1}{d\eta} + \frac{1}{\beta\eta}\frac{d(\eta f_2)}{d\eta} = 0$ (544)

**Page 192, 12th line:** The Reynolds shear stress decreases inversely...

**Page 192:** With Eqs. (554a-d) the momentum equation (549) is simplified to the ODE

**Page 200, 15th line:** ... propositions:

**Page 210, 19th line:** ...relation, which is stated as:

**Page 216, figure 9.20:** c) d)

**Page 217, 6th line:** ... (647a-c)

**11th line:** ... (648a,b)

**Page 222, 8th line:** Next, it must be proven that...

**Page 224, 5th line:** $f_{21} = \ldots$ (684)

**Page 226, 13th line:** ...turbulence, the DQTM, was derived, contains....

**Page 234:** Applying these two equations to (720) and dividing the resulting equation by $u^{*2}$ yields

**Page 237, 2nd line:** $\ldots = \sqrt{\frac{\beta}{4-\beta}}$,

**Page 240:** Table 9.3: $g_1 = 1/2\langle 1 + \sqrt{(1-\chi/\chi)}\tan\{\arctan[\sqrt{\chi/(1-\chi)}]\eta\}\rangle$

**Page 248, Fig. Capt.:** The flow was calculated with a LES dynamic...

**Page 249, 9th line :** ... Reynolds shear stress, $\tau_{21}$,

**10th line:** $\tau_{tot} = \tau_0 + \tau_{21} = \mu\frac{d\overline{u}_1}{x_2} - \rho\overline{u_2' u_1'}$. (782a,b)

**Page 251, 12th line:** $\tau_{21} = \ldots$ (796a,b)

**15th line:** In terms of the self-similar coordinate ...

**Page 258, 2nd last line:** ... in earlier equations) $'$:=d/d$\psi$ , lead to ...

**Page 260, 4th line:** In Eqs. (727a,b), replacing $f_1(1)$ by $f_1(0)$ and substituting these two equations into (844a), it follows that

**Page 270, 20th line:** ...olds numbers up to 35,259,000, which is the highest Re value...

**Page 271, 23th line:** .....kind of flows why in this book a deficit power law...

**Page 279, Capt. Fig. 9.55:** ....inverse Reynolds number].

**Page 310, 13th line:** . . . (1015a-c)

**Page 310:** Replace in Eqs. (1013a,b) $\widehat{u}_{i\vec{n}}$ by $\widehat{u}_{i,\vec{n}}$

**Page 311, 5th line :** . . . , (1018a,b)

**Page 313, 10th line:** . . .very close to "1" (see Sect. 10.4) and power law exponents close to -3, are expected to occur.

**22nd & 23rd line:** . . . *of modeling turbulent fields.*

**Page 322:** $$\ldots - \lambda_2 \left[ \int_{\Gamma} x^2 p(x)\, \mathrm{d}x - \sigma^2 \right], \qquad (1066)$$

**Page 323:** $$p(x) = A\ \exp\left(-\frac{x^2}{2\sigma^2}\right) \qquad (1067)$$

**Page 325:** $$\Phi[P] = k \frac{1 - \int_0^\infty P(x)\mathrm{d}x}{q-1} - \ldots \qquad (1080)$$

**Page 334, in the figure:** . . .

**Page 335:** A comparison of Eq. (1114a) with the maximum value $q$ reveals Eq. (1114c).

**Page 343:** (a) $E_\varepsilon = E$ is the sum of energy contributions $E_n$: $E = \sum_{n=0}^{N} E_n$ (see Eq. (1011f))

**Page 346:** [6]Temperatures of turbulence were also proposed by Chorin (1994) and Castaign (1996).

**Page 347, 13th line:** .. and Tsallis et al. (1995)

**Page 350:** Chorin, A.J., Vorticity and Turbulence. Applied Mathematical Sciences 103. Springer edition, New York, Inc. (1994), ISBN 0-387-94197-5

**Page 362:** These limit the translational variance for size-scaled eddies to the distance of periodic length $2\lambda$.

**Page 364, 13th line:** . . . (1181a-c)

**Page 365, 26th line:** Still a further quantity, that diverges as $s_{1c}$ is approached, may. . .

**Page 366, 18th line:** Following Eq. (1187a-c), the compressibility diverges. . .

**24th line:** . . . (1193a-c)

**Page 367, 3rd line:** . . . (1194a-c)

**6th line:** . . . (1195a-c)

**16th line** . . . (1197a-c)

**Page 368, 1st line:** . . . (1198a-c)

**Page 370, 12th line:** . . . , (1210)

**Page 376, 19th line:** . . . (1228a-c)

**Page 376, 21st line:** . . . (1229a-c)

**Page 378, 17th line:** . . . (1241a,b)

**Page 378, 19th line:** . . . (1242a,b)

**Page 378, 21st line:** . . . (1243a-c)

**Page 379, 27th line:** . . . , (1245a,b)

**Page 383, 8th line:** . . . (1256a-d)

**Page 384, 14th line:** . . . (1261a,b)

**Page 388, 1st line:** . . . domains with disordered and the white with ordered spins.

**Page 396, 2nd last line:** . . .proportion is, which in analogy, we call the *marvelous*. . .

**Page 401:** Song, F., Karniadakis,. . .

**Page 413:** . . . (D.45a-c)

**Page 415:** . . . (E.7a-c)

# Appendices

## Appendix A: Normalization of Probability Distribution

Appendices A–C mainly follow recipes from Shlesinger et al. (1986).

The proof for correct normalization is started with formula (313)

$$\sum_{s=-b^N}^{b^N} p(s) = \frac{1}{2}\frac{a^{N+1}-a^N}{a^{N+1}-1}\left[\sum_{n=0}^{N}\frac{1}{a^n}\left(\delta_{s,b^n}+\delta_{s,-b^n}\right)\right]. \tag{A.1}$$

Because of symmetric behavior of this sum, it follows that

$$\sum_{s=-b^N}^{b^N} p(s) = \frac{a^{N+1}-a^N}{a^{N+1}-1}\left[\sum_{n=0}^{N}\frac{1}{a^n}\right]. \tag{A.2}$$

By complete induction one easily verifies that

$$\sum_{n=0}^{N} a^n = \frac{1-a^{N+1}}{1-a}. \tag{A.3}$$

Then with the reciprocal value, in analogy it follows that

$$\sum_{n=0}^{N}\frac{1}{a^n} = \sum_{n=0}^{N}\left(\frac{1}{a}\right)^n = \frac{1-\left(\frac{1}{a}\right)^{N+1}}{1-\frac{1}{a}} = \frac{1-\frac{1}{a^{N+1}}}{1-\frac{1}{a}} = \frac{a^{N+1}-1}{a^{N+1}-a^N}. \tag{A.4a – d}$$

Substituting (A.4d) into (A.2) one obtains

The original version of this chapter was revised. A correction to this chapter can be found at https://doi.org/10.1007/978-3-030-26033-0_14



P. W. Egolf, K. Hutter, *Nonlinear, Nonlocal and Fractional Turbulence*,
https://doi.org/10.1007/978-3-030-26033-0

$$\sum_{s=-b^N}^{b^N} p(s) = 1. \tag{A.5}$$

As a result, it is proven that the proposed probability distribution is normalized.

## Appendix B: The Variance of Lévy Flight Processes

The variance of a Lévy flight process is

$$\langle s^2 \rangle = \sum_{s=-\infty}^{s=+\infty} p(s)s^2. \tag{B.1}$$

By substituting the right-hand side of Eq. (315) into (B.1) it follows that

$$\langle s^2 \rangle = \frac{a-1}{2a}\left[\sum_{s=-\infty}^{s=+\infty}\sum_{n=0}^{\infty}\frac{1}{a^n}\left(\delta_{s,b^n}+\delta_{s,-b^n}\right)s^2\right]. \tag{B.2}$$

In this expression, the sums are exchanged

$$\langle s^2 \rangle = \frac{a-1}{2a}\left[\sum_{n=0}^{\infty}\sum_{s=-\infty}^{s=+\infty}\frac{1}{a^n}\left(\delta_{s,b^n}+\delta_{s,-b^n}\right)s^2\right]. \tag{B.3}$$

Evaluating the right-hand side leads to

$$\begin{aligned}\langle s^2 \rangle &= \frac{a-1}{2a}\left\{\sum_{n=0}^{\infty}\frac{1}{a^n}\left[b^{2n}+(-b)^{2n}\right]\right\} = \frac{a-1}{a}\left(\sum_{n=0}^{\infty}\frac{b^{2n}}{a^n}\right)\\ &= \frac{a-1}{a}\left[\sum_{n=0}^{\infty}\left(\frac{b^2}{a}\right)^n\right].\end{aligned} \tag{B.4a – c}$$

If $b$ is equal or larger than the root of $a$, it immediately follows that the variance is infinite

$$\langle s^2 \rangle = \frac{a-1}{a}\left[\sum_{n=0}^{\infty}\left(\frac{b^2}{a}\right)^n\right] \to \infty, \quad b \geq \sqrt{a}. \tag{B.5a, b}$$

This infinite variance is an important characteristics of Lévy flights.

## Appendix C: The Structure Function

The structure function is the Fourier transform of the probability function (see Shlesinger et al. 1986)

$$\lambda(k) = \sum_{s=-\infty}^{+\infty} p(s) \exp(iks). \tag{C.1}$$

Introducing Eq. (315) one obtains

$$\lambda(k) = \sum_{s=-\infty}^{+\infty} \frac{a-1}{2a} \left[ \sum_{n=0}^{\infty} \frac{1}{a^n} (\delta_{s,b^n} + \delta_{s,-b^n}) \right] e^{iks}. \tag{C.2}$$

Changing the order of the sums leads to

$$\lambda(k) = \frac{a-1}{2a} \left[ \sum_{n=0}^{\infty} \frac{1}{a^n} \sum_{s=-\infty}^{+\infty} (\delta_{s,b^n} + \delta_{s,-b^n}) e^{isk} \right]. \tag{C.3}$$

By applying a further evaluation, it follows that

$$\lambda(k) = \frac{a-1}{a} \left[ \sum_{n=0}^{\infty} \frac{1}{a^n} \frac{e^{ib^n k} + e^{-ib^n k}}{2} \right]. \tag{C.4}$$

which may be rewritten as

$$\lambda(k) = \frac{a-1}{a} \left[ \sum_{n=0}^{\infty} \frac{1}{a^n} \cos(b^n k) \right]. \tag{C.5}$$

This function is called “Weierstrass function.” Now a renormalization transformation is applied to the structure function. We start by splitting of the term $n = 0$ (this corresponds to a renormalization group transformation)

$$\lambda(k) = \frac{a-1}{a} \left[ \cos(k) + \sum_{n=1}^{\infty} \frac{1}{a^n} \cos(b^n k) \right]. \tag{C.6}$$

With new indices, namely

$$l = n - 1. \tag{C.7}$$

one obtains

$$\lambda(k) = \frac{a-1}{a}\left[\cos(k) + \sum_{l=0}^{\infty}\frac{1}{a^{l+1}}\cos\left(b^{l+1}k\right)\right]. \tag{C.8}$$

This is rearranged to become

$$\begin{aligned}\lambda(k) &= \frac{a-1}{a}\left\langle \cos(k) + \frac{1}{a}\left\{\sum_{l=0}^{\infty}\frac{1}{a^{l}}\cos\left[b^{l}(bk)\right]\right\}\right\rangle \\ &= \frac{a-1}{a}\cos(k) + \frac{1}{a}\frac{a-1}{a}\left\langle \cos(bk) + \left\{\sum_{l=1}^{\infty}\frac{1}{a^{l}}\cos\left[b^{l}(bk)\right]\right\}\right\rangle,\end{aligned} \tag{C.8a, b}$$

and as a result the two functional relations follow,

$$\lambda(k) = \frac{1}{a}\lambda(bk) + \frac{a-1}{a}\cos(k), \tag{C.9}$$

and

$$\lambda(bk) = a\,\lambda(k) - (a-1)\cos(k). \tag{C.10}$$

## Appendix D: Circular Mean Velocity Profile of Plane Turbulent Poiseuille Flows

This proof is presented in Egolf and Weiss (2000) in a very brief and mathematical rigorous way. In their article numerous formulas are prepared, e.g., in Taylor series approximations and then in a final merging together process, with additionally taking the limit to infinite Reynolds number Re $\to \infty$ or $\kappa \to \infty$, respectively, as limit of the Bessel function velocity profile the circle profile results. In this appendix, we present an extended version of this proof, and because of didactic reasons much more details are outlined. The above described limit is not taken in a final procedure, it is applied in several stages (only where it is actually allowed) and at the same time different order approximations are kept in the limiting processes to be applied at later stages. As such, this procedure is mathematically less satisfying, but has the great advantage that the reader has easier access to the ideas of this mathematical proof.

We start with some preparatory work. With the help of Eqs. (838a) and (841), we obtain

$$\psi = \frac{\sqrt{\beta}}{\alpha}\eta. \tag{D.1}$$

Furthermore, by combining Eq. (D.1) with (844a), it follows that

$$\psi = \frac{2\kappa}{\sqrt{\beta}}\eta. \tag{D.2}$$

With this relation Eq. (867) is rewritten as

$$g_1(\eta) = \frac{\eta}{\sqrt{\beta}} \frac{J_{\kappa-1}\left(\frac{2\kappa}{\sqrt{\beta}}\eta\right)}{J_\kappa\left(\frac{2\kappa}{\sqrt{\beta}}\eta\right)}; \tag{D.3}$$

this is the formula with which we shall now study the averaged stream-wise velocity profile.

It shall be demonstrated that if Re and $\kappa$ tend to infinity (or $\alpha \to 0$ and $\beta \to 4$), for $|\eta| < 1$[1]), the averaged velocity profile (D.3) tends point-wise toward the circular profile (826a).

Abramowitz and Stegun (1984) list the asymptotic relation of the Bessel function for large order, viz,

$$J_\kappa(\kappa \ \mathrm{sech}\gamma) \propto \frac{\exp\{\kappa[\tanh(\gamma) - \gamma]\}}{\sqrt{2\pi\kappa \tanh(\gamma)}}, \qquad \gamma > 0, \qquad \kappa \to \infty. \tag{D.4}$$

To adapt this asymptotic formula to our notation, by comparison of the arguments of (D.3) and (D.4), we set

$$\frac{2\eta}{\sqrt{\beta}} = \mathrm{sech}(\gamma_1), \tag{D.5}$$

where we added the index 1 to the variable $\gamma$.
Recall the definition

$$\mathrm{sech}(\gamma) = \frac{1}{\cosh(\gamma)}. \tag{D.6}$$

With the aid of (D.4) to (D.6), one obtains in the same limit

$$J_{\kappa-1}\left(\frac{2\kappa}{\sqrt{\beta}}\eta\right) = J_{\kappa-1}[\kappa \, \mathrm{sech}(\gamma_1)], \qquad \gamma_1 > 0, \qquad \kappa \to \infty. \tag{D.7}$$

Next, we demand that

[1]Due to Eq. (804a), without any loss of information, it may be assumed that $\eta \geq 0$.

$$J_{\kappa-1}[(\kappa-1)\mathrm{sech}(\gamma_2)] = J_{\kappa-1}[\kappa\,\mathrm{sech}(\gamma_1)]. \tag{D.8}$$

Therefore, we state that there exists a $\gamma_2$ such that

$$(\kappa-1)\mathrm{sech}(\gamma_2) = \kappa\,\mathrm{sech}(\gamma_1); \tag{D.9}$$

where with (D.5), (D.9), and (844a) it follows that

$$\mathrm{sech}(\gamma_2) = \frac{\kappa}{\kappa-1}\mathrm{sech}(\gamma_1) = \frac{\kappa}{\kappa-1}\frac{2\eta}{\sqrt{\beta}} = \frac{2\eta}{\sqrt{\beta}}\frac{\beta}{\beta-2\alpha}. \tag{D.10a – c}$$

The next step is somewhat delicate because one could imagine that the term (D.10c) is larger than unity. However, at this stage we request the freely chosen $\eta$ to be so small that (D.10c) becomes smaller than unity. This procedure is allowed, because in the limit as $\kappa \to \infty$, that we will apply later, this term will then tend to

$$\mathrm{sech}(\gamma_2) = \mathrm{sech}(\gamma_1) = 2\eta/\sqrt{\beta} = \eta \leq 1. \tag{D.11a – d}$$

(see (D.5 and D.10a)). Moreover, during this process $\eta$ can then be extended also up to the value equal to one.
Next, we apply (D.4) twice to obtain

$$\begin{aligned} J_\kappa(\kappa\,\mathrm{sech}\gamma_1) &\propto \frac{\exp\{\kappa[\tanh(\gamma_1)-\gamma_1]\}}{\sqrt{2\pi\kappa\ \tanh(\gamma_1)}}, \qquad \kappa\to\infty \\ J_{\kappa-1}[(\kappa-1)\mathrm{sech}\gamma_2] &\propto \frac{\exp\{(\kappa-1)[\tanh(\gamma_2)-\gamma_2]\}}{\sqrt{2\pi(\kappa-1)\tanh(\gamma_2)}}, \qquad \kappa\to\infty. \end{aligned} \tag{D.12a, b}$$

These two equations are inserted into Eq. (D.3), which yields

$$g_1(\eta) = \frac{\eta}{\sqrt{\beta}}\frac{\dfrac{\exp\{(\kappa-1)[\tanh(\gamma_2)-\gamma_2]\}}{\sqrt{2\pi(\kappa-1)\tanh(\gamma_2)}}}{\dfrac{\exp\{\kappa[\tanh(\gamma_1)-\gamma_1]\}}{\sqrt{2\pi\kappa\ \tanh(\gamma_1)}}}, \qquad \kappa\to\infty, \tag{D.13}$$

or

$$\begin{aligned} g_1(\eta) &= \frac{\eta}{\sqrt{\beta}}\exp\{(\kappa-1)[\tanh(\gamma_2)-\gamma_2]-\kappa[\tanh(\gamma_1)-\gamma_1]\} \\ &\times\frac{\sqrt{\kappa\ \tanh(\gamma_1)}}{\sqrt{(\kappa-1)\tanh(\gamma_2)}}, \qquad \kappa\to\infty, \end{aligned} \tag{D.14}$$

respectively. This can be rearranged as

$$g_1(\eta) = \frac{\eta}{\sqrt{\beta}} \exp\left\{ \kappa \left[ \underbrace{\tanh(\gamma_2) - \tanh(\gamma_1)}_{F_A} + \underbrace{(\gamma_1 - \gamma_2)}_{F_B} \right] - \left[ \underbrace{\tanh(\gamma_2)}_{F_C} - \underbrace{\gamma_2}_{F_D} \right] \right\} \underbrace{\sqrt{\frac{\kappa}{\kappa - 1} \frac{\tanh(\gamma_1)}{\tanh(\gamma_2)}}}_{F_E}, \quad \kappa \to \infty. \tag{D.15}$$

In the following sections, the functions $F_A$ to $F_E$ will be further elaborated. Therefore, we rewrite (D.15) in the form

$$g_1(\eta) = \frac{\eta}{\sqrt{\beta}} \exp\{[\kappa(F_A + F_B)] - F_C + F_D\} F_E, \tag{D.16}$$

as indicted with the following auxiliary functions

$$\begin{aligned} F_A &= \tanh(\gamma_2) - \tanh(\gamma_1), \\ F_B &= \gamma_1 - \gamma_2, \\ F_C &= \tanh(\gamma_2), \\ F_D &= \gamma_2, \\ F_E &= \sqrt{\frac{\kappa}{\kappa - 1} \frac{\tanh(\gamma_1)}{\tanh(\gamma_2)}}. \end{aligned} \tag{D.17a – e}$$

To start this procedure, first we study the auxiliary function $F_A$. From Eq. (D10a), it follows that

$$\frac{\mathrm{sech}(\gamma_1)}{\mathrm{sech}(\gamma_2)} = \frac{\kappa - 1}{\kappa} = 1 - \frac{1}{\kappa}. \tag{D.18a, b}$$

This can be rearranged to obtain

$$\mathrm{sech}(\gamma_1) = \mathrm{sech}(\gamma_2) - \frac{1}{\kappa} \mathrm{sech}(\gamma_2). \tag{D.19}$$

Furthermore, it is well known that

$$\tanh(\gamma) = \frac{\sinh(\gamma)}{\cosh(\gamma)} \tag{D.20}$$

and

$$\cosh^2(\gamma) - \sinh^2(\gamma) = 1. \tag{D.21}$$

Inserting Eq. (D.21) into (D.20) yields

$$\tanh(\gamma) = \frac{\sqrt{\cosh^2(\gamma) - 1}}{\cosh(\gamma)}. \tag{D.22}$$

With Eq. (D.6) this becomes

$$\tanh(\gamma) = \sqrt{1 - \mathrm{sech}^2(\gamma)}. \tag{D.23}$$

With this equation, we transform Eq. (D.17a) into

$$F_A = \sqrt{1 - \mathrm{sech}^2(\gamma_2)} - \sqrt{1 - \mathrm{sech}^2(\gamma_1)}. \tag{D.24}$$

With the square of Eq. (D.19), we rewrite the second root in the following manner:

$$\sqrt{1 - \mathrm{sech}^2(\gamma_1)} = \sqrt{1 - \left[1 - \frac{2}{\kappa} + \frac{1}{\kappa^2}\right]\mathrm{sech}^2(\gamma_2)} \tag{D.25}$$

and drop the $O(\kappa^{-2})$ term in the root leading to

$$\sqrt{1 - \mathrm{sech}^2(\gamma_1)} = \sqrt{1 - \mathrm{sech}^2(\gamma_2) + \frac{2}{\kappa}\mathrm{sech}^2(\gamma_2)}. \tag{D.26}$$

This expression is described by a Taylor series approximation of only first order at the position $1 - \mathrm{sech}^2(\gamma_2)$. This is allowed because we can choose $\kappa$ so large that the second term in (D.26) in the root becomes sufficiently small. The result of this calculus is

$$\sqrt{1 - \mathrm{sech}^2(\gamma_1)} = \sqrt{1 - \mathrm{sech}^2(\gamma_2)} + \frac{1}{\kappa}\frac{\mathrm{sech}^2(\gamma_2)}{\sqrt{1 - \mathrm{sech}^2(\gamma_2)}}. \tag{D.27}$$

Equation (D.27) is inserted into (D.24), which leads to

$$F_A = -\frac{1}{\kappa}\frac{\mathrm{sech}^2(\gamma_2)}{\sqrt{1 - \mathrm{sech}^2(\gamma_2)}} + O(\kappa^{-2}). \tag{D.28}$$

This intermediate result will be applied in the final operation of calculating (D.15).

The second auxiliary computation is devoted to $F_B$. To calculate the $\gamma$'s, a further auxiliary calculation is now outlined. With the basic definitions

$$\sinh(\gamma) = \frac{\exp(\gamma) - \exp(-\gamma)}{2} \tag{D.29}$$

and

$$\cosh(\gamma) = \frac{\exp(\gamma) + \exp(-\gamma)}{2}, \tag{D.30}$$

it follows that

$$\exp(\gamma) = \sinh(\gamma) + \cosh(\gamma). \tag{D.31}$$

Inserting this equation into $\gamma = \log_e[\exp(\gamma)]$ yields

$$\gamma = \log_e[\sinh(\gamma) + \cosh(\gamma)]. \tag{D.32}$$

Applying relation (D.21) transforms this equation to

$$\gamma = \log_e\left[\cosh(\gamma) + \sqrt{\cosh^2(\gamma) - 1}\right]. \tag{D.33}$$

Finally, (D.33) with (D.6) leads to

$$\gamma = \log_e\left[\frac{1}{\mathrm{sech}(\gamma)} + \sqrt{\frac{1}{\mathrm{sech}^2(\gamma)} - 1}\right]. \tag{D.34}$$

Applying (D.34) twice, it follows that

$$\begin{aligned} \exp(\kappa F_B) &= \exp[\kappa(\gamma_1 - \gamma_2)] \\ &= \left[1 + \frac{1}{\kappa}\frac{1}{\sqrt{1 - \mathrm{sech}^2(\gamma_2)}} + \mathrm{O}(\kappa^{-2})\right]^{\kappa}. \end{aligned} \qquad \text{(D.35a, b)}$$

In the limit $\kappa \to \infty$, Eq. (D.35b) is identical to

$$\exp\left(\kappa F_B\right) = \exp\left[\frac{1}{\sqrt{1-\operatorname{sech}^2(\gamma_2)}}\right]. \tag{D.36}$$

Now it follows that

$$\exp\left[\kappa(F_A+F_B)\right] = \exp\left[\frac{1}{\sqrt{1-\operatorname{sech}^2(\gamma_2)}} - \frac{\operatorname{sech}^2(\gamma_2)}{\sqrt{1-\operatorname{sech}^2(\gamma_2)}}\right]. \quad \kappa \to \infty \tag{D.37}$$

This is identical to

$$\exp\left[\kappa(F_A+F_B)\right] = \exp\left(\sqrt{1-\operatorname{sech}^2(\gamma_2)}\right). \tag{D.38}$$

With Eq. (D.23) this leads to

$$\exp\left[\kappa(F_A+F_B)\right] = \exp\left[\tanh\left(\gamma_2\right)\right] = \exp\left(F_C\right). \tag{D.39}$$

Now (D.16) reduces to

$$g_1(\eta) = \frac{\eta}{\sqrt{\beta}} \exp\left(F_D\right) F_E. \tag{D.40}$$

Furthermore, we let the function $F_D = \gamma_2$ for a later elaboration and, now, have a closer look at function $F_E$. After (D.17e) there is

$$F_E = \sqrt{\frac{\kappa}{\kappa-1}} \sqrt{\frac{\tanh\left(\gamma_1\right)}{\tanh\left(\gamma_2\right)}}. \tag{D.41}$$

With help of (D.23), this may be rewritten to yield

$$F_E = \sqrt{\frac{\kappa}{\kappa-1}} \left[\frac{1-\operatorname{sech}(\gamma_1)}{1-\operatorname{sech}(\gamma_2)}\right]^{1/4}. \tag{D.42}$$

By applying Eq. (D.19) it follows that

$$F_E = \sqrt{\frac{\kappa}{\kappa-1}} \left[\frac{1-\left(1-\frac{1}{\kappa}\right)\operatorname{sech}(\gamma_2)}{1-\operatorname{sech}(\gamma_2)}\right]^{1/4}. \tag{D.43}$$

Now, it becomes clear that in the Re $\to \infty$ limit

$$F_E = 1, \qquad \kappa \to \infty. \tag{D.44}$$

If we, now, substitute the result (D.44) into (D.40), the simpler relation

$$g_1(\eta) = \frac{\eta}{\sqrt{\beta}} \exp(F_D) = \frac{\eta}{\sqrt{\beta}} \exp(\gamma_2) = \frac{\eta}{2} \exp(\gamma_2), \qquad \kappa \to \infty \tag{D.45a – c}$$

follows. This result denotes the infinite Reynolds number limit for $\beta = 4$.

Now, Eq. (D.34) is substituted into (D.45c), which leads to

$$g_1(\eta) = \frac{\eta}{2} \frac{1}{\text{sech}(\gamma_2)} \left(1 + \sqrt{1 - \text{sech}^2(\gamma_2)}\right). \tag{D.46}$$

If now Eq. (D.11c) is inserted, the final result is

$$g_1(\eta) = \frac{1 + \sqrt{1 - \eta^2}}{2}, \tag{D.47}$$

which is identical to the circular profile of plane turbulent Poiseuille flow at infinite Reynolds number (compare with Eq. (826a)), q.e.d.

## Appendix E: Fourier Transform for $q$-Generalized Energy Spectrum of Turbulent Flows

This proof is a review of work from Egolf and Hutter (2018) with some minor amendments.

To determine the energy-enstrophy spectrum of *2-d* isotropic turbulence, we apply a Fourier transform to the generalized energy $\breve{E}_\varepsilon(\vec{x})$,

$$\widehat{\breve{E}}_\varepsilon(\vec{k}) = \frac{1}{2\pi} \frac{1}{Z_{ES\varepsilon}} \int_{-\infty}^{\infty} \int_{-\infty}^{\infty} \breve{E}_\varepsilon(\vec{x})\, P\left[\breve{E}_\varepsilon(\vec{x})\right] e^{-i\left(\vec{k}\,\vec{x}\right)} \mathrm{d}\vec{x}. \tag{E.1}$$

This is identical to

$$\widehat{\breve{E}}_\varepsilon(\vec{k}) = \frac{1}{2\pi} \frac{1}{Z_{ES\varepsilon}} \int_{-\infty}^{\infty} \int_{-\infty}^{\infty} \breve{E}_\varepsilon(x_1, x_2) P\left[\breve{E}_\varepsilon(x_1, x_2)\right] e^{-i(k_1 x_1 + k_2 x_2)} \mathrm{d}x_1 \mathrm{d}x_2, \tag{E.2}$$

or

$$\widehat{\breve{E}}_{\varepsilon}(\vec{k}) = \frac{1}{2\pi}\frac{1}{Z_{ES\varepsilon}} \int_{-\infty}^{\infty}\int_{-\infty}^{\infty} \breve{E}_{\varepsilon}(x_1,x_2)e^{-i(k_1x_1+k_2x_2)} \times \left[e^{-\beta_q \breve{E}_{\varepsilon}(x_1,x_2)}\right]^q \mathrm{d}x_1\mathrm{d}x_2, \tag{E.3}$$

respectively. Here, the escort probability distribution function (1075a,b) was inserted. This yields

$$\widehat{\breve{E}}_{\varepsilon}(\vec{k}) = \frac{1}{2\pi}\frac{1}{Z_{ES\varepsilon}} \int_{-\infty}^{\infty}\int_{-\infty}^{\infty} \breve{E}_{\varepsilon}(x_1,x_2)e^{-i(k_1x_1+k_2x_2)}e^{-q\beta_q \breve{E}_{\varepsilon}(x_1,x_2)}\mathrm{d}x_1\mathrm{d}x_2. \tag{E.4}$$

It is clear that in the context of this proof also the enstrophy, Eq. (1012a,b), requires a generalization by fractional calculus methods. Therefore, we introduce the modified enstrophy,

$$\widetilde{\Omega}_{\varepsilon}^{(q)} = \sum_{i,j=1}^{2} \left|\frac{\partial^q \widetilde{u}_i}{\partial x_j^q}\right|^2, \qquad \mathrm{D}_j^q := \frac{\partial^q}{\partial x_j^q}, \tag{E.5a, b}$$

with the fractional derivatives given by their two most common notations. Notice that, to perform this proof, we will not discuss which of the various fractional derivatives is best suited for this special application; this will be left for future clarification. The employed property of transformation of a fractional derivative of a certain order to a power law exponent with a value of exactly this order is valid for practically all versions of fractional derivatives, e.g., the Fourier fractional derivative (see Herrmann 2014) and Riemann–Liouville fractional derivative (Herrmann 2011). It is straightforward to see that, with the help of Eqs. (1011f) and (1018b), this generalization leads to the following modification of Eq. (1021a):

$$\widetilde{E}_{\varepsilon}^{(q)} = \left(\alpha + \beta k^{2q}\right)\widehat{\widetilde{u}}^{\,2} = \left(\alpha + \beta k^{2q}\right)\left[\sum_{l=n}^{N} \widehat{u}_l^{\,2}\right], \tag{E.6a, b}$$

where a possibly occurring constant in front of $k^{2q}$ is absorbed in $\beta$.

Now, an important question is, whether in the fractal case the subspace conserving energy and enstrophy still remains a quadratic form. It is expected that in phase space this could be a fractal subset (strange attractor). However, in Sect. 10.11, we prove that even if the total energy presents itself in Euclidian form that it is based on fractal features with self-similarity. Therefore, in agreement with Kraichnan we write

$$\widehat{\breve{u}}^2 = a_{ik}\tilde{x}_i\tilde{x}_k = a_{11}\tilde{x}_1^2 + a_{12}\tilde{x}_1\tilde{x}_2 + a_{21}\tilde{x}_2\tilde{x}_1 + a_{22}\tilde{x}_2^2, \qquad a_{12} = a_{21} \qquad \text{(E.7a – c)}$$

Then a rotational main axis transformation ($\tilde{x}_i \to \widehat{x}_i$) is applied and the coordinates are compressed or stretched in such a manner that relation (E.7a) simplifies to

$$\widehat{\breve{u}}^2 = \widehat{x}_1^2 + \widehat{x}_2^2, \qquad \text{(E.8)}$$

where in the following for simplicity, the convex curved overheads will be dropped. In Sect. 10.11, we have already outlined a motivation and proof of this relation. Notice that this does not imply a decoupling of Fourier modes. It is evident that the sum of the replicas for each spatial component inserted into the square quantities of Eq. (E.8) describes self interaction and interaction mode coupling. Now, it follows that

$$\breve{E}_\varepsilon = \left(\alpha + \beta k^{2q}\right)\left(x_1^2 + x_2^2\right). \qquad \text{(E.9)}$$

This is inserted into Eq. (E.4) and yields

$$\begin{aligned}
\widehat{\breve{E}}_\varepsilon(\vec{k}) &= \frac{1}{2\pi}\frac{1}{Z_{ES\varepsilon}}\int_{-\infty}^{\infty}\int_{-\infty}^{\infty}\left(\alpha + \beta k^{2q}\right)\left(x_1^2 + x_2^2\right)\\
&\times e^{-i(k_1x_1+k_2x_2)}e^{-q\beta_q\left(\alpha+\beta k^{2q}\right)\left(x_1^2+x_2^2\right)}\mathrm{d}x_1\mathrm{d}x_2 =\\
&\frac{1}{2\pi}\frac{1}{Z_{ES\varepsilon}}\int_{-\infty}^{\infty}\int_{-\infty}^{\infty}\left(\alpha + \beta k^{2q}\right)x_1^2 e^{-i(k_1x_1+k_2x_2)}\\
&\times e^{-q\beta_q\left(\alpha+\beta k^{2q}\right)\left(x_1^2+x_2^2\right)}\mathrm{d}x_1\mathrm{d}x_2 +\\
&\frac{1}{2\pi}\frac{1}{Z_{ES\varepsilon}}\int_{-\infty}^{\infty}\int_{-\infty}^{\infty}\left(\alpha + \beta k^{2q}\right)x_2^2 e^{-i(k_1x_1+k_2x_2)}\\
&\times e^{-q\beta_q\left(\alpha+\beta k^{2q}\right)\left(x_1^2+x_2^2\right)}\mathrm{d}x_1\mathrm{d}x_2 =\\
&\frac{1}{2\pi}\frac{1}{Z_{ES\varepsilon}}\int_{-\infty}^{\infty}\int_{-\infty}^{\infty}\left(\alpha + \beta k^{2q}\right)x_1^2 e^{-ik_1x_1}e^{-ik_2x_2}\\
&\times e^{-q\beta_q\left(\alpha+\beta k^{2q}\right)x_1^2}e^{-q\left(\alpha+\beta k^{2q}\right)x_2^2}\mathrm{d}x_1\mathrm{d}x_2 +\\
&\frac{1}{2\pi}\frac{1}{Z_{ES\varepsilon}}\int_{-\infty}^{\infty}\int_{-\infty}^{\infty}\left(\alpha + \beta k^{2q}\right)x_2^2 e^{-ik_1x_1}e^{-ik_2x_2}\\
&\times e^{-q\beta_q\left(\alpha+\beta k^{2q}\right)x_1^2}e^{-q\left(\alpha+\beta k^{2q}\right)x_2^2}\mathrm{d}x_1\mathrm{d}x_2.
\end{aligned} \qquad \text{(E.10a – c)}$$

An advantage is that it is possible to separate the double integrals in $x_1$ and $x_2$

$$\widehat{E}_{\varepsilon}(\vec{k}) = \frac{1}{Z_{ES\varepsilon}}\left[\frac{1}{\sqrt{2\pi}}\int\limits_{-\infty}^{\infty}(\alpha+\beta k^{2q})x_1^2 e^{-ik_1x_1}e^{-q\beta_q\left(\alpha+\beta k^{2q}\right)x_1^2}\mathrm{d}x_1\right.$$
$$\left.\times\frac{1}{\sqrt{2\pi}}\int\limits_{-\infty}^{\infty}e^{-ik_2x_2}e^{-q\left(\alpha+\beta k^{2q}\right)x_2^2}\mathrm{d}x_2\right]$$
$$+\frac{1}{Z_{ES\varepsilon}}\left[\frac{1}{\sqrt{2\pi}}\int\limits_{-\infty}^{\infty}(\alpha+\beta k^{2q})x_2^2 e^{-ik_2x_2}e^{-q\beta_q\left(\alpha k^{2q}+\beta\right)x_2^2}\mathrm{d}x_2\right.$$
$$\left.\times\frac{1}{\sqrt{2\pi}}\int\limits_{-\infty}^{\infty}e^{-ik_1x_1}e^{-q\left(\alpha+\beta k^{2q}\right)x_1^2}\mathrm{d}x_1\right]. \tag{E.11}$$

In this expression, the various integrals are solved one after the other. The simpler two integrals are

$$I_i^{(2)} = \frac{1}{\sqrt{2\pi}}\int\limits_{-\infty}^{\infty}e^{-ik_ix_i}e^{-q\left(\alpha+\beta k^{2q}\right)x_i^2}\mathrm{d}x_i, \qquad i\in\{1,2\}. \tag{E.12}$$

With the abbreviation

$$\gamma = 2q(\alpha+\beta k^{2q}), \tag{E.13}$$

it follows that

$$I_i^{(2)} = \frac{1}{\sqrt{2\pi}}\int\limits_{-\infty}^{\infty}e^{-ik_ix_i-\frac{\gamma}{2}x_i^2}\mathrm{d}x_i, \qquad i\in\{1,2\}. \tag{E.14}$$

Furthermore, a quadratic expansion is performed

$$I_i^{(2)} = \frac{1}{\sqrt{2\pi}}e^{-\frac{1}{2\gamma}k_i^2}\int\limits_{-\infty}^{\infty}e^{-\frac{\gamma}{2}\left[x_i^2+\frac{2i}{\gamma}k_ix_i+\left(\frac{ik_i}{\gamma}\right)^2\right]}\mathrm{d}x_i, \qquad i\in\{1,2\}, \tag{E.15}$$

which is identical to

$$I_i^{(2)} = \frac{1}{\sqrt{2\pi}}e^{-\frac{1}{2\gamma}k_i^2}\int\limits_{-\infty}^{\infty}e^{-\frac{\gamma}{2}\left(x_i+\frac{ik_i}{\gamma}\right)^2}\mathrm{d}x_i, \qquad i\in\{1,2\}. \tag{E.16}$$

A second abbreviation is introduced by the expression

$$y_i = x_i + \frac{ik_i}{\gamma} \quad \Rightarrow \quad \mathrm{d}y_i = \mathrm{d}x_i, \qquad i \in \{1,2\}, \tag{E.17a, b}$$

which simplifies Eq. (E.16) to

$$I_i^{(2)} = \frac{1}{\sqrt{2\pi}} e^{-\frac{1}{2\gamma}k_i^2} \int_{-\infty}^{\infty} e^{-\frac{\gamma}{2}y_i^2} \mathrm{d}y_i, \qquad i \in \{1,2\}. \tag{E.18}$$

In the pocket book of mathematics of Bronstein and Semendjajew (1981), the following formula is listed for $a > 0$:

$$\int_0^{\infty} e^{-a^2x^2} \mathrm{d}x = \sqrt{\frac{\pi}{2a}} \quad \Rightarrow \quad \int_{-\infty}^{\infty} e^{-\frac{\gamma}{2}y_i^2} \mathrm{d}y_i = \sqrt{\frac{2\pi}{\gamma}}, \tag{E.19a, b}$$

which is directly inserted into (E.18). This yields the final result for this integration problem

$$I_i^{(2)} = \frac{1}{\gamma^{1/2}} e^{-\frac{1}{2\gamma}k_i^2}, \qquad i \in \{1,2\}. \tag{E.20}$$

The two slightly larger integrals are

$$I_i^{(1)} = \frac{1}{\sqrt{2\pi}} \int_{-\infty}^{\infty} (\alpha + \beta k^{2q}) x_i^2 e^{-ik_ix_i} e^{-q(\alpha+\beta k^{2q})x_i^2} \mathrm{d}x_i, \qquad i \in \{1,2\} \tag{E.21}$$

With abbreviation (E.13) this transforms to

$$I_i^{(1)} = \frac{1}{2\sqrt{2\pi}} \frac{\gamma}{q} \int_{-\infty}^{\infty} x_i^2 e^{-ik_ix_i - \frac{\gamma}{2}x_i^2} \mathrm{d}x_i, \qquad i \in \{1,2\}. \tag{E.22}$$

and with (E.17) the result is

$$I_i^{(1)} = \frac{1}{2\sqrt{2\pi}} \frac{\gamma}{q} e^{-\frac{1}{2\gamma}k_i^2} \int_{-\infty}^{\infty} \left(y_i - \frac{ik_i}{\gamma}\right)^2 e^{-\frac{\gamma}{2}y_i^2} \mathrm{d}y_i, \qquad i \in \{1,2\}. \tag{E.23}$$

By working on the parenthesis term of the integrand, this integral is separated into three integrals

$$I_i^{(1)} = \frac{1}{2\sqrt{2\pi}} \times \frac{\gamma}{q} e^{-\frac{1}{2\gamma}k_i^2} \left( \int_{-\infty}^{\infty} y_i^2 e^{-\frac{\gamma}{2}y_i^2} \mathrm{d}y_i - 2i\frac{k_i}{\gamma} \int_{-\infty}^{\infty} y_i e^{-\frac{\gamma}{2}y_i^2} \mathrm{d}y_i - \frac{k_i^2}{\gamma^2} \int_{-\infty}^{\infty} e^{-\frac{\gamma}{2}y_i^2} \mathrm{d}y_i \right). \quad \text{(E.24)}$$

The second integral, because of its antisymmetric nature, is zero; so Eq. (E.24) simplifies to

$$I_i^{(1)} = \frac{1}{2\sqrt{2\pi}} \frac{\gamma}{q} e^{-\frac{1}{2\gamma}k_i^2} \int_{-\infty}^{\infty} y_i^2 e^{-\frac{\gamma}{2}y_i^2} \mathrm{d}y_i - \frac{1}{2\sqrt{2\pi}} \frac{k_i^2}{q\gamma} e^{-\frac{1}{2\gamma}k_i^2} \int_{-\infty}^{\infty} e^{-\frac{\gamma}{2}y_i^2} \mathrm{d}y_i. \quad \text{(E.25)}$$

Bronstein and Semendjajew (1981) list the following result:

$$\int_0^{\infty} x^2 e^{-a^2x^2} \mathrm{d}x = \frac{\sqrt{\pi}}{4a^3} \quad \Rightarrow \quad \int_{-\infty}^{\infty} y_i^2 e^{-\frac{\gamma}{2}y_i^2} \mathrm{d}y_i = \sqrt{2\pi} \frac{1}{\gamma^{3/2}}. \quad \text{(E.26a, b)}$$

With (E.19) and (E.26) it is concluded that

$$I_i^{(1)} = \frac{1}{2q} \left( \frac{1}{\gamma^{1/2}} - \frac{k_i^2}{\gamma^{3/2}} \right) e^{-\frac{1}{2\gamma}k_i^2}. \quad \text{(E.27)}$$

Equation (E.11) can be written as

$$\widehat{E}_\varepsilon(\vec{k}) = \frac{1}{Z_{ES\varepsilon}} \left( I_1^{(1)} \cdot I_2^{(2)} + \mathrm{I}_2^{(1)} \cdot \mathrm{I}_1^{(2)} \right). \quad \text{(E.28)}$$

Substituting the partial solutions of the integrals (E.20) and (E.27) into (E.28), the result is

$$\begin{aligned} \widehat{E}_\varepsilon(\vec{k}) = \frac{1}{Z_{ES\varepsilon}} \bigg[ & \frac{1}{2q} \left( \frac{1}{\gamma^{1/2}} - \frac{k_1^2}{\gamma^{3/2}} \right) e^{-\frac{1}{2\gamma}k_1^2} \\ & \times \frac{1}{\gamma^{1/2}} e^{-\frac{1}{2\gamma}k_2^2} + \frac{1}{2q} \left( \frac{1}{\gamma^{1/2}} - \frac{k_2^2}{\gamma^{3/2}} \right) e^{-\frac{1}{2\gamma}k_2^2} \frac{1}{\gamma^{1/2}} e^{-\frac{1}{2\gamma}k_2^2} \bigg], \end{aligned} \quad \text{(E.29)}$$

which simplifies to

$$\widehat{E}_\varepsilon(\vec{k}) = \frac{1}{Z_{ES\varepsilon}} \frac{1}{q} \left[ \left( \frac{1}{\gamma} - \frac{1}{2} \frac{k^2}{\gamma^2} \right) \right] e^{-\frac{1}{2\gamma}k^2}, \qquad k = \sqrt{k_1^2 + k_2^2}. \quad \text{(E.30a, b)}$$

For $\alpha \ll \beta$ and $k \ll \beta(2\sqrt{\beta q})^{1/(1-q)}$, respectively, $\beta \ll \alpha$ and $k \ll 2\sqrt{\alpha q}$, the second term in the square brackets is neglected. This leads to the following expression:

$$\widehat{\widetilde{E}}_{\varepsilon}(\vec{k}) = \frac{1}{Z_{ES\varepsilon}} \frac{1}{q\gamma} e^{-\frac{1}{2\gamma}k^2}. \tag{E.31}$$

With a Taylor series expansion of the exponential function, respecting only the leading term, by applying the same argument again, one concludes that

$$\widehat{\widetilde{E}}_{\varepsilon}(\vec{k}) = \frac{1}{Z_{ES\varepsilon}} \frac{1}{q\gamma}. \tag{E.32}$$

Substituting Eq. (E.13) into Eq. (E.32) leads to

$$\widehat{\widetilde{E}}_{\varepsilon}(\vec{k}) = \frac{1}{2} \frac{1}{Z_{ES\varepsilon}} \frac{1}{q^2} \frac{1}{\alpha + \beta k^{2q}}, \tag{E.33}$$

which is the proof for the energy intensity spectrum (1109), q.e.d.

For $q = \beta_q = Z_{ES\varepsilon} = 1$ from (E.33) it follows that

$$\widehat{\widetilde{E}}_{\varepsilon}(\vec{k}) = \frac{1}{2} \frac{1}{\alpha + \beta k^2}, \tag{E.34}$$

called the Kraichnanian energy-enstrophy intensity spectrum (1022), q.e.d.

## References


Abramowitz, M., Stegun, I.A.: Pocketbook of Mathematical Functions. Verlag Harri Deutsch, Thun (1984). ISBN 3-87144-818-4

Bronstein, I.N., Semendjajew, K.A.: Taschenbuch der Mathematik. Edition Harri Deutsch, Thun (1981). ISBN 3-87-144-492-8 (in German)

Egolf, P.W., Hutter, K.: Tsallis extended thermodynamics applied to 2-d turbulence: Lévy statistics and q-fractional generalized Kraichnanian energy and enstrophy spectra. J. Entropy. **20**, 109 (2018)

Egolf, P.W., Weiss, D.A.: Difference-quotient turbulence model: analytical solutions for the core region of plane Poiseuille flow. Phys. Rev E. **62**(1-A), 553 (2000)

Herrmann, R.: Fractional Calculus. World Scientific, New Jersey (2011). ISBN 13-978-981-4340-24-3

Herrmann, R.: Fraktionale Infinitesimalrechnung, 2nd edn. Books on Demand, Norderstedt (2014). ISBN 978-3-7357-4109-7 (in German)

Shlesinger, M.F., Klafter, J., West, B.J.: Lévy walks with applications to turbulence and chaos. Physica. **140A**, 212 (1986)

# Author Index





P. W. Egolf, K. Hutter, *Nonlinear, Nonlocal and Fractional Turbulence*,
https://doi.org/10.1007/978-3-030-26033-0

**Y**



**Z**

# Subject Index





P. W. Egolf, K. Hutter, *Nonlinear, Nonlocal and Fractional Turbulence*,
https://doi.org/10.1007/978-3-030-26033-0

**E**

**M**

**N**

**O**

**T**

**Y**



**Z**

The manufacturer's authorised representative in the EU is Springer Nature Customer Service Centre GmbH, Europaplatz 3, 69115 Heidelberg, Germany. If you have any concerns regarding our products, please contact ProductSafety@springernature.com

Printed and bound by CPI Group (UK) Ltd, Croydon, CR0 4YY
05/10/2026
02228124-0019